LALLEMENT—Semigroups and Combinatorial Applications
LAMB—Elements of Soliton Theory
LAY—Convex Sets and Their Applications
LINZ—Theoretical Numerical Analysis: An Introduction to Advanced Techniques
LOVELOCK and RUND—Tensors, Differential Forms, and Variational Principles
MARTIN—Nonlinear Operators and Differential Equations in Banach Spaces
MELZAK—Companion to Concrete Mathematics
MELZAK—Invitation to Geometry
NAYFEH—Perturbation Methods
NAYFEH and MOOK—Nonlinear Oscillations
ODEN and REDDY—An Introduction to the Mathematical Theory of Finite Elements
PASSMAN—The Algebraic Structure of Group Rings
PETRICH—Inverse Semigroups
PIER—Amenable Locally Compact Groups
PRENTER—Splines and Variational Methods
RIBENBOIM—Algebraic Numbers
RICHTMYER and MORTON—Difference Methods for Initial-Value Problems,
            2nd Edition
RIVLIN—The Chebyshev Polynomials
ROCKAFELLAR—Network Flows and Monotropic Optimization
RUDIN—Fourier Analysis on Groups
SAMELSON—An Introduction to Linear Algebra
SCHUMAKER—Spline Functions: Basic Theory
SHAPIRO—Introduction to the Theory of Numbers
SIEGEL—Topics in Complex Function Theory
        Volume 1—Elliptic Functions and Uniformization Theory
        Volume 2—Automorphic Functions and Abelian Integrals
        Volume 3—Abelian Functions and Modular Functions of Several Variables
STAKGOLD—Green's Functions and Boundary Value Problems
STOKER—Differential Geometry
STOKER—Nonlinear Vibrations in Mechanical and Electrical Systems
STOKER—Water Waves
TURÁN—On A New Method of Analysis and Its Applications
WHITHAM—Linear and Nonlinear Waves
WOUK—A Course of Applied Functional Analysis
ZAUDERER—Partial Differential Equations of Applied Mathematics

AMENABLE LOCALLY
COMPACT GROUPS

AMENABLE LOCALLY COMPACT GROUPS

JEAN-PAUL PIER
Centre Universitaire of Luxembourg

A Wiley-Interscience Publication
JOHN WILEY & SONS
New York • Chichester • Brisbane • Toronto • Singapore

Copyright © 1984 by John Wiley & Sons, Inc.

All rights reserved. Published simultaneously in Canada.

Reproduction or translation of any part of this work beyond that permitted by Section 107 or 108 of the 1976 United States Copyright Act without the permission of the copyright owner is unlawful. Requests for permission or further information should be addressed to the Permissions Department, John Wiley & Sons, Inc.

Library of Congress Cataloging in Publication Data
Pier, Jean-Paul, 1933–
   Amenable locally compact groups.

   (Pure and applied mathematics)
   "A Wiley-Interscience publication."
   Bibliography: p.
   Includes indexes.
   1. Locally compact groups. I. Title. II. Series:
Pure and applied mathematics (John Wiley & Sons)

QA387.P55   1984   512'.2   84-7366
ISBN 0-471-89390-0

Printed in the United States of America

10 9 8 7 6 5 4 3 2 1

*To Christiane,
Anne-Marie, and Benoît*

PREFACE

Over the past decade the theory of amenable groups has continued to grow and many properties related to various mathematical topics have been discovered, with most recent results scattered throughout the literature. Although the end of the theory's development does not seem to be in sight, I wrote this book to give an expository report on the actual situation, and to present a detailed investigation on the major features of the subject. In the first sections I try to give a reasonably self-contained account, relying only on the fundamental properties of harmonic analysis. Chapter 2 presents the principal characterizations of amenable locally compact groups with the basic amenable groups discussed in Chapter 3. Chapters 4 and 5 provide more detailed coverage of the properties discussed in Chapter 2. The final chapter gives some supplementary description of amenable groups and the various directions into which amenability properties have been generalized. It is my hope that this text may facilitate further studies on the phenomenon of amenability.

I would like to express my profound gratitude to Professor Pierre Eymard, University of Nancy, who introduced me to this subject. I also thank Professor Peter Gerl, University of Salzburg, for his comments. I acknowledge the kind help and patience of my colleagues from the Mathematical Seminar of my institute, especially the assistance of Marco Charpentier, who read a large part of the manuscript and contributed valuable criticism. Finally I thank the Luxembourg Ministry of Cultural Affairs for its financial support.

JEAN-PAUL PIER

Luxembourg
April 1984

CONTENTS

1. **Preliminaries**   1
   1. Introduction, 1
   2. Background and Notations, 6
   3. General Considerations on Means and Averaging Processes, 23

2. **Fundamental Characterizations of Amenable Groups**   30
   4. Invariant Means on Various Function Spaces, 30
   5. Flows and Fixed Point Properties, 42
   6. Convergence to Invariance, 48
   7. Structural Properties, 61
   8. Amenability and Unitary Group Representations, 72
   9. Amenability and Convolution Operators, 82
   10. Existence of Approximate Units, 90
   11. Cohomological Properties, 98

3. **The Class of Amenable Groups**   112
   12. Examples of Amenable Groups, 112
   13. Stability Properties of Amenability, 117
   14. The Structure of Amenable Groups, 122

4. **Complements on the Characteristic Properties of Amenable Groups**   140
   15. Quasi-Invariance Properties, 140
   16. Supplementary Structural Properties, 157
   17. Supplementary Unitary Group Representation Properties, 182
   18. Supplementary Convolution Properties, 191
   19. Fourier Algebra Properties, 208

5. Complements on Invariant Means Properties 229

20. Existence of Invariant Means, 229
21. Kernels and Ranges of Invariant Means, 246
22. The Sets of Invariant Means, 264

6. The Phenomenon of Amenability 309

23. Complements on the Class of Amenable Groups, 309
24. Generalized Notions of Amenability, 342

Bibliography 377

Notation Index 405

Author Index 411

Subject Index 415

AMENABLE LOCALLY COMPACT GROUPS

1
PRELIMINARIES

1. INTRODUCTION

The etymological origin of the term "mean" is the middle of a segment. If a_1 and a_2 are real numbers, their arithmetical mean value $(a_1 + a_2)/2$ is the midpoint of the interval $[a_1, a_2]$. More generally, for given real numbers $a_1, \ldots, a_n$, one considers the arithmetical mean value

$$\frac{\sum_{i=1}^{n} a_i}{n} = \frac{a_1 + \cdots + a_n}{n}.$$

If these numbers occur with the frequencies $k_1, \ldots, k_n$, the mean value is given by

$$\frac{\sum_{i=1}^{n} k_i a_i}{\sum_{i=1}^{n} k_i} = \frac{k_1}{\sum_{i=1}^{n} k_i} a_1 + \cdots + \frac{k_n}{\sum_{i=1}^{n} k_i} a_n.$$

So in general, for the real numbers $a_1, \ldots, a_n$ affected with the weights $\alpha_1, \ldots, \alpha_n \in \mathbf{R}_+^*$ such that $\sum_{i=1}^{n} \alpha_i = 1$, one defines the mean value or average value $a = \sum_{i=1}^{n} \alpha_i a_i$; then

$$\inf\{a_1, \ldots, a_n\} \leq a \leq \sup\{a_1, \ldots, a_n\}.$$

The latter inequalities constitute the weakest conditions to be fulfilled by a mean value.

In 1904, extending Riemann integration, Lebesgue [344] considered bounded real-valued functions defined on $\mathbf{R}$ admitting integrals that satisfy the following six conditions:

(a) For all a, b, h, $\int_a^b f(x)\, dx = \int_{a+h}^{b+h} f(x - h)\, dx$; that is, the integral is translation-invariant.
(b) For all a, b, c, $\int_a^b f(x)\, dx + \int_b^c f(x)\, dx + \int_c^a f(x)\, dx = 0$.

(c) $\int_a^b [f(x) + \varphi(x)]\, dx = \int_a^b f(x)\, dx + \int_a^b \varphi(x)\, dx$.
(d) If $f \geq 0$ and $b > a$, then also $\int_a^b f(x)\, dx \geq 0$.
(e) $\int_0^1 1\, dx = 1$.
(f) If f_n increases to f, then the integral of f_n converges to the integral of f.

Whereas the first five conditions are independent, Lebesgue asked whether the last property may be independent of the other ones. He observed that it suffices to define the integral for characteristic functions. For any bounded subset E of $\mathbf{R}$ one has to determine a number $m(E) \in \mathbf{R}_+$ called the measure of E such that the following properties hold:

(a') Two bounded subsets, that may be superposed by translations, have equal measures (m is invariant by translations).
(b') The measure of a bounded subset, that is a finite or countable union of pairwise disjoint subsets, is the sum of the measures of these subsets (m is σ-additive).
(c') The interval $[0, 1]$ has measure 1 (m is normalized on $[0, 1]$).

Note that (a) is equivalent to (a') and (e) is equivalent to (c'); the conditions (c) and (f) imply (b'). The measure of the interval determined by a and b ($a < b$) is the length $b-a$, whether the interval contains a and b or not. If E is a bounded subset of $\mathbf{R}$, one considers any finite or countable union of intervals in which E is included. The sums of the lengths of these intervals admit a greatest lower bound which is called the outer measure $m_e(E)$ of E. Consider a bounded interval A containing E and define the inner measure of E to be $m_i(E) = m(A) - m_e(A \setminus E)$. If $m_i(E) = m_e(E)$, E is said to be Lebesgue measurable; the common value $m(E)$ satisfies the conditions (a'), (b'), (c'). In particular, among the Lebesgue measurable subsets are the bounded subsets of $\mathbf{R}$ that are Borel subsets (i.e., obtained from open subsets when one iterates the operations of countable unions and differences of subsets).

The notion of measurability is extended to unbounded subsets. Also for every $\mathbf{R}^n$ ($n \in \mathbf{N}^*$) there exists a Lebesgue measure that is invariant by isometries, σ-additive, and normalized on a specific bounded Lebesgue measurable subset, for instance the unit ball centered at the origin. As not all bounded subsets of $\mathbf{R}^n$ are Lebesgue measurable, in 1914 Hausdorff [253] raised the problem of associating to any bounded subset of $\mathbf{R}^n$ a number $m(E) \in \mathbf{R}_+$ such that

(1) $m(E_1) = m(E_2)$ if E_1 and E_2 may be superposed by isometries (m is invariant by isometries).
(2) $m(E_1 \cup E_2) = m(E_1) + m(E_2)$ whenever $E_1 \cap E_2 = \emptyset$ (m is finitely additive).
(3) $m(E_0) = 1$ for a specific E_0 (m is normalized on E_0).

Hausdorff solved the problem negatively for the case $n \geq 3$ by proving via the axiom of choice the existence of a partition $\{A, B, C, D\}$ of the unit sphere S_2 in $\mathbf{R}^3$ where D is countable. By appropriate rotations of 120° [resp. 180°] he succeeded in superposing A, B, C [resp. A and $B \cup C$]. If m did exist for $E_0 = S_2$, a contradiction would arise because then $m(D) = 0$ and $m(S_2) = 1$, $m(A) = m(B) = m(C) = \frac{1}{3}$, $2m(B) = m(B \cup C) = m(A) = \frac{1}{2}$. No solution exists either if E_0 is chosen to be a unit cube in $\mathbf{R}^n$ for $n \geq 3$.

In 1923 these questions were taken up by Banach [28]. He answered Hausdorff's problem affirmatively in case $n = 1$ [resp. $n = 2$] and the measure is normalized on a segment [resp. a unit square]; he thus succeeded in extending the Lebesgue integral to a nonnegative, finitely additive functional over all bounded functions of 1 or 2 variables that is invariant under isometries. Further investigations were made by Banach and Tarski [29] and the name of *Hausdorff-Banach-Tarski paradox* became attached to this phenomenon on $\mathbf{R}^n$ having opposite aspects for the cases $n = 1, 2$, and $n \geq 3$. Banach and Tarski say that two subsets of a euclidian space are equivalent for finite [resp. countable] decompositions if they may be decomposed into a partition formed by the same finite number of [resp. countably many] subsets that are pairwise congruent (i.e., isometric). If $n \geq 1$, any pair of subsets of $\mathbf{R}^n$ having interior points are equivalent for countable decompositions. If $n \geq 3$, but not if $n = 1, 2$, any pair of bounded subsets of $\mathbf{R}^n$ having interior points are equivalent for finite decompositions. Considering two bounded subsets A and B in $\mathbf{R}^n$ ($n \geq 3$) for which $A \subset B$ and A as well as $B \setminus A$ have nonvoid interiors, one concludes that there cannot exist a nonzero, nonnegative, finitely additive measure on all subsets of $\mathbf{R}^n$ that is invariant by isometries. Let us quote the comment of Banach and Tarski on the existence or nonexistence of *paradoxical decompositions*.

Le rôle que joue (l'axiome de Zermelo) dans nos raisonnements nous semble mériter l'attention. Envisageons, en effet, les deux théorèmes suivants qui résultent de nos recherches:

Deux polyèdres arbitraires sont équivalents par décomposition finie;

deux polygones différents, dont l'un est contenu dans l'autre, ne sont jamais équivalents par décomposition finie.

Or, on ne sait démontrer aucun de ces deux théorèmes sans faire appel à l'axiome du choix: ni le premier qui semble peut-être paradoxal, ni le second qui est en plein accord avec l'intuition. De plus, en analysant leurs démonstrations, on peut constater que l'axiome du choix intervient dans la démonstration du premier théorème sous une forme bien plus restreinte que dans le cas du second.

The problem may be stated in terms of the action of the isometry group on $\mathbf{R}^n$ and one studies in general the action of a transformation group G on an arbitrary set S in view of finding nonzero, nonnegative, finitely additive measures living on S that are normalized on a given subset A of S and remain invariant under the action of G. In 1929 von Neumann [414] observed that the dichotomy of the Hausdorff–Banach–Tarski paradox is not due to the difference of the structures of the corresponding euclidian spaces, but to the difference of the structures of the isometry groups. Studying Hausdorff's proof thoroughly he noticed that if $n \geq 3$, the rotation group contains the free group on two generators, whereas that is not the case for $n = 2$.

Hence the group itself plays the important role in the game, and the problem may be studied essentially in case $S = G$ and also $A = G$. The aim is to prove the existence of a finitely additive probability measure μ on G that is G-invariant, that is, for any subset E in G and any $a \in G$, $\mu(aE) = \mu(E)$. To μ one associates canonically a linear functional M on the vector space $l^\infty(G)$ of bounded complex-valued functions defined on G by putting $M(1_E) = \mu(E)$, where 1_E is the characteristic function of E; M verifies the conditions:

$$(\forall f \in l^\infty(G))f \geq 0 \Rightarrow M(f) \geq 0; \qquad M(1_G) = 1.$$

These properties are equivalent to the following one: For every bounded real-valued function f defined on G,

$$\inf f \leq M(f) \leq \sup f;$$

so M may be called a mean. Thus the problem consists in showing the existence of a mean M that is G-invariant, that is, for all $f \in l^\infty(G)$ and $a \in G$, $M(f) = M(g)$ with $g(x) = f(ax)$, $x \in G$.

In order to simplify the construction of the invariant Haar measure on a compact group, von Neumann [415] had the idea to consider almost periodic functions. This theory had been inaugurated by Bohr in 1924–1925 and exposed more systematically in 1932 [39]. The complex-valued function f on a group G is called almost periodic if, for every $\varepsilon > 0$, there exists a partition $\{A_1, \ldots, A_n\}$ of G such that, for all $x, y \in A_i$ ($i = 1, \ldots, n$), $c \in G$, $d \in G$, one has $|f(cxd) - f(cyd)| < \varepsilon$. If f is an almost periodic function, one may determine $M(f) \in \mathbf{C}$ such that, for every $\varepsilon > 0$, there exist $a_1, \ldots, a_n \in G$ with $|M(f) - (1/n)\sum_{i=1}^{n} f(ca_i d)| < \varepsilon$ whenever $c, d \in G$; $f \mapsto M(f)$ defines a unique mean on the space of almost periodic functions that is invariant for left and right translations as well as for inversion on G. Von Neumann's proof is an elementary one. Weil [552] specified the link existing between the space of almost periodic continuous functions and the space of continuous functions on a compact group. To any topological group G one may associate a compact group K and a continuous homomorphism h of G onto a dense subgroup of K; a complex-valued continuous function f on G is almost periodic if and only if there exists a complex-valued continuous function g on K such that $f = g \circ h$. The invariant mean on the space of almost periodic continuous functions defined on G is induced by the normalized Haar integral on the space of continuous functions defined on the compactification K.

INTRODUCTION

In particular, a continuous function f on $\mathbf{R}$ is almost periodic if and only if, for every $\varepsilon > 0$, there exists $T > 0$ such that in any interval of length T there is a point t satisfying $\sup\{|f(x) - f(x+t)| : x \in \mathbf{R}\} < \varepsilon$. The Bohr integral mean M is given by

$$M(f) = \lim_{T \to \infty} \frac{1}{2T} \int_{-T}^{T} f(x)\, dx.$$

We are interested in the existence of invariant means on spaces of Haar measurable functions on a locally compact group that are bigger than the space of almost periodic continuous functions, for instance the space of bounded uniformly continuous, bounded continuous or essentially bounded measurable functions.

The theory may be carried out on topological (not necessarily locally compact) groups, semigroups, transformation groups, and homogeneous spaces, but the fundamental ideas appear in the most striking manner on *locally compact groups*, and we focus on that situation.

The class of all locally compact Hausdorff groups G for which invariant means may be considered on $L^\infty(G, \lambda)$, λ being a left Haar measure, is formed by groups that are sufficiently far away from being free—an idea going back to von Neumann. It contains all abelian and more generally all solvable groups, as well as all compact groups. We should mention that von Neumann [414] was surprised to see abelian and finite groups sitting in one and the same interesting class of discrete groups.

It turns out that the class may be characterized by a great deal of other properties that do not appear to be related to the existence of invariant means at first glance. Various properties concern algebras of functions on the groups and reflect aspects of locally compact abelian groups or compact groups for which certain of these conditions may be established by more classical or more direct proofs. They generally express some kind of invariance or quasi-invariance. As a matter of fact, from different viewpoints many characterizations appear to be much more interesting than invariant-means properties. Originally these groups had been called "meßbar" by von Neumann [414], later they were termed groups with full Banach mean value [187]. But this well-behaving class deserves the name given to them by Day [112]—The groups not only admit an invariant mean, they are indeed *amenable*!

Basic facts about analysis and harmonic analysis are discussed in Section 2; general properties of means are stated in Section 3. Chapter Two gives a first insight into various equivalent characterizations of amenable groups. In Chapter Three we list examples of amenable groups and study the class of amenable groups. Chapters Four and Five give further details on necessary and/or sufficient conditions for locally compact Hausdorff groups to be amenable. Finally complementary questions are dealt with in Chapter Six.

We do not attempt to obtain the utmost general refinements of the statements. Also, we try to focus on amenability properties and do not intend to treat more particular or more general situations. For a global reference on the classification of locally compact Hausdorff groups one may consult Palmer [425]. Recent summaries on the theory of amenable groups are Greenleaf [226], Eymard [175], Bondar–Milnes [44], and Pier [436].

The comments made at the end of certain sections provide information on interesting results that could not appear in the core of the text. The notes are intended to trace historical developments, and we apologize in advance for any unavoidable errors or omissions in these citations. Within the sections, formulas are listed by one number; in a reference to another section, it is preceded by the number of that section. Numbers between brackets concern the bibliography.

2. BACKGROUND AND NOTATIONS

We recall some basic facts; for more information one should consult the main reference sources listed at the beginning of the Bibliography. The reader interested in a particular section may proceed immediately to that point and switch back occasionally to have a look at the corresponding part of this Section.

If $n \in \mathbf{N}^*$, S_n is the n-dimensional unit sphere; $\mathbf{T}$ is the torus $\mathbf{R}/\mathbf{Z}$. For any set S, we denote by $\mathfrak{P}(S)$ [resp. $\mathfrak{F}(S)$] the family of all [resp. all nonvoid finite] subsets of S. The cardinal of S is designated by card S; $\mathfrak{c}$ is the cardinality of the continuum. Finite sets are considered to be countable. In general, a net is supposed to be directed by $\prec$. Throughout the book we assume the axiom of choice to hold whenever necessary.

A. Function Spaces

The characteristic function of a subset A in a set is denoted by 1_A. Symbols representing complex-valued function spaces are equipped with subscripts $+$ or $\mathbf{R}$ if only nonnegative-valued or real-valued functions are considered. The support of a complex-valued function f is denoted by supp f.

If S is a topological space, we denote by $\mathfrak{B}(S)$ the class of all Borel subsets. We write $\mathscr{C}(S)$ for the Banach space of all complex-valued bounded continuous functions on S; if $f \in \mathscr{C}(S)$,

$$\|f\| = \sup\{|f(x)| : x \in S\}.$$

Any locally compact space is assumed to be a Hausdorff space. If S is a locally compact space, we denote by $\mathfrak{K}(S)$ the family of all nonvoid compact subsets in S. We write $\mathscr{K}(S)$ for the space of all complex-valued continuous (bounded) functions on S having compact support. Also $\mathscr{C}_0(S)$ is the Banach space of complex-valued (bounded) functions f on S that vanish at infinity, that is, for any $\varepsilon > 0$, there exists a compact subset K of S such that $|f(x)| < \varepsilon$ whenever $x \in S \setminus K$.

Every completely regular space S is homeomorphic with a dense subset of a compact space βS in such a way that any $f \in \mathscr{C}(S)$ may be extended to a unique element of $\mathscr{C}(\beta S)$; βS is the Stone–Čech compactification of S ([DS] IV.6.22).

The topological space is said to be completely separable if it admits a countable open basis.

B. Topological Vector Spaces

1. If X is a real or complex locally convex Hausdorff *topological vector space*, we denote by X' [resp. X^*] its algebraic [resp. topological] dual space consisting of all [continuous] linear functionals on X; if $\xi \in X$ and $F \in X'$, we write $\langle \xi, F \rangle$ for $F(\xi)$. We always assume X to be canonically embedded into $X^{*\prime}$. The weak topology, or w-topology, on X is the $\sigma(X, X^*)$-topology, that is, the weakest topology on X for which each $F \in X^*$ is continuous on X. If $\xi \in X$, the sets

$$U(\xi; F_1, \ldots, F_n; \varepsilon) = \{\eta \in X : |\langle \eta, F_i \rangle - \langle \xi, F_i \rangle| < \varepsilon; i = 1, \ldots, n\}$$

where $\varepsilon \in \mathbf{R}^*_+$ and $\{F_1, \ldots, F_n\}$ is a finite subset of X^*, form a basis of open neighborhoods of ξ in the w-topology of X. The weak-$*$-topology or w$*$-topology on X^* is the $\sigma(X^*, X)$-topology, that is, the weakest topology on X^* for which each $\xi \in X$ is continuous on X^*. If $F \in X^*$, the sets

$$V(F; \xi_1, \ldots, \xi_n; \varepsilon) = \{L \in X^* : |\langle \xi_i, L \rangle - \langle \xi_i, F \rangle| < \varepsilon; i = 1, \ldots, n\}$$

where $\varepsilon \in \mathbf{R}^*_+$ and $\{\xi_1, \ldots, \xi_n\}$ is a finite subset of X, form a basis of open neighborhoods of F in the w$*$-topology of X^*.

If X and Y are normed spaces, $\mathscr{L}(X, Y)$ is the normed space of all continuous (i.e., bounded) linear operators from X into Y; for $T \in \mathscr{L}(X, Y)$,

$$\|T\| = \sup\left\{\frac{\|T\xi\|_Y}{\|\xi\|_X} : \xi \in X, \xi \neq 0\right\}.$$

If Y is a Banach space, so is $\mathscr{L}(X, Y)$ ([DS] II.3.8). The space of all continuous endomorphisms of the normed space X is denoted by $\mathscr{L}(X)$; $\mathscr{L}(X)$ and X^* are Banach spaces in case X is. If X is a normed space and $T \in \mathscr{L}(X)$, we consider the adjoint operator $T^\sim$ defined by

$$\langle \xi, T^\sim(\eta) \rangle = \langle T(\xi), \eta \rangle; \xi \in X, \eta \in X^*.$$

We denote by $\mathscr{L}_{\mathbf{R}}(X, Y)$ [resp. $X^*_{\mathbf{R}}$] the space of all continuous linear operators [resp. functionals] corresponding to the underlying real spaces. If X, Y are partially ordered, $\mathscr{L}_+(X, Y)$ [resp. X^*_+] is the set of all elements F in $\mathscr{L}(X, Y)$ [resp. X^*] that are nonnegative, that is, $F(f) \geq 0$ whenever $f \in X, f \geq 0$.

We denote the convex hull of a subset E in a topological vector space by co E. If X is a locally convex Hausdorff topological vector space and $E \subset X$, then the closed convex hull $\overline{\text{co}}\, E$ of E coincides with the w-closure of co E ([B EVT] Chap. IV, Section 2, No. 3, corollaire 2). The weakly closed convex hull of a weakly compact subset of a Banach space is also weakly compact ([DS] V.6.4).

2. Let A be a an *algebra* over $\mathbf{C}$; A is called involutive algebra if it admits an involution, that is, a mapping $x \mapsto x^*$ of A into itself such that, for all $x, y \in A$ and $\alpha \in \mathbf{C}, (x^*)^* = x, (x + y)^* = x^* + y^*, (xy)^* = y^*x^*$, and $(\alpha x)^* = \bar{\alpha} x^*$. The set of hermitian elements $x = x^*$ in A is denoted by $\mathbf{A}_h$. If A admits a unit u, an element x of A is called unitary if $x^*x = u = xx^*$, that is, $x^* = x^{-1}$. The unitary elements form the unitary group A_u in A. If A is a normed algebra, we suppose that $\|x^*\| = \|x\|$ whenever $x \in A$. Any subalgebra of A that is closed under involution is termed $*$-subalgebra of A.

If A is a *Banach algebra*, A_1 is the unit sphere in A, and B is a dense linear subspace of A, then $\overline{B \cap A_1} = A_1$. As a matter of fact, if $x \in A_1$ and $\varepsilon > 0$, there exists $y \in B$ such that $\|y\| > 0$ and $\|x - y\| < \varepsilon/2$; for $z = y/\|y\| \in B \cap A_1$, we have

$$\|x - z\| \leq \|x - y\| + \left\|y - \frac{y}{\|y\|}\right\| = \|x - y\| + |\|y\| - 1|$$

$$= \|x - y\| + |\|y\| - \|x\|| \leq 2\|x - y\| < \varepsilon.$$

Any normed [resp. involutive, Banach] algebra A over $\mathbf{C}$ may be extended to a normed [resp. involutive, Banach] algebra $\tilde{A}$ that consists of all elements (x, α) in $A \times \mathbf{C}$ with $\|(x, \alpha)\|_{\tilde{A}} = \|x\|_A + |\alpha|$ and admits the unit $u = (0_A, 1)$, multiplication being defined by $(x, \alpha)(y, \beta) = (xy + \alpha y + \beta x, \alpha\beta)$ for $(x, \alpha), (y, \beta) \in A \times \mathbf{C}$ [and $(x, \alpha)^* = (x^*, \bar{\alpha})$ for $(x, \alpha) \in A \times \mathbf{C}$] ([HR] C.3). The spectrum of an element x in A is the set sp x consisting of all $\alpha \in \mathbf{C}$ such that $x - \alpha u$ is not invertible in $\tilde{A}$; sp x is nonvoid and compact. The spectral radius of x is

$$r(x) = \sup\{|\alpha| : \alpha \in \text{sp}\, x\} = \lim_{n \to \infty} \|x^n\|^{1/n}; \quad r(x) \leq \|x\|.$$

The involutive Banach algebra A is said to be a C^*-algebra if $\|x^*x\| = \|x\|^2$ whenever $x \in A$.

3. If A is a *commutative Banach algebra* with unit, we consider its structure space $\hat{A}$ consisting of all (continuous) multiplicative linear functionals [characters] on A. For $x \in A$, the (Fourier–) Gelfand transform $\hat{x}$ is defined by

$$\hat{x}(\chi) = \chi(x),$$

$\chi \in \hat{A}$. On $\hat{A}$ one considers the Gelfand topology, that is, the weakest topology for which all the functions $\hat{x}$ ($x \in A$) are continuous; $\hat{A}$ is compact, and $x \mapsto \hat{x}$ is a homomorphism of A onto a subalgebra of $\mathscr{C}(\hat{A})$. In particular, if A is a commutative C^*-algebra with unit, A and $\mathscr{C}(\hat{A})$ are isometrically isomorphic ([HR] C.21,25,28). Any compact space X may be identified with $\overline{\mathscr{C}(X)}$ ([HR] C.32); in particular, if S is a locally compact space, the compact space βS may be identified with $\overline{\mathscr{C}(\beta S)}$, that is, with $\overline{\mathscr{C}(S)}$.

4. We say that a complex vector space E admits a *prehilbert structure* if there exists a function $f: E \times E \to \mathbf{C}$ that is linear in the first variable, hermitian (i.e., $f(x, y) = \overline{f(y, x)}$ whenever $x, y \in E$), and nonnegative-definite (i.e., $f(x, x) \geq 0$ whenever $x \in E$). The mapping $x \mapsto f(x, x)^{1/2}$ is a seminorm on E. Let $N = \{x \in E : f(x, x) = 0\}$; then E/N is a normed space with $\|\dot{x}\| = f(x, x)^{1/2}$, $x \in \dot{x}$, and $\dot{x} \in E/N$. The completion of E/N is a Hilbert space.

5. If $\mathscr{H}$ is a *Hilbert space*, $\mathscr{L}(\mathscr{H})$ may be endowed with the strong [resp. weak, ultrastrong, ultraweak] topology defined by the seminorms $T \mapsto \|T\xi\|$, $\xi \in \mathscr{H}$ [resp. $T \mapsto |(T\xi|\eta)|, \xi, \eta \in \mathscr{H}$;

$$T \mapsto \left(\sum_{i=1}^{\infty} \|T\xi_i\|^2\right)^{1/2}, \xi_1, \xi_2, \ldots \in \mathscr{H}, \sum_{i=1}^{\infty} \|\xi_i\|^2 < \infty; \quad T \mapsto \left|\sum_{i=1}^{\infty} (T\xi_i|\eta_i)\right|,$$

$$\xi_1, \xi_2, \ldots \in \mathscr{H}, \eta_1, \eta_2, \ldots \in \mathscr{H}, \sum_{i=1}^{\infty} \|\xi_i\|^2 < \infty, \sum_{i=1}^{\infty} \|\eta_i\|^2 < \infty\bigg].$$

On bounded subsets of $\mathscr{L}(\mathscr{H})$ the strong [resp. weak] topology coincides with the ultrastrong [resp. ultraweak] topology ([140] Chapitre I, Section 3). For $T \in \mathscr{L}(\mathscr{H})$ we consider the adjoint operator $T^* \in \mathscr{L}(\mathscr{H})$ defined by $(T\xi|\eta) = (\xi|T^*\eta), \xi \in \mathscr{H}, \eta \in \mathscr{H}$.

C. Measures and Integration

1. Let X be a locally compact space. We denote by $\mathscr{M}(X)$ the space of all complex (Radon) measures on X and by $\mathscr{M}^1(X)$ the Banach space of all complex bounded measures on X; we have $\mathscr{M}^1(X) = \mathscr{C}_0(X)^*$. Subscripts + [resp. **R**] indicate that we consider nonnegative [resp. real] measures. We denote by $\operatorname{supp}\mu$ the support of the measure μ. Let $\mathscr{M}_c^1(X)$ be the dense subset of $\mathscr{M}^1(X)$ formed by the measures in $\mathscr{M}^1(X)$ having compact support; $\mathscr{M}_f^1(X)$ consists of the measures in $\mathscr{M}^1(X)$ having finite support. We denote by $M^1(X)$ the convex set formed by the probability measures on X, that is, all $\mu \in \mathscr{M}_+^1(X)$ for which $\|\mu\| = \langle 1_X, \mu\rangle = \int_X d\mu(x) = 1$. The Dirac measure $\delta_a \in M^1(X)$ at $a \in X$ is defined by $\langle \varphi, \delta_a \rangle = \varphi(a)$ for $\varphi \in \mathscr{C}_0(X)$. We denote by $D^1(X)$ the set of Dirac measures on X. Let also $M^0(X) = \{\mu \in \mathscr{M}^1(X) : \langle 1_X, \mu \rangle = 0\}$.

2. Assume now that $\mu \in \mathcal{M}_+^1(X)$. We write $\mathcal{S}(X, \mu)$ for the space of all complex-valued μ-measurable simple functions on X. We consider the space $\mathcal{L}^\infty(X, \mu)$ of all complex-valued μ-measurable functions f on X for which $|f|$ is essentially bounded; $L^\infty(X, \mu)$ is the set of equivalence classes in $\mathcal{L}^\infty(X, \mu)$ where one identifies $f_1, f_2 \in \mathcal{L}^\infty(X, \mu)$ for which ess sup $|f_1 - f_2| = 0$. Then $L^\infty(X, \mu)$ is equipped with the norm defined by $\|f\|_\infty = $ ess sup $|f|$, $f \in L^\infty(X, \mu)$. In particular, if X is discrete, we write $l^\infty(X)$ for the space of complex-valued bounded functions defined on X.

For $p \in [1, \infty[$, $\mathcal{L}^p(X, \mu)$ is the space of all complex-valued μ-measurable functions f on X such that $\int_X |f|^p = \int_X |f(x)|^p d\mu(x) < \infty$; $L^p(X, \mu)$ is the set of equivalence classes in $L^p(X, \mu)$ where one identifies $f_1, f_2 \in \mathcal{L}^p(X, \mu)$ for which $\int_X |(f_1 - f_2)(x)|^p d\mu(x) = 0$. Then $L^p(X, \mu)$ is equipped with the norm defined by $\|f\|_p = (\int_X |f(x)|^p d\mu(x))^{1/p}$, $f \in L^p(X, \mu)$.

All spaces $L^p(X, \mu)$ $(1 \le p \le \infty)$ are Banach spaces. If $p \in]1, \infty[$, we define $p' \in]1, \infty[$ by $1/p + 1/p' = 1$; we also put $1' = \infty$. If $1 \le p < \infty$, $L^p(X)^* = L^{p'}(X)$; the duality is given by the relation

$$\langle f, g \rangle = \int_X f(x) g(x) \, d\mu(x)$$

for $f \in L^p(X)$, $g \in L^{p'}(X)$. In case $1 \le p < \infty$, $f \in L^p(X)$ and $g \in L^{p'}(X)$, we also put

$$(f|g) = \langle f, \bar{g} \rangle.$$

In particular, $L^2(X)^* = L^2(X)$ and $L^2(X)$ is a Hilbert space for $(.|.)$. If $1 \le p < \infty$, we define $P^p(X, \mu) = \{f \in L_+^p(X, \mu) : \|f\|_p = 1\}$. Let $L^0(X, \mu) = \{f \in L^1(X, \mu) : \int f d\mu = 0\}$.

If X is discrete and $1 \le p < \infty$, we also write $l^p(X)$ instead of $L^p(X)$.

D. Topological Groups

1. If A and B are nonvoid subsets of a *group* G, one considers $AB = \{ab : a \in A, b \in B\}$. We denote by e the identity element of G; $a^0 = e$ for any $a \in G$. For every $f : G \to \mathbf{C}$, we define

$$\check{f}(x) = f(x^{-1}), \tilde{f}(x) = \overline{f(x^{-1})},$$

$x \in G$; if $f : G \to \mathbf{C}$ and $a \in G$, we put

$$_a f(x) = f(ax), f_a(x) = f(xa),$$

$x \in G$. If $f : G \to \mathbf{C}$ and $a \in G$, we also consider $L_a(f) = {}_{a^{-1}} f$; for all a, $b \in G$, we have $L_{ab} f = L_a(L_b f)$. If X is a set of complex-valued [resp. real-valued] functions on G, we denote by $\mathcal{N}(X)$ the complex [resp. real] linear span of $\{f - {}_a f : f \in X, a \in G\}$.

2. Any *topological group* G is assumed to be T_0, hence also completely regular ([HR I] p. 83). We denote by $\mathscr{RUC}(G)$ [resp. $\mathscr{LUC}(G)$] the Banach space consisting of the complex-valued bounded functions f on G that are right [resp. left] uniformly continuous, that is, for every $\varepsilon > 0$, there exists a neighborhood U of e such that $|f(x) - f(x')| < \varepsilon$ for all $x, x' \in G$ satisfying $x'x^{-1} \in U$ [resp. $x^{-1}x' \in U$]. The property signifies that $\|_{x'}f - _xf\| < \varepsilon$ [resp. $\|f_{x'} - f_x\| < \varepsilon$] for all $x, x' \in G$ such that $x'x^{-1} \in U$ [resp. $x^{-1}x' \in U$]. For $f \in l^\infty(G)$, we have $f \in \mathscr{RUC}(G)$ [resp. $f \in \mathscr{LUC}(G)$] if and only if the mapping $x \mapsto _xf$ [resp. $x \mapsto f_x$] from G into $l^\infty(G)$ is right [resp. left] uniformly continuous. For any $f \in \mathscr{RUC}(G)$ [resp. $f \in \mathscr{LUC}(G)$] and any $a \in G$, we have $_af, f_a \in \mathscr{RUC}(G)$ [resp. $_af, f_a \in \mathscr{LUC}(G)$]; we say that $\mathscr{RUC}(G)$ and $\mathscr{LUC}(G)$ are (translation) invariant. We also consider the Banach space $\mathscr{UC}(G) = \mathscr{RUC}(G) \cap \mathscr{LUC}(G)$ of all complex-valued bounded uniformly continuous functions on G. The mappings $f \mapsto \check{f}$ and $f \mapsto \tilde{f}$ define bijections from $\mathscr{RUC}(G)$ onto $\mathscr{LUC}(G)$, from $\mathscr{LUC}(G)$ onto $\mathscr{RUC}(G)$ and from $\mathscr{UC}(G)$ onto itself.

3. We now consider a *locally compact group* G. Recall that it is assumed to be a Hausdorff space. We denote by G_d the group G equipped with the discrete topology. The locally compact group G admits a left Haar measure (defined uniquely up to a positive multiplicative constant) denoted by λ or $d\lambda(x)$ or dx; λ is a nonnegative nonzero measure that is left invariant, that is, for all $f \in L^1(G) = L^1(G, \lambda)$ and $a \in G$, $\int _af = \int f$. If A is a λ-measurable subset of G, we put $|A| = \lambda(A)$. If $0 < |A| < \infty$, we also consider the mapping

$$\xi_A : x \mapsto \frac{1}{|A|} 1_A$$

defined on G. For every open subset U in G, $|U| > 0$. For every λ-measurable subset A of G and every $a \in G$, $|aA| = |A|$. We denote by $\mathfrak{H}(G)$ the family of Haar measurable subsets of G and by $\mathfrak{K}_0(G)$ the family of compact subsets of G having nonzero measure. On the locally compact group G the Haar measure is finite if and only if G is compact; in that case, we always assume the Haar measure to be normalized, that is, $\int_G d\lambda(x) = |G| = 1$. If G is discrete, a Haar measure is given by $|A| = \operatorname{card} A$ for $A \subset G$.

There exists a continuous homomorphism Δ of the locally compact group G into the multiplicative group $\mathbf{R}_+^*$, called modular function, such that, for all $f \in L^1(G)$ and $a \in G$,

$$\int f(xa) \Delta(a) \, dx = \int f(x) \, dx$$

and

$$\int f(x^{-1}) \Delta(x^{-1}) \, dx = \int f(x) \, dx.$$

For $f: G \to \mathbf{C}$ and $a \in G$, we define $R_a f = \Delta(a^{-1}) f_{a^{-1}}$. Then for all $a, b \in G$, we have $R_{ab} f = R_a(R_b f)$. If $f \in L^1(G)$ and $a \in G$, $\|L_a f\|_1 = \|R_a f\|_1 = \|f\|_1$. In case $\Delta = 1_G$, the group G is called unimodular; locally compact abelian groups, discrete groups, and compact groups are unimodular.

If $f \in L^p(G)$, $1 \leq p < \infty$, the mapping $x \mapsto {}_x f$ of G into $L^p(G)$ is right uniformly continuous, whereas the mapping $x \mapsto f_x$ of G into $L^p(G)$ is continuous ([HR] 20.4): Given $\varepsilon > 0$, there exists a neighborhood U of e in G such that $\|_x f - {}_{x'} f\|_p < \varepsilon$ whenever $x, x' \in G$ with $xx'^{-1} \in U$; given $\varepsilon > 0$ and $x \in G$, there exists a neighborhood V of x in G such that $\|f_x - f_{x'}\|_p < \varepsilon$ whenever $x' \in V$.

We consider the closed linear subspace $\mathcal{M}_a^1(G)$ of $\mathcal{M}^1(G)$ consisting of all bounded measures on G that are absolutely continuous with respect to λ; $\mathcal{M}_a^1(G) = \{f\lambda : f \in L^1(G)\}$ may be identified with $L^1(G)$. We denote by $\mathcal{M}_s^1(G)$ the closed linear subspace of all bounded measures on G that are singular with respect to the Haar measure. The closed linear subspace $\mathcal{M}_d^1(G)$ of $\mathcal{M}^1(G)$ consists of all purely discontinuous measures $\mu = \sum_{i=1}^\infty \alpha_i \delta_{a_i}$, where $\alpha_i \in \mathbf{C}$, $a_i \in G$ for any $i \in \mathbf{N}^*$, and $\sum_{i=1}^\infty |\alpha_i| < \infty$. If G is nondiscrete, for every $\mu \in \mathcal{M}^1(G)$, there exists a unique decomposition $\mu = \mu^a + \mu^s + \mu^d$ where $\mu^a \in \mathcal{M}_a^1(G)$, $\mu^s \in \mathcal{M}_s^1(G)$, and $\mu^d \in \mathcal{M}_d^1(G)$; also $\|\mu\| = \|\mu^a\| + \|\mu^s\| + \|\mu^d\|$. If G is discrete, $\mathcal{M}^1(G) = \mathcal{M}_d^1(G) = \mathcal{M}_a^1(G)$, and $\mathcal{M}_s^1(G) = \{0\}$.

On $\mathcal{M}^1(G)$ a Banach algebra structure is defined by the convolution product: If $\mu_1, \mu_2 \in \mathcal{M}^1(G)$,

$$\langle \varphi, \mu_1 * \mu_2 \rangle = \int_G \varphi(z) \, d(\mu_1 * \mu_2)(z)$$

$$= \int_G \int_G \varphi(xy) \, d\mu_1(x) \, d\mu_2(y), \varphi \in \mathscr{C}_0(G);$$

$$\mu_1 * \mu_2 \in \mathcal{M}^1(G), \|\mu_1 * \mu_2\| \leq \|\mu_1\| \|\mu_2\|$$

and $\operatorname{supp}(\mu_1 * \mu_2) \subset \operatorname{supp} \mu_1 \operatorname{supp} \mu_2$. If $\mu \in \mathcal{M}^1(G)$ and $n \in \mathbf{N}^*$, we write μ^{*n} for the convolution product of n factors μ; let also $\mu^{*0} = \delta_e$.

If $\mu \in \mathcal{M}^1(G)$, $g \in \mathscr{L}^p(G)$, $1 \leq p < \infty$ [resp. $p = \infty$], $\mu * g(x) = \int g(y^{-1}x) \, d\mu(y)$, $g * \mu(x) = \int g(xy^{-1}) \Delta(y^{-1}) \, d\mu(y)$ is defined for [locally] μ-almost every $x \in G$. Via Hölder's inequality we may consider $\mu * g, g * \mu \in L^p(G)$; also $\|\mu * g\|_p \leq \|\mu\| \|g\|_p$ ([HR] 20.12-13). If $f \in L^1(G)$, $g \in L^p(G)$, $1 \leq p < \infty$ [resp. $p = \infty$], $f * g(x) = \int f(y) g(y^{-1} x) \, dy = \int f(xy) g(y^{-1}) \, dy$ for [locally] λ-almost every $x \in G$. We consider $f * g \in L^p(G)$ and have $\|f * g\|_p \leq \|f\|_1 \|g\|_p$. For complex-valued functions f, g defined on G we write

$$L_f g(x) = f * g(x) = \int f(xy) g(y^{-1}) \, dy = \left[\int {}_x f(y) \check{g}(y) \, dy \right]$$

BACKGROUND AND NOTATIONS

($x \in G$) whenever the expression makes sense. Hölder's inequality implies that, if $f \in L^p(G)$, $g \in L^{p'}(G)$ with $1 < p < \infty$, and $1/p + 1/p' = 1$, then $f * \check{g} \in \mathscr{C}_0(G)$ and $\|f * \check{g}\| \leq \|f\|_p \|g\|_{p'}$. If $f \in L^1(G)$, $g \in L^\infty(G)$, then $f * g \in \mathscr{RUC}(G)$ and $\|f * g\| \leq \|f\|_1 \|g\|_\infty$; also $g * \check{f} \in \mathscr{LUC}(G)$ and $\|g * \check{f}\| \leq \|f\|_1 \|g\|_\infty$ ([HR] 20.16). If $f \in \mathscr{LUC}(G)$ and $\mu \in \mathscr{M}^1(G)$, then $\mu * f \in \mathscr{LUC}(G)$; in particular, if $f \in \mathscr{K}(G)$ and $\mu \in \mathscr{M}_c^1(G)$, we have $\mu * f \in \mathscr{K}(G)$.

Let $f, g \in \mathscr{L}^1(G)$; then

$$f * g(x) = \int f(y) g(y^{-1}x) \, dy = \int f(xy) g(y^{-1}) \, dy$$

$$= \int f(y^{-1}) g(yx) \Delta(y^{-1}) \, dy = \int f(xy^{-1}) g(y) \Delta(y^{-1}) \, dy$$

for λ-almost every $x \in G$. Also $L^1(G)$, equipped with the convolution product, is a Banach subalgebra of $\mathscr{M}^1(G)$, called the group algebra of G. It is commutative if and only if G is abelian. From the general property of Banach algebras considered in B.2 we deduce that the closed semigroup $M^1(G)$ [resp. $P^1(G), M_d^1(G)$] admits the dense subsemigroup $M_c^1(G)$ [resp. $P^1(G) \cap \mathscr{K}(G), M_f^1(G)$].

If $\mu \in \mathscr{M}^1(G)$, we define $d\mu^*(x) = \overline{d\mu(x^{-1})}$. For $\varphi \in \mathscr{C}_0(G)$, we have $\langle \varphi, \mu^* \rangle = \overline{\langle \tilde{\varphi}, \mu \rangle}$. The mapping $\mu \mapsto \mu^*$ is an involutive isometric automorphism of $\mathscr{M}^1(G)$. In particular, for $a \in G$, $\delta_a^* = \delta_{a^{-1}}$. Consider $f \in L^1(G)$ and let $\mu = f\lambda$. Then for any $\varphi \in \mathscr{C}_0(G)$,

$$\langle \varphi, \mu^* \rangle = \int \overline{\tilde{\varphi}(x) \, d\mu(x)} = \int \varphi(x^{-1}) \overline{f(x)} \, dx = \int \varphi(x) \tilde{f}(x) \Delta(x^{-1}) \, dx$$

hence $\mu^* = (f\lambda)^* = (\check{\Delta} \tilde{f})\lambda$. In general, for any $f: G \to \mathbf{C}$, we define

$$f^* = \check{\Delta} \tilde{f} = (\Delta \tilde{f})^{\check{}}.$$

The mapping $f \mapsto f^*$ defines an involutive isometric automorphism on $L^1(G)$ and maps $P^1(G)$ onto $P^1(G)$.

E. Convolution Formulas

We list here a collection of formulas which may be checked readily for a locally compact group G.

If $f: G \to \mathbf{C}$ and $a \in G$,

$$({}_a f)^{\check{}} = \check{f}_{a^{-1}}, \, (f_a)^{\check{}} = {}_{a^{-1}} \check{f}, \, (L_a f)^* = R_{a^{-1}} f^*. \tag{1}$$

If the following expressions make sense for $f: G \mapsto \mathbf{C}$, $g: G \mapsto \mathbf{C}$, $a \in G$, we have the corresponding identities:

$$\delta_a * f = {}_{a^{-1}}f = L_a f \qquad f * \delta_a = R_a f, \tag{2}$$

$$(f * g)^{\vee} = \check{g} * \check{f}, \tag{3}$$

$${}_a(f * g) = ({}_a f) * g, \tag{4}$$

$$(f * g)_a = f * (g_a), \tag{5}$$

$$f * (L_a g) = (R_a f) * g, \tag{6}$$

$$(f_a) * g = \Delta(a^{-1}) f * ({}_a g), \tag{7}$$

$$({}_a f)^* * g = f^* * ({}_{a^{-1}} g), \tag{8}$$

$$({}_a f | g) = (f|_{a^{-1}} g) = f * \tilde{g}(a), \quad (g_a | f) = f^* * g(a). \tag{9}$$

In particular, for $f \in L^2(G)$, $a \in G$,

$$({}_a f | f) = f * \tilde{f}(a), \quad (f_a | f) = f^* * f(a) \tag{10}$$

hence

$$f * \tilde{f}(e) = f^* * f(e) = \|f\|_2^2. \tag{11}$$

If $\varphi, \psi \in L^1(G)$ and $f \in L^\infty(G)$,

$$(\varphi | f * \tilde{\psi}) = (\varphi * \psi | f) = (\psi | \varphi^* * f). \tag{12}$$

If $\mu \in \mathcal{M}^1(G)$ [resp. $\mu \in \mathcal{M}^1_\mathbf{R}(G)$] and $f \in L^p(G)$, $g \in L^{p'}(G)$ with $1 < p < \infty$, $1/p + 1/p' = 1$ or $p = 1$, $p' = \infty$,

$$(f | \mu * g) = (\mu^* * f | g)$$

$$[\text{resp. } \langle f, \mu * g \rangle = \langle \mu^* * f, g \rangle]. \tag{13}$$

F. Approximate Units

We begin by recalling standard definitions.

The normed algebra A is said to have *left* [resp. *right*] *approximate units* (*bounded* by $c \in \mathbf{R}^*_+$) if, given $a \in A$ and $\varepsilon > 0$, there exists $u \in A$ (with $\|u\| \leq c$) such that $\|a - ua\| < \varepsilon$ [resp. $\|a - au\| < \varepsilon$]. It is said to have *multiple left* [resp. *right*] *approximate units* (*bounded* by $c \in \mathbf{R}^*_+$) if, given $a_1, \ldots, a_n \in A$, $n \in \mathbf{N}^*$, and $\varepsilon > 0$, there exists $u \in A$ (with $\|u\| \leq c$) such

BACKGROUND AND NOTATIONS

that $\|a_j - ua_j\| < \varepsilon$ [resp. $\|a_j - a_j u\| < \varepsilon$] whenever $j = 1, \ldots, n$; the latter condition says that there exists a net $(u_i)_{i \in I}$ in A (bounded by c) such that ($\|u_i\| \leq c$ whenever $i \in I$ and) $\lim_i \|a - u_i a\| = 0$ [resp. $\lim_i \|a - au_i\| = 0$] whenever $a \in A$.

For the sake of completeness we add the proofs of two important general results.

Proposition 2.1. *Let A be a normed algebra. If A admits left [resp. right] approximate units bounded by $c \in \mathbf{R}_+^*$, then it admits multiple approximate left [resp. right] units bounded by c.*

Proof (of the left-hand version). We may add a formal unit u to the abstract algebra A. Let $a_1, \ldots, a_n \in A$ and $\varepsilon > 0$; we put $\delta = \varepsilon((1 + c)^{n+1} + \sup\{\|a_j\| : j = 1, \ldots, n\})^{-1}$. By assumption there exists $u_1 \in A$ such that $\|u_1\| \leq c$ and $\|(u - u_1)a_1\| = \|a_1 - u_1 a_1\| < \delta$; we may determine inductively $u_1, \ldots, u_n \in A$ such that, for every $j = 1, \ldots, n$, $\|u_j\| \leq c$ and

$$\|(u - u_j)(u - u_{j-1}) \cdots (u - u_1)a_i\|$$
$$= \|(u - u_{j-1}) \cdots (u - u_1)a_i - u_j(u - u_{j-1}) \cdots (u - u_1)a_i\| < \delta.$$

Now let

$$v = u - (u - u_n) \cdots (u - u_1) = u_1 + \cdots + u_n + \cdots + (-1)^{n+1} u_n \cdots u_1.$$

Then for every $j = 1, \ldots, n$,

$$\|a_j - va_j\|$$
$$= \|(u - v)a_j\| = \|((u - u_n) \cdots (u - u_{j+1}))((u - u_j) \cdots (u - u_1)a_j)\|$$
$$\leq (1 + c)^{n-j}\delta < (1 + c)^n \delta.$$

There also exists $w \in A$ such that $\|w\| \leq c$ and $\|v - wv\| < \delta$. Thus for every $j = 1, \ldots, n$,

$$\|a_j - wa_j\| \leq \|a_j - va_j\| + \|(v - wv)a_j\| + \|w(a_j - va_j)\|$$
$$\leq \|a_j - va_j\| + \|v - wv\| \|a_j\| + \|w\| \|a_j - va_j\|$$
$$< (1 + c)^n \delta + \|a_j\|\delta + c(1 + c)^n \delta = \left((1 + c)^{n+1} + \|a_j\|\right)\delta \leq \varepsilon. \quad \square$$

Proposition 2.2. *Let G be a locally compact group.*
  (a) *Let $p \in]1, \infty[$. Given $f_1, \ldots, f_n \in L^p(G)$ and $\varepsilon \in \mathbf{R}_+^*$, there exists $\varphi = \varphi^* \in P^1(G) \cap \mathcal{K}(G)$ such that $\|\varphi * f_i - f_i\|_p < \varepsilon$ whenever $i = 1, \ldots, n$.*

(b) *The Banach algebra $L^1(G)$ admits hermitian multiple left and right approximate units belonging to $P^1(G) \cap \mathcal{K}(G)$.*

The element $\varphi \in \mathcal{K}(G)$ [resp. the approximate units] may be chosen in order to have their supports in a preassigned compact neighborhood U of e.

Proof. Let $p \in [1, \infty[$ and consider $f_1, \ldots, f_n \in L^p(G)$. [If $p = 1$, let $f_{n+i} = f_i^* \in L^1(G)$ for $i = 1, \ldots, n$]. Given $\varepsilon > 0$, there exists a compact neighborhood U' of e such that $U' \subset U$ and $\|_{y^{-1}}f_j - f_j\|_p < \varepsilon$ whenever $y \in U'$ and $j = 1, \ldots, n$ [resp. $j = 1, \ldots, 2n$]. Let V be a compact symmetric neighborhood of e in G such that $V^2 \subset U'$ and let $\varphi = \xi_V^* * \xi_V \in P^1(G)$;

$$\varphi = \varphi^* = \xi_V^* * \xi_V \in \mathscr{LUC}(G) \text{ (Section 2D.3),}$$

and

$$\operatorname{supp} \varphi \subset V^2 \subset U' \subset U.$$

(a) Let $p \in]1, \infty[$. For every $i = 1, \ldots, n$, Hölder's inequality implies that, for all $g \in L^{p'}(G)$ with $1/p + 1/p' = 1$,

$$|\langle \varphi * f_i - f_i, g \rangle| = \left| \int \left(\int \varphi(y) f_i(y^{-1}x) \, dy - f_i(x) \int \varphi(y) \, dy \right) g(x) \, dx \right|$$

$$\leq \left| \int \left(\int (_{y^{-1}}f_i - f_i)(x) g(x) \, dx \right) \varphi(y) \, dy \right|$$

$$\leq \int \|_{y^{-1}}f_i - f_i\|_p \|g\|_{p'} \varphi(y) \, dy$$

$$\leq \varepsilon \|g\|_{p'}.$$

Hence

$$\|\varphi * f_i - f_i\|_p = \sup \left\{ |\langle \varphi * f_i - f_i, g \rangle| : g \in L^{p'}(G), \|g\|_{p'} \leq 1 \right\} \leq \varepsilon.$$

(b) Let $p = 1$. For every $j = 1, \ldots, 2n$,

$$\|\varphi * f_j - f_j\|_1 = \int |\varphi * f_j(x) - f_j(x)| \, dx$$

$$= \int \left| \int \varphi(y) f_j(y^{-1}x) \, dy - f_j(x) \int \varphi(y) \, dy \right| dx$$

$$\leq \int \int |f_j(y^{-1}x) - f_j(x)| \varphi(y) \, dy \, dx$$

$$= \int \left(\int |f_j(y^{-1}x) - f_j(x)| \, dx \right) \varphi(y) \, dy < \varepsilon$$

hence for $i = 1, \ldots, n$,

$$\|\varphi * f_i - f_i\|_1 < \varepsilon$$

and also, by isometry of the involution,

$$\|f_i * \varphi - f_i\|_1 = \|\varphi^* * f_i^* - f_i^*\|_1 = \|\varphi * f_{n+i} - f_{n+i}\|_1 < \varepsilon. \qquad \square$$

G. Banach Modules, Invariant Function Spaces and Arens Multiplication

1. A discrete semigroup S is said to *act* on a set X (from the left) if there exists a mapping

$$(x, \xi) \mapsto x\xi$$

$$S \times X \to X$$

such that $(xy)\xi = x(y\xi)$ whenever $x, y \in S$ and $\xi \in X$; S is a *transformation semigroup* on X. If $A \subset S$ and $Y \subset X$ we say that Y is A-*invariant* in case $x\xi \in Y$ whenever $x \in A$ and $\xi \in Y$.

Let A be a complex Banach algebra and let G be a topological group. We say that the complex Banach space X constitutes a *left Banach A-module* [resp. *a left Banach G-module*] if there exists a mapping

$$(a, \xi) \mapsto a\xi$$

$$A \times X \to X$$

$$[\text{resp. } (x, \xi) \mapsto x\xi$$

$$G \times X \to X]$$

having the following properties:

(i) X is a left A-module, that is,

$$(\forall a \in A)(\forall b \in A)(\forall \xi \in X)(a + b)\xi = a\xi + b\xi,$$

$$(\forall a \in A)(\forall \xi \in X)(\forall \eta \in X)a(\xi + \eta) = a\xi + a\eta,$$

$$(\forall \alpha \in \mathbb{C})(\forall a \in A)(\forall \xi \in X)(\alpha a)\xi = \alpha(a\xi) = a(\alpha\xi),$$

$$(\forall a \in A)(\forall b \in A)(\forall \xi \in X)(ab)\xi = a(b\xi)$$

[resp. X is a left G-module, that is,

$$(\forall x \in G)(\forall \xi \in X)(\forall \eta \in X)x(\xi + \eta) = x\xi + x\eta,$$

$$(\forall \alpha \in \mathbf{C})(\forall x \in G)(\forall \xi \in X)\alpha(x\xi) = x(\alpha\xi),$$

$$(\forall x \in G)(\forall y \in G)(\forall \xi \in X)(xy)\xi = x(y\xi)].$$

(ii) The mapping $(a, \xi) \mapsto a\xi$ is continuous, that is,

$$A \times X \to X$$

$$(\exists k \in \mathbf{R}_+^*)(\forall a \in A)(\forall \xi \in X)\|a\xi\|_X \leq k\|a\|_A\|\xi\|_X$$

[resp. $(\exists k \in \mathbf{R}_+^*)(\forall x \in G)(\forall \xi \in X)\|x\xi\|_X \leq k\|\xi\|_X$ and, for every $\xi \in X$, the mapping $\genfrac{}{}{0pt}{}{x \mapsto x\xi}{G \to X}$ is continuous].

Notice that one does not require that $u\xi = u = \xi u$ [resp. $e\xi = \xi = \xi e$] for $\xi \in X$ in case A has a unit u [resp. X is a G-module]. In the same way, one defines a *right Banach A-module* [resp. *a right Banach G-module*]. The (two-sided) *Banach A-module* [resp. *Banach G-module*] X is a left and right Banach A-module [resp. Banach G-module] such that

$$(\forall a \in A)(\forall b \in A)(\forall \xi \in X)(a\xi)b = a(\xi b)$$

$$\left[\text{resp. } (\forall x \in G)(\forall y \in G)(\forall \xi \in X)(x\xi)y = x(\xi y)\right].$$

Every Banach algebra A may be regarded as a Banach A-module (with $k = 1$). If A is a Banach algebra and X is a left [resp. right] Banach A-module, we may define in a canonical way a right [resp. left] Banach A-module structure on X^* by putting

$$\langle \xi, Fa \rangle = \langle a\xi, F \rangle \; [\text{resp. } \langle \xi, aF \rangle = \langle \xi a, F \rangle]$$

for $a \in A$, $\xi \in X$, $F \in X^*$; then

$$\|Fa\| = \sup\{|\langle a\xi, F \rangle| : \xi \in X, \|\xi\|_X \leq 1\} \leq \|F\|k\|a\|_A$$

$$\left[\text{resp. } \|aF\| = \sup\{|\langle \xi a, F \rangle| : \xi \in X, \|\xi\|_X \leq 1\} \leq \|F\|k\|a\|_A\right].$$

If G is a locally compact group and X is a left [resp. right] Banach G-module, we define similarly a *right* [resp. *left*] *dual G-module structure* on X^* by putting

$$\langle \xi, Fx \rangle = \langle x\xi, F \rangle \; [\text{resp. } \langle \xi, xF \rangle = \langle \xi x, F \rangle],$$

BACKGROUND AND NOTATIONS   19

$x \in G$, $\xi \in X$, $F \in X^*$; for every $F \in X^*$, $x \mapsto Fx$ [resp. $x \mapsto xF$] is a continuous mapping of G into X^* equipped with the $\sigma(X^*, X)$-topology.

If X and Y are left Banach A-modules, an *A-multiplier* or *A-centralizer* is any $T \in \mathscr{L}(X, Y)$ such that $T(a\xi) = aT(\xi)$ whenever $a \in A$ and $\xi \in X$. In particular, if A is a Banach algebra, a multiplier of A is any $T \in \mathscr{L}(A)$ such that $T(ab) = aT(b)$ whenever $a, b \in A$.

2. Let G be a locally compact group and let $p \in [1, \infty]$. For any closed $*$-subalgebra A of $\mathscr{M}^1(G)$, the convolution product defines a left Banach A-module structure on $L^p(G)$; $\mu f = \mu * f$, $\|\mu * f\|_p \leq \|\mu\| \|f\|_p$ whenever $\mu \in A$ and $f \in L^p(G)$. If $p \in [1, \infty[$, the dual space $L^{p'}(G)$ is a right Banach A-module with $f\mu = \overline{\mu^* * \bar{f}}$ for $\mu \in A$, $f \in L^{p'}(G)$; notice that, for every $g \in L^p(G)$, by (13) we have

$$\langle g, f\mu \rangle = \langle \mu g, f \rangle = \langle \mu * g, f \rangle = \left(\mu * g | \bar{f} \right) = \left(g | \mu^* * \bar{f} \right) = \langle g, \overline{\mu^* * \bar{f}} \rangle.$$

If $p \in [1, \infty[$, the $\mathscr{M}_d^1(G)$-module structure determines a left Banach G-module structure; notice that, for $f \in L^p(G)$, $x \mapsto xf = \delta_x * f = L_x f$ is a continuous mapping of G into $L^p(G)$.

If G is a locally compact group, $X = L^\infty(G)$, $\mathscr{S}(G), \mathscr{C}(G), \mathscr{RUC}(G)$, $\mathscr{LUC}(G), \mathscr{UC}(G)$ are examples of subspaces of $L^\infty(G)$ that are closed under complex conjugation and admit the unit 1_G for the pointwise multiplication; all but possibly $\mathscr{S}(G)$ are Banach subspaces and even Banach subalgebras.

Definition 2.3. *Let G be a locally compact group, X a subspace of $L^\infty(G)$ that is closed under complex conjugation and E a subset of $\mathscr{M}^1(G)$. We say that X is E-invariant if $\mu * f \in X$ whenever $f \in X$ and $\mu \in E$.*

Note that $\mathscr{M}^1(G)$ [resp. $L^1(G), \mathscr{M}_d^1(G)$]-invariance is equivalent to $M^1(G)$ [resp. $P^1(G), M_d^1(G)$]-invariance and also, if X is closed, to $M_c^1(G)$ [resp. $P^1(G) \cap \mathscr{K}(G), M_f^1(G)$]-invariance; $M_f^1(G)$-invariance is $D^1(G)$-invariance.

Definition 2.4. *Let G be a locally compact group and let X be a subspace of $L^\infty(G)$. We say that X is left invariant if it is $D^1(G)$-invariant, that is, $_af \in X$ whenever $f \in X$ and $a \in G$; we say that X is topologically left invariant if it is $P^1(G)$-invariant, that is, $\varphi * f \in X$ whenever $f \in X$ and $\varphi \in P^1(G)$.*

3. We now indicate some important *examples* of structures of the types just defined.

(a) $L^\infty(G)$ is $\mathscr{M}^1(G)$-invariant; $L^\infty(G)$ is a left Banach $\mathscr{M}^1(G)$-module.
(b) $L^\infty(G)$, $\mathscr{S}(G)$, $\mathscr{C}(G)$, $\mathscr{RUC}(G)$, $\mathscr{LUC}(G)$, and $\mathscr{UC}(G)$ are left invariant; $L^\infty(G)$, $\mathscr{C}(G)$, $\mathscr{RUC}(G)$, $\mathscr{LUC}(G)$, and $\mathscr{UC}(G)$ are left Banach $\mathscr{M}_d^1(G)$-modules.
(c) $L^\infty(G)$, $\mathscr{C}(G)$, $\mathscr{RUC}(G)$, $\mathscr{LUC}(G)$, and $\mathscr{UC}(G)$ are topologically left invariant; they form left Banach $L^1(G)$-modules.

As a matter of fact, for every $\varphi \in L^1(G)$ and every $f \in L^\infty(G)$, $\varphi * f \in \mathscr{R}\mathscr{U}\mathscr{C}(G)$ (Section 2D.3). So topological left invariance is established for $L^\infty(G)$, $\mathscr{C}(G)$, and $\mathscr{R}\mathscr{U}\mathscr{C}(G)$. For any $\varphi \in L^1(G)$ and any $f \in \mathscr{L}\mathscr{U}\mathscr{C}(G)$, we have $\varphi * f \in \mathscr{L}\mathscr{U}\mathscr{C}(G)$. Indeed, given $\varepsilon > 0$, there exists a neighborhood U of e in G such that for all $x, x' \in G$ satisfying $x^{-1}x' \in U$, the inequality $\|f_x - f_{x'}\| < \varepsilon$ holds; hence we have

$$|\varphi * f(x) - \varphi * f(x')| = \left|\int \varphi(y)(f(y^{-1}x) - f(y^{-1}x'))\, dy\right|$$

$$\leq \int |\varphi(y)||f(y^{-1}x) - f(y^{-1}x')|\, dy < \varepsilon \|\varphi\|_1.$$

We conclude that $\mathscr{L}\mathscr{U}\mathscr{C}(G)$ and $\mathscr{U}\mathscr{C}(G)$ are topologically left invariant.

Considering the mapping $f \mapsto \check{f}$ on $L^\infty(G)$ one deduces from (b), (c) and (1), (3) that the spaces $Y = L^\infty(G)$, $\mathscr{S}(G)$, $\mathscr{C}(G)$, $\mathscr{L}\mathscr{U}\mathscr{C}(G)$, $\mathscr{R}\mathscr{U}\mathscr{C}(G)$, and $\mathscr{U}\mathscr{C}(G)$ are *right invariant*, that is, $f_a \in Y$ whenever $f \in Y$, $a \in G$ and that $Y = L^\infty(G)$, $\mathscr{C}(G)$, $\mathscr{L}\mathscr{U}\mathscr{C}(G)$, $\mathscr{R}\mathscr{U}\mathscr{C}(G)$, $\mathscr{U}\mathscr{C}(G)$ are *topologically right invariant*, that is, $f * \check{\varphi} \in Y$ whenever $f \in Y$, $\varphi \in L^1(G)$.

4. Let G be a locally compact group, A a Banach $*$-subalgebra of $\mathscr{M}^1(G)$ and X a Banach subspace of $L^\infty(G)$ that is closed under complex conjugation and A-invariant. On X we consider the right Banach A-module structure defined by

$$f\mu = \overline{\mu^* * \bar{f}}$$

for $(\mu, f) \in A \times X$. If $F \in X^*$, $f \in X$, $\mu \in A$, we have

$$|\langle \overline{\mu^* * \bar{f}}, F\rangle| \leq \|F\|\, \|\mu^*\|\, \|f\|_\infty = \|F\|\, \|\mu\|\, \|f\|_\infty.$$

We may put

$$\langle \mu, F * f\rangle = \langle f\mu, F\rangle = \langle \overline{\mu^* * \bar{f}}, F\rangle;$$

$F * f \in A^*$ and

$$\|F * f\|_{A^*} \leq \|F\|\, \|f\|_\infty.$$

If $A = L^1(G)$, for $F \in X^*$, $f \in X$, we have $F * f \in L^1(G)^* = L^\infty(G)$. Then for all $a \in G$, $f \in X$, $F \in X^*$ and $\varphi \in L^1(G)$, (8) implies that

$$\langle \varphi, {}_a(F * f)\rangle = \langle {}_{a^{-1}}\varphi, F * f\rangle = \langle \overline{({}_{a^{-1}}\varphi)^* * \bar{f}}, F\rangle$$

$$= \langle \overline{\varphi^* * ({}_a\bar{f})}, F\rangle = \langle \varphi, F * ({}_a f)\rangle,$$

BACKGROUND AND NOTATIONS

hence
$$_a(F \star f) = F \star (_a f). \tag{14}$$

In the particular case $A = \mathcal{M}_d^1(G) = l^1(G_d)$, for $F \in X^*$, $f \in X$, we have $F \star f \in l^1(G_d)^* = l^\infty(G_d)$. So for all $a \in G$, $f \in X$, $F \in X^*$,

$$\langle \delta_a, F \star f \rangle = \langle \overline{\delta_a^* \star \bar{f}}, F \rangle = \langle \delta_{a^{-1}} \star f, F \rangle = \langle _a f, F \rangle,$$

that is,
$$F \star f(a) = f(_a f). \tag{15}$$

Definition 2.5. *Let G be a locally compact group, A a Banach $\star$-subalgebra of $\mathcal{M}^1(G)$ and X a Banach subspace of $L^\infty(G)$ that is closed under complex conjugation and A-invariant. We say that X is A-introverted if $F \star f \in X$ whenever $F \in X^*$ and $f \in X$.*

Consider the following examples of introverted spaces:

(a) $X = \mathcal{RUC}(G)$, $A = \mathcal{M}_d^1(G)$. By (15), for all $F \in X^*$, $f \in X$, $x, x' \in G$,

$$|F \star f(x) - F \star f(x')| = |F(_x f - _{x'} f)| \leq \|F\| \|_x f - _{x'} f\|.$$

As $f \in \mathcal{RUC}(G)$, we have also $F \star f \in \mathcal{RUC}(G)$.

(b) $X = L^\infty(G)$, $A = L^1(G)$. For all $F \in X^*$ and $f \in X$, we have $F \star f \in L^1(G)^* = L^\infty(G)$.

(c) $X = \mathcal{RUC}(G)$, $A = L^1(G)$. Let $F \in X^*$ and $f \in X$. If $x, x' \in G$, by (14),

$$\|_x(F \star f) - _{x'}(F \star f)\| = \|F \star (_x f - _{x'} f)\| \leq \|F\| \|_x f - _{x'} f\|.$$

The fact that $f \in \mathcal{RUC}(G)$ implies that $F \star f \in \mathcal{RUC}(G)$.

If X is any A-introverted subspace of $L^\infty(G)$, for $F_1, F_2 \in X^*$ and $f \in X$, we have

$$|\langle F_2 \star f, F_1 \rangle| \leq \|F_1\| \|F_2 \star f\| \leq \|F_1\| \|F_2\| \|f\|.$$

So we may put
$$\langle f, F_1 \odot F_2 \rangle = \langle F_2 \star f, F_1 \rangle;$$

$F_1 \odot F_2 \in X^*$ and
$$\|F_1 \odot F_2\| \leq \|F_1\| \|F_2\|.$$

For all $F_1 \in X^*$, $F_2 \in X^*$, $\alpha \in \mathbf{C}$, we have $\alpha(F_1 \odot F_2) = (\alpha F_1) \odot F_2 = F_1 \odot (\alpha F_2)$; for all $F_1, F_2, F_3 \in X^*$, $(F_1 + F_2) \odot F_3 = (F_1 \odot F_3) + (F_2 \odot F_3)$ and $F_1 \odot (F_2 + F_3) = (F_1 \odot F_2) + (F_1 \odot F_3)$.

Let A now be a Banach $*$-subalgebra of $\mathcal{M}^1(G)$ containing $L^1(G)$ and consider an A-introverted subspace X of $L^\infty(G)$. For all $\mu \in A$, $f \in X$, $F \in X^*$, we have $F \star f \in L^\infty(G)$ and also $F \star (f\mu) \in L^\infty(G)$. For every $\varphi \in L^1(G)$, by (13) we obtain

$$\langle \varphi, F \star (f\mu) \rangle = \langle (f\mu)\varphi, F \rangle = \langle f(\mu * \varphi), F \rangle = \langle \mu * \varphi, F \star f \rangle$$
$$= (\mu * \varphi | \overline{F \star f}) = (\varphi | \mu^* * (\overline{F \star f})) = (\varphi | \overline{(F \star f)\mu}) = \langle \varphi, (F \star f)\mu \rangle;$$

therefore

$$F \star (f\mu) = (F \star f)\mu. \tag{16}$$

Finally we show that if X is an $L^1(G)$-introverted subspace of $L^\infty(G)$, then X^* is a Banach algebra associated to X; $\odot$ is an *Arens multiplication*. Notice that, for all $F_1 \in X^*$, $F_2 \in X^*$, $f \in X$,

$$(F_1 \odot F_2) \star f = F_1 \star (F_2 \star f) \tag{17}$$

as, for every $\varphi \in L^1(G)$, by (16),

$$\langle \varphi, (F_1 \odot F_2) \star f \rangle = \langle f\varphi, F_1 \odot F_2 \rangle = \langle F_2 \star (f\varphi), F_1 \rangle$$
$$= \langle (F_2 \star f)\varphi, F_1 \rangle = \langle \varphi, F_1 \star (F_2 \star f) \rangle.$$

For all $F_1, F_2, F_3 \in X^*$ and $f \in X$, by (17) we have

$$\langle f, (F_1 \odot F_2) \odot F_3 \rangle = \langle F_3 \star f, F_1 \odot F_2 \rangle = \langle F_2 \star (F_3 \star f), F_1 \rangle$$
$$= \langle (F_2 \odot F_3) \star f, F_1 \rangle = \langle f, F_1 \odot (F_2 \odot F_3) \rangle$$

hence $(F_1 \odot F_2) \odot F_3 = F_1 \odot (F_2 \odot F_3)$.

The Arens multiplication of $L^\infty(G)^*$ coincides with the convolution product on the subspace $L^1(G)$. As a matter of fact, for all $\varphi, \psi \in L^1(G)$ and $f \in L^\infty(G)$, by (12),

$$\langle f, \varphi \odot \psi \rangle = \langle \psi \star f, \varphi \rangle = \langle \varphi, \psi \star f \rangle = \langle f\varphi, \psi \rangle$$
$$= \langle \overline{\varphi^* * \bar{f}}, \psi \rangle = (\psi | \varphi^* * \bar{f}) = (\varphi * \psi | \bar{f}) = \langle f, \varphi * \psi \rangle.$$

NOTES

General references for approximate units are Reiter [455], and Doran and Wichmann [142]. Wichmann [554] gave a proof of 2.1 due to Saeki. For 2.2 see Loomis [370].

In our context Banach modules were studied by Johnson [291]. General constructions performed by Arens [14] allow the introduction of a multiplication in the bidual space of an arbitrary Banach algebra that extends the multiplication of the algebra. The general notion of introversion was defined by Wong [560] for subspaces of $L_\mathbf{R}^\infty(G)$. He transposed previous applications of the Arens multiplication due to Day [115] concerning the set of means on $l_\mathbf{R}^\infty(S)$ for a discrete semigroup S.

3. GENERAL CONSIDERATIONS ON MEANS AND AVERAGING PROCESSES

We indicate some general properties of means, which constitute the fundamental notion of amenability, and on averaging processes from which means originate.

A. Basic Properties of Means

If X is any vector space of real- or complex-valued functions defined on a set S such that $1_S \in X$, a *mean* M on X is a *state* on X, that is, a normalized nonnegative functional: $M \in X'$, $M(1_S) = 1$ and $M(f) \geq 0$ whenever $f \in X_+$. We denote by $\mathfrak{M}(X)$ the convex set of all means on X.

Restricting ourselves to the type of functions to be considered mainly in the sequel, we formulate an equivalent definition.

Definition 3.1. *If G is a locally compact group, let X be a real [resp. complex] subspace of $L_\mathbf{R}^\infty(G)$ [resp. $L^\infty(G)$] with $1_G \in X$ [that is closed under complex conjugation]. A real [resp. complex] linear functional M on X is called a mean on X if the following property holds:*

  (i) $(\forall f \in X_\mathbf{R}) \operatorname{ess\,inf} f \leq M(f) \leq \operatorname{ess\,sup} f.$

Note that if one of the inequalities holds for every $f \in X_\mathbf{R}$, so does the other one. Assume that, for instance, $M(f) \leq \operatorname{ess\,sup} f$ whenever $f \in X_\mathbf{R}$; then

$$-M(f) = M(-f) \leq \operatorname{ess\,sup}(-f) = -\operatorname{ess\,inf} f$$

and $\operatorname{ess\,inf} f \leq M(f)$.

If M is a mean on the complex space X, (i) implies that, for every $f \in X_\mathbf{R}$, $M(f) \in \mathbf{R}$. For $f = f_1 + if_2 \in X$ with $f_1, f_2 \in X_\mathbf{R}$, we have

$$M(\bar{f}) = M(f_1 - if_2) = M(f_1) - iM(f_2)$$
$$= \overline{M(f_1) + iM(f_2)} = \overline{M(f)}.$$

Proposition 3.2. *Let G be a locally compact group and let X be a real [resp. complex] subspace of $L_\mathbf{R}^\infty(G)$ [resp. $L^\infty(G)$] with $1_G \in X$ [that is closed under*

complex conjugation]. *A real* [*resp. complex*] *linear functional M on X is a mean on X if and only if any pair of the following conditions hold*:
  (ii) $(\forall f \in X_+) M(f) \geq 0$, *that is, M is nonnegative*.
  (iii) $M(1_G) = 1$.
  (iv) $\|M\| = 1$.

Proof. More precisely we show that (i) $\Rightarrow$ [(ii) and (iii) and (iv)], [(ii) and (iii)] $\Rightarrow$ (i), [(ii) and (iv)] $\Rightarrow$ (iii), [(iii) and (iv)] $\Rightarrow$ (ii).

$$(i) \Rightarrow (ii)$$

Applying (i) to $f \in X_+$ we obtain $0 \leq \operatorname{ess\,inf} f \leq M(f)$.

$$(i) \Rightarrow (iii)$$

Applying (i) to 1_G we obtain $1 \leq M(1_G) \leq 1$.

$$(i) \Rightarrow (iv)$$

If $f \in X$, there exists $\alpha \in \mathbf{R}$ [resp. $\alpha \in \mathbf{C}$] such that $|\alpha| = 1$ and $|M(f)| = \alpha M(f) = M(\alpha f)$.

In the real case, (i) implies that $|M(f)| \leq \|\alpha f\|_\infty = \|f\|_\infty$; $\|M\| \leq 1$. In the complex case, as X is closed under complex conjugation, we may consider the decomposition $\alpha f = f_1 + if_2$, where $f_1, f_2 \in X_\mathbf{R}$. We have $M(f_1), M(f_2) \in \mathbf{R}$ and $M(f_1) + iM(f_2) = M(\alpha f) \geq 0$; hence $M(f_2) = 0$ and $M(f_1) \geq 0$. By (i), $M(f_1) \leq \|f_1\|_\infty$ and we conclude that

$$|M(f)| = M(f_1) \leq \|f_1\|_\infty \leq \|\alpha f\|_\infty = \|f\|_\infty;$$

$\|M\| \leq 1$.
As (i) $\Rightarrow$ (iii), we have precisely $\|M\| = 1$.

$$[(ii) \text{ and } (iii)] \Rightarrow (i)$$

For any $f \in X_\mathbf{R}$, $(\operatorname{ess\,inf} f) 1_G \leq f \leq (\operatorname{ess\,sup} f) 1_G$. From (ii), (iii) and the linearity of M we deduce that

$$\operatorname{ess\,inf} f = (\operatorname{ess\,inf} f) M(1_G) \leq M(f) \leq (\operatorname{ess\,sup} f) M(1_G)$$

$$= \operatorname{ess\,sup} f.$$

$$[(ii) \text{ and } (iv)] \Rightarrow (iii)$$

From (ii) we obtain $M(1_G) \geq 0$; (iv) implies that $M(1_G) \leq 1$.
Let $f \in X$. There exists $\alpha \in \mathbf{C}$ such that $|\alpha| = 1$ and $M(\alpha f) = \alpha M(f) = |M(f)|$. If $f_1 = \mathcal{R}e\,\alpha f$, $f_1 \in X_\mathbf{R}$, $M(f_1) = M(\alpha f)$ and $f_1 \leq |\alpha f| \leq \|f\|_\infty 1_G$.

Hence by linearity of M and (ii), $|M(f)| = M(f_1) \leq \|f\|_\infty M(1_G)$ and therefore $1 = \|M\| = M(1_G)$.

$$[(iii) \text{ and } (iv)] \Rightarrow (ii)$$

Consider $f \in X_\mathbf{R}$ such that $0 \leq f \leq 1_G$. Let $g = 2f - 1_G$, hence $\|g\|_\infty \leq 1$. If $M(g) = \alpha + i\beta$ for $\alpha, \beta \in \mathbf{R}$, then for any $\gamma \in \mathbf{R}$, by (iii) and (iv),

$$(\beta + \gamma)^2 \leq |\alpha + i(\beta + \gamma)|^2$$

$$= |M(g + i\gamma 1_G)|^2 \leq \|g + i\gamma 1_G\|_\infty^2 \leq 1 + \gamma^2; \qquad 2\beta\gamma \leq 1 - \beta^2.$$

As γ may be chosen arbitrarily, we must have $\beta = 0$, $M(g) = \alpha \in \mathbf{R}$. By (iv), $|M(g)| \leq \|g\|_\infty \leq 1$, so by (iii) and the linearity of M we have

$$M(f) = \tfrac{1}{2}M(1_G + g) = \tfrac{1}{2}(1 + M(g)) \geq 0.$$

The statement now follows from the homogeneity of M. □

If G is a locally compact group and X_1, X_2 are subspaces of $L_\mathbf{R}^\infty(G)$, [resp. $L^\infty(G)$ and are closed under complex conjugation] with $1_G \in X_1 \subset X_2$, then obviously any mean on X_2 may be restricted to a mean on X_1. If X is a subspace of $L_\mathbf{R}^\infty(G)$ [resp. $L^\infty(G)$ and is closed under complex conjugation] such that $1_G \in X$, then by (iv) any mean on X is continuous.

Proposition 3.3. *Let G be a locally compact group and let X be a subspace of $L_\mathbf{R}^\infty(G)$ [resp. $L^\infty(G)$ that is closed under complex conjugation] with $1_G \in X$. Then the set $\mathfrak{M}(X)$ consisting of the means on X is w^*-compact in X^*; the sets $P^1(G)$ and $P^1(G) \cap \mathcal{K}(G)$ are w^*-dense in $\mathfrak{M}(X)$.*

Proof. By Alaoglu's theorem, the unit ball of X^* is w^*-compact ([HR] B.25); as $\mathfrak{M}(X)$ is a w^*-closed subset of this ball by Proposition 3.2, it is w^*-compact too.

If $\varphi \in P^1(G)$, for any $f \in X_+$, $\int f\varphi \geq 0$ and $\int 1_G \varphi = 1$; hence φ verifies properties (ii), (iii) of Proposition 3.2 and $P^1(G) \subset \mathfrak{M}(X)$. Let $\mathfrak{M}_1(X)$ be the w^*-closure of $P^1(G)$ in $\mathfrak{M}(X)$. If $\mathfrak{M}_1(X)$ were different from $\mathfrak{M}(X)$, we could pick $M_0 \in \mathfrak{M}(X) \setminus \mathfrak{M}_1(X)$. As $\mathfrak{M}_1(X)$ is convex, the Hahn–Banach theorem ([B EVT] Chap. II, 5, Section 5, No. 2) implies the existence of $f \in X_\mathbf{R}$ such that, for any $N \in \mathfrak{M}_1(X)$,

$$\langle f, N \rangle \leq 1 \qquad (1)$$

whereas

$$\langle f, M_0 \rangle > 1. \qquad (2)$$

Then

$$\operatorname{ess\,sup} f \leq 1 \qquad (3)$$

because otherwise there would exist a measurable subset A of G such that $0 < |A| < \infty$ and $f(x) > 1$ locally for almost every $x \in A$; one would then have $\langle f, \xi_A \rangle = 1/|A| \int_A f(x)\,dx > 1$ and contradict (1). But (3) implies that $\langle f, M_0 \rangle \leq 1$ which contradicts (2).

As $P^1(G) \cap \mathscr{K}(G)$ is dense in $P^1(G)$ (Section 2D.3), then also $P^1(G) \cap \mathscr{K}(G)$ is w*-dense in $\mathfrak{M}(X)$. $\square$

Corollary 3.4. *Let G be a locally compact group and let X be a subspace of $\mathscr{C}_\mathbf{R}(G)$ [resp. $\mathscr{C}(G)$ that is closed under complex conjugation] with $1_G \in X$. Then the sets $M_d^1(G)$ and $M_f^1(G)$ are w*-dense in $\mathfrak{M}(X)$.*

Proof. As $\mathscr{C}_\mathbf{R}(G)$ [resp. $\mathscr{C}(G)$] may be considered to be a subspace of $l_\mathbf{R}^\infty(G)$ [resp. $l^\infty(G)$], the statement follows from Proposition 3.3 applied to G_d with $M_d^1(G) = M^1(G_d) = P^1(G_d)$. $\square$

Note that, for $X = \mathscr{C}(G), \mathscr{C}_\mathbf{R}(G)$, the corollary follows also from the Kreĭn–Mil'man theorem saying that $\overline{\operatorname{co}}\{\delta_x : x \in G\}$ is the w*-compact convex set $\mathfrak{M}(X)$ ([HR] B.30).

Let S be any topological space and consider a commutative C^*-algebra X of complex-valued functions defined on S such that $1_S \in X$. Let $\hat{X}$ be the structure space of X; via the Gelfand transform, X may be identified with $\mathscr{C}(\hat{X})$ (Section 2B.3) and therefore X^* may be identified with $\mathscr{M}^1(\hat{X})$, the correspondence being ruled by

$$\langle \hat{f}, \hat{F} \rangle = \langle f, F \rangle \qquad (4)$$

for $f \in X$, $F \in X^*$, $\hat{f} \in \mathscr{C}(\hat{X})$, $\hat{F} \in \mathscr{M}^1(\hat{X})$. In particular, $\mathfrak{M}(X)$ may be identified with $M^1(\hat{X})$ via (4). By the Hahn–Jordan decomposition ([HS] 19.13) any measure on $\hat{X}$ is a finite linear combination of probability measures on $\hat{X}$; therefore any $F \in X^*$ is a finite linear combination of elements in $\mathfrak{M}(X)$.

Proposition 3.5. *Let G be a locally compact group and let X be a topologically left invariant Banach subspace of $L^\infty(G)$ that is closed under complex conjugation and contains 1_G. Then $\mathfrak{M}(X)$ is a semigroup for the Arens multiplication.*

Proof. For every $\varphi \in L^1(G)$, $\varphi * 1_G = \overline{(\int \varphi)}1_G$ and, for every $M \in \mathfrak{M}(X)$, we have

$$\langle \varphi, M * 1_G \rangle = \langle \overline{\varphi^* * 1_G}, M \rangle = \left(\int \varphi \right) \langle 1_G, M \rangle = \int \varphi$$

hence $M * 1_G = 1_G$. If $M_1, M_2 \in \mathfrak{M}(X)$,

$$\langle 1_G, M_1 \odot M_2 \rangle = \langle M_2 * 1_G, M_1 \rangle = \langle 1_G, M_1 \rangle = 1.$$

As X^* is a Banach algebra for $\odot$, $\|M_1 \odot M_2\| \leq \|M_1\| \|M_2\| = 1$; then $\|M_1 \odot M_2\| = 1$. We deduce from Proposition 3.2 that $M_1 \odot M_2 \in \mathfrak{M}(X)$. □

B. Averaging Processes

Let G be a locally compact group and let X be a Banach subspace of $\mathscr{C}(G)$ that is closed under complex conjugation and contains 1_G. Proposition 3.3 and Corollary 3.4 show that $P^1(G)$ and $M_d^1(G)$ are both dense in the set of means on X; all means are approximated by averaging processes. We now establish analogous facts replacing functionals by endomorphisms.

If $p \in [1, \infty]$, to $\mu \in M^1(G)$ we associate the *convolution operator*

$$L_\mu : f \mapsto \mu * f$$

on $L^p(G)$; $L_\mu \geq 0$ and, for every $f \in L^p(G)$, $\|\mu * f\|_p \leq \|\mu\| \|f\|_p = \|f\|_p$, hence $\|L_\mu\| \leq 1$ and $L_\mu \in \mathscr{L}(L^p(G))$. We denote by L_G the convex hull of the set of isometric operators L_x, that is, $L_{\delta_x}(x \in G)$ on $L^1(G)$. If $f \in L^1(G)$ [resp. $f \in \mathscr{C}(G)$] we consider in $L^1(G)$ [resp. $\mathscr{C}(G)$] the sets

$$L_G(f) = \overline{\mathrm{co}}\{L_x f : x \in G\}$$

and

$$\mathscr{L}_G(f) = \left\{ \sum_{i=1}^n \alpha_i L_{x_i} f : \alpha_i \in \mathbf{C}, x_i \in G; \; i = 1, \ldots, n; \; \sum_{i=1}^n \alpha_i = 1; \; n \in \mathbf{N}^* \right\}^-.$$

Proposition 3.6. *Let G be a locally compact group. If $f \in L^1(G)$ or $f \in \mathscr{RUC}(G)$, $L_G(f) = \{\varphi * f : \varphi \in P^1(G)\}^-$ [resp. $\mathscr{L}_G(f) = \{\varphi * f : \varphi \in L^1(G), \int \varphi = 1\}^-$].*

Proof. We put $\|\!|\cdot|\!\| = \|\cdot\|_1$ or $\|\!|\cdot|\!\| = \|\cdot\|_\infty$.

(a) We establish $L_G(f) \subset \{\varphi * f : \varphi \in P^1(G)\}^-$ [resp. $\mathscr{L}_G(f) \subset \{\varphi * f : \varphi \in L^1(G), \int \varphi = 1\}^-$].

Let $x_1, \ldots, x_n \in G$, $\alpha_1, \ldots, \alpha_n \in \mathbf{R}_+^*$ [resp. $\alpha_1, \ldots, \alpha_n \in \mathbf{C}^*$] with $\sum_{i=1}^n \alpha_i = 1$, and $\varepsilon \in \mathbf{R}_+^*$; we put $\varepsilon' = \varepsilon \left(\sum_{i=1}^n |\alpha_i|\right)^{-1}$. As $L^1(G)$ admits approximate units belonging to $P^1(G)$ by Proposition 2.2, if $f \in L^1(G)$, there exists $\psi \in P^1(G)$ such that $\|\psi * f - f\|_1 < \varepsilon'$. If $f \in \mathscr{RUC}(G)$, there exists a compact neighborhood U of e in G such that $|f(y^{-1}x) - f(x)| < \varepsilon'$ whenever $x \in G$

and $y \in U$; then for $\psi = \xi_U \in P^1(G)$ and every $x \in G$,

$$|(\psi * f - f)(x)| = \left| \frac{1}{|U|} \int_U f(y^{-1}x)\, dy - \frac{1}{|U|} \int_U f(x)\, dy \right|$$

$$\leq \frac{1}{|U|} \int_U |f(y^{-1}x) - f(x)|\, dy < \varepsilon',$$

that is, $\|\psi * f - f\| < \varepsilon'$.

The function $\varphi = \sum_{i=1}^n \alpha_i {}_{x_i}\psi$ belongs to $L^1_+(G)$ [resp. $L^1(G)$] and $\int \varphi = 1$. Then by (2.4), we have

$$\left\| \varphi * f - \sum_{i=1}^n \alpha_i {}_{x_i} f \right\| = \left\| \sum_{i=1}^n \alpha_i(({}_{x_i}\psi) * f - {}_{x_i}f) \right\| = \left\| \sum_{i=1}^n \alpha_i {}_{x_i}(\psi * f - f) \right\|$$

$$\leq \sum_{i=1}^n |\alpha_i| \|\|\psi * f - f\|\| < \varepsilon.$$

(b) We establish $\{\varphi * f : \varphi \in P^1(G)\}^- \subset L_G(f)$ [resp. $\{\varphi * f : \varphi \in L^1(G), \int \varphi = 1\}^- \subset \mathscr{L}_G(f)$].

Let $\varphi \in L^1_+(G)$ [resp. $\varphi \in L^1(G)$] with $\int \varphi = 1$ and $\varepsilon \in \mathbf{R}^*_+$; we put $\delta = \varepsilon(2\|\|f\|\| + \|\varphi\|_1)^{-1}$. There exists a compact subset K in G such that $\int_{G \setminus K} |\varphi| \leq \delta$. As $f \in \mathscr{RUC}(G)$ or, in case $f \in L^1(G)$, the mapping $x \mapsto {}_x f$ is right uniformly continuous (Section 2D.3), there exists an open neighborhood V of e in G such that, for all $z, z' \in G$ with $z \in z'V$, $\|\|_{z^{-1}}f - {}_{z'^{-1}}f\|\| < \delta$. We may determine a subset $\{y_1, \ldots, y_n\}$ in G such that $K \subset \bigcup_{i=1}^n y_i V$ and $\|\|_{y^{-1}}f - {}_{y_i^{-1}}f\|\| < \delta$ whenever $y \in (y_i V) \cap K$; $i = 1, \ldots, n$. We put $y_0 = e$, $A_0 = G \setminus K$, $A_1 = y_1 V \cap K$ and define inductively $A_i = y_i V \cap (K \setminus \bigcup_{j=1}^{i-1} A_j)$ for $i = 2, \ldots, n$. If $i = 0, 1, \ldots, n$, we also put $\alpha_i = \int_{A_i} \varphi(y)\, dy$; $\alpha_i \in \mathbf{R}_+$ [resp. $\alpha_i \in \mathbf{C}$]. Then $\sum_{i=0}^n \alpha_i = 1$; for almost every $x \in G$,

$$\left| \varphi * f(x) - \sum_{i=0}^n \alpha_i L_{y_i} f(x) \right| = \left| \int_G \varphi(y) f(y^{-1}x)\, dy - \sum_{i=0}^n \alpha_i f(y_i^{-1}x) \right|$$

$$= \left| \sum_{i=0}^n \int_{A_i} \varphi(y)(f(y^{-1}x) - f(y_i^{-1}x))\, dy \right|$$

$$\leq \sum_{i=0}^n \int_{A_i} |\varphi(y)| |f(y^{-1}x) - f(y_i^{-1}x)|\, dy.$$

We obtain [by making use of Fubini's theorem in the case where $f \in L^1(G)$]

$$\left\| \varphi * f - \sum_{i=0}^{n} \alpha_i L_{y_i} f \right\| \leq 2\|\|f\|\| \int_{A_0} |\varphi(y)| \, dy + \delta \sum_{i=1}^{n} \int_{A_i} |\varphi(y)| \, dy$$

$$\leq 2\|\|f\|\|\delta + \delta\|\varphi\|_1 = (2\|\|f\|\| + \|\varphi\|_1)\delta = \varepsilon. \quad \square$$

Corollary 3.7. *Let G be a locally compact group and let X be a Banach subspace of $\mathscr{RUC}(G)$. Then X is left invariant if and only if it is topologically left invariant.*

Proof. The statement follows from the equality $L_G(f) = \{\varphi * f : \varphi \in P^1(G)\}^-$ established in Proposition 3.6 for every $f \in \mathscr{RUC}(G)$. $\quad \square$

NOTES

Following von Neumann's fundamental investigations, the first systematic study on means (for bounded functions on discrete semigroups) was Day [115]. He established 3.3 in the discrete case. Hulanicki [268] proved the general property relying on results concerning linear lattices due to Namioka [410]. As noted in Section 2, Wong [560] applied Arens multiplication to the theory of means.

A demonstration of 3.6 applying to bounded right uniformly continuous bounded functions is due to Wong [560]. For the proof concerning the group algebra we refer to Reiter [453] and Rindler [468]; also see Lau [337].

2
FUNDAMENTAL CHARACTERIZATIONS OF AMENABLE GROUPS

In this chapter we study the principal characterizations of amenable locally compact groups listing related properties under the same heading; more details appear in Chapters Four and Five. We intend to give here a bird's eye view of the interplay between the various properties. All these characterizations may be used to produce examples of amenable groups. The reader should look at Section 12A to see that at least for compact groups and abelian groups amenability may be checked easily.

4. INVARIANT MEANS ON VARIOUS FUNCTION SPACES

If G is locally compact group, we may consider various $*$-subalgebras A of $\mathcal{M}^1(G)$ and also various subspaces X of $L^\infty(G)$ that contain 1_G, are closed under complex conjugation and under the action of A, that is, $\mu * f \in X$ whenever $\mu \in A$ and $f \in X$ (Section 2G). We are interested in means M on X that are invariant under the action of the convex semigroup $A^1 = A \cap M^1(G)$, that is, $M(\mu * f) = M(f)$ whenever $\mu \in A^1$ and $f \in X$. We say that M is an A^1-*invariant mean*. Obviously an invariant mean on a space is also invariant on subspaces. The remarkable point is that a large amount of existence properties of invariant means are equivalent. They give the first characterizations of amenable locally compact groups. We finally state a useful criterion expressing existence of invariant means.

A. Basic Definitions

Consider the interesting standard cases.

Definition 4.1. *Let G be a locally compact group and let X be a real [resp. complex] subspace of $L^\infty_\mathbf{R}(G)$ [resp. $L^\infty(G)$] with $1_G \in X$ that is [closed under complex conjugation and] left invariant. A mean M on X is called left invariant mean if*

$$(\forall f \in X)(\forall a \in G)\ M(_a f) = M(f).$$

This definition makes sense for $X = L^\infty(G)$, $\mathscr{S}(G)$, $\mathscr{C}(G)$, $\mathscr{RUC}(G)$, $\mathscr{LUC}(G)$, $\mathscr{UC}(G)$ (Section 2G.3); it signifies that M is $M^1_f(G)$-invariant.

Definition 4.2. *A locally compact group G is called amenable group (F: groupe moyennable; G: mittelbare Gruppe; R: аменабельная группа) if it admits a left invariant mean on $L^\infty(G)$.*

We add some more definitions concerning the standard situations.

Definition 4.3. *Let G be a locally compact group and let X be a real [resp. complex] subspace of $L^\infty_\mathbf{R}(G)$ [resp. $L^\infty(G)$] with $1_G \in X$ that is [closed under complex conjugation and] topologically left invariant. A mean M on X is called topologically left invariant mean if*

$$(\forall f \in X)(\forall \varphi \in P^1(G))\ M(\varphi * f) = M(f).$$

This definition makes sense for $X = L^\infty(G)$, $\mathscr{C}(G)$, $\mathscr{RUC}(G)$, $\mathscr{LUC}(G)$, $\mathscr{UC}(G)$ (Section 2G.3).

Definition 4.4. *Let G be a locally compact group and let X be a real [resp. complex] subspace of $L^\infty_\mathbf{R}(G)$ [resp. $L^\infty(G)$] with $1_G \in X$ [that is closed under complex conjugation]. A mean M on X is called right invariant if X is right invariant and*

$$(\forall f \in X)(\forall a \in G)\ M(f_a) = M(f).$$

A mean M on X is called topologically right invariant if X is topologically right invariant and

$$(\forall f \in X)(\forall \varphi \in P^1(G))\ M(f * \check\varphi) = M(f).$$

Right invariant means may be considered on $L^\infty(G)$, $\mathscr{S}(G)$, $\mathscr{C}(G)$, $\mathscr{RUC}(G)$, $\mathscr{LUC}(G)$, and $\mathscr{UC}(G)$. Topologically right invariant means may be considered on $L^\infty(G)$, $\mathscr{C}(G)$, $\mathscr{LUC}(G)$, $\mathscr{RUC}(G)$, $\mathscr{UC}(G)$. (Section 2G.3).

Definition 4.5. *Let G be a locally compact group and let X be a real [resp. complex] subspace of $L^\infty_\mathbf{R}(G)$ [resp. $L^\infty(G)$] with $1_G \in X$ [that is closed under* complex conjugation]. If X is left and right invariant, a left invariant and right invariant mean on X is called invariant mean. If X is topologically left and right invariant, a topologically left invariant and topologically right invariant mean on X is called topologically invariant mean.

The convex set of left invariant [resp. right invariant, invariant, topologically left invariant, topologically right invariant, topologically invariant] means on X is denoted by LIM(X) [resp. RIM(X), IM(X), TLIM(X), TRIM(X), TIM(X)]; it is w*-compact by Proposition 3.3.

Proposition 4.6. *Let G be a locally compact group and let X be a subspace of $L^\infty(G)$ with $1_G \in X$ that is closed under complex conjugation and is left invariant [resp. right invariant, left and right invariant, topologically left invariant, topologically right invariant, topologically left and right invariant]. Then the following properties are equivalent*:

(i) LIM(X) $\neq \emptyset$ [resp. RIM(X) $\neq \emptyset$, IM(X) $\neq \emptyset$, TLIM(X) $\neq \emptyset$, TRIM(X) $\neq \emptyset$, TIM(X) $\neq \emptyset$].

(ii) LIM($X_\mathbf{R}$) $\neq \emptyset$ [resp. RIM($X_\mathbf{R}$) $\neq \emptyset$, IM($X_\mathbf{R}$) $\neq \emptyset$, TLIM($X_\mathbf{R}$) $\neq \emptyset$, TRIM($X_\mathbf{R}$) $\neq \emptyset$, TIM($X_\mathbf{R}$) $\neq \emptyset$].

Proof. Obviously (i) $\Rightarrow$ (ii). To establish the reverse implication, associate to an element M of LIM($X_\mathbf{R}$) [resp. RIM($X_\mathbf{R}$), IM($X_\mathbf{R}$), TLIM($X_\mathbf{R}$), TRIM($X_\mathbf{R}$), TIM($X_\mathbf{R}$)] the functional M_1 on X defined by

$$M_1(f + ig) = M(f) + iM(g),$$

$f, g \in X_\mathbf{R}$. Then M_1 is an element of LIM(X) [resp. RIM(X), IM(X), TLIM(X), TRIM(X), TIM(X)]. $\square$

Thus, as far as existence properties are concerned, it makes no difference to consider the real or the complex case.

Proposition 4.7. *Let G be a locally compact group and let X be a subspace of $L^\infty(G)$ with $1_G \in X$ that is closed under complex conjugation; let also $\check{X} = \{\check{f}: f \in X\}$ and consider the mapping*

$$\Upsilon: F \mapsto \check{F}$$

$$X^* \to \check{X}^*$$

defined by $\langle f, \check{F} \rangle = \langle \check{f}, F \rangle$ for $f \in X$. If X is left [resp. right] invariant, then $\check{X}$ is right [resp. left] invariant and Υ maps bijectively LIM(X) [resp. RIM(X)] onto RIM($\check{X}$) [resp. LIM($\check{X}$)]; if X is topologically left [resp. topologically right] invariant, then $\check{X}$ is topologically right [resp. topologically left] invariant and Υ maps bijectively TLIM(X) [resp. TRIM(X)] onto TRIM(X) [resp. TLIM(X)].

Proof. Obviously Υ maps bijectively $\mathfrak{M}(X)$ onto $\mathfrak{M}(\check{X})$. Note then that, for $f \in X$ and $a \in G$, $(\check{f}_a)\check{} = {}_{a^{-1}}f$ by (2.1); also for $f \in X$ and $\varphi \in P^1(G)$, $(\check{f} * \check{\varphi})\check{} = \varphi * f$ by (2.3). $\square$

Corollary 4.8. *Let G be a locally compact group and let $X = L^\infty(G)$, $\mathscr{S}(G)$, $\mathscr{C}(G)$, $\mathscr{U}\mathscr{C}(G)$ [resp. $X = L^\infty(G)$, $\mathscr{C}(G)$, $\mathscr{U}\mathscr{C}(G)$]. There exists a bijection of* LIM(X) *onto* RIM(X) *[resp.* TLIM(X) *onto* TRIM(X)*].*

Proof. The statement concerns particular cases of Proposition 4.7. $\square$

If G is a locally compact group, $\mathscr{L}\mathscr{U}\mathscr{C}(G)\check{} = \mathscr{R}\mathscr{U}\mathscr{C}(G)$ (Section 2D.2); so Proposition 4.7 affords, for instance, a bijection of LIM($\mathscr{R}\mathscr{U}\mathscr{C}(G)$) onto RIM($\mathscr{L}\mathscr{U}\mathscr{C}(G)$). Proposition 4.7 illustrates the following general fact: For most invariance properties of amenable groups it suffices to establish one-sided versions.

B. Main Characterizations

Proposition 4.9. *If G is a locally compact group, any left invariant mean on $\mathscr{S}(G)$ may be extended uniquely to a left invariant mean on $L^\infty(G)$.*

Proof. Any left invariant mean M on $\mathscr{S}(G)$ may be extended to a unique $M_1 \in L^\infty(G)^*$ with $\|M_1\| = 1$ and $M_1(1_G) = 1$ ([HR] B.11); by Proposition 3.2, M_1 is a mean on $L^\infty(G)$.

Let $f \in L^\infty(G)$ and $a \in G$. For every $\varepsilon > 0$, there exists $g \in \mathscr{S}(G)$ such that $\|g - f\|_\infty < \varepsilon/2$. Thus we have

$$|M_1({}_af - f)| \le |M_1({}_a(f - g))| + |M_1({}_ag - g)| + |M_1(g - f)|$$

$$\le 2\|f - g\|_\infty < \varepsilon.$$

As $\varepsilon > 0$ is arbitrary, $M_1({}_af - f) = 0$ and $M_1 \in \text{LIM}(L^\infty(G))$. $\square$

We now introduce constants measuring the deviation of a mean from left invariance [resp. topological left invariance].

Definition 4.10. *If G is a locally compact group, X is a subspace of $L^\infty(G)$ with $1_G \in X$ that is closed under complex conjugation and left invariant [resp. topologically left invariant], for $M \in \mathfrak{M}(X)$, let*

$$\alpha(M, X) = \sup\{|M({}_xf - f)| : f \in X, \|f\|_\infty \le 1, x \in G\}$$

$$\left[\text{resp. } \beta(M, X) = \sup\{|M(\varphi * f - f)| : f \in X, \|f\|_\infty \le 1, \varphi \in P^1(G)\}\right].$$

We have $0 \leq \alpha(M, X) \leq 2$; $0 \leq \beta(M, X) \leq 2$. The mean M is left invariant [resp. topologically left invariant] on X if and only if $\alpha(M, X) = 0$ [resp. $\beta(M, X) = 0$].

Proposition 4.11. *Let G be a locally compact group, X a subspace of $L^\infty(G)$ with $1_G \in X$ that is closed under complex conjugation, left invariant and topologically left invariant. For every $M \in \mathfrak{M}(X)$, $\alpha(M, X) \leq 2\beta(M, X)$.*

Proof. Consider a fixed element $\varphi_0 \in P^1(G)$. For any $f \in X$ and any $a \in G$, $R_{a^{-1}}\varphi_0 \in P^1(G)$ and by (2.6) $\varphi_0 * (_a f) = (R_{a^{-1}}\varphi_0) * f$; hence if $\|f\|_\infty \leq 1$,

$$|M(_a f - f)| \leq |M(_a f - \varphi_0 * (_a f))| + |M((R_{a^{-1}}\varphi_0) * f - f)| \leq 2\beta(M, X)$$

and then also $\alpha(M, X) \leq 2\beta(M, X)$. □

Corollary 4.12. *Let G be a locally compact group, X a subspace of $L^\infty(G)$ with $1_G \in X$ that is closed under complex conjugation, left invariant and topologically left invariant. Then $\mathrm{TLIM}(X) \subset \mathrm{LIM}(X)$.*

Proof. The statement is an immediate consequence of Proposition 4.11. □

Proposition 4.13. *Let G be a locally compact group, X a closed subspace of $\mathscr{RUC}(G)$ with $1_G \in X$ that is closed under complex conjugation and is left invariant. Then for every $M \in \mathfrak{M}(X)$, $\beta(M, X) \leq \alpha(M, X)$.*

Proof. Recall that, by Corollary 3.7, X is left invariant if and only if it is topologically left invariant. Let $f \in X$ and $\varphi \in P^1(G)$. If $\varepsilon > 0$, by Proposition 3.6, there exist $\alpha_1, \ldots, \alpha_n \in \mathbf{R}_+$ with $\sum_{i=1}^n \alpha_i = 1$ and $a_1, \ldots, a_n \in G$ such that

$$\left\| \varphi * f - \sum_{i=1}^n \alpha_i {}_{a_i}f \right\| < \varepsilon.$$

Then also

$$|M(\varphi * f - f)| = \left| M(\varphi * f) - \sum_{i=1}^n \alpha_i M(f) \right|$$

$$\leq \left| M\left(\varphi * f - \sum_{i=1}^n \alpha_i {}_{a_i}f\right) \right|$$

$$+ \left| \sum_{i=1}^n \alpha_i M(_{a_i}f - f) \right|$$

$$< \varepsilon + \sum_{i=1}^n \alpha_i \alpha(M, X)\|f\| = \varepsilon + \alpha(M, X)\|f\|.$$

As $\varepsilon > 0$ is arbitrary, we obtain $\beta(M, X) \leq \alpha(M, X)$. □

Corollary 4.14. *If G is a locally compact group,* $\mathrm{LIM}(\mathcal{RUC}(G)) = \mathrm{TLIM}(\mathcal{RUC}(G))$ *and* $\mathrm{LIM}(\mathcal{UC}(G)) = \mathrm{TLIM}(\mathcal{UC}(G))$.

Proof. The statement follows from Corollary 4.12 and Proposition 4.13. □

Lemma 4.15. *Let G be a locally compact group and let $M \in \mathrm{TLIM}(\mathcal{UC}(G))$. Then for all $f \in \mathcal{LUC}(G)$, $M(\varphi * f) = M(\psi * f)$ whenever $\varphi, \psi \in P^1(G)$.*

Proof. Let (φ_i) be a net of approximate units in $L^1(G)$. For every $\varphi \in P^1(G)$,

$$\lim_i \|\varphi * \varphi_i * f - \varphi * f\| \leq \lim_i \|\varphi * \varphi_i - \varphi\|_1 \|f\| = 0.$$

As $f \in \mathcal{LUC}(G)$, $\varphi * f \in \mathcal{UC}(G)$ and, for every $i \in I$, $\varphi_i * f \in \mathcal{UC}(G)$ (Section 2G.3). As $\|M\| = 1$ and M is topologically left invariant on $\mathcal{UC}(G)$, we obtain

$$M(\varphi * f) = \lim_i M(\varphi * \varphi_i * f) = \lim_i M(\varphi_i * f).$$

Therefore the number $M(\varphi * f)$ is independent of the choice of $\varphi \in P^1(G)$. □

Lemma 4.16. *Let G be a locally compact group and let $M \in \mathrm{TRIM}(\mathcal{UC}(G))$. Then for every $f \in \mathcal{RUC}(G)$, $M(f * \check{\varphi}) = M(f * \check{\psi})$ whenever $\varphi, \psi \in P^1(G)$.*

Proof. The statement follows from Corollary 4.8, Lemma 4.15 and (2.3). □

Proposition 4.17. *Let G be a locally compact group. If $M \in \mathrm{TLIM}(\mathcal{UC}(G))$ [resp. $M \in \mathrm{TIM}(\mathcal{UC}(G))$] and U is an open relatively compact subset of G, there exists $N \in \mathrm{TLIM}(L^\infty(G))$ [resp. $N \in \mathrm{TIM}(L^\infty(G))$] defined by*

$$N(f) = \frac{|U|}{|U^{-1}|} M(\xi_U * f * \xi_U),$$

$f \in L^\infty(G)$. Any topologically invariant mean on $\mathcal{UC}(G)$ may be extended to a topologically invariant mean on $L^\infty(G)$.

Proof. Let $\theta = (|U|/|U^{-1}|)\xi_U = (1/|U^{-1}|)1_U$; then $\check{\theta} \in P^1(G) \cap L^\infty(G)$. For any $f \in L^\infty(G)$, $\xi_U * (f * \theta) \in \mathcal{RUC}(G)$ and $(\xi_U * f) * \theta \in \mathcal{LUC}(G)$ (Section 2D.3). We consider

$$N(f) = M(\xi_U * f * \theta) = \frac{|U|}{|U^{-1}|} M(\xi_U * f * \xi_U),$$

$f \in L^\infty(G)$. Obviously $N(f) \geq 0$ whenever $f \in L^\infty_+(G)$; also $\xi_U * 1_G * \theta = 1_G$ and $N(1_G) = 1$. Therefore we have $N \in \mathfrak{M}(L^\infty(G))$.

By Lemma 4.15 [and Lemma 4.16], for all $f \in L^\infty(G)$, $\varphi \in P^1(G)$, as $f * \theta \in \mathscr{L}\mathscr{U}\mathscr{C}(G)$ [and $\xi_U * f \in \mathscr{R}\mathscr{U}\mathscr{C}(G)$], we have

$$N(\varphi * f) = M(\xi_U * (\varphi * f) * \theta) = M((\xi_U * \varphi) * f * \theta)$$
$$= M(\xi_U * f * \theta) = N(f)$$

[and $N(f * \check{\varphi}) = M(\xi_U * (f * \check{\varphi}) * \theta) = M(\xi_U * f * (\check{\theta} * \varphi)\check{\ }) = M(\xi_U * f * \theta) = N(f)$].

In particular, if U is symmetric, one obtains an extension of $M \in \text{TIM}(\mathscr{U}\mathscr{C}(G))$ to an element of $\text{TIM}(L^\infty(G))$. □

Proposition 4.18. *Let G be a locally compact group. If $\text{RIM}(\mathscr{L}\mathscr{U}\mathscr{C}(G)) \neq \varnothing$, then $\text{IM}(\mathscr{U}\mathscr{C}(G)) \neq \varnothing$.*

Proof. Let $M \in \text{RIM}(\mathscr{L}\mathscr{U}\mathscr{C}(G))$. If $f \in \mathscr{R}\mathscr{U}\mathscr{C}(G)$, the mapping $f' : x \mapsto M(_x f)$ belongs to $\mathscr{R}\mathscr{U}\mathscr{C}(G)$. By Corollary 4.8 we have $\check{M} \in \text{LIM}(\mathscr{R}\mathscr{U}\mathscr{C}(G))$ and we may now define

$$N(f) = \check{M}(f')$$

for $f \in \mathscr{U}\mathscr{C}(G)$. Obviously $N(1_G) = 1$ and $N(f) \geq 0$ whenever $f \in \mathscr{U}\mathscr{C}_+(G)$; hence $N \in \mathfrak{M}(\mathscr{U}\mathscr{C}(G))$.

If $f \in \mathscr{U}\mathscr{C}(G), a, x \in G$, we have $_x f \in \mathscr{L}\mathscr{U}\mathscr{C}(G)$, $M(_x f) = M(_x f_a)$, hence $(f_a)' = f'$ and

$$N(f_a) = \check{M}((f_a)') = \check{M}(f') = N(f).$$

Moreover $_a f \in \mathscr{U}\mathscr{C}(G)$,

$$(_a f)'(x) = M(_x(_a f)) = M(_{ax} f) = f'(ax) = {_a f'(x)}$$

and

$$N(_a f) = \check{M}((_a f)') = \check{M}(_a f') = \check{M}(f') = N(f).$$

Therefore N is an invariant mean on $\mathscr{U}\mathscr{C}(G)$. □

Theorem 4.19. *Let G be a locally compact group. The following properties are equivalent and characterize amenability of G:*

(a) $\text{LIM}(X) \neq \varnothing$ *if $X = \mathscr{S}(G)$ or X is a left invariant subspace of $L^\infty(G)$ that contains $\mathscr{U}\mathscr{C}(G)$ and is closed under complex conjugation.*

(b) $\text{RIM}(X) \neq \varnothing$ *if $X = \mathscr{S}(G)$ or X is a right invariant subspace of $L^\infty(G)$ that contains $\mathscr{U}\mathscr{C}(G)$ and is closed under complex conjugation.*

(c) $\text{IM}(X) \neq \varnothing$ *if $X = \mathscr{S}(G)$ or X is an invariant subspace of $L^\infty(G)$ that contains $\mathscr{U}\mathscr{C}(G)$ and is closed under complex conjugation.*

(d) TLIM(X) ≠ ∅ if X is a topologically left invariant subspace of $L^\infty(G)$ that contains $\mathcal{UC}(G)$ and is closed under complex conjugation.
(e) TRIM(X) ≠ ∅ if X is a topologically right invariant subspace of $L^\infty(G)$ that contains $\mathcal{UC}(G)$ and is closed under complex conjugation.
(f) TIM(X) ≠ ∅ if X is a topologically invariant subspace of $L^\infty(G)$ that contains $\mathcal{UC}(G)$ and is closed under complex conjugation.

The corresponding conditions for real-valued functions are also equivalent.

Proof.

1. Observe first that invariance properties of means are transmitted by heredity onto subspaces. Recall also that LIM($\mathcal{UC}(G)$) = TLIM($\mathcal{UC}(G)$) by Corollary 4.14.

2. The extensions of left invariant means properties to larger spaces are insured by the following implications:

$$[\text{LIM}(\mathcal{S}(G)) \neq \varnothing] \Rightarrow [\text{LIM}(L^\infty(G)) \neq \varnothing] \quad \text{(Proposition 4.9)}$$

and

$$[\text{TLIM}(\mathcal{UC}(G)) \neq \varnothing] \Rightarrow [\text{TLIM}(L^\infty(G)) \neq \varnothing] \quad \text{(Proposition 4.17)}$$

$$[\text{TLIM}(L^\infty(G)) \neq \varnothing] \Rightarrow [\text{LIM}(L^\infty(G)) \neq \varnothing] \quad \text{(Corollary 4.12)}.$$

3. The equivalence of left-hand properties with right-hand properties is due to Proposition 4.7.

4. The equivalence of one-sided invariance properties with the apparently stronger bilateral properties is due to the following implications:

$$[\text{RIM}(\mathcal{LUC}(G)) \neq \varnothing] \Rightarrow [\text{IM}(\mathcal{UC}(G)) \neq \varnothing] \quad \text{(Proposition 4.18)}$$

$$[\text{IM}(\mathcal{UC}(G)) \neq \varnothing] \Rightarrow [\text{TIM}(\mathcal{UC}(G)) \neq \varnothing]$$

(Corollary 4.14 and Proposition 4.7)

$$[\text{TIM}(\mathcal{UC}(G)) \neq \varnothing] \Rightarrow [\text{TIM}(L^\infty(G)) \neq \varnothing] \quad \text{(Proposition 4.17)}$$

$$[\text{TIM}(L^\infty(G)) \neq \varnothing] \Rightarrow [\text{IM}(L^\infty(G)) \neq \varnothing]$$

(Corollary 4.12 and Proposition 4.7).

5. The equivalence with the conditions on real-valued functions is established in Proposition 4.6. □

The following corollary characterizes amenability in terms of the original description of the phenomenon.

Corollary 4.20. *A locally compact group G is amenable if and only if it admits a finitely additive probability measure μ on the σ-algebra $\mathfrak{H}(G)$ of Haar measurable subsets in G, that is left invariant [resp. right invariant; invariant], that is, for every $A \in \mathfrak{H}(G)$, one has $\mu(aA) = \mu(A)$ whenever $a \in G$ [resp. $\mu(Ab) = \mu(A)$ whenever $b \in G$; $\mu(aAb) = \mu(A)$ whenever $a, b \in G$].*

Proof. The corollary states the characterization of the amenability of G via the condition $\text{LIM}(\mathscr{S}_{\mathbf{R}}(G)) \neq \varnothing$ [resp. $\text{RIM}(\mathscr{S}_{\mathbf{R}}(G)) \neq \varnothing$, $\text{IM}(\mathscr{S}_{\mathbf{R}}(G)) \neq \varnothing$]. □

Note that if G is a noncompact amenable group, the measure μ given by Corollary 4.20 is not σ-additive because otherwise it would constitute a finite Haar measure.

Observe also that in Theorem 4.19 several implications are independent of the hypothesis of local compactness; they remain valid for arbitrary topological groups. Therefore, one may call a topological group G amenable if $\mathscr{C}(G)$ admits a left invariant mean. Unless the contrary is specified, *amenable group* will always signify that *the group is assumed to be locally compact*. Notice right now that at least in the discrete case one may define amenability for a semigroup S. If $f \in l^\infty(S)$ and $a \in S$, consider $_af \colon \begin{smallmatrix} S \to \mathbf{C} \\ x \mapsto f(ax) \end{smallmatrix}$. The semigroup S is called amenable if there exists a mean M on $l^\infty(S)$ such that $M(_af) = M(f)$ whenever $f \in l^\infty(S)$ and $a \in S$.

C. Consequences

Proposition 4.21. *If the group G admits locally compact group structures for two topologies $\mathscr{T}_1$ and $\mathscr{T}_2$ such that $\mathscr{T}_1$ is finer than $\mathscr{T}_2$, then amenability of $(G, \mathscr{T}_1)$ implies amenability of $(G, \mathscr{T}_2)$. In particular, a locally compact group G, for which G_d is amenable, is itself amenable.*

Proof. It suffices to notice that, as $\mathscr{C}(G, \mathscr{T}_2) \subset \mathscr{C}(G, \mathscr{T}_1)$, we have $\text{LIM}(\mathscr{C}(G, \mathscr{T}_1)) \subset \text{LIM}(\mathscr{C}(G, \mathscr{T}_2))$. □

We add some supplementary characterizations of amenability; with respect to Propositions 4.6 and 4.7 we may restrict ourselves to the complex case and also to one-sided versions.

Proposition 4.22. *A locally compact group G is amenable if and only if there exists an $M^1(G)$-invariant mean M on $L^\infty(G)$, that is, $M(\mu * f) = M(f)$ whenever $f \in L^\infty(G)$ and $\mu \in M^1(G)$.*

Proof. The condition is obviously sufficient because the mean belongs to $\text{TLIM}(L^\infty(G))$. Conversely, if $M \in \text{TLIM}(L^\infty(G))$, choose an arbitrary $\varphi_0 \in P^1(G)$. For all $f \in L^\infty(G)$, $\mu \in M^1(G)$, we have $\varphi_0 * \mu \in P^1(G)$, $\mu * f \in L^\infty(G)$ and then also

$$M(f) = M((\varphi_0 * \mu) * f) = M(\varphi_0 * (\mu * f)) = M(\mu * f). \qquad \square$$

We state some characterizations of amenability that constitute formal weakenings of the standard properties.

Proposition 4.23. *A locally compact group G is amenable if and only if there exists a mean M on $\mathscr{RUC}(G)$ [resp. $\mathscr{UC}(G)$] such that, for every $f \in \mathscr{RUC}(G)$ [resp. every $f \in \mathscr{UC}(G)$] and every element d of a dense subset D of G, $M(_d f) = M(f)$ holds.*

Proof. Assume that the condition holds. Let $f \in \mathscr{RUC}(G)$ [resp. $f \in \mathscr{UC}(G)$] and $a \in G$. As f is right uniformly continuous, for $\varepsilon > 0$ there exists $d \in D$ such that $\|_a f - {_d f}\| < \varepsilon$; thus for the mean M given by hypothesis, we have

$$|M(_a f) - M(f)| \leq |M(_a f - {_d f})| + |M(_d f - f)| \leq \|_a f - {_d f}\| < \varepsilon.$$

As $\varepsilon > 0$ is arbitrary, we conclude that $M(_a f) = M(f)$. □

Lemma 4.24. *Let G be a locally compact group, A a Banach $*$-subalgebra of $\mathscr{M}^1(G)$, $A^1 = A \cap M^1(G)$ and X an A-invariant subspace of $L^\infty(G)$. If B is a dense linear subspace of A and $M \in \mathfrak{M}(X)$ such that $M(\mu * f) = M(f)$ whenever $f \in X$, $\mu \in B \cap A^1$, then also $M(\mu * f) = M(f)$ whenever $f \in X$, $\mu \in A^1$.*

Proof. If A_1 is the unit sphere of A, $\overline{B \cap A_1} = A_1$ (Section 2B.2); so also $\overline{B \cap A^1} = A^1$. Given $f \in X$, $\mu \in A^1$ and $\varepsilon \in \mathbf{R}_+^*$, there exists $\mu_1 \in B \cap A^1$ such that $\|\mu - \mu_1\| \|f\|_\infty < \varepsilon$, hence $\|\mu * f - \mu_1 * f\|_\infty < \varepsilon$. By hypothesis we have

$$|M(\mu * f - f)| \leq |M((\mu - \mu_1) * f)| + |M(\mu_1 * f - f)|$$

$$\leq \|(\mu - \mu_1) * f\|_\infty < \varepsilon.$$

As $\varepsilon \in \mathbf{R}_+^*$ is arbitrary, $M(\mu * f) = M(f)$. □

Proposition 4.25. *A locally compact group G is amenable if and only if any one of the following conditions holds*:
  (a) *On $L^\infty(G)$ there exists a mean M such that $M(\mu * f) = M(f)$ whenever $f \in L^\infty(G)$ and $\mu \in M_c^1(G)$.*
  (b) *On every topologically left invariant subspace X of $L^\infty(G)$ that contains $\mathscr{UC}(G)$ and is closed under complex conjugation there exists a mean M such that $M(\varphi * f) = M(f)$ whenever $f \in X$ and $\varphi \in P^1(G) \cap \mathscr{K}(G)$.*

Proof. The statement follows readily from Proposition 4.22, Theorem 4.19, and Lemma 4.24. □

If G is an amenable group, every closed subalgebra X of $L^\infty(G)$ that contains $\mathscr{UC}(G)$, is closed under complex conjugation and left invariant, admits a left invariant mean. As we noted in Section 3A, to X we associate the structure space $\hat{X}$; if $f \in X$ and $a \in G$, let $a \cdot \hat{f} = \widehat{L_a f}$. The mapping $M \mapsto \hat{M}$ identifies $\mathrm{LIM}(X)$ with the set of probability measures on $\hat{X}$ that are G-invariant; for every $\hat{f} \in \mathscr{C}(\hat{X})$ and every $a \in G$,

$$\langle a \cdot \hat{f}, \hat{M} \rangle = \langle \widehat{L_a f}, \hat{M} \rangle = \langle L_a f, M \rangle = \langle f, M \rangle = \langle \hat{f}, \hat{M} \rangle.$$

We complete these considerations by making a final remark on the structures of the set of left invariant means and the set of topologically left invariant means in particular cases.

Proposition 4.26. *Let G be an amenable group.*
(a) *Then $\mathrm{LIM}(L^\infty(G))$ is a right ideal in $\mathfrak{M}(L^\infty(G))$.*
(b) *Let $X = L^\infty(G)$ or $\mathcal{RUC}(G)$. If $F \in X^*$ such that $\langle \varphi * f, F \rangle = \langle f, F \rangle$ whenever $f \in X$, $\varphi \in P^1(G)$, and $M_1 \in \mathfrak{M}(X)$, then $M_1 \odot F = F$. Also $\mathrm{TLIM}(X)$ is an ideal in $\mathfrak{M}(X)$ with $M_1 \odot M_2 = M_2$ whenever $M_1 \in \mathfrak{M}(X)$ and $M_2 \in \mathrm{TLIM}(X)$. For $N_1, N_2 \in \mathrm{TLIM}(X)$, $N_1 \neq N_2$, one has $N_1 \odot N_2 \neq N_2 \odot N_1$.*

Proof. (a) If $M_1 \in \mathrm{LIM}(L^\infty(G))$, $M_2 \in \mathfrak{M}(L^\infty(G))$ and $f \in L^\infty(G)$, $a \in G$, by (2.14), we have

$$\langle {}_a f, M_1 \odot M_2 \rangle = \langle M_2 * ({}_a f), M_1 \rangle = \langle {}_a (M_2 * f), M_1 \rangle$$
$$= \langle M_2 * f, M_1 \rangle = \langle f, M_1 \odot M_2 \rangle.$$

(b) Given $f \in X$, we have $F * f \in L^\infty(G)$ and, for every $\psi \in P^1(G)$,

$$\langle \psi, F * f \rangle = \langle f\psi, F \rangle = \langle f, F \rangle = \langle \psi, F(f) 1_G \rangle;$$

as $P^1(G)$ generates $L^1(G)$, we conclude that $F * f = F(f) 1_G \in L^\infty(G)$. Hence

$$\langle f, M_1 \odot F \rangle = \langle F * f, M_1 \rangle = \langle F(f) 1_G, M_1 \rangle = F(f)$$
$$= \langle f, F \rangle,$$

that is, $M_1 \odot F = F$.

In particular $M_1 \odot M_2 = M_2$ and $\mathrm{TLIM}(X)$ is a left ideal in $\mathfrak{M}(X)$. On the other hand, for all $\varphi \in P^1(G)$, by (2.16) we have

$$\langle f\varphi, M_2 \odot M_1 \rangle = \langle M_1 * (f\varphi), M_2 \rangle = \langle (M_1 * f)\varphi, M_2 \rangle$$
$$= \langle M_1 * f, M_2 \rangle = \langle f, M_2 \odot M_1 \rangle$$

hence $M_2 \odot M_1 \in \mathrm{TLIM}(X)$ and $\mathrm{TLIM}(X)$ is a right ideal. If $N_1, N_2 \in \mathrm{TLIM}(X)$ with $N_1 \neq N_2$, we obtain

$$N_1 \odot N_2 = N_2 \neq N_1 = N_2 \odot N_1. \qquad \square$$

D. The Dixmier Criterion

We establish a criterion that insures the existence of invariant means via the Hahn–Banach theorem, a definitely nonconstructive procedure.

Lemma 4.27. *Let G be a locally compact group. Let X be a subspace of $L^\infty_{\mathbf{R}}(G)$ such that $1_G \in X$ and Y a subspace of X. Then the following properties are equivalent:*
  (i) $(\exists M \in \mathfrak{M}(X))(\forall f \in Y)\, M(f) = 0$.
  (ii) $(\forall f \in Y)\, \operatorname{ess\,sup} f \geq 0$.

Proof. (i) $\Rightarrow$ (ii)
By the definition of means, for every $f \in Y$,

$$0 = M(f) \leq \operatorname{ess\,sup} f.$$

(ii) $\Rightarrow$ (i)

Let $M_0(h) = 0$ for $h \in Y$ and $N(f) = \operatorname{ess\,sup} f$ for $f \in X$. If $h \in Y$, $M_0(h) \leq N(h)$; N is a sublinear functional on X. By the Hahn–Banach theorem ([HS] 14.9) the linear functional M_0 on Y may be extended to a linear functional M on X such that, for every $f \in X$, $M(f) \leq \operatorname{ess\,sup} f$. Therefore M is a mean on X. For every $f \in Y$, $M(f) = 0$. □

Note that in the preceding lemma necessarily $1_G \notin Y$.

Lemma 4.28. *Let G be a locally compact group, X a subspace of $L^\infty_{\mathbf{R}}(G)$ with $1_G \in X$, A a $*$-subalgebra of $\mathscr{M}^1(G)$. Assume X to be A-invariant and let S be a subset of $A \cap M^1(G)$. Then there exists a mean M on X such that $M(\mu * f) = M(f)$ whenever $f \in X$ and $\mu \in S$ if and only if $\operatorname{ess\,sup} \sum_{i=1}^n (f_i - \mu_i * f_i) \geq 0$ whenever $f_1, \ldots, f_n \in X$, $\mu_1, \ldots, \mu_n \in S$, $n \in \mathbf{N}^*$.*

Proof. One applies Lemma 4.27 to the real linear subspace Y of X generated by the functions $f_i - \mu_i * f_i$, where $f_i \in X$, $\mu_i \in S$, $i = 1, \ldots, n$, $n \in \mathbf{N}^*$. □

Proposition 4.29. *Amenability of a locally compact group G is characterized by each of the following conditions:*
  (a) *If $X = \mathscr{S}_{\mathbf{R}}(G)$ or X is a left invariant subspace of $L^\infty_{\mathbf{R}}(G)$ containing $\mathscr{U\!C}_{\mathbf{R}}(G)$, then $\operatorname{ess\,sup} \sum_{i=1}^n (f_i - {}_{a_i}f_i) \geq 0$ whenever $f_1, \ldots, f_n \in X$, $a_1, \ldots, a_n \in G$, $n \in \mathbf{N}^*$.*
  (b) *If $X = \mathscr{R\!U\!C}_{\mathbf{R}}(G)$ or $\mathscr{U\!C}_{\mathbf{R}}(G)$ and D is a dense subset of G, then $\operatorname{ess\,sup} \sum_{i=1}^n (f_i - {}_{a_i}f_i) \geq 0$ whenever $f_1, \ldots, f_n \in X$, $a_1, \ldots, a_n \in D$, $n \in \mathbf{N}^*$.*

(c) *If X is a topologically left invariant subspace of $L_\mathbf{R}^\infty(G)$ containing $\mathscr{UC}_\mathbf{R}(G)$, then* $\operatorname{ess\,sup} \sum_{i=1}^n (f_i - \varphi_i * f_i) \geq 0$ *whenever* $f_1, \ldots, f_n \in X$, $\varphi_1, \ldots, \varphi_n \in P^1(G)$, $n \in \mathbf{N}^*$.

(d) *If X is a topologically left invariant subspace of $L_\mathbf{R}^\infty(G)$ containing $\mathscr{UC}_\mathbf{R}(G)$, then* $\operatorname{ess\,sup} \sum_{i=1}^n (f_i - \varphi_i * f_i) \geq 0$ *whenever* $f_1, \ldots, f_n \in X$, $\varphi_1, \ldots, \varphi_n \in P^1(G) \cap \mathscr{K}(G)$, $n \in \mathbf{N}^*$.

(e) $\operatorname{ess\,sup} \sum_{i=1}^n (f_i - \mu_i * f_i) \geq 0$ *whenever* $f_1, \ldots, f_n \in L_\mathbf{R}^\infty(G)$, $\mu_1, \ldots, \mu_n \in M^1(G)$, $n \in \mathbf{N}^*$.

(f) $\operatorname{ess\,sup} \sum_{i=1}^n (f_i - \mu_i * f_i) \geq 0$ *whenever* $f_1, \ldots, f_n \in L_\mathbf{R}^\infty(G)$, $\mu_1, \ldots, \mu_n \in M_c^1(G)$, $n \in \mathbf{N}^*$.

Proof. The statement follows from Lemma 4.28 and Theorem 4.19, Proposition 4.23, Proposition 4.22, Proposition 4.25. □

With respect to Theorem 4.19 amenability of a locally compact group is also characterized by right-hand versions and bilateral versions of Proposition 4.29.

NOTES

As we pointed out in Section 1, von Neumann [414] inaugurated the study of invariant means for the space of bounded functions on a discrete group. The existence of invariant means for discrete semigroups and groups was widely investigated by Day [114, 115]. Reiter [452] observed the equivalence [LIM($L^\infty(G)$) ≠ ∅] ⇔ [LIM($\mathscr{C}(G)$) ≠ ∅]. Hulanicki [268] introduced the notion of topological invariance and considered $M^1(G)$-invariant means. He showed that every topologically left invariant mean on $\mathscr{C}(G)$ may be extended to a topologically left invariant mean on $L^\infty(G)$ and that every topologically left invariant mean on $L^\infty(G)$ is left invariant; he also established 4.22. Greenleaf [226] studied invariance of means extensively and organized the subject. He proved that the existence of a left invariant mean on $\mathscr{UC}(G)$ implies the existence of a topologically left invariant mean on $L^\infty(G)$. Namioka [412] established the implication [LIM($L^\infty(G)$) ≠ ∅] ⇒ [TLIM($L^\infty(G)$) ≠ ∅]. Greenleaf used Namioka's argument to simplify an earlier proof showing that on $\mathscr{UC}(G)$ a left invariant mean is also topologically left invariant.

Day [115] proved 4.26 in the discrete case. Granirer [210] noticed the noncommutativity of TLIM($\mathscr{RUC}(G)$) if the set is not reduced to a singleton. For the general version of 4.26 see Wong [560].

Dixmier [138] proved his criterion for semigroups. Hulanicki [268] transposed it to topological invariance.

5. FLOWS AND FIXED POINT PROPERTIES

The existence of an invariant mean is a fixed point property. In the framework of the theory of flows, we consider at present a general fixed point property that is shown to be equivalent to the particular case concerning the existence of a left invariant mean on the vector space of right uniformly continuous bounded real-valued functions defined on a topological group.

A. Flows

If G is a topological group and Z is a compact space, the pair (G, Z) is called a *flow* in case G is a *transformation group* on Z, that is, there exists a continuous mapping

$$(x, \zeta) \mapsto x\zeta$$

$$G \times Z \to Z$$

such that $x_1(x_2\zeta) = (x_1 x_2)\zeta$ for all $x_1, x_2 \in G$, $\zeta \in Z$ and $e\zeta = \zeta$ for all $\zeta \in Z$. For a given $x \in G$, the mapping $\zeta \mapsto x\zeta$ is a homeomorphism from Z onto Z. The continuity of the transformation group mapping is equivalent to the separate continuity in each variable, a classical result due to Ellis ([158] Theorem 1). If $f \in \mathscr{C}(Z)$, for $x \in G$, we define

$$fx : \zeta \mapsto f(x\zeta)$$

$$Z \to \mathbf{C};$$

$\|fx\| = \|f\|$ and $fx \in \mathscr{C}(Z)$. Given $\zeta \in Z$ and $\varepsilon \in \mathbf{R}_+^*$, there exist neighborhoods U_ζ of e in G and V_ζ of ζ in Z such that $|f(x\eta) - f(\eta)| < \varepsilon$ whenever $x \in U_\zeta$ and $\eta \in V_\zeta$. As Z is compact, one can determine $\zeta_1, \ldots, \zeta_n \in Z$ such that $Z = \cup_{i=1}^n V_{\zeta_i}$. Consider the neighborhood $U = \cap_{i=1}^n U_{\zeta_i}$ of e in G. We have then $|f(x\zeta) - f(\zeta)| < \varepsilon$ for all $x \in U$ and $\zeta \in Z$; so for all $\zeta \in Z$ and all $x, y \in G$ with $xy^{-1} \in U$,

$$|f(x\zeta) - f(y\zeta)| = |f((xy^{-1})(y\zeta)) - f(y\zeta)| < \varepsilon,$$

that is, the mapping $x \mapsto fx$ is right uniformly continuous on G.
$$G \to \mathscr{C}(Z)$$

For each $\zeta \in Z$, we consider the *orbit* $O_\zeta = \{x\zeta : x \in G\}$ of ζ in Z. The flow (G, Z) is said to be *minimal* if every orbit is dense in Z. If Y is a closed subset of Z that is G-invariant, that is, $x\eta \in Y$ whenever $x \in G$, $\eta \in Y$, we may consider the *subflow* (G, Y) of (G, Z). A flow is minimal if and only if it does not admit a proper subflow. Two points ζ, ζ' in Z are called *proximal* if there exists a net (x_i) in G such that $\lim_i x_i\zeta = \lim_i x_i\zeta'$; the flow (G, Z) is called *proximal flow* if any pair of points in Z is proximal.

To the action of G on Z we associate the action of G on $\mathscr{M}^1(Z)$ defined by

$$\langle f, x\mu \rangle = \langle fx, \mu \rangle$$

for $f \in \mathscr{C}(Z)$, $\mu \in \mathscr{M}^1(Z)$, $x \in G$. As $M^1(Z)$ is obviously G-invariant, we may consider the flow $(G, M^1(Z))$ for the w*-topology of $M^1(Z)$. If $\zeta \in Z$ and $x \in G$, $f \in \mathscr{C}(Z)$, we have

$$f(x\zeta) = (fx)(\zeta) = \langle fx, \delta_\zeta \rangle = \langle f, x\delta_\zeta \rangle.$$

Therefore, we may identify ζ with δ_ζ and consider (G, Z) to be a subflow of $(G, M^1(Z))$. The flow (G, Z) is called *strongly proximal flow* if the flow $(G, M^1(Z))$ is proximal.

Lemma 5.1. *A flow (G, Z) is strongly proximal if and only if, for every $\mu \in M^1(Z)$, there exist a net (x_i) in G and $\zeta \in Z$ such that $\delta_\zeta = \lim_i x_i\mu \in \overline{O_\mu}$.*

Proof. (a) Assume that (G, Z) is strongly proximal. If $\mu \in M^1(Z)$ and $\eta \in Z$, there exists a net (x_i) in G such that $\lim_i x_i\mu = \lim_i x_i\delta_\eta$. It suffices to put $\zeta = \lim_i x_i\eta$ for (G, Z) identified with a subflow of $(G, M^1(Z))$.

(b) Let $\mu, \nu \in M^1(Z)$ and $\sigma = \frac{1}{2}(\mu + \nu) \in M^1(Z)$. If the condition holds, there exist a net (x_i) in G and $\zeta \in Z$ such that $\lim_i x_i\sigma = \delta_\zeta$. For a subnet (x_{i_j}) of (x_i) we have $\lim_j x_{i_j}\mu = \mu_1$ and $\lim_j x_{i_j}\nu = \nu_1$ in $M^1(Z)$; then $\delta_\zeta = \frac{1}{2}(\mu_1 + \nu_1)$. As δ_ζ is an extreme point in $M^1(Z)$ ([D] 2.5.2, 2.5.5(ii)), we must have $\mu_1 = \nu_1 = \delta_\zeta$; the proximality of $M^1(Z)$ is established. □

If E is a locally convex Hausdorff topological real vector space and Z is a compact convex subset of E, a flow (G, Z) is called *affine flow* in the case, for every $x \in G$, the mapping $\zeta \mapsto x\zeta$ is affine, that is, for all $\zeta_1, \zeta_2 \in Z$ and $\alpha \in [0, 1]$, $x(\alpha\zeta_1 + (1 - \alpha)\zeta_2) = \alpha x\zeta_1 + (1 - \alpha)x\zeta_2$. Every nonempty closed convex G-invariant subset Y of Z gives rise to an affine subflow (G, Y) of (G, Z). The affine flow (G, Z) is called *irreducible affine flow* if it admits no proper affine subflow.

Lemma 5.2. *Let Z be a compact convex subset of a locally convex Hausdorff topological real vector space E and let (G, Z) be an irreducible affine flow. If Y is the closure of the set of extreme points in Z, then (G, Y) is the unique minimal subflow of (G, Z). Moreover (G, Z) is strongly proximal and (G, Y) is a minimal strongly proximal flow.*

Proof. (a) Let S be the set of extreme points of Z. If $\zeta \in S$, $x \in G$ and $x\zeta = \alpha\zeta_1 + (1 - \alpha)\zeta_2$ for $\zeta_1, \zeta_2 \in Z$, $\alpha \in [0, 1]$, then

$$\zeta = x^{-1}(\alpha\zeta_1 + (1 - \alpha)\zeta_2) = \alpha x^{-1}\zeta_1 + (1 - \alpha)x^{-1}\zeta_2.$$

Necessarily $\alpha = 0$ or 1, that is, $x\zeta \in S$. Therefore S and Y are G-invariant; (G, Y) is a subflow.

(b) Let Z_1 be a closed subset of Z such that (G, Z_1) is a minimal subflow of (G, Z); as $(G, \overline{co}Z_1)$ is an affine subflow and (G, Z) is irreducible, $\overline{co}Z_1 = Z$. Then necessarily $S \subset Z_1$ ([B EVT] II, Section 7, No. 1, proposition 2, corollaire) and also $Y \subset Z_1$. Hence (G, Y) is the unique minimal subflow of (G, Z).

FLOWS AND FIXED POINT PROPERTIES

(c) Any element ν in $M^1(Z)$ admits a barycenter $g(\nu) = \int_Z \zeta \, d\nu(\zeta)$ in Z, that is,

$$\int_Z \langle \zeta, \eta \rangle \, d\nu(\zeta) = \langle g(\nu), \eta \rangle$$

whenever $\eta \in E^*$. If $\mu \in M^1(Z)$ and B is the set of all barycenters of the elements of the G-invariant set $\overline{O_\mu}$, then $\overline{co}B$ is a closed convex G-invariant subset of Z that coincides with $g(\overline{co}\,\overline{O_\mu})$ ([B INT] Chap. IV, Section 7, No. 1).

Since (G, Z) is irreducible, $\overline{co}B = Z$ and then again as in (b) we have $S \subset B$. If $\eta \in S$, δ_η is the unique element in $M^1(Z)$ admitting η as its barycenter ([B INT] Chap. IV, Section 7, No. 2, proposition 3, corollaire). As $g(\overline{co}\,\overline{O_\mu}) = \overline{co}B$, necessarily $\delta_\eta \in \overline{O_\mu}$. Lemma 5.1 implies then that (G, Z) is strongly proximal. □

B. The Fixed Point Characterization of Amenability

Definition 5.3. *A topological group G is said to have the fixed point property (FP) if for every affine flow (G, Z) there exists a point ζ in Z that is invariant under the action of G, that is, $x\zeta = \zeta$ whenever $x \in G$.*

The characterization of amenability used in the following theorem is the existence of a left invariant mean on the space of right uniformly continuous bounded real-valued functions; it applies to topological nonnecessarily locally compact groups.

Theorem 5.4. *If G is a locally compact group, the following conditions are equivalent*:
  (i)   *G is amenable.*
  (ii) *G has the fixed point property (FP).*
  (iii) *Every flow (G, Z) admits a G-invariant probability measure μ, that is, for all $f \in \mathscr{C}(Z)$, $x \in G$, $\langle f, \mu \rangle = \langle fx, \mu \rangle$ with $fx(\zeta) = f(x\zeta)$, $\zeta \in Z$.*
  (iv) *Every minimal strongly proximal flow (G, Y) is trivial, that is, Y is a singleton.*

Proof. [LIM($\mathscr{RUC}_\mathbf{R}(G)) \neq \varnothing$] $\Rightarrow$ (ii)

Let (G, Z) be an affine flow on a compact convex subset Z of a locally convex Hausdorff topological real vector space E.

If $f \in E^*$ and $\zeta \in Z$, we consider the mapping

$$f^\zeta : x \mapsto \langle x\zeta, f \rangle$$

$$G \to \mathbf{R}.$$

By the considerations made in A, we have $f^\zeta \in \mathscr{RUC}_\mathbf{R}(G)$.

We embed E into the algebraic dual $(E^*)'$ of E^* with the topology $\sigma((E^*)', E^*)$; since Z is compact in E, it is closed in $(E^*)'$. Let ζ_0 be a fixed element in Z. For every $M \in \mathfrak{M}(\mathcal{RUC}_R(G))$, the mapping

$$g \mapsto M(g^{\zeta_0})$$

$$E^* \to \mathbf{R}$$

belongs to $(E^*)'$, that is, there exists an element $\theta(M)$ in $(E^*)'$ such that the duality relation

$$\langle g, \theta(M) \rangle = M(g^{\zeta_0})$$

holds for every $g \in E^*$. In particular, if $M = \sum_{i=1}^n \alpha_i \delta_{x_i}$, where $\alpha_1, \ldots, \alpha_n \in \mathbf{R}_+^*$ with $\sum_{i=1}^n \alpha_i = 1$ and $x_1, \ldots, x_n \in G$, convexity and G-invariance of Z imply that $\theta(M) \in Z$. The mapping θ from $\mathfrak{M}(\mathcal{RUC}_R(G))$ equipped with the w*-topology into $(E^*)'$ equipped with the $\sigma((E^*)', E^*)$-topology is continuous. Since $M_f^1(G)$ is w*-dense in $\mathfrak{M}(\mathcal{RUC}(G))$ by Corollary 3.4 and Z is closed in $(E^*)'$, we conclude that $\theta(M) \in Z$ for every $M \in \mathfrak{M}(\mathcal{RUC}_R(G))$.

If $a \in G$, we consider

$$\Lambda_a : \zeta \mapsto a\zeta$$

$$Z \to Z.$$

Let $M \in \mathrm{LIM}(\mathcal{RUC}_R(G))$. For all $a \in G$, $g \in E^*$, we have

$$\langle g, a\theta(M) \rangle = \langle g \circ \Lambda_a, \theta(M) \rangle = M\big((g \circ \Lambda_a)^{\zeta_0}\big).$$

But

$$(g \circ \Lambda_a)^{\zeta_0} : x \mapsto \langle x\zeta_0, g \circ \Lambda_a \rangle = \langle ax\zeta_0, g \rangle = g(ax\zeta_0)$$

$$G \to \mathbf{R}$$

so $M((g \circ \Lambda_a)^{\zeta_0}) = M(g^{\zeta_0}) = \langle g, \theta(M) \rangle$, hence $a\theta(M) = \theta(M)$ for every $a \in G$, that is, $\theta(M)$ is a fixed point under the action of G.

$$\text{(ii)} \Rightarrow [\mathrm{LIM}(\mathcal{RUC}_R(G)) \neq \varnothing]$$

By Proposition 3.3 the set $\mathfrak{M}(\mathcal{RUC}_R(G))$ is convex and w*-compact in $\mathcal{RUC}_R(G)^*$. We define the flow $(G, \mathfrak{M}(\mathcal{RUC}_R(G)))$ by putting

$$\langle f, xM \rangle = \langle _x f, M \rangle = \langle L_{x^{-1}} f, M \rangle$$

for $x \in G$, $M \in \mathfrak{M}(\mathcal{RUC}_R(G))$ and $f \in \mathcal{RUC}_R(G)$. If $x, y \in G$, $M \in \mathfrak{M}(\mathcal{RUC}_R(G))$ and $f \in \mathcal{RUC}_R(G)$,

$$\langle f, x(yM) \rangle = \langle L_{y^{-1}}(L_{x^{-1}} f), M \rangle = \langle L_{(xy)^{-1}} f, M \rangle = \langle f, xyM \rangle.$$

It suffices now to establish the continuity of the mapping

$$(x, M) \mapsto xM$$

$$G \times \mathfrak{M}(\mathcal{RUC}_{\mathbf{R}}(G)) \to \mathfrak{M}(\mathcal{RUC}_{\mathbf{R}}(G)).$$

Let $(x_0, M_0) \in G \times \mathfrak{M}(\mathcal{RUC}_{\mathbf{R}}(G))$ and $\varepsilon \in \mathbf{R}_+^*$. Given $f \in \mathcal{RUC}_{\mathbf{R}}(G)$, there exists a neighborhood V of x_0 in G such that $\|_x f - _{x_0} f\| < \varepsilon/2$ whenever $x \in V$. Consider the w*-neighborhood $W = \{M \in \mathcal{RUC}_{\mathbf{R}}(G)^* : |\langle_{x_0} f, M - M_0\rangle| < \varepsilon/2\}$ of M_0. Then for every $(x, M) \in V \times W$,

$$|\langle f, xM\rangle - \langle f, x_0 M_0\rangle| \leq |\langle_x f - _{x_0} f, M\rangle| + |\langle_{x_0} f, M - M_0\rangle|$$

$$\leq \|_x f - _{x_0} f\| + |\langle_{x_0} f, M - M_0\rangle| < \varepsilon.$$

By hypothesis there exists $M_1 \in \mathfrak{M}(\mathcal{RUC}_{\mathbf{R}}(G))$ that is fixed under the action of G, that is, for every $f \in \mathcal{RUC}_{\mathbf{R}}(G)$ and every $x \in G$,

$$\langle_x f, M_1\rangle = \langle f, xM_1\rangle = \langle f, M_1\rangle$$

hence $M_1 \in \mathrm{LIM}(\mathcal{RUC}_{\mathbf{R}}(G))$.

$$(\mathrm{ii}) \Rightarrow (\mathrm{iii})$$

One applies (FP) to the affine flow $(G, M^1(Z))$.

$$(\mathrm{iii}) \Rightarrow (\mathrm{iv})$$

By hypothesis, the affine flow $(G, M^1(Y))$ admits a fixed point $\mu \in M^1(Y)$. Then Lemma 5.1 implies the existence of $\eta \in Y$ such that $\mu = \delta_\eta$ and η is a fixed point in Y. Since (G, Y) is minimal, $Y = \{\eta\}$.

$$(\mathrm{iv}) \Rightarrow (\mathrm{ii})$$

Let (G, Z) be an affine flow. By Zorn's lemma it admits an irreducible affine subflow; by Lemma 5.2 the latter admits a minimal subflow (G, Y) that is strongly proximal. So Y reduces to $\{\eta\}$ and then η is a fixed point for G in Z. □

NOTES

Fixed point properties in relation to amenability were considered for the first time by Day [116, 117]. His proofs concerning topological semigroups show that, for a locally compact group G, $[\mathrm{LIM}(\mathcal{C}(G)) \neq \emptyset] \Rightarrow (FP)$ and that the converse holds if G is compact or discrete. Furstenberg [191] studied the class of groups having (FP). Rickert [462] proved the equivalence (i) $\Leftrightarrow$ (ii) in 5.4.

The theory of flows was developed mainly by Ellis [159]. An exhaustive study on the subject is Glasner [203]; it contains the proof of 5.4 and the foregoing lemmas.

6. CONVERGENCE TO INVARIANCE

Proposition 3.3 establishes that, for a locally compact group G, $P^1(G)$ is w*-dense in the set of all means on $L^\infty(G)$. Hence amenability of G, that is, existence of a left invariant mean on $L^\infty(G)$ may be characterized by the existence of an asymptotically left invariant net in $P^1(G)$. We begin by examining this convergence to invariance; we then study other powerful properties of almost invariance.

A. Day's Asymptotical Invariance Properties

We first state a general definition.

Definition 6.1. *Let G be a locally compact group and let S be a subset of $M^1(G)$. We say that a net (μ_i) in $M^1(G)$ is asymptotically S-invariant if $\lim_i \|\mu * \mu_i - \mu_i\| = 0$ whenever $\mu \in S$.*

We introduce supplementary definitions that, as a matter of fact, characterize amenability.

Definition 6.2. *Let G be a locally compact group. For $p \in [1, \infty[$, the group is said to have property (D_p) if there exists a net (φ_i) in $P^p(G)$ such that $\lim_i \|_a\varphi_i - \varphi_i\|_p = 0$ whenever $a \in G$. The group is said to have property (D) [resp. (D')] if there exists a net (φ_i) in $P^1(G)$ that is asymptotically $P^1(G)$-invariant [resp. asymptotically $M^1(G)$-invariant].*

Lemma 6.3. *Let G be a locally compact group and let S be a subset of $M^1(G)$. There exists a net (φ_i) in $P^1(G)$ such that, for every $\mu \in S$, $(\mu * \varphi_i - \varphi_i)$ converges to 0 in the strong topology of $L^1(G)$ if and only if there exists a net (ψ_i) in $P^1(G)$ such that, for every $\mu \in S$, $(\mu * \psi_i - \psi_i)$ converges to 0 in the w-topology of $L^1(G)$.*

Proof. To every $\mu \in S$ we associate the operator $N_\mu: f \mapsto \mu * f - f$ on $L^1(G)$. Let also

$$N_S: f \mapsto \left(N_\mu(f)\right)_{\mu \in S}$$

$$L^1(G) \to [L^1(G)]^S.$$

The w-topology on the product $[L^1(G)]^S$ is the product of the w-topologies on

$L^1(G)$ ([B EVT] II, Section 6, No. 6, proposition 8). As $P^1(G)$ is convex, the closures of $N_S(P^1(G))$ in $[L^1(G)]^S$ for the strong topology and the w-topology coincide (Section 2B.1). □

Lemma 6.4. *A locally compact group G has property* (D_1) [*resp.* $(D), (D')$] *if and only if it has the following property*: (Δ_1) [*resp.* $(\Delta), (\Delta')$]. *There exists a net* (φ_i) *in* $P^1(G)$ *such that, for every* $a \in G$ [*resp. every* $\varphi \in P^1(G)$, *every* $\mu \in M^1(G)$], *the net* $({}_a\varphi_i - \varphi_i)$ [*resp.* $(\varphi * \varphi_i - \varphi_i), (\mu * \varphi_i - \varphi_i)$] *converges to 0 in the w-topology of* $L^1(G)$.

Proof. The equivalences concern particular instances of Lemma 6.3 □

Lemma 6.5. *Let G be a locally compact group and let S be a subset of* $M^1(G)$. *Then the following properties are equivalent*:
 (i) *There exists a mean M on* $L^\infty(G)$ *such that, for every* $f \in L^\infty(G)$ *and every* $\mu \in S$, $M(\mu^* * f) = M(f)$.
 (ii) *There exists a net* (φ_i) *in* $P^1(G)$ *such that, for every* $\mu \in S$, *the net* $(\mu * \varphi_i - \varphi_i)$ *converges to 0 in the w-topology* $\sigma(L^1(G), L^\infty(G))$.

More precisely, if (i) *holds, there exists a net* (φ_i) *in* $P^1(G)$ *converging to M in the w*-topology such that, for every* $\mu \in S$, $(\mu * \varphi_i - \varphi_i)$ *converges to 0 in the w-topology; if* (ii) *holds, a subnet of* (φ_i) *converges in the w*-topology to a mean M on* $L^\infty(G)$ *satisfying* (i).

Proof. (i) ⇒ (ii)
As $P^1(G)$ is w*-dense in $\mathfrak{M}(L^\infty(G))$ by Proposition 3.3, there exists a net (φ_i) in $P^1(G)$ such that, for every $f \in L^\infty(G)$ and every $\mu \in S$,

$$\lim_i \langle \varphi_i, \mu^* * f - f \rangle = M(\mu^* * f - f) = 0;$$

hence by (2.13) we have

$$\lim_i \langle \mu * \varphi_i - \varphi_i, f \rangle = 0.$$

(ii) ⇒ (i)

By Proposition 3.3 again the net (φ_i) admits a subnet (φ_{i_j}) converging to a mean M in the w*-topology of $L^\infty(G)$. Hence for every $f \in L^\infty(G)$ and every $\mu \in S$, by (2.13) also we have

$$\langle \mu^* * f - f, M \rangle = \lim_j \langle \varphi_{i_j}, \mu^* * f - f \rangle = \lim_j \langle \mu * \varphi_{i_j} - \varphi_{i_j}, f \rangle = 0. \quad □$$

Proposition 6.6. *The amenability of a locally compact group is characterized by the properties* $(D_1), (D), (D'), (\Delta_1), (\Delta), (\Delta')$.

Proof. The statement follows from Theorem 4.19, Proposition 4.22, Lemma 6.5, and Lemma 6.4. □

Proposition 6.7. *Amenability of a locally compact group G is characterized by each of the following conditions:*

(i) *There exists a net $(\mu_i)_{i \in I}$ in $M^1(G)$ that is asymptotically $M^1(G)$-invariant.*

(ii) *There exists a net $(\mu_i)_{i \in I}$ in $M^1(G)$ such that, for every compact subset K of G, $\lim_i \|\mu * \mu_i - \mu_i\| = 0$ uniformly for every $\mu \in M_c^1(G)$ with $\operatorname{supp}\mu \subset K$.*

(iii) *There exists a net $(\psi_i)_{i \in I}$ in $P^1(G)$ such that, for every compact subset K of G, $\lim_i \|\mu * \psi_i - \psi_i\|_1 = 0$ uniformly for every $\mu \in M_c^1(G)$ with $\operatorname{supp}\mu \subset K$.*

Proof. By Proposition 6.6, (D') characterizes amenability of G.

$$(D') \Rightarrow (i) \text{ Trivial}$$

$$(i) \Rightarrow (D')$$

Choose $\varphi \in P^1(G)$ and let $\varphi_i = \mu_i * \varphi \in P^1(G)$, $i \in I$. Then for every $\mu \in M^1(G)$ and every $i \in I$, $\|\mu * \varphi_i - \varphi_i\|_1 \leq \|\mu * \mu_i - \mu_i\|$ and $\lim_i \|\mu * \varphi_i - \varphi_i\|_1 = 0$.

$$(D') \Rightarrow (iii)$$

Let $(\varphi_i)_{i \in I}$ be a net in $P^1(G)$ such that $\lim_i \|\nu * \varphi_i - \varphi_i\|_1 = 0$ whenever $\nu \in M^1(G)$. Choose $\varphi \in P^1(G)$ and, for every $i \in I$, let $\psi_i = \varphi * \varphi_i \in P^1(G)$; so we have

$$\|\nu * \psi_i - \psi_i\|_1 \leq \|(\nu * \varphi) * \varphi_i - \varphi_i\|_1 + \|\varphi * \varphi_i - \varphi_i\|_1$$

and therefore by hypothesis

$$\lim_i \|\nu * \psi_i - \psi_i\|_1 = 0. \qquad (1)$$

Given $\varepsilon > 0$, there exists a neighborhood U of e in G such that, for every $z \in U$, $\|_z\varphi - \varphi\|_1 < \varepsilon/2$; hence for every $i \in I$, by (2.4)

$$\|_z\psi_i - \psi_i\|_1 = \|(_z\varphi - \varphi) * \varphi_i\|_1 \leq \|_z\varphi - \varphi\|_1 < \frac{\varepsilon}{2}.$$

Let $\mu \in M_c^1(G)$ such that $\operatorname{supp}\mu \subset K$. There exist $a_1, \ldots, a_n \in G$ such that $K^{-1} \subset \cup_{j=1}^n a_j U$. By (1) there exists $i_0 \in I$ such that, for every $i \in I$ with $i \succ i_0$ and every $j = 1, \ldots, n$,

$$\|_{a_j}\psi_i - \psi_i\|_1 = \|\delta_{a_j^{-1}} * \psi_i - \psi_i\|_1 < \frac{\varepsilon}{2}.$$

CONVERGENCE TO INVARIANCE

Hence for all $i \in I$ with $i \succ i_0$ and $j \in \{1, \ldots, n\}$, $z \in U$,

$$\|a_{jz}\psi_i - \psi_i\|_1 = \|_z(a_j\psi_i - \psi_i) + (_z\psi_i - \psi_i)\|_1$$

$$\leq \|a_j\psi_i - \psi_i\|_1 + \|_z\psi_i - \psi_i\|_1 < \frac{\varepsilon}{2} + \frac{\varepsilon}{2} = \varepsilon,$$

that is, $\|_{y^{-1}}\psi_i - \psi_i\|_1 < \varepsilon$ for every $y \in K$; then also

$$\|\mu * \psi_i - \psi_i\|_1 = \int_G \left| \int_G \psi_i(y^{-1}x)\, d\mu(y) - \psi_i(x) \right| dx$$

$$= \int_G \left| \int_G \psi_i(y^{-1}x)\, d\mu(y) - \psi_i(x) \int_G d\mu(y) \right| dx$$

$$\leq \int_G \int_G |\psi_i(y^{-1}x) - \psi_i(x)|\, dx\, d\mu(y)$$

$$= \int_K \left(\int_G |\psi_i(y^{-1}x) - \psi_i(x)|\, dx \right) d\mu(y)$$

$$= \int_K \|_{y^{-1}}\psi_i - \psi_i\|_1\, d\mu(y) < \varepsilon.$$

(iii) $\Rightarrow$ (ii) Trivial

(ii) $\Rightarrow$ (i)

This implication follows directly from the fact that $\overline{M_c^1(G)} = M^1(G)$. $\square$

B. Reiter's Conditions

Proposition 6.6 and Proposition 6.7 readily show the importance of the next definitions for the study of amenability, at least in case $p = 1$.

Definition 6.8. *Let G be a locally compact group. If $p \in [1, \infty[$, one considers the following property:*

(P_p) [resp. (P_p^*)] *For every compact subset K [resp. every finite subset F] of G and every $\varepsilon \in \mathbf{R}_+^*$, there exists $\varphi \in P^p(G)$ such that $\|_a\varphi - \varphi\|_p < \varepsilon$ whenever $a \in K$ [resp. $a \in F$].*

We introduce two constants related to these properties.

Definition 6.9. *If G is a locally compact group and $p \in [1, \infty[$, we define*

$$\gamma_p = \sup \left\{ \inf \left\{ \sup \{\|_a\varphi - \varphi\|_p : a \in K\} : \varphi \in P^p(G) \right\} : K \in \mathfrak{K}(G) \right\}$$

and

$$\gamma_p^* = \sup\left\{\inf\left\{\sup\{\|_a\varphi - \varphi\|_p : a \in F\} : \varphi \in P^p(G)\right\} : F \in \mathfrak{F}(G)\right\}.$$

For any $p \in [1, \infty[$, the property (P_p) [resp. (P_p^*)] signifies that $\gamma_p = 0$ [resp. $\gamma_p^* = 0$].

Before we go on, we state an elementary fact: If $0 \le a \le b$ and $1 \le s < \infty$, we have

$$(b - a)^s \le b^s - a^s. \tag{2}$$

It suffices to notice that the real function f defined on $\mathbf{R}_+$ by $f(x) = x^s - a^s - (x - a)^s$ is increasing and $f(a) = 0$.

If $p \in [1, \infty[$, $\varphi \in P^p(G)$, $a \in G$, by (2)

$$\|_a\varphi - \varphi\|_p^p = \int |\varphi(ax) - \varphi(x)|^p\, dx \le \int |\varphi(ax)^p - \varphi(x)^p|\, dx$$

$$\le \int \varphi(ax)^p\, dx + \int \varphi(x)^p\, dx = 2,$$

hence

$$0 \le \gamma_p^* \le \gamma_p \le 2^{1/p}. \tag{3}$$

Proposition 6.10. *In a locally compact group G the properties (P_1) and (P_1^*) are equivalent.*

Proof. Obviously (P_1) implies (P_1^*). Assume now that (P_1^*) holds.

Let K be a compact subset of G and $\varepsilon \in \mathbf{R}_+^*$. We consider the compact subset $K_1 = K^{-1} \cup \{e\}$ and $\delta = \varepsilon/10$. Let $\varphi \in P^1(G)$. There exists a neighborhood U of e such that $\|_{y^{-1}}\varphi - \varphi\|_1 < \delta$ whenever $y \in U$. Since K_1 is compact, there exists a compact neighborhood V of e such that $b^{-1}Vb \subset U$ for every $b \in K_1$ ([HR] 4.9). Hence for every $b \in K_1$ and every $z \in V$, $\|_{b^{-1}z^{-1}b}\varphi - \varphi\|_1 < \delta$. Let $\psi = \xi_V \in P^1(G)$. For every $b \in K_1$,

$$\|\psi * _{b^{-1}}\varphi - _{b^{-1}}\varphi\|_1 = \int_G \left|\int_G \psi(z)\varphi(b^{-1}z^{-1}x)\, dz - \int_G \psi(z)\varphi(b^{-1}x)\, dz\right| dx$$

$$\le \int_G \psi(z)\left(\int_G |\varphi(b^{-1}z^{-1}bx) - \varphi(x)|\, dx\right) dz$$

$$= \int_V \psi(z)\|_{b^{-1}z^{-1}b}\varphi - \varphi\|_1\, dz < \delta. \tag{4}$$

There exists a compact subset C in G such that $\int_{G\setminus C}\varphi(x)\, dx < \delta$. Consider the compact subset $C_1 = K_1^{-1}C$. For every $b \in K_1$, we have $bK_1^{-1}C \supset C$, and then

$$\int_{G \setminus C_1} \varphi(b^{-1}x)\, dx = \int_{G \setminus bK_1^{-1}C} \varphi(x)\, dx \leq \int_{G \setminus C} \varphi(x)\, dx < \delta. \tag{5}$$

There exists a compact neighborhood W of e in G such that $\|_x\psi^* - \psi^*\|_1 < \delta$ for every $x \in W$. Hence by (2.2) and (2.1) also $\|\psi * \delta_x - \psi\|_1 = \|R_x\psi - \psi\|_1 < \delta$ for every $x \in W$. Let W' be an open neighborhood of e such that $W'W'^{-1} \subset W$. As C_1 is compact, there exist $c_1, \ldots, c_m \in C_1$ such that $C_1 \subset \bigcup_{i=1}^m W'c_i$. For every $i = 1, \ldots, m$, let $A_i = W'c_i$, then $A_iA_i^{-1} \subset W$. Thus we may determine a finite partition $\{B_j : j = 1, \ldots, n\}$ formed by Borel subsets for C_1 such that $B_jB_j^{-1} \subset W$ whenever $j = 1, \ldots, n$. For every $j = 1, \ldots, n$, choose $b_j \in B_j$. So if $x \in B_j$,

$$\|\psi * \delta_x - \psi * \delta_{b_j}\|_1 = \|(\psi * \delta_{xb_j^{-1}} - \psi) * \delta_{b_j}\|_1 = \|\psi * \delta_{xb_j^{-1}} - \psi\|_1 < \delta. \tag{6}$$

By (P_1^*) there exists $\varphi' \in P^1(G)$ such that

$$\|\delta_{b_j} * \varphi' - \varphi'\|_1 = \|_{b_j^{-1}}\varphi' - \varphi'\|_1 < \delta \tag{7}$$

for every $j = 1, \ldots, n$.

For every $b \in K_1$, by (2.4) and (4) we have

$$\|_{b^{-1}}(\varphi * \varphi') - \psi *_{b^{-1}}(\varphi * \varphi')\|_1 \leq \|_{b^{-1}}\varphi - (\psi *_{b^{-1}}\varphi)\|_1 < \delta; \tag{8}$$

if $\varphi'' = {}_{b^{-1}}\varphi \in P^1(G)$, by (5), (6), (7) also

$$\|\psi * \varphi'' * \varphi' - \psi * \varphi'\|_1 = \int_G |\psi * \varphi'' * \varphi'(z) - \psi * \varphi'(z)|\, dz$$

$$= \int_G \left| \int_G \int_G \psi(yx^{-1}) \Delta(x^{-1}) \varphi''(x) \varphi'(y^{-1}z)\, dy\, dx - \psi * \varphi'(z) \right| dz$$

$$= \int_G \left| \int_G \varphi''(x) \psi * \delta_x * \varphi'(z)\, dx - \left(\int_G \varphi''(x)\, dx \right) \psi * \varphi'(z) \right| dz$$

$$\leq \int_{G \setminus C_1} \varphi''(x) \|\psi * \delta_x * \varphi' - \psi * \varphi'\|_1\, dx$$

$$+ \sum_{j=1}^n \int_{B_j} \varphi''(x) \|\psi * \delta_x * \varphi' - \psi * \delta_{b_j} * \varphi'\|_1\, dx$$

$$+ \sum_{j=1}^n \int_{B_j} \varphi''(x) \|\psi * \delta_{b_j} * \varphi' - \psi * \varphi'\|_1\, dx$$

$$< 2\delta + \delta + \delta = 4\delta. \tag{9}$$

Now $\varphi * \varphi' \in P^1(G)$. Let $a \in K$. Since $a^{-1} \in K_1$ and $e \in K_1$, (8) and (9) imply that

$$\|_a(\varphi * \varphi') - \varphi * \varphi'\|_1 \leq \|_a(\varphi * \varphi') - \psi * (_a(\varphi * \varphi'))\|_1$$

$$+ \|\psi * (_a\varphi) * \varphi' - \psi * \varphi'\|_1 + \|\psi * \varphi' - \psi * \varphi * \varphi'\|_1$$

$$+ \|\psi * \varphi * \varphi' - \varphi * \varphi'\|_1 < \delta + 4\delta + 4\delta + \delta = \varepsilon. \quad \square$$

Proposition 6.11. *If G is a locally compact group, there exists a mean M on $L^\infty(G)$ such that $\alpha(M, L^\infty(G)) \leq \gamma_1$.*

Proof. To each compact subset K in G and each $\varepsilon \in \mathbf{R}_+^*$ we associate the nonvoid subset

$$A_{K,\varepsilon} = \{\varphi \in P^1(G) : (\forall a \in K) \|_a\varphi - \varphi\|_1 < \gamma_1 + \varepsilon\}.$$

For each $\varphi \in A_{K,\varepsilon}$, we consider $F_\varphi \in L^\infty(G)^*$ defined by

$$\langle f, F_\varphi \rangle = \int\int f(xy)\varphi(x)\varphi(y)\,dx\,dy,$$

$f \in L^\infty(G)$. Let $B_{K,\varepsilon} = \{F_\varphi : \varphi \in A_{K,\varepsilon}\}$; the w*-closure $\overline{B_{K,\varepsilon}}$ of $B_{K,\varepsilon}$ is contained in $\mathfrak{M}(L^\infty(G))$. For all $f \in L^\infty(G)$, $\varphi \in A_{K,\varepsilon}$, $a \in K$, we have

$$|\langle _{a^{-1}}f - f, F_\varphi \rangle| = \left|\int\int (f(a^{-1}xy) - f(xy))\varphi(x)\varphi(y)\,dx\,dy\right|$$

$$= \left|\int\int f(xy)(\varphi(ax) - \varphi(x))\varphi(y)\,dx\,dy\right|$$

$$\leq \|f\|_\infty \|_a\varphi - \varphi\|_1 \leq \|f\|_\infty(\gamma_1 + \varepsilon).$$

If $K_1, \ldots, K_n \in \mathfrak{K}(G)$ and $\varepsilon_1, \ldots, \varepsilon_n \in \mathbf{R}_+^*$, let $K = K_1 \cup \cdots \cup K_n$ and $\varepsilon = \inf\{\varepsilon_1, \ldots, \varepsilon_n\}$; $A_{K,\varepsilon} \neq \emptyset$ and $B_{K,\varepsilon} \neq \emptyset$. Therefore $\cap_{i=1}^n \overline{B_{K_i,\varepsilon_i}} \neq \emptyset$. As $\mathfrak{M}(L^\infty(G))$ is w*-compact by Proposition 3.3, we must have

$$\bigcap_{\substack{K \in \mathfrak{K}(G) \\ \varepsilon \in \mathbf{R}_+^*}} \overline{B_{K,\varepsilon}} \neq \emptyset.$$

For any mean M in this latter intersection, $\alpha(M, L^\infty(G)) \leq \gamma_1$ holds. $\quad \square$

Proposition 6.12. *Amenability of a locally compact group G is characterized by (P_1) and (P_1^*).*

Proof. By Proposition 6.10, (P_1) and (P_1^*) are equivalent.
If (P_1) holds, then Proposition 6.11 implies that $\text{LIM}(L^\infty(G)) \neq \emptyset$.

If G is amenable, by Proposition 6.7, given a compact subset K of G and $\varepsilon > 0$, there exists $\varphi \in P^1(G)$ such that $\|\mu * \varphi - \varphi\|_1 < \varepsilon$ whenever $\mu \in M^1(G)$ with $\operatorname{supp} \mu \subset K^{-1}$. Hence, in particular by (2.2), for every $a \in K$, we have

$$\|{}_a\varphi - \varphi\|_1 = \|\delta_{a^{-1}} * \varphi - \varphi\|_1 < \varepsilon. \qquad \square$$

C. Consequences

Lemma 6.13. *Let G be a locally compact group. If $A \subset G$ and $p \in [1, \infty[$, let*

$$(R_{A,p}) \quad (\forall \varepsilon \in \mathbf{R}_+^*)(\exists \varphi \in P^p(G))(\forall a \in A)\|{}_a\varphi - \varphi\|_p < \varepsilon,$$

$$(R'_{A,p}) \quad (\forall \varepsilon \in \mathbf{R}_+^*)(\exists \varphi \in P^p(G) \cap \mathcal{K}(G))(\forall a \in A)\|{}_a\varphi - \varphi\|_p < \varepsilon.$$

We have the following properties:
1. $(\forall p \in [1, \infty[) (\forall A \in \mathfrak{P}(G)) (R_{A,p}) \Leftrightarrow (R'_{A,p})$.
2. Given $A \subset G$, if $(R_{A,p})$ [resp. $(R'_{A,p})$] holds for one $p \in [1, \infty[$, then it holds for every $p \in [1, \infty[$.

Proof. 1. We show that $(R_{A,p})$ implies $(R'_{A,p})$.
If $\varepsilon \in \mathbf{R}_+^*$, let $\delta = \varepsilon(3 + \varepsilon)^{-1}$. By assumption there exists $\varphi \in P^p(G)$ such that $\|{}_a\varphi - \varphi\|_p < \delta$ for every $a \in A$. Since $\mathcal{K}(G)$ is dense in $L^p(G)$, there exists $\varphi' \in \mathcal{K}_+(G)$ such that $\|\varphi' - \varphi\|_p < \delta$. Hence $\|\varphi'\|_p > \|\varphi\|_p - \delta = 1 - \delta > 0$ and, for every $a \in A$,

$$\|{}_a\varphi' - \varphi'\|_p \leq \|{}_a(\varphi' - \varphi)\|_p + \|{}_a\varphi - \varphi\|_p + \|\varphi - \varphi'\|_p < 3\delta.$$

Now let $\psi = \varphi'/\|\varphi'\|_p \in P^p(G) \cap \mathcal{K}(G)$. For every $a \in A$,

$$\|{}_a\psi - \psi\|_p = \frac{\|{}_a\varphi' - \varphi'\|_p}{\|\varphi'\|_p} < \frac{3\delta}{1 - \delta} = \varepsilon.$$

2. The statement is now a consequence of the following properties:

(a) If $q, r \in [1, \infty[$ and $r \leq q$, $(R_{A,r})$ implies $(R_{A,q})$.
(b) If $p \in [1, \infty[$, $(R_{A,2p})$ implies $(R_{A,p})$.

We proceed to give the proofs of these implications.
(a) Let $\varepsilon > 0$. By hypothesis there exists $\varphi \in P^r(G)$ such that, for every $a \in A$, $\|{}_a\varphi - \varphi\|_r < \varepsilon^{q/r}$. Let $\psi = \varphi^{r/q} \in P^q(G)$. For any $a \in A$, by (2),

have

$$\|_a\psi - \psi\|_q = \left(\int \left(|_a\psi(x) - \psi(x)|^{q/r}\right)^r dx\right)^{1/q}$$

$$\leq \left(\int |_a\varphi(x) - \varphi(x)|^r dx\right)^{1/q} = \|_a\varphi - \varphi\|_r^{r/q} < \varepsilon.$$

(b) Let $\varepsilon > 0$. By hypothesis there exists $\varphi \in P^{2p}(G)$ such that, for every $a \in A$, $\|_a\varphi - \varphi\|_{2p} < \varepsilon/2$. Let $\psi = \varphi^2 \in P^p(G)$; for any $a \in A$, by the Cauchy–Schwarz inequality we obtain

$$\|_a\psi - \psi\|_p = \left(\int |_a\psi(x) - \psi(x)|^p dx\right)^{1/p}$$

$$= \left(\int |_a\varphi(x) + \varphi(x)|^p |_a\varphi(x) - \varphi(x)|^p dx\right)^{1/p}$$

$$\leq \left(\int |_a\varphi(x) + \varphi(x)|^{2p} dx\right)^{1/2p} \left(\int |_a\varphi(x) - \varphi(x)|^{2p} dx\right)^{1/2p}$$

$$\leq (\|_a\varphi\|_{2p} + \|\varphi\|_{2p})\frac{\varepsilon}{2} = \varepsilon. \qquad \square$$

Theorem 6.14. *A locally compact group is amenable if and only if, for at least one $p \in [1, \infty[$ (for every $p \in [1, \infty[$) any one of the following properties holds: $(D_p), (P_p), (P_p^*), [\gamma_p = 0], [\gamma_p^* = 0]$.*

Proof. Apply the second part of Lemma 6.13 to Proposition 6.6 and Proposition 6.12. $\qquad \square$

Corollary 6.15. *A locally compact group G is amenable if and only if, for at least one $p \in [1, \infty[$ (for every $p \in [1, \infty[$), any one of the following properties holds:*

(a) *There exist a net (φ_i) in $P^p(G) \cap \mathcal{K}(G)$ such that $\lim_i \|_a\varphi_i - \varphi_i\|_p = 0$ whenever $a \in G$.*

(b) $\sup \{\inf \{\sup \{\|_a\varphi - \varphi\|_p : a \in K\} : \varphi \in P^p(G) \cap \mathcal{K}(G)\} : K \in \mathfrak{K}(G)\} = 0$.

(c) $\sup \{\inf \{\sup \{\|_a\varphi - \varphi\|_p : a \in F\} : \varphi \in P^p(G) \cap \mathcal{K}(G)\} : F \in \mathfrak{F}(G)\} = 0$.

Proof. The statement is a consequence of Theorem 6.14 and the first part of Lemma 6.13. $\qquad \square$

Lemma 6.16. *If G is a locally compact group, let $X_0(G)$ be the set of all $f \in L_\mathbb{R}^\infty(G)$ such that $\inf \{\|\varphi * f\|_\infty : \varphi \in P^1(G)\} = 0$. The following properties*

hold:
(a) For any $f \in L_\mathbf{R}^\infty(G)$ and any $a \in G$, $f - {}_af \in X_0(G)$.
(b) For any $g \in \mathscr{S}_\mathbf{R}(G) \cap X_0(G)$, ess sup $g \geq 0$.

Proof. (a) Let A be a measurable subset of G such that $0 < |A| < \infty$. If $n \in \mathbf{N}^*$, we consider $\varphi_n = (1/n|A|)\sum_{k=1}^n 1_{a^k A} \in P^1(G)$. For $n \in \mathbf{N}^*$ and locally almost every $x \in G$,

$$\varphi_n^* * (f - {}_af)(x) = \int \varphi_n(y^{-1})\Delta(y^{-1})(f(y^{-1}x) - f(ay^{-1}x))\,dy$$

$$= \frac{1}{n|A|}\sum_{k=1}^n \int_{a^k A} (f(yx) - f(ayx))\,dy$$

$$= \frac{1}{n|A|}\left(\int_{aA} f(yx)\,dy - \int_{a^{n+1}A} f(yx)\,dy\right);$$

hence

$$\|\varphi_n^* * (f - {}_af)\|_\infty \leq \frac{2}{n}\|f\|_\infty.$$

Therefore $f - {}_af \in X_0(G)$.

(b) Let $g \in \mathscr{S}_\mathbf{R}(G)$ such that ess sup $g \geq 0$ does not hold. We must have $\alpha = $ ess sup $g < 0$. If $\varphi \in P^1(G)$, for locally almost every $x \in G$,

$$\varphi^* * g(x) = \int \varphi(y)g(yx)\,dy \leq \alpha \int \varphi(y)\,dy = \alpha < 0;$$

then $g \notin X_0(G)$. $\square$

Proposition 6.17. *If G is a locally compact group, the following properties are equivalent*:
(i) G is amenable.
(ii) The set $Y_0(G)$ of all $f \in \mathscr{S}_\mathbf{R}(G)$ for which $\inf\{\|\varphi * f\|_\infty : \varphi \in P^1(G)\} = 0$ constitutes a real vector space.
(iii) For all $\varphi_1, \varphi_2 \in P^1(G)$, $\inf\{\|\varphi_1 * \psi_1 - \varphi_2 * \psi_2\|_1 : \psi_1, \psi_2 \in P^1(G)\} = 0$.
(iv) For all $\varphi_1, \varphi_2 \in P^1(G)$, $\inf\{\|(\varphi_1 - \varphi_2) * \psi\|_1 : \psi \in P^1(G)\} = 0$.

Proof. (i) $\Rightarrow$ (iv)
The amenable group G satisfies (Δ) by Proposition 6.6. There exists a net (ψ_i) in $P^1(G)$ such that

$$\lim_i \|\varphi_1 * \psi_i - \psi_i\|_1 = \lim_i \|\varphi_2 * \psi_i - \psi_i\|_1 = 0$$

whenever $\varphi_1, \varphi_2 \in P^1(G)$. Then also $\lim_i \|(\varphi_1 - \varphi_2) * \psi_i\|_1 = 0$.

(iv) $\Rightarrow$ (iii) Trivial

(iii) $\Rightarrow$ (ii)

It suffices to prove that $Y_0(G)$ is closed for addition. Let $f_1, f_2 \in Y_0(G)$ and $\varepsilon \in \mathbf{R}_+^*$. We put $\delta = \varepsilon(2 + \|f_2\|_\infty)^{-1}$. There exist $\varphi_1, \varphi_2 \in P^1(G)$ such that $\|\varphi_1^* * f_1\|_\infty < \delta$ and $\|\varphi_2^* * f_2\|_\infty < \delta$. By hypothesis there exist $\psi_1, \psi_2 \in P^1(G)$ such that $\|\psi_1^* * \varphi_1^* - \psi_2^* * \varphi_2^*\|_1 = \|\varphi_1 * \psi_1 - \varphi_2 * \psi_2\|_1 < \delta$. Let $\varphi = \psi_1^* * \varphi_1^* \in P^1(G)$. We have

$$\|\varphi * (f_1 + f_2)\|_\infty \leq \|\psi_1^* * \varphi_1^* * f_1\|_\infty$$

$$+ \|(\psi_1^* * \varphi_1^* - \psi_2^* * \varphi_2^*) * f_2\|_\infty + \|\psi_2^* * \varphi_2^* * f_2\|_\infty$$

$$\leq \|\varphi_1^* * f_1\|_\infty + \|\psi_1^* * \varphi_1^* - \psi_2^* * \varphi_2^*\|_1 \|f_2\|_\infty + \|\varphi_2^* * f_2\|_\infty$$

$$\leq (2 + \|f_2\|_\infty)\delta = \varepsilon.$$

(ii) $\Rightarrow$ (i)

As $Y_0(G)$ is a real vector space, Lemma 6.16 (a) insures that the real vector space $\mathcal{N}(\mathcal{S}_\mathbf{R}(G))$ belongs to $Y_0(G)$. Then Lemma 6.16 (b) and Proposition 4.29 imply amenability of G. $\square$

D. The Glicksberg–Reiter Property

We establish a characterization of amenability in terms of limits of averaging operators.

Let G be a locally compact group. If $f \in L^1(G)$, we consider the closed convex subset $L_G(f)$ of $L^1(G)$ defined in Section 3B. Let also R_G be the convex hull of the set of isometric operators $R_x (x \in G)$ on $L^1(G)$ and

$$R_G(f) = \overline{\operatorname{co}}\{R_x f : x \in G\},$$

$f \in L^1(G)$. For every $f \in L^1(G)$, by (2.1) involution on $L^1(G)$ induces an isometry of $L_G(f)$ onto $R_G(f^*)$. We next introduce a notation for the distances of the null function to $L_G(f)$ and $R_G(f)$ ($f \in L^1(G)$) in $L^1(G)$.

Definition 6.18. *Let G be a locally compact group. If $f \in L^1(G)$, we define $d_l(f) = \inf\{\|g\|_1 : g \in L_G(f)\}$ and $d_r(f) = \inf\{\|g\|_1 : g \in R_G(f)\}$.*

For any $f \in L^1(G)$, by (2.1) we have

$$d_l(f) = d_r(f^*). \tag{10}$$

Moreover $|\int f| = |\int \varphi * f| \leq \int |\varphi * f|$ whenever $\varphi \in P^1(G)$; therefore we deduce from Proposition 3.6 that $|\int f| \leq d_l(f)$.

Definition 6.19. *Let G be a locally compact group. We consider the following Glicksberg–Reiter properties:*

$$(GR_l) \quad (\forall f \in L^1(G)) \, d_l(f) = \left|\int f\right|;$$

$$(GR_r) \quad (\forall f \in L^1(G)) \, d_r(f) = \left|\int f\right|.$$

Lemma 6.20. *Let G be a locally compact group satisfying (GR_l) [resp. (GR_r)] and let $f_1, \ldots, f_n \in L^0(G)$. Then for every $\varepsilon > 0$, there exists $T \in L_G$ [resp. $T \in R_G$] such that $\|T(f_i)\|_1 < \varepsilon$ whenever $i = 1, \ldots, n$.*

Proof. By (2.1) it suffices to consider, for instance, the left-hand version.

If $n = 1$, the property is trivially verified by (GR_l). Assume that it holds for $f_1, \ldots, f_{n-1}$, that is, there exists $T' \in L_G$ such that $\|T'(f_i)\|_1 < \varepsilon$ whenever $i = 1, \ldots, n - 1$. Since $T'(f_n) \in L^0(G)$, by (GR_l) again there exists $T'' \in L_G$ such that $\|T''(T'(f_n))\|_1 < \varepsilon$. Then $T = T'' \circ T' \in L_G$ and, for every $i = 1, \ldots, n - 1$, also $\|T(f_i)\|_1 = \|T'' \circ T'(f_i)\|_1 \leq \|T'(f_i)\|_1 < \varepsilon$. □

Theorem 6.21. *Amenability of a locally compact group G is characterized by (GR_l) and by (GR_r).*

Proof. For every $f \in L^1(G)$, $\int f = \int f^*$ and involution maps $L_G(f)$ isometrically onto $R_G(f^*)$. Thus we deduce from (10) that (GR_l) and (GR_r) are equivalent. Recall that (P_1) characterizes amenability by Proposition 6.12.

$$(P_1) \Rightarrow (GR_l)$$

Let $f \in L^1(G)$. If $\|f\|_1 = 0$, we have trivially $d_l(f) = |\int f|$. We now consider the case $\|f\|_1 > 0$. If $\varepsilon \in \mathbf{R}_+^*$, there exists a compact subset K in G such that $\int_{G \setminus K} |f^*| < \varepsilon/3$; (P_1) implies the existence of $\varphi \in P^1(G)$ such that $\|_{y^{-1}}\varphi - \varphi\|_1 < \varepsilon/3\|f\|_1$ whenever $y \in K$. We obtain via involution

$$\left\|\varphi * * f - \left(\int f\right)\varphi^*\right\|_1 = \left\|f^* * \varphi - \left(\int f^*\right)\varphi\right\|_1$$

$$= \int_G \left|\int_G f^*(y)(\varphi(y^{-1}x) - \varphi(x)) \, dy\right| dx$$

$$\leq \int_{G \setminus K} + \int_K \|_{y^{-1}}\varphi - \varphi\|_1 |f^*(y)| \, dy$$

$$< 2\frac{\varepsilon}{3} + \frac{\varepsilon}{3\|f\|_1} \int_K |f^*(y)| \, dy \leq \varepsilon$$

hence

$$\|\varphi^* * f\|_1 \leq \left|\int f\right| \|\varphi^*\|_1 + \varepsilon = \left|\int f\right| + \varepsilon.$$

With respect to Proposition 3.6 we have $d_l(f) \leq |\int f| + \varepsilon$. Since $\varepsilon \in \mathbf{R}^*_+$ is chosen arbitrarily, $d_l(f) \leq |\int f|$.

$$(GR_r) \Rightarrow (P_1)$$

Let K be a compact subset of G and let $\varepsilon \in \mathbf{R}^*_+$. Consider a fixed element φ in $P^1(G)$. There exists a neighborhood U of e in G such that $\|_y\varphi - \varphi\|_1 < \varepsilon/2$ for every $y \in U$. Consider $a_1, \ldots, a_n \in K$ such that $K \subset \cup_{i=1}^n Ua_i$. By Lemma 6.20 there exists $T \in R_G$ such that, for any $i = 1, \ldots, n$, $\|T(_{a_i}\varphi - \varphi)\|_1 < \varepsilon/2$. Let now $\psi = T(\varphi) \in P^1(G)$. For any $a \in K$, there exist $i \in \{1, \ldots, n\}$ and $y \in U$ such that $a = ya_i$. Then we have

$$\|_a\psi - \psi\|_1 \leq \|_a\psi - _{a_i}\psi\|_1 + \|_{a_i}\psi - \psi\|_1$$

$$= \|_{a_i}(_y\psi - \psi)\|_1 + \|_{a_i}\psi - \psi\|_1$$

$$= \|_y\psi - \psi\|_1 + \|_{a_i}\psi - \psi\|_1 = \|T(_y\varphi - \varphi)\|_1 + \|T(_{a_i}\varphi - \varphi)\|_1$$

$$\leq \|_y\varphi - \varphi\|_1 + \|T(_{a_i}\varphi - \varphi)\|_1 < \frac{\varepsilon}{2} + \frac{\varepsilon}{2} = \varepsilon. \qquad \square$$

As for most characteristic properties of amenability we consider merely one-sided versions.

Corollary 6.22. *The locally compact group G is amenable if and only if, for every $f \in \mathcal{K}(G)$, $d_l(f) = |\int f|$.*

Proof. We show that the condition implies (GR_l).

For any $f \in L^1(G)$ and any $\varepsilon \in \mathbf{R}^*_+$, there exists $g \in \mathcal{K}(G)$ such that $\|g - f\|_1 < \varepsilon/3$. By hypothesis there exists $T \in L_G$ such that

$$\|T(g)\|_1 < \left|\int g\right| + \frac{\varepsilon}{3} \leq \left|\int f\right| + \|g - f\|_1 + \frac{\varepsilon}{3} < \left|\int f\right| + 2\frac{\varepsilon}{3}.$$

Then we have

$$\|T(f)\|_1 \leq \|T(g)\|_1 + \|T(g - f)\|_1 < \left|\int f\right| + 2\frac{\varepsilon}{3} + \|g - f\|_1 < \left|\int f\right| + \varepsilon.$$

As $\varepsilon \in \mathbf{R}^*_+$ may be chosen arbitrarily, $d_l(f) \leq |\int f|$.

Corollary 6.23. *The locally compact group G is amenable if and only if, for every $f \in L^0(G) \cap \mathcal{K}(G)$, $d_l(f) = 0$.*

Proof. The property is necessary by Proposition 6.21. We show that it implies the condition of Corollary 6.22.

Let $g \in \mathcal{K}(G)$ and $\varepsilon \in \mathbf{R}_+^*$; we put $\alpha = \int g$ and choose $\varphi \in P^1(G) \cap \mathcal{K}(G)$. Let $k = g - \alpha\varphi \in L^0(G) \cap \mathcal{K}(G)$. By hypothesis there exists $T \in L_G$ such that $\|T(k)\|_1 < \varepsilon$. We have

$$\|T(g)\|_1 \leq \|T(k)\|_1 + \|T(\alpha\varphi)\|_1 < \varepsilon + |\alpha|\|\varphi\|_1 = \varepsilon + |\alpha|.$$

Since $\varepsilon \in \mathbf{R}_+^*$ may be chosen arbitrarily, $d_l(g) \leq |\alpha| = |\int g|$. □

NOTES

Asymptotical invariance properties were introduced by Day [112]. He proved amenability of a discrete group to be equivalent to the existence of a net (φ_i) of functions (with finite support) satisfying (D_1) [114, 115]. Namioka [411] gave a generalization for arbitrary topological groups. Hulanicki [268] showed that (D), (Δ), (D'), and (Δ') characterize amenability of a locally compact group; also see Day [119]. For 6.7 we make reference to Wong [563]; these properties were also investigated by Day [122].

Property (P_1) was introduced and studied extensively by Reiter. Initially he had focused his interest on locally compact abelian groups [450]. He was the first to observe that, for a general locally compact group G, (P_1) is equivalent to the existence of a left invariant mean on $\mathscr{C}(G)$ and $L^\infty(G)$ [452]. The proof of 6.11 is part of a generalization due to Derighetti [126] of Reiter's [453] demonstration. Hulanicki [268] showed the equivalence $(D_1) \Leftrightarrow (P_1)$. Anker [10, 11] established the elegant proof of the implication $(P_1^*) \Rightarrow (P_1)$ bypassing the considerations on invariant means made in Reiter's original demonstration [454].

Dieudonné [136] was the first to consider the general property (P_p); he noticed the implication $(P_1) \Rightarrow (P_p)$. Reiter [451] showed that $(P_2) \Rightarrow (P_1)$. The idea of 6.13.2 is due to Stegeman [517]. Emerson [167] established 6.16–17.

The proof of 6.21 was given by Reiter [452, 453]. For the implication $[\text{RIM}(\mathscr{C}(G)) \neq \varnothing] \Rightarrow (GR_r)$ he adapted a general abstract statement, obtained by Glicksberg [205], to a weakly continuous representation of the group on a Banach space via contracting operators.

7. STRUCTURAL PROPERTIES

In this section we characterize amenability by properties of the locally compact group itself that reflect in a certain sense the structure of the group. A locally compact group is amenable if and only if it admits measurable subsets having finite nonzero measures that are not substantially modified by translations.

A. Følner's Conditions and Leptin's Conditions

We establish the equivalence of amenability for a locally compact group with each of the properties defined in 7.1 concerning (a) Følner conditions, (b) weak Følner conditions, (c) strong Følner conditions, and (d) Leptin conditions.

Definition 7.1. *Let G be a locally compact group. We consider the following properties*:
   (a) (F) [*resp.* (F^*)] *If K is a compact subset* [*resp. F is a finite subset*] *of G and $\varepsilon > 0$, there exists a measurable subset U in G such that $0 < |U| < \infty$ and $|aU \triangle U|/|U| < \varepsilon$ for every $a \in K$* [*resp. every $a \in F$*].
   (b) (WF) *If K is a compact subset of G and $\delta, \varepsilon > 0$, there exist measurable subsets U and N in G such that $0 < |U| < \infty$, $|N| < \delta$, and $|aU \triangle U|/|U| < \varepsilon$ for every $a \in K \setminus N$.*
   (c) (SF) [*resp.* (SF^*)] *If K is a compact subset* [*resp. F is a finite subset*] *of G and $\varepsilon > 0$, there exists a measurable subset U in G such that $0 < |U| < \infty$ and $|KU \triangle U|/|U| < \varepsilon$* [*resp. $|FU \triangle U|/|U| < \varepsilon$*].
   (d) (L) [*resp.* (L^*)] *If K is a compact subset* [*resp. F is a finite subset*] *of G and $\varepsilon > 0$, there exists a measurable subset U in G such that $0 < |U| < \infty$ and $|KU|/|U| < 1 + \varepsilon$* [*resp. $|FU|/|U| < 1 + \varepsilon$*].

We add the subscript e in the conditions (SF), (L) [*resp.* (SF^*), (L^*)] if we suppose that e is an element of K [*resp.* F]. We add the subscript $_0$ in the conditions (SF), (L) if we suppose that $\overset{\circ}{K} \neq \varnothing$; both subscripts signify that K is a neighborhood of e.

We state some simple general formulas. Let G be a locally compact group. If A is a measurable subset of G and $a \in G$,

$$|aA \triangle A| = |aA| + |A| - 2|aA \cap A|$$

hence

$$|aA \triangle A| = 2(|A| - |aA \cap A|). \qquad (1)$$

If A and B are nonvoid measurable subsets of G such that BA is measurable, then

$$|A \cap BA| + |A \setminus BA| = |A| \leq |BA| = |BA \setminus A| + |BA \cap A|$$

hence

$$|A \setminus BA| \leq |BA \setminus A|$$

and

$$|BA \triangle A| \leq 2|BA \setminus A|. \qquad (2)$$

B. The Pointwise Characterizations

Lemma 7.2. *Let G be a locally compact group and let $\psi \in P^1(G) \cap \mathscr{S}(G)$. There exist measurable subsets $A_1, \ldots, A_n$ in G with $A_n \subset \cdots \subset A_2 \subset A_1$, $0 < |A_n|, |A_1| < \infty$ and $\alpha_1, \ldots, \alpha_n \in \mathbf{R}_+^*$ with $\sum_{i=1}^{n} \alpha_i = 1$ such that $\psi = \sum_{i=1}^{n} \alpha_i \xi_{A_i}$ and*

$$\|_{a^{-1}}\psi - \psi\|_1 = \sum_{i=1}^{n} \alpha_i \frac{|aA_i \triangle A_i|}{|A_i|}$$

whenever $a \in G$. If $\psi = \check{\psi}$, all the sets A_i ($i = 1, \ldots, n$) may be assumed to be symmetric.

Proof. There exist $\beta_1, \ldots, \beta_n \in \mathbf{R}_+^*$ such that $0 < \beta_1 < \cdots < \beta_n$ and pairwise disjoint measurable subsets $B_1, \ldots, B_n$ in G such that $0 < |B_j| < \infty$ whenever $j = 1, \ldots, n$ and $\psi = \sum_{j=1}^{n} \beta_j 1_{B_j} \in L^\infty(G)$. For $i = 1, \ldots, n$, let $A_i = \cup_{j=i}^{n} B_j$. Then $A_n \subset \cdots \subset A_2 \subset A_1$, $0 < |A_n|, |A_1| < \infty$ and

$$\psi = \beta_1 1_{A_1} + (\beta_2 - \beta_1) 1_{A_2} + \cdots + (\beta_n - \beta_{n-1}) 1_{A_n}.$$

Let $\alpha_1 = \beta_1 |A_1|$ and $\alpha_i = (\beta_i - \beta_{i-1})|A_i|$ for $i = 2, \ldots, n$. We then have $\psi = \sum_{i=1}^{n} \alpha_i \xi_{A_i}$ and $1 = \|\psi\|_1 = \sum_{i=1}^{n} \alpha_i$.

If $i, j = 1, \ldots, n$, either $A_i \subset A_j$ or $A_j \subset A_i$ so, for $a \in G$, either $aA_i \setminus A_i \subset aA_j$ or $A_j \setminus aA_j \subset A_i$. In both cases $(aA_i \setminus A_i) \cap (A_j \setminus aA_j) = \varnothing$. Let $A = \cup_{i=1}^{n} (A_i \setminus aA_i)$. For every $i = 1, \ldots, n$, $aA_i \setminus A_i \subset G \setminus A$. Now

$$\|_{a^{-1}}\psi - \psi\|_1 = \int_G \left| \sum_{i=1}^{n} \alpha_i (\xi_{aA_i}(x) - \xi_{A_i}(x)) \right| dx.$$

If $x \in A$, there exists $i = 1, \ldots, n$ such that $x \in A_i \setminus aA_i$ and $x \notin aA_j \setminus A_j$ whenever $j = 1, \ldots, n$ with $j \neq i$; hence $1_{A_i}(x) - 1_{aA_i}(x) = 1$ and $1_{A_j}(x) - 1_{aA_j}(x) \geq 0$. If $x \in G \setminus A$, then $x \notin A_i \setminus aA_i$ whenever $i = 1, \ldots, n$, so $1_{aA_i}(x) - 1_{A_i}(x) \geq 0$. Therefore we have

$$\|_{a^{-1}}\psi - \psi\|_1 = \int_A \sum_{i=1}^{n} \frac{\alpha_i}{|A_i|} (1_{A_i}(x) - 1_{aA_i}(x)) \, dx$$

$$+ \int_{G \setminus A} \sum_{i=1}^{n} \frac{\alpha_i}{|A_i|} (1_{aA_i}(x) - 1_{A_i}(x)) \, dx$$

$$= \sum_{i=1}^{n} \frac{\alpha_i}{|A_i|} \|1_{aA_i} - 1_{A_i}\|_1 = \sum_{i=1}^{n} \alpha_i \frac{|aA_i \triangle A_i|}{|A_i|}. \quad \square$$

Theorem 7.3. *Amenability of a locally compact group G is characterized by each of the properties* $(F), (F^*), (WF)$.

Proof. Recall that by Proposition 6.12 the properties (P_1) and (P_1^*) characterize amenability.

$$(P_1) \Rightarrow (WF)$$

Let a compact subset K of G and $\delta, \varepsilon \in \mathbf{R}_+^*$ be given. In case $|K| = 0$, one chooses $N = K$ and the verification is trivial. Assume now that $|K| > 0$.

Let $\alpha = \delta\varepsilon(\delta\varepsilon + 3|K|)^{-1} \in]0, 1[$. By hypothesis there exists $\varphi \in P^1(G)$ such that, for any $x \in K$, $\|_{x^{-1}}\varphi - \varphi\|_1 < \alpha$. By density there exists $\theta \in \mathscr{S}_+(G)$ such that $\|\varphi - \theta\|_1 < \alpha$, hence $\|\theta\|_1 > 1 - \alpha > 0$. Let $\psi = \theta/\|\theta\|_1 \in P^1(G) \cap \mathscr{S}(G)$. For any $x \in K$,

$$\|_{x^{-1}}\psi - \psi\|_1 \leq \frac{\|_{x^{-1}}\theta - \theta\|_1}{\|\theta\|_1} \leq \frac{\|_{x^{-1}}(\theta - \varphi)\|_1 + \|_{x^{-1}}\varphi - \varphi\|_1 + \|\varphi - \theta\|_1}{\|\theta\|_1}$$

$$< \frac{3\alpha}{1 - \alpha} = \frac{\delta\varepsilon}{|K|}.$$

We now choose the measurable subsets $A_1, \ldots, A_n$ in G and $\alpha_1, \ldots, \alpha_n \in \mathbf{R}_+^*$ associated to ψ by Lemma 7.2; $\sum_{i=1}^n \alpha_i = 1$ and, for every $x \in K$,

$$\sum_{i=1}^n \alpha_i \frac{|xA_i \triangle A_i|}{|A_i|} < \frac{\delta\varepsilon}{|K|}.$$

By integration over K, we obtain

$$\sum_{i=1}^n \alpha_i \int_K \frac{|xA_i \triangle A_i|}{|A_i|} \, dx < \delta\varepsilon.$$

As the left-hand side is a convex linear combination, there must exist $i_0 \in \{1, \ldots, n\}$ such that $\int_K |xA_{i_0} \triangle A_{i_0}|/|A_{i_0}| \, dx < \delta\varepsilon$. It suffices now to put

$$N = \left\{ x \in K : \frac{|xA_{i_0} \triangle A_{i_0}|}{|A_{i_0}|} \geq \varepsilon \right\}$$

and $U = A_{i_0}$.

$$(WF) \Rightarrow (F)$$

Let K be a compact subset of G and let $\varepsilon \in \mathbf{R}_+^*$. Consider a compact neighborhood V of e containing K and let $W = V^2$. For every $a \in K$ and every $x \in V$, $ax \in W \cap aW$; hence

$$|W \cap aW| \geq |aV| = |V|. \tag{3}$$

STRUCTURAL PROPERTIES

We apply (WF) to $\delta = |V|/2$, $\varepsilon/2$ and W; there exist measurable subsets U and N in G such that $0 < |U| < \infty$, $N \subset W$, $|N| < \delta$ and $|yU \triangle U| < (\varepsilon/2)|U|$ whenever $y \in W \setminus N$.

For every $a \in K$, by (3)

$$2\delta = |V| \le |W \cap aW| \le |(W \setminus N) \cap a(W \setminus N)| + |N| + |aN|$$

$$< |(W \setminus N) \cap a(W \setminus N)| + 2\delta.$$

So we must have $(W \setminus N) \cap a(W \setminus N) \ne \emptyset$ and there exist $y, z \in W \setminus N$ such that $a = yz^{-1}$. Then

$$|aU \triangle U| = |yz^{-1}U \triangle U| = |z^{-1}U \triangle y^{-1}U|$$

$$\le |z^{-1}U \triangle U| + |y^{-1}U \triangle U| = |U \triangle zU| + |U \triangle yU| < \varepsilon|U|.$$

$(F) \Rightarrow (F^*)$ Trivial

$(F^*) \Rightarrow (P_1^*)$

Let F be a finite subset of G and let $\varepsilon \in \mathbf{R}_+^*$. By hypothesis there exists a measurable subset U of G such that $0 < |U| < \infty$ and $|aU \triangle U| < \varepsilon|U|$ whenever $a \in F$. Then also

$$\|{}_a\xi_U - \xi_U\|_1 = \frac{|a^{-1}U \triangle U|}{|U|} = \frac{|U \triangle aU|}{|U|} < \varepsilon$$

for every $a \in F$. □

Proposition 7.4. *Let G be an amenable group, K [resp. F] a compact [resp. finite] subset of G and $\varepsilon \in \mathbf{R}_+^*$. Then the measurable subset U given by (F) [resp. (F^*)] may be supposed to be compact.*

Proof. By (F) [resp. F^*] there exists a measurable subset U of G such that $0 < |U| < \infty$ and $|aU \triangle U|/|U| < \varepsilon/6$ whenever $a \in K$ [resp. $a \in F$]. Regularity of Haar measure implies the existence of a compact subset U_1 in U such that $|U \setminus U_1| < \inf\{|U|/2, (\varepsilon/6)|U|\}$. Then $|U_1| > 0$, $|U| = |U_1| + |U \setminus U_1| < |U_1| + |U|/2$, $|U|/|U_1| < 2$ and, for every $a \in K$ [resp. $a \in F$],

$$\frac{|aU_1 \triangle U_1|}{|U_1|} \le \frac{|U|}{|U_1|}\left(\frac{|aU \triangle U|}{|U|} + \frac{|a(U \setminus U_1)|}{|U|} + \frac{|U \setminus U_1|}{|U|}\right)$$

$$< 2 \cdot 3 \cdot \frac{\varepsilon}{6} = \varepsilon.$$ □

C. Global Characterizations

We begin by stating a general *covering lemma*.

Lemma 7.5. *Let V be a relatively compact neighborhood of a point in a locally compact group G. Then there exist a subset T of G and $k \in \mathbf{N}^*$ such that $G = TV$ and at most k of the sets tV ($t \in T$) have a nonvoid common intersection.*

Proof. (a) We show that if the lemma holds for some relatively compact neighborhood V of a point in G, it holds for every relatively compact neighborhood W of a point in G.

There exist $a_1, \ldots, a_m \in G$ such that $V = \cup_{i=1}^m a_i W = W'$. Let $S = \{ta_i : t \in T; \ i = 1, \ldots, m\}$; then $G = TV = SW$. As W' is relatively compact, there exist $b_1, \ldots, b_n \in G$ such that $W' \subset \cup_{j=1}^n Vb_j$. At most kmn the sets sW ($s \in S$) have a nonvoid common intersection. Otherwise there would exist $a \in G$ and $i_0 \in \{1, \ldots, m\}$ with a belonging to more than kn sets $ta_{i_0}W$ ($t \in T$), hence to more than kn sets tVb_j ($t \in T$; $j = 1, \ldots, n$); then there would exist $j_0 \in \{1, \ldots, n\}$ with $ab_{j_0}^{-1}$ belonging to more than k sets tV ($t \in T$) and we would come to a contradiction.

(b) By (a) we may now assume that V is an open symmetric relatively compact neighborhood of e. We consider the open subgroup $H = \cup_{n=1}^\infty V^n$ of G. We show that if the lemma holds for H, then it holds for G.

As H is an open and closed subgroup of G, V is an open relatively compact neighborhood of e in H. Let T' be a subset of H and $k' \in \mathbf{N}^*$ such that $H = T'V$ and at most k' of the sets $t'V$ ($t' \in T'$) have a nonvoid common intersection. For every $a \in G$, $aH = (aT')V$ and at most k' of the sets uV ($u \in aT'$) admit a nonvoid common intersection. If A is a system of representatives for the left cosets of H in G, then $G = AT'V$. As these cosets are pairwise disjoint, at most k' of the sets vV ($v \in AT'$) have a nonvoid common intersection.

(c) By (b) we may assume that $G = \cup_{n=1}^\infty V^n$, where V is an open symmetric relatively compact neighborhood of e in G. If G is compact, the lemma is trivially verified. Suppose now that G is noncompact.

Let $t_1 = e$. As G is noncompact, $\overline{V^2} \neq V$; choose $t_2 \in \overline{V^2} \setminus V$. If $\overline{V^2} \setminus \cup_{i=1}^2 t_i V \neq \emptyset$, let $t_3 \in \overline{V^2} \setminus \cup_{i=1}^2 t_i V$. Continuing this procedure one can find $t_1, t_2, \ldots, t_{n_2} \in G$ such that $\overline{V^2} \subset \cup_{i=1}^{n_2} t_i V$ and $t_j \notin \cup_{i=1}^{j-1} t_i V$ for $j = 2, \ldots, n_2$.

If $\overline{V^3} \setminus \cup_{i=1}^{n_2} t_i V \neq \emptyset$, let $t_{n_2+1} \in \overline{V^3} \setminus \cup_{i=1}^{n_2} t_i V$. Going on with this procedure, one determines $t_1, \ldots, t_{n_p} \in G$ such that $\overline{V^p} \subset \cup_{i=1}^{n_p} t_i V$ and $t_j \notin \cup_{i=1}^{j-1} t_i V$ for $j = 2, \ldots, n_p$. As G is noncompact, the construction gives rise to an infinite sequence $(t_i)_{i \in \mathbf{N}^*}$ in G such that $G = TV$ for $T = \{t_i : i \in \mathbf{N}^*\}$.

Consider an open symmetric neighborhood W of e such that $W^2 \subset V$. Choose any subset S of T with $x \in \cap_{s \in S} sV$. Then for every $s \in S$, one has $s \in xV$, $sW \subset xVW$ and
$$SW \subset xVW. \tag{4}$$

If, for $i, j \in \mathbf{N}^*$, there exists $y \in t_i W \cap t_j W$, then $t_j \in t_i W^2 \subset t_i V$, and similarly $t_i \in t_j V$. Owing to the definition of (t_i) we necessarily have $i = j$; hence the sets tW ($t \in T$) are pairwise disjoint. Then by (4), (card S) $|W| \leq |VW|$, card $S \leq |VW|/|W|$; also $k = |VW|/|W|$ is independent of the choice of S and at most k of the sets $t_i V$ ($i \in I$) have a nonvoid common intersection. □

Lemma 7.6. *Let G be an amenable group and let K be a compact neighborhood of e in G. By Proposition 7.4, for every $n \in \mathbf{N}^*$, there exists a compact subset U_n of G such that $0 < |U_n|$ and $|aU_n \triangle U_n|/|U_n| < 1/n$ whenever $a \in K' = KKK^{-1}$. Given $\varepsilon > 0$, let $\delta = \varepsilon/2(1 + \varepsilon)$. There exists a sequence (V_n) of compact subsets in G such that, for every $n \in \mathbf{N}^*$, $0 < |V_n|$, $V_n \subset U_n$, $|U_n \setminus V_n| \leq \delta |U_n|$, and $\limsup_{n \to \infty} |KV_n|/|V_n| \leq 1 + \varepsilon$.*

Proof. We start by fixing some notations. For all $n \in \mathbf{N}^*$, $x \in G$, $\alpha \in \mathbf{R}_+^*$, we consider the measurable subsets

$$A(x, n) = K' \setminus U_n x^{-1},$$

$$B(n, \alpha) = \{x \in U_n : |A(x, n)| > \alpha\}$$

and let

$$\alpha_n = \inf\{\alpha \in \mathbf{R}_+^* : |B(n, \alpha)| \leq \delta |U_n|\}.$$

(a) We show that $\lim_{n \to \infty} \alpha_n = 0$.

If the equality does not hold, there exist $\beta \in \mathbf{R}_+^*$ and a subsequence (α_{n_m}) of (α_n) such that $\beta < \alpha_{n_m}$ whenever $m \in \mathbf{N}^*$, hence

$$|B(n_m, \beta)| > \delta |U_{n_m}| \tag{5}$$

whenever $m \in \mathbf{N}^*$. If $y \in K'$ and $m \in \mathbf{N}^*$, we consider the measurable subset

$$C(y, m) = \{x \in B(n_m, \beta) : y \in A(x, n_m)\};$$

then by Fubini's theorem

$$\int_{B(n_m, \beta)} |A(x, n_m)| \, dx = \int_{B(n_m, \beta)} \int_{K'} 1_{A(x, n_m)}(y) \, dy \, dx$$

$$= \int_{K'} |C(y, m)| \, dy. \tag{6}$$

For every $x \in C(y, m)$, $y \notin U_{n_m} x^{-1}$, thus $yC(y, m) \cap U_{n_m} = \emptyset$ and

$$(yU_{n_m}) \cap U_{n_m} = y(U_{n_m} \setminus C(y, m)) \cap U_{n_m}.$$

But as $y \in K'$,

$$\lim_{m \to \infty} \frac{|yU_{n_m} \cap U_{n_m}|}{|U_{n_m}|} = 1;$$

so

$$1 = \lim_{m \to \infty} \frac{|y(U_{n_m} \setminus C(y,m)) \cap U_{n_m}|}{|U_{n_m}|} \leq \liminf_{m \to \infty} \frac{|y(U_{n_m} \setminus C(y,m))|}{|U_{n_m}|}$$

$$= \liminf_{m \to \infty} \frac{|U_{n_m}| - |C(y,m)|}{|U_{n_m}|} = 1 - \limsup_{m \to \infty} \frac{|C(y,m)|}{|U_{n_m}|},$$

that is, $\lim_{m \to \infty} |C(y,m)|/|U_{n_m}| = 0$. By Lebesgue's dominated convergence theorem we have then

$$\lim_{m \to \infty} \int_{K'} \frac{|C(y,m)|}{|U_{n_m}|} dy = 0.$$

Thus (6) and (5) imply that

$$0 = \lim_{m \to \infty} \frac{1}{|U_{n_m}|} \int_{B(n_m, \beta)} |A(x, n_m)| \, dx$$

$$> \liminf_{m \to \infty} \frac{1}{|U_{n_m}|} \beta \delta |U_{n_m}| = \beta \delta.$$

We come to a contradiction.

(b) Via the isomorphism $x \mapsto x^{-1}$ on G, the covering Lemma 7.5 insures the existence of a subset T in G and $k \in \mathbf{N}^*$ such that $G = KT$ and at most k of the sets Kt ($t \in T$) admit a nonvoid common intersection.

For every $n \in \mathbf{N}^*$, let $\beta_n = \alpha_n + 1/n$, $W_n = U_n \setminus B(n, \beta_n)$, and $T_n = \{t \in T : Kt \cap W_n \neq \emptyset\}$; since W_n is relatively compact, T_n must be finite. To every $t \in T_n$ we associate $z_t \in Kt \cap W_n$; then

$$Kt \subset K'z_t \tag{7}$$

and

$$KW_n \subset KKT_n \subset \bigcup_{t \in T_n} KKK^{-1} z_t = \bigcup_{t \in T_n} K'z_t. \tag{8}$$

For $t \in T_n$,

$$\Delta(z_t) \leq \gamma \Delta(t) \tag{9}$$

where $\gamma = \sup\{\Delta(x) : x \in K\}$. If $x \in W_n$, we have $|K' \setminus U_n x^{-1}| \leq \beta_n$ hence

$$|K'x \setminus U_n| \leq \beta_n \Delta(x). \tag{10}$$

We obtain

$$\sum_{t \in T_n} |Kt \cap U_n| = \sum_{t \in T_n} \int 1_{U_n}(y) 1_{Kt}(y) \, dy = \int 1_{U_n}(y) \left(\sum_{t \in T_n} 1_{Kt}(y) \right) dy$$

$$\leq k|U_n|. \tag{11}$$

If $n \in \mathbf{N}^*$ and $t \in T_n$, (7), (10), and (9) imply that

$$|Kt \setminus U_n| \leq |K'z_t \setminus U_n| \leq \beta_n \Delta(z_t) \leq \beta_n \gamma \Delta(t) \tag{12}$$

and then by (11)

$$k|U_n| \geq \sum_{t \in T_n} |Kt \cap U_n| \geq (|K| - \beta_n \gamma) \sum_{t \in T_n} \Delta(t).$$

As $\lim_{n \to \infty} \beta_n = 0$ and $|K| > 0$, there exists $n_0 \in \mathbf{N}^*$ such that, for every $n \geq n_0$, $|K| - \beta_n \gamma > 0$ and

$$\sum_{t \in T_n} \Delta(t) \leq \frac{k|U_n|}{|K| - \beta_n \gamma};$$

there also exists a compact subset V_n in G such that $V_n \subset W_n$ and $|W_n \setminus V_n| < \delta|U_n|$. By (8) and (12) we have

$$|KV_n| \leq \sum_{t \in T_n} |K'z_t \setminus U_n| + |U_n| \leq \beta_n \gamma \sum_{t \in T_n} \Delta(t) + |U_n|$$

$$\leq \left(\frac{k\beta_n \gamma}{|K| - \beta_n \gamma} + 1 \right) |U_n|.$$

But for every $n \in \mathbf{N}^*$, $\alpha_n < \beta_n$, hence $|B(n, \beta_n)| \leq \delta|U_n|$ and

$$|W_n| = |U_n| - |B(n, \beta_n)| \geq (1 - \delta)|U_n|,$$

$$|V_n| = |W_n| - |W_n \setminus V_n| > (1 - 2\delta)|U_n| > 0.$$

As $\lim_{n \to \infty} \beta_n = 0$, we finally obtain

$$\limsup_{n \to \infty} \frac{|KV_n|}{|V_n|} \leq \limsup_{n \to \infty} \left(\frac{k\beta_n \gamma}{|K| - \beta_n \gamma} + 1 \right) \frac{1}{1 - 2\delta} = \frac{1}{1 - 2\delta} = 1 + \varepsilon. \quad \square$$

Lemma 7.7. *Let G be a locally compact group and suppose that, for a compact subset K in G and $\varepsilon \in \,]0,1[$, there exists a measurable subset U in G such that $0 < |U| < \infty$ and $|KU \triangle U|/|U| < \varepsilon$. Then there exists a compact subset V in G such that $0 < |V|$, $V \subset U$, $|U \setminus V| < \varepsilon |U|$, and $|KV \triangle V|/|V| < 4\varepsilon/(1-\varepsilon)$.*

Proof. Note the measurability of the subsets $A = KU \cup U = (KU \triangle U) \cup U$, $B = KU \cap U = A \setminus (KU \triangle U)$, $C = KU \setminus U = A \setminus U$, $D = KU = B \cup C$. By regularity of Haar measure there exists a compact subset V of U such that $|V| > 0$ and $|U \setminus V| < \varepsilon |U|$. We have

$$|U| < |V| + \varepsilon |U|, \quad \frac{|U|}{|V|} < \frac{1}{1-\varepsilon};$$

$$|KV \triangle V| \le |KV \triangle KU| + |KU \triangle U| + |U \triangle V|$$
$$= |KU \setminus KV| + |KU \triangle U| + |U \setminus V|$$

whereas

$$|KU \setminus KV| = |KU| - |KV| \le |KU| - |V| \le |KU \triangle U| + |U| - |V|.$$

Therefore we obtain

$$\frac{|KV \triangle V|}{|V|} \le 2 \frac{|U|}{|V|} \left(\frac{|KU \triangle U|}{|U|} + \frac{|U \setminus V|}{|U|} \right) < 2 \frac{1}{1-\varepsilon}(\varepsilon + \varepsilon) = \frac{4\varepsilon}{1-\varepsilon}. \quad \square$$

Lemma 7.8. *Let G be a locally compact group and suppose that, for every compact subset K of G and every $\varepsilon > 0$, there exists a compact subset U of G such that $0 < |U|$ and $|KU \triangle U|/|U| < \varepsilon$. Then for every compact subset K of G and every $\varepsilon > 0$, there exists a compact subset U' of G such that $0 < |U'|$ and $|KU'|/|U'| < 1 + \varepsilon$.*

Proof. Let K be a compact subset of G and let $\varepsilon > 0$. Consider $K' = K \cup \{e\}$. By hypothesis there exists a compact subset U' of G such that $0 < |U'|$ and

$$\frac{|K'U'|}{|U'|} - 1 = \frac{|K'U' \setminus U'|}{|U'|} = \frac{|K'U' \triangle U'|}{|U'|} < \varepsilon.$$

Then also

$$\frac{|KU'|}{|U'|} \le \frac{|K'U'|}{|U'|} < 1 + \varepsilon. \quad \square$$

Theorem 7.9. *Amenability of a locally compact group G is characterized by each of the properties $(SF), (SF^*), (L), (L^*), (SF_e), (SF_e^*), (L_e), (L_e^*), (SF_0), (L_0), (SF_{e,0}), (L_{e,0})$.*

STRUCTURAL PROPERTIES

Proof. Diagram of the proof:

$$(F) \Rightarrow (L_{e,0}) \Rightarrow (L_e) \Rightarrow (SF_e) \Rightarrow (SF) \Rightarrow (SF^*) \Leftarrow (SF_e^*)$$

$$(F^*) \Uparrow$$

$$(SF_{e,0}) \Rightarrow (SF_0) \Rightarrow (L_0) \Rightarrow (L) \Rightarrow (L^*) \Rightarrow (L_e^*)$$

with $(F) \Rightarrow (L_{e,0})$ going down (⇓) and $(L_e^*) \Uparrow (SF_e^*)$ going up.

Recall that (F) and (F^*) characterize amenability by Theorem 7.3. The implications $(SF) \Rightarrow (SF^*)$ and $(L) \Rightarrow (L^*) \Rightarrow (L_e^*)$ are trivial. The implications $(L_e) \Rightarrow (SF_e)$, $(L_{e,0}) \Rightarrow (SF_{e,0})$, $(L_e^*) \Rightarrow (SF_e^*)$, $(SF_0) \Rightarrow (L_0)$ are immediate.

$$(F) \Rightarrow (L_{e,0})$$

This implication follows from Lemma 7.6.

$$(L_0) \Rightarrow (L) \quad [\text{resp. } (L_{e,0}) \Rightarrow (L_e)]$$

If K is a compact subset of G [such that $e \in K$], consider a compact neighborhood V of e containing K. By hypothesis and Lemma 7.8, given $\varepsilon > 0$, there exists a compact subset U in G such that $0 < |U|$ and $|VU|/|U| < 1 + \varepsilon$; then a fortiori $|KU|/|U| < 1 + \varepsilon$.

$$(SF_{e,0}) \Rightarrow (SF_0) \quad [\text{resp. } (SF_e) \Rightarrow (SF), (SF_e^*) \Rightarrow (SF^*)]$$

If K is a subset of G that is compact and has nonvoid interior [resp. is compact, finite], consider an arbitrary compact neighborhood V of e and let $K' = K \cup V$ [resp. $K' = K \cup \{e\}$]. By hypothesis and Lemma 7.7, given $\varepsilon > 0$ there exists a compact subset U in G such that $0 < |U|$ and $|K'U \triangle U|/|U| < \varepsilon/2$. Then by (2)

$$\frac{|KU \triangle U|}{|U|} \leq 2\frac{|KU \setminus U|}{|U|} \leq 2\frac{|K'U \setminus U|}{|U|} = 2\frac{|K'U \triangle U|}{|U|} < \varepsilon.$$

$$(SF^*) \Rightarrow (F^*)$$

Consider a finite subset F in G and $\varepsilon > 0$; let $F_1 = F \cup F^{-1} \cup \{e\}$. By hypothesis there exists a measurable subset U of G such that $0 < |U| < \infty$ and $|F_1 U \triangle U|/|U| < \varepsilon/2$. For any $a \in F$,

$$|aU \triangle U| = |aU \setminus U| + |U \setminus aU| = |aU \setminus U| + |a^{-1}U \setminus U|$$

$$\leq 2|F_1 U \setminus U| = 2|F_1 U \triangle U| < \varepsilon |U|. \quad \square$$

Remark 7.10. In Theorem 7.9 the proof of the implication $(F) \Rightarrow (SF_e^*)$ may be carried out simply without making use of the covering Lemma 7.5.

As a matter of fact, let F be a finite subset of G such that $e \in F$ and let $\varepsilon > 0$; there exists a measurable subset U of G such that $0 < |U| < \infty$ and $|aU \triangle U| < (\varepsilon/\text{card } F)|U|$ whenever $a \in F$. Therefore

$$|FU \triangle U| = |FU \setminus U| = \left| \bigcup_{a \in F} (aU \setminus U) \right| \leq \sum_{a \in F} |aU \setminus U|$$

$$< \text{card } F \frac{\varepsilon}{\text{card } F} |U| = \varepsilon |U|.$$

Proposition 7.11. *In the properties characterizing amenability of the locally compact group G given by Theorem 7.9 the measurable subset U may be supposed to be compact.*

Proof. The statement is a consequence of Lemmas 7.7 and 7.8. □

NOTES

Originally Følner [187] showed by combinatorial methods that, for any discrete group, (F^*) is equivalent to the existence of a left invariant mean on the real-valued bounded functions on the group. Namioka [411] noticed that, for a general locally compact group, (F^*) follows from (D_1). Hulanicki [268] simplified Namioka's demonstration. The proof of $(P_1) \Rightarrow (WF)$ given in 7.2-3 is a modification due to Greenleaf [226] of an argument developed by Namioka [411]. A direct proof of the implication $(F^*) \Rightarrow [\text{LIM}(\mathscr{S}_\mathbf{R}(G)) \neq \varnothing]$ in case of discrete countable groups G may be found in Mycielski [409].

The covering lemma was first established by Emerson and Greenleaf [169] via structure-theoretical methods; the proof given in 7.5 is due to Milnes and Bondar [389]. Leptin introduced (L) in [350]; he showed that $(L) \Rightarrow (F)$ [353]. Strong Følner conditions were studied by Emerson and Greenleaf [169]; the proof of the crucial implication $(F) \Rightarrow (L_{e,0})$ given in 7.6 is due to these authors. Making use of the theory of invariant capacities, Moulin Ollagnier and Pinchon [404] established the implication $(FP) \Rightarrow (L)$ bypassing the covering lemma.

8. AMENABILITY AND UNITARY GROUP REPRESENTATIONS

We begin this section by summarizing some general properties of the theory of weak containment for unitary representations introduced by Fell [181, 182]; another reference is Dixmier [D]. We then state Godement's deep characterization of amenability.

A. General Considerations on Weak Containment

If A is an involutive algebra over $\mathbf{C}$, a linear functional f on A is said to be a *positive functional* if $f(x^*x) \geq 0$ whenever $x \in A$. Then for all $x, y \in A$, $f((x + y)^*(x + y)) \geq 0$ and therefore $f(y^*x) = \overline{f(x^*y)}$. So we define a prehilbert structure on A by putting $(x|y) = f(y^*x)$ for x, $y \in A$; the Cauchy–Schwarz inequality is $|f(y^*x)|^2 \leq f(x^*x)f(y^*y)$ for x, $y \in A$. In particular, if A admits a unit u, for every $x \in A$, $f(x^*) = \overline{f(x)}$ and $|f(x)|^2 \leq f(u)f(x^*x)$. If A is an involutive Banach algebra with unit u such that $\|u\| = 1$, then every positive functional f on A is continuous and $\|f\| = f(u)$.

If $\mathcal{H}$ is a Hilbert space, $\mathcal{L}(\mathcal{H})$ is a C^*-algebra, Φ^* being the adjoint operator of $\Phi \in \mathcal{L}(\mathcal{H})$. For every $\Phi \in \mathcal{L}(\mathcal{H})$, $\|\Phi\| = \sup\{|(\Phi\xi|\xi)| : \xi \in \mathcal{H}, \|\xi\| = 1\}$ ([HS] 16.41, 16.55). If Φ is an element of the real subalgebra $\mathcal{L}(\mathcal{H})_h$ formed by all hermitian elements in $\mathcal{L}(\mathcal{H})$, then $(\Phi\xi|\xi) \in \mathbf{R}$ for every $\xi \in \mathcal{H}$ and also $\mathrm{sp}\,\Phi \subset \mathbf{R}$ ([HR] C.34). The (hermitian) operator $\Phi \in \mathcal{L}(\mathcal{H})$ is said to be *positive-definite* if $(\Phi\xi|\xi) \in \mathbf{R}_+$ whenever $\xi \in \mathcal{H}$. For every $\Phi \in \mathcal{L}(\mathcal{H})_h$, we consider $\Phi^+(\xi) = \sup\{(\Phi\xi|\xi), 0\}$, $\Phi^-(\xi) = \sup\{(-\Phi\xi|\xi), 0\}$, $\xi \in \mathcal{H}$; then $(\Phi\xi|\xi) = \Phi^+(\xi) - \Phi^-(\xi)$ for every $\xi \in \mathcal{H}$. Also let $\|\!|\Phi^\pm|\!\| = \sup\{\Phi^\pm(\xi) : \xi \in \mathcal{H}, \|\xi\| \leq 1\}$; $\|\!|\Phi^\pm|\!\| \leq \|\Phi\|$. If $\Phi \in \mathcal{L}(\mathcal{H})$, the following properties are equivalent: (i) Φ is positive-definite; (ii) Φ is hermitian and $\mathrm{sp}\,\Phi \subset \mathbf{R}_+$; (iii) there exists $\Phi_0 \in \mathcal{L}(\mathcal{H})_h$ such that $\Phi_0^2 = \Phi$. The positive-definite operator Φ_0 is denoted by $\Phi^{1/2}$ and called the square root of Φ ([B TS] chap. I, Section 6, No. 8, proposition 13). For every $\Phi \in \mathcal{L}(\mathcal{H})$, $\Phi^*\Phi$ is positive-definite.

If A is an involutive algebra, a *representation* T of A admitting the Hilbert space $\mathcal{H} = \mathcal{H}_T$ as its representation space is a $*$-homomorphism $T : A \mapsto \mathcal{L}(\mathcal{H})$, that is, for all x, $y \in A$ and $\alpha \in \mathbf{C}$, one has $T_{x+y} = T_x + T_y$, $T_{\alpha x} = \alpha T_x$, $T_{xy} = T_x T_y$, $T_x^* = T_{x^*}$. If A is a C^*-algebra, $\|T\| \leq 1$ and T is automatically continuous ([HR] 21.22). The representation T is said to be *nondegenerate* if $\{T_x \xi : x \in A, \xi \in \mathcal{H}\}^- = \mathcal{H}$. The vector $\xi \in \mathcal{H}$ is said to be *cyclic* for T if $\{T_x \xi : x \in A\}^- = \mathcal{H}$. Two representations T and T' with representation spaces $\mathcal{H}_T$ and $\mathcal{H}_{T'}$ are called *equivalent* if there exists an isomorphism Ψ of $\mathcal{H}_T$ onto $\mathcal{H}_{T'}$ such that $\Psi \circ T_x \circ \Psi^{-1} = T'_x$ whenever $x \in A$; Ψ is an intertwining operator. If T is a representation of A, for every $\xi \in \mathcal{H}_T$,

$$f : x \mapsto (T_x \xi | \xi)$$

is a positive functional on A; note that, for every $x \in A$,

$$f(x^*x) = (T_{x^*x}\xi|\xi) = (T_x\xi|T_x\xi) \geq 0.$$

We say that f is a positive functional *associated* to T. For any $y \in A$, $f' : x \mapsto f(y^*xy)$ is also a positive functional associated to T because

$$(T_{y^*x^*xy}\xi|\xi) = (T_{xy}\xi|T_{xy}\xi) \geq 0$$

whenever $x \in A$ and $\xi \in \mathcal{H}_T$.

Let A be a closed $*$-subalgebra of $\mathcal{L}(\mathcal{H})$. If $\xi_1, \ldots, \xi_n \in \mathcal{H}$, consider

$$\Psi : \Phi \mapsto \sum_{i=1}^n (\Phi\xi_i|\xi_i)$$

$$A_h \to \mathbf{R};$$

Ψ is an element of the dual A_h^* of the real Banach space A_h and Ψ is positive, that is, $\langle \Phi, \Psi \rangle \in \mathbf{R}_+$ whenever Φ is a positive-definite operator in A_h. We denote by $N(A)$ the set of all positive elements in A_h^* that are of the type

$$\Psi : \Phi \mapsto \sum_{i=1}^{n} (\Phi \xi_i | \xi_i)$$

where $\xi_1, \ldots, \xi_n \in \mathcal{H}$ and $\sum_{i=1}^{n} \|\xi_i\|^2 \leq 1$; then $\|\Psi\| \leq 1$.

Lemma 8.1. *Let $\mathcal{H}$ be a Hilbert space and let A be a closed $*$-subalgebra of $\mathcal{L}(\mathcal{H})$. Then the set of positive elements Ψ in A_h^* such that $\|\Psi\| \leq 1$ is the w^*-closure of $N(A)$.*

Proof. (a) The polar set

$$N(A)^0 = \{ \Phi \in A_h : (\forall \Psi \in N(A)) \langle \Phi, \Psi \rangle \geq -1 \}$$

of $N(A)$ is the set of all $\Phi \in A_h$ such that $\|\|\Phi^-\|\| \leq 1$.

In view of establishing this statement notice that, if $\Phi \in N(A)^0$,

$$-1 \leq \inf \{ (\Phi \xi | \xi) : \xi \in \mathcal{H}, \|\xi\| \leq 1 \}$$

hence

$$\sup \{ (-\Phi \xi | \xi) : \xi \in \mathcal{H}, \|\xi\| \leq 1 \} \leq 1$$

and $\|\|\Phi^-\|\| \leq 1$. Conversely let $\Phi \in A_h$ with $\|\|\Phi^-\|\| \leq 1$; then for all $\xi_1, \ldots, \xi_n \in \mathcal{H}$ with $\sum_{i=1}^{n} \|\xi_i\|^2 \leq 1$, we have $\sum_{i=1}^{n} \Phi^-(\xi_i) \leq \sum_{i=1}^{n} \|\xi_i\|^2 \leq 1$ and, as also $\sum_{i=1}^{n} \Phi^+(\xi_i) \geq 0$,

$$\sum_{i=1}^{n} (\Phi \xi_i | \xi_i) = \sum_{i=1}^{n} \Phi^+(\xi_i) - \sum_{i=1}^{n} \Phi^-(\xi_i) \geq -1.$$

(b) The bipolar set

$$N(A)^{00} = \{ \Psi \in A_h^* : (\forall \Phi \in N(A)^0) \langle \Phi, \Psi \rangle \geq -1 \}$$

of $N(A)$ is the set of all positive elements Ψ in A_h^* such that $\|\Psi\| \leq 1$.

As a matter of fact, let $\Psi \in N(A)^{00}$. If $\Phi \in A_h$ with $\|\Phi\| \leq 1$, we have $\|\|(-\Phi)^-\|\| = \|\|\Phi^+\|\| \leq 1$ and $\|\|\Phi^-\|\| \leq 1$; hence (a) implies that $-\Phi \in N(A)^0$, $\Phi \in N(A)^0$, so $\langle -\Phi, \Psi \rangle \geq -1$, that is, $\langle \Phi, \Psi \rangle \leq 1$ and also $\langle \Phi, \Psi \rangle \geq -1$. Therefore $|\langle \Phi, \Psi \rangle| \leq 1$ and finally $\|\Psi\| \leq 1$. Moreover if $\Phi \in A_h$ is positive-definite and $\alpha \in \mathbf{R}_+^*$, then $\alpha \Phi$ is positive-definite. Hence $\|\|(\alpha \Phi)^-\|\| = 0$ and, by (a), $\alpha \Phi \in N(A)^0$, that is, $\langle \alpha \Phi, \Psi \rangle \geq -1$, $\langle \Phi, \Psi \rangle \geq -1/\alpha$. As

$\alpha \in \mathbf{R}_+^*$ may be chosen arbitrarily, $\langle \Phi, \Psi \rangle \geq 0$. Conversely let Ψ be a positive element in A_h^* with $\|\Psi\| \leq 1$. By (a), for any $\Phi \in N(A)^0$, $\Phi + id_{\mathcal{H}}$ is positive-definite, therefore $\langle \Phi, \Psi \rangle \geq -1$.

(c) We now apply the bipolar theorem ([B EVT] Chap. II, Section 6, No. 3, théorème 1): As $N(A)$ is convex and $0 \in N(A)$, $N(A)^{00}$ is the w*-closure of $N(A)$ in A_h^*. □

Lemma 8.2. *Let A be a C*-algebra. If $\mathcal{T} = \{T^{(i)} : i \in I\}$ is a family of representations of A and S is a particular representation of A, then the following conditions are equivalent:*

(i) $\cap_{i \in I} \mathcal{K}er\, T^{(i)} \subset \mathcal{K}er\, S$.

(ii) *Every positive functional on A associated to S is the w*-limit of sums of positive functionals associated to elements of $\mathcal{T}$.*

(iii) *If ξ is a cyclic vector for the representation S, the positive functional $f: x \mapsto (S_x \xi | \xi)$ is the w*-limit of sums of positive functionals associated to elements of $\mathcal{T}$.*

Proof. (i) ⇒ (ii)

We may suppose each $T^{(i)}$ ($i \in I$) to be nondegenerate. We put $T = \oplus_{i \in I} T^{(i)}$ on $\mathcal{H} = \oplus_{i \in I} \mathcal{H}_{T^{(i)}}$. As we may always pass to the quotient space by $\cap_{i \in I} \mathcal{K}er\, T^{(i)}$, we assume that $\cap_{i \in I} \mathcal{K}er\, T^{(i)} = \{0\}$. So T is injective and we identify A with the closed *-subalgebra $T_A = \{T_x : x \in A\}$ of $\mathcal{L}(\mathcal{H})$. Then the hypothesis and Lemma 8.1 applied to T_A yield that any positive functional associated to S is the w*-limit of sums of positive functionals associated to T; therefore (ii) holds.

$$(ii) \Rightarrow (i)$$

If $a \in \cap_{i \in I} \mathcal{K}er\, T^{(i)}$, for every $i \in I$ and every $\xi \in \mathcal{H}_{T^{(i)}}$, $(T_{a^*a}^{(i)} \xi | \xi) = (T_a^{(i)} \xi | T_a^{(i)} \xi) = 0$; hence for every positive functional f associated to an element of $\mathcal{T}$, $f(a^*a) = 0$. Then by (ii), for every $\zeta \in \mathcal{H}_S$, $\|S_a \zeta\|^2 = (S_a \zeta | S_a \zeta) = (S_{a^*a} \zeta | \zeta) = 0$, that is, $a \in \mathcal{K}er\, S$.

$$(ii) \Rightarrow (iii) \quad \text{Trivial}$$

$$(iii) \Rightarrow (ii)$$

If $S = 0$, the implication is trivial. Otherwise let $\eta \in \mathcal{H}_S, \eta \neq 0$ and consider the positive functional g associated to S that is defined by $g(x) = (S_x \eta | \eta)$, $x \in A$. Given $\varepsilon > 0$, let $\varepsilon_1 = (\|\eta\|^2 + \varepsilon)^{1/2} - \|\eta\|$. There exists $y \in A$ such that $\|S_y \xi - \eta\| < \varepsilon_1$. Notice that, by hypothesis, the positive functional f' defined by $f'(x) = f(y^*xy)$, $x \in A$, is the w*-limit of sums of positive functionals associated to elements of $\mathcal{T}$; moreover for each $x \in A$,

$$|g(x) - f'(x)| = |(S_x\eta|\eta) - (S_{xy}\xi|S_y\xi)|$$

$$\leq |(S_x\eta|\eta - S_y\xi)| + |(S_x(\eta - S_y\xi)|S_y\xi)|$$

$$\leq \|S_x\| \|\eta\|\varepsilon_1 + \|S_x\|\varepsilon_1(\|\eta\| + \varepsilon_1) = (\varepsilon_1^2 + 2\varepsilon_1\|\eta\|)\|S_x\|$$

$$= \varepsilon \|S_x\|. \qquad \square$$

Definition 8.3. *Let A be a C^*-algebra, $\mathcal{T}$ a family of representations of A, and S a particular representation of A. If $\mathcal{T}$ and S satisfy the equivalent conditions of Lemma 8.2, we say that S is weakly contained in $\mathcal{T}$. The support of a representation T is the set of all representations of A that are weakly contained in $\{T\}$.*

Let G be a locally compact group. A *unitary representation T* of G admitting the Hilbert space $\mathcal{H} = \mathcal{H}_T$ as its representation space is a homomorphism $T: x \mapsto T_x$ of G into the unitary group of $\mathcal{L}(\mathcal{H})$; $T_e = id_\mathcal{H}$ and $T_{x^{-1}} = T_x^*$ for any $x \in G$. The unitary representation T is called *continuous* if it is continuous with respect to the *strong topology* of $\mathcal{L}(\mathcal{H})$, that is, the mapping $x \mapsto \|T_x\xi\|$ is continuous whenever $\xi \in \mathcal{H}$, or equivalently, with respect to the *weak topology* of $\mathcal{L}(\mathcal{H})$, that is, the mapping $x \mapsto (T_x\xi|\eta)$ is continuous whenever $\xi, \eta \in \mathcal{H}$ ([HR] 22.9). If T is continuous, then for all $\xi, \eta \in \mathcal{H}$, the mappings $x \mapsto T_x\xi$ and $x \mapsto (T_x\xi|\eta)$ are left uniformly continuous on G ([HR] 22.20).
$$G \to \mathcal{H} \qquad G \to \mathbf{C}$$

We consider the important particular case of the unitary representation of the locally compact group G given by the *left regular representation L* of G on the Hilbert space $L^2(G)$, that is,

$$x \mapsto L_x f = {}_{x^{-1}}f = \delta_x * f$$

$$G \to L^2(G)$$

for $f \in L^2(G)$; the left regular representation is continuous (Section 2D.3). Notice that, for all $f, g \in L^2(G)$ and $x \in G$,

$$(L_x f|g) = ({}_{x^{-1}}f|g) = (f|{}_x g) = (f|L_{x^{-1}}g) = (f|L_x^{-1}g).$$

By (2.10) we have

$$(L_x \tilde{f}|\tilde{f}) = (\tilde{f}|L_x f) = (L_{x^{-1}}\tilde{f}|f) = f * \tilde{f}(x). \qquad (1)$$

Let G be a locally compact group. If T is a continuous unitary representation of G on the Hilbert space $\mathcal{H}$ and μ is any element in $\mathcal{M}^1(G)$, we put

$$T_\mu = \int_G T_x \, d\mu(x) \in \mathcal{L}(\mathcal{H}).$$

In this way we define a representation of $\mathcal{M}^1(G)$ associated to T which we denote also by T: For $\xi, \eta \in \mathcal{H}$, $\mu \in \mathcal{M}^1(G)$,

$$(T_\mu \xi | \eta) = \int_G (T_x \xi | \eta) \, d\mu(x);$$

$T_\mu^* = T_{\mu^*}$. The restriction to $L^1(G)$ of this representation of $\mathcal{M}^1(G)$ is nondegenerate. More precisely, we obtain a bijection of the set of all continuous unitary representations of G onto the set of all nondegenerate representations of $L^1(G)$ ([D] 13.3.1, 13.3.4). In particular, this correspondence associates to the left regular representation L of G the *left regular representation L of $L^1(G)$* defined by

$$L_f : g \mapsto f * g$$

$$L^2(G) \to L^2(G)$$

for $f \in L^1(G)$. Notice that, if $f \in L^1(G)$ and $g, h \in L^2(G)$,

$$\int (L_x g | h) f(x) \, dx = \iint g(x^{-1}y) \overline{h(y)} f(x) \, dx \, dy$$

$$= \int f * g(y) \overline{h(y)} \, dy$$

$$= (f * g | h).$$

If T is a nondegenerate representation of the involutive algebra A [resp. a continuous unitary representation of the locally compact group G], the following properties are equivalent: The only closed subspaces of $\mathcal{H}_T$, that are invariant under T, are $\{0\}$ and $\mathcal{H}_T$; the only elements of $\mathcal{L}(\mathcal{H}_T)$ that commute with all $T_x (x \in A)$ [resp. all $T_x (x \in G)$] are the scalar operators $\alpha \, \mathrm{id}_{\mathcal{H}_T}$ ($\alpha \in \mathbf{C}$); every $\xi \in \mathcal{H}_T \setminus \{0\}$ is a cyclic vector for T ([HR] 21.30). If these equivalent conditions hold, T is said to be *topologically irreducible*. The *dual* of the involutive algebra A is the set of all (equivalence classes of) topologically irreducible nonzero representations of A. Via Lemma 8.2 we may define a topology on the dual of a C^*-algebra A by calling a nonvoid subset $\mathcal{S}$ closed if it consists of all (equivalence classes of) topologically irreducible representations of A that are weakly contained in $\mathscr{S}$. This nonnecessarily Hausdorff topology is the *Fell topology*.

Let now $\mathbf{A}$ be an involutive Banach algebra admitting approximate units and let $\mathscr{S}$ be a subset of its dual. For $x \in A$, we put

$$\|x\|_{\mathscr{S}} = \sup\{\|T_x\| : T \in \mathscr{S}\};$$

$\|\ \|$ is a seminorm on A. Let $N_{\mathscr{S}} = \{x \in A : \|x\|_{\mathscr{S}} = 0\}$. The completion of $A/N_{\mathscr{S}}$ is a C^*-algebra ([D] 2.7.2). If $\mathscr{S}$ is the dual of A, the associated C^*-algebra is called *enveloping C^* algebra* of A. In particular, if $A = L^1(G)$ for a locally compact group G, we denote by $C^*_{\mathscr{S}}(G)$ the completion of $A/N_{\mathscr{S}}$ and by $C^*(G)$ the corresponding enveloping C^*-algebra. There exists a canonical bijection from the dual of $C^*(G)$ onto the set of (equivalence classes of) topologically irreducible continuous unitary representations of G ([D] 18.1.1). The latter set, equipped with the topology induced from the dual of $C^*(G)$, is the *dual* (object) $\hat{G}$ of G. If $S \in \hat{G}$ and $\mathscr{T} \subset \hat{G}$ we also say that S is weakly contained in $\mathscr{T}$ if S, considered as an element of the dual of $C^*(G)$, is weakly contained in $\mathscr{T}$, considered as a subset of the dual of $C^*(G)$. By Lemma 8.2, S is weakly contained in $\mathscr{T}$ if and only if $C^*_S(G)\ [= C^*_{\{S\}}(G)] \subset C^*_{\mathscr{T}}(G)$. The *support* of S is again the set of all elements of $\hat{G}$ that are weakly contained in $\{S\}$. The *reduced dual* $\hat{G}_r$ of G is the support of the left regular representation L of G.

Let G be a locally compact group. We say that a continuous function $\varphi: G \to \mathbf{C}$ is *positive-definite* if, for all $n \in \mathbf{N}^*$, $a_1, \ldots, a_n \in G$ and $\alpha_1, \ldots, \alpha_n \in \mathbf{C}$,

$$\sum_{i,j=1}^{n} \alpha_i \overline{\alpha_j} \varphi(a_i^{-1} a_j) \geq 0.$$

We denote by $P(G)$ the cone of all continuous positive-definite functions defined on G. In particular, $1_G \in P(G)$. For every $\varphi \in P(G)$, we have $\varphi(e) \geq 0$, $\tilde{\varphi} = \varphi$, and $\|\varphi\| = \varphi(e)$ ([D] 13.4.3), hence $P(G) \subset \mathscr{C}(G)$. Moreover $P(G)$ is closed for the pointwise multiplication ([HR] 32.9). On $\{\varphi \in P(G) : \varphi(e) = 1\}$ the w*-topology $\sigma(L^\infty(G), L^1(G))$ coincides with the topology of uniform convergence on compacta ([D] 13.5.2). A complex-valued function φ defined on G is an element of $P(G)$ if and only if there exist a continuous unitary representation T of G and $\xi \in \mathscr{H}_T$ such that $\varphi(x) = (T_x \xi | \xi)$ for every $x \in G$ ([HR] 32.8; [D] 13.4.5). The continuous unitary representation T of G is weakly contained in a set $\mathscr{S}$ of such representations if and only if any positive-definite function associated to T is the uniform limit on compacta of sums of positive-definite functions associated to elements of $\mathscr{S}$. By (1) we have $f * \tilde{f} \in P(G)$ whenever $f \in L^2(G)$. By a very important result due to Godement, if G is a locally compact group and $\varphi \in P(G) \cap L^2(G)$, then there exists $\psi \in P(G) \cap L^2(G)$ such that $\varphi = \psi * \tilde{\psi}$ ([208] théorème 17; [D] 13.8.6). We say that $\mu \in \mathscr{M}^1(G)$ is a *positive-definite measure* on G if $\langle f * \tilde{f}, \mu \rangle \geq 0$

whenever $f \in \mathcal{K}(G)$; a positive-definite measure is necessarily hermitian ([D] 13.7.2).

Finally recall that, if G is a locally compact abelian group, all continuous unitary representations of G have dimension 1 ([HR] 22.17); they constitute *characters*, that is, continuous homomorphisms of G into the torus $\mathbf{T}$. The Fell topology on $\hat{G}$ is the w*-topology $\sigma(L^\infty(G), L^1(G))$ or equivalently the topology of uniform convergence on compacta for the characters; $\hat{G}$ constitutes a locally compact abelian group called the *character group* of G ([D] 18.1.6; [HR] 23.15).

B. Godement's Conditions

Let G be a locally compact group. For every $\varphi \in P^2(G)$ and every $a \in G$, by (2.10) we have

$$\|_a\varphi \pm \varphi\|_2^2 = (_a\varphi \pm \varphi|_a\varphi \pm \varphi) = \|_a\varphi\|^2 + \|\varphi\|^2 \pm 2(_a\varphi|\varphi)$$

$$= 2(1 \pm (_a\varphi|\varphi)) = 2(1 \pm \varphi * \tilde{\varphi}(a)). \qquad (2)$$

Lemma 8.4. *Let G be a locally compact group. Let $X = L^2(G)$, $X_1 = P^2(G)$ or $X = \mathcal{K}(G)$, $X_1 = P^2(G) \cap \mathcal{K}(G)$. Then the following conditions are equivalent:*

(i) *For every $\varepsilon > 0$ and every compact subset K [resp. every finite subset F] in G, there exists $\varphi \in X$ such that $|1 - \varphi * \tilde{\varphi}(a)| < \varepsilon$ whenever $a \in K$ [resp. $a \in F$].*

(ii) *For every $\varepsilon > 0$ and every compact subset K [resp. every finite subset F] in G, there exists $\varphi \in X_1$ such that $0 \le 1 - \varphi * \tilde{\varphi}(a) < \varepsilon$ whenever $a \in K$ [resp. $a \in F$].*

Proof. Assume that (i) holds. Given $\varepsilon > 0$ and a compact subset K [resp. a finite subset F] in G, let $\varepsilon' = \varepsilon(\varepsilon + 2)^{-1}$ and $K' = K \cup \{e\}$ [resp. $F' = F \cup \{e\}$]. By hypothesis, there exists $\varphi \in X$ such that $|1 - \varphi * \tilde{\varphi}(a)| < \varepsilon'$ whenever $a \in K'$ [resp. $a \in F'$]. In particular, by (2.11), $|1 - \|\varphi\|_2^2| = |1 - \varphi * \tilde{\varphi}(e)| < \varepsilon'$ and $\|\varphi\|_2^2 > 1 - \varepsilon' > 0$. Let $\psi = |\varphi|/\|\varphi\|_2$; $\psi \in X_1$. For every $a \in K$ [resp. every $a \in F$], by (2.10), $|\varphi| * |\varphi|\tilde{\,}(a) = (_a|\varphi|\,|\varphi|) \le \|\varphi\|_2^2$ and

$$0 \le 1 - \frac{|\varphi| * |\varphi|\tilde{\,}(a)}{\|\varphi\|_2^2} \le \frac{\|\varphi\|_2^2 - |\varphi * \tilde{\varphi}(a)|}{\|\varphi\|_2^2}$$

$$\le \frac{|1 - \|\varphi\|_2^2|}{\|\varphi\|_2^2} + \frac{|1 - \varphi * \tilde{\varphi}(a)|}{\|\varphi\|_2^2} < 2\frac{\varepsilon'}{1 - \varepsilon'} = \varepsilon,$$

that is, $0 \le 1 - \psi * \tilde{\psi}(a) < \varepsilon$. □

Proposition 8.5. *A locally compact group G is amenable if and only if any one of the following properties holds*:

(G) [*resp.* (G^*)]. *On every compact* [*resp. finite*] *subset of G, the function* 1_G *is the uniform limit of functions of the form* $\varphi * \check{\varphi}$, *where* $\varphi \in L^2(G)$, $\mathcal{K}(G)$, $P^2(G)$, *or* $P^2(G) \cap \mathcal{K}(G)$.

Proof. The statement follows from Corollary 6.15, (2) and Lemma 8.4. □

Definition 8.6. *Let G be a locally compact group. We define*

$$\sigma = \sup\left\{\inf\left\{\sup\left\{1 - \varphi * \check{\varphi}(a) : a \in K\right\} : \varphi \in P^2(G)\right\} : K \in \mathfrak{K}(G)\right\},$$

$$\sigma^* = \sup\left\{\inf\left\{\sup\left\{1 - \varphi * \check{\varphi}(a) : a \in F\right\} : \varphi \in P^2(G)\right\} : F \in \mathfrak{F}(G)\right\}.$$

Obviously $\sigma^* \leq \sigma$; by (2.10), $0 \leq \sigma^*$. By (2) we have $\gamma_2^* = (2\sigma^*)^{1/2}$ and $\gamma_2 = (2\sigma)^{1/2}$.

Proposition 8.7. *A locally compact group G is amenable if and only if* $\sigma = 0$ *or equivalently* $\sigma^* = 0$.

Proof. The statement is an immediate consequence of Proposition 8.5 as the conditions are reformulations of (G) and (G^*). □

Proposition 8.8. *A locally compact group G is amenable if and only if every $f \in P(G)$ is a uniform limit, on every compact* [*resp. every finite*] *subset in G, of functions of the form* $\varphi * \check{\varphi}$, *where* $\varphi \in L^2(G)$ *or* $\varphi \in \mathcal{K}(G)$.

Proof. As $1_G \in P(G)$, the condition yields amenability by Proposition 8.5.

Assume now that G is amenable. By Proposition 8.5, on every compact [resp. finite] subset in G, 1_G is a uniform limit of functions belonging to $P(G) \cap \mathcal{K}(G)$. Then also $f = f 1_G$ is such a limit. Now recall that every element of $P(G) \cap L^2(G)$ is of the form $\varphi * \check{\varphi}$, where $\varphi \in L^2(G)$, and $\mathcal{K}(G)$ is dense in $L^2(G)$. □

Theorem 8.9. *Let G be a locally compact group. Each of the following properties characterizes amenability*:

(i) *The dual $\hat{G}$ of G coincides with the reduced dual $\hat{G}_r$ of G, that is,* $C_L^*(G) = C^*(G)$.

(i') *The trivial representation $i_G : x \mapsto \mathrm{id}_{L^2(G)}$ of G is an element of the reduced dual $\hat{G}_r$ of G.*

(ii) *For every positive-definite measure μ in $\mathcal{M}^1(G)$ and every $f \in P(G)$,* $\langle f, \mu \rangle \geq 0$.

(ii') *For every positive-definite measure μ in $\mathcal{M}^1(G)$, $\int d\mu \geq 0$.*

Proof. (i) ⇒ (i') Trivial

If G is amenable, Proposition 8.8 and (1) imply that $\hat{G} = \hat{G}_r$. If (i') holds, 1_G is a uniform limit on compacta of sums of positive-definite functions associated

NOTES 81

to L; by (1) such functions are of the type $\varphi * \tilde{\varphi}$, for $\varphi \in L^2(G)$. The sums belong themselves to $P(G) \cap L^2(G)$ and therefore are of the same type. In particular, (G) holds, and amenability follows from Proposition 8.5.

If G is amenable, (ii) is an immediate consequence of the definition of positive-definite measures on G as, by Proposition 8.8, on every compact subset, f is the uniform limit of functions of the form $\varphi * \tilde{\varphi}$, where $\varphi \in \mathcal{K}(G)$. The implication (ii) $\Rightarrow$ (ii') is due to the fact that $1_G \in P(G)$.

$$(\text{ii}') \Rightarrow (\text{i}')$$

We show that 1_G is contained in the convex cone C of all functions in $P(G)$ that are uniform limits on compacta of functions of the form $\varphi * \tilde{\varphi}$, where $\varphi \in L^2(G)$; C is w*-closed in $L^\infty(G)$. As $0 \in C$, by the bipolar theorem it suffices to show that 1_G belongs to the bipolar set C^{00} of C ([B EVT] Chap. II, Section 6, No. 3, théorème 1).

Let $g \in L^1(G) \cap C^0$; for every $\varphi \in \mathcal{K}(G)$, $\mathcal{R}e \langle \varphi * \tilde{\varphi}, g \rangle \geq 0$, C being a cone ([B EVT] Chap. II, Section 6, No. 3, proposition 4). As $(\varphi * \tilde{\varphi})\tilde{\ } = \varphi * \tilde{\varphi}$,

$$\int \varphi * \tilde{\varphi}(x) g^*(x)\, dx = \int \varphi * \tilde{\varphi}(x) \overline{g(x^{-1})} \Delta(x^{-1})\, dx = \int \overline{\varphi * \tilde{\varphi}(x) g(x)}\, dx;$$

we obtain

$$\langle \varphi * \tilde{\varphi}, g + g^* \rangle \geq 0$$

so $g + g^*$ is a positive-definite measure on G. Then by hypothesis $\int (g + g^*) \geq 0$, that is, $\int (g + \bar{g}) \geq 0$; thus $\mathcal{R}e \langle 1_G, g \rangle = \mathcal{R}e \int g \geq 0$, that is, $1_G \in C^{00}$. □

NOTES

The original work on weak containment was done by Fell [181, 182]. We do not intend to trace the history of positive-definite functions up to Weil, Gelfand and Raĭkov, and Godement; [HR II] gives precise accounts. An interesting historical survey on the development of the theory of positive-definite functions is Stewart [519].

Godement [208] proved the equivalence of (G) with property (ii') of 8.9; for the demonstrations of the other equivalence properties in 8.9 we refer to Dixmier [D]. Reiter [451] noticed that $(P_1) \Leftrightarrow (G)$. Hulanicki [267, 268] established the implication $(F) \Rightarrow (G)$ and Guichardet [240] showed the equivalence of (D_2) with the weak containment characterizations of amenability. See Gilbert [199] for the equivalence $(G) \Leftrightarrow (G^*)$.

9. AMENABILITY AND CONVOLUTION OPERATORS

We state a collection of characterizations of amenability for a locally compact group G in terms of the operators on $L^p(G)$ $(1 < p < \infty)$ that consist in left convolution by measures.

A. General Considerations on Convolutors

Definition 9.1. *Let G be a locally compact group and $p \in [1, \infty]$. We consider the Banach subspace $Cv^p(G)$ of $\mathscr{L}(L^p(G))$ formed by all convolutors of $L^p(G)$, that is, all $T \in \mathscr{L}(L^p(G))$ such that*

$$T(f * g) = (Tf) * g$$

whenever $f, g \in \mathscr{K}(G)$.

If G is a locally compact group and $p \in [1, \infty]$, we denote by $\|\cdot\|_{Cv^p}$ the norm of $\mathscr{L}(L^p(G))$. If $\mu \in \mathscr{M}^1(G)$ and $f \in L^p(G)$, we consider

$$L_\mu : f \mapsto \mu * f$$
$$L^p(G) \to L^p(G);$$

we have

$$\|L_\mu\|_{Cv^p} \leq \|\mu\| \tag{1}$$

(Section 2D.3). Let now $\mu \in \mathscr{M}^1_+(G)$. As $\|\mu * 1_G\| = \|\mu\|$,

$$\|L_\mu\|_{Cv^\infty} = \|\mu\|; \tag{2}$$

for every $\varphi \in P^1(G)$, $\|\mu * \varphi\|_1 = \|\mu\|$, hence also

$$\|L_\mu\|_{Cv^1} = \|\mu\|. \tag{3}$$

For any $\mu \in \mathscr{M}^1(G)$, L_μ is a convolutor on $L^1(G)$. If $1 \leq p < \infty$, by density of $\mathscr{K}(G)$ in $L^p(G)$, $Cv^p(G)$ is a Banach subalgebra of $\mathscr{L}(L^p(G))$.

If G is a locally compact group, $f \in \mathscr{K}(G)$, $\mu \in \mathscr{M}(G)$, we may also define

$$L_\mu f(x) = \langle (f_x)\check{\ }, \mu \rangle = \int_G f(y^{-1}x) \, d\mu(y),$$

$x \in G$. We show that $L_\mu f$ is continuous. Let $x \in G$ and let V be a fixed compact neighborhood of e in G. For every $x' \in xV$,

$$\left| L_\mu f(x) - L_\mu f(x') \right| \leq \int_G \left| f(y^{-1}x) - f(y^{-1}x') \right| d|\mu|(y)$$

$$= \int_{xV(\text{supp} f)^{-1}} \left| f(y^{-1}x) - f(y^{-1}x') \right| d|\mu|(y).$$

If $|\mu|(xV(\operatorname{supp} f)^{-1}) = 0$, $L_\mu f(x) = L_\mu f(x')$. If $|\mu|(xV(\operatorname{supp} f)^{-1}) > 0$, given $\varepsilon > 0$, there exists a compact neighborhood U of e in V such that

$$|f(z) - f(z')| < \varepsilon \bigl(|\mu|\bigl(xV(\operatorname{supp} f)^{-1}\bigr)\bigr)^{-1}$$

whenever $z, z' \in G$ with $z^{-1}z' \in U$. Then we have also

$$\bigl|L_\mu f(x) - L_\mu f(x')\bigr| < \varepsilon$$

for $x' \in xU$.

If G is a locally compact group, for all $f, g \in \mathscr{K}(G)$ and $\mu \in \mathscr{M}(G)$, by Fubini's theorem,

$$\langle f * g, \mu \rangle = \int_G \left(\int_G f(y) g(y^{-1}x) \, dy \right) d\mu(x)$$

$$= \int_G f(y) \left(\int_G \check{g}(x^{-1}y) \, d\mu(x) \right) dy = \langle f, L_\mu \check{g} \rangle. \qquad (4)$$

Proposition 9.2. *Let G be a locally compact group and let $p \in]1, \infty[$. For any $T \in Cv^p(G)$, there exists $\mu \in \mathscr{M}(G)$ such that $T(g) = L_\mu g$ for every $g \in \mathscr{K}(G)$. In case T is a nonnegative operator, we have $\mu \in \mathscr{M}_+(G)$ and*

$$\|T\|_{Cv^p} = \sup \left\{ \langle f * g, \mu \rangle : f, g \in \mathscr{K}_+(G), \|f\|_{p'} \leq 1, \|g\|_p \leq 1 \right\}$$

with $1/p + 1/p' = 1$. In particular, in the case that T is hermitian,

$$\|T\|_{Cv^2} = \sup \left\{ \langle f * \check{f}, \mu \rangle : f \in \mathscr{K}_+(G), \|f\|_2 = 1 \right\}.$$

Proof. (a) By Proposition 2.2 there exists a net $(u_i)_{i \in I}$ in $P^1(G) \cap \mathscr{K}(G)$ such that $\lim_i \|u_i * g - g\|_p = 0$ whenever $g \in L^p(G)$. For every $i \in I$, $T(u_i) \in L^p(G)$; hence also $\mu_i = T(u_i) \in \mathscr{M}(G)$. In particular, if $T \geq 0$, we have $\mu_i \in \mathscr{M}_+(G)$ for every $i \in I$.

(b) We show that, for every $h \in \mathscr{K}(G)$, $\sup \{|\langle h, \mu_i \rangle| : i \in I\} < \infty$.

It suffices to verify the statement for any $h \in \mathscr{K}(G)$ that is of the type $f * g$, where $f, g \in \mathscr{K}(G)$. As a matter of fact, if $h \in \mathscr{K}(G)$, on any open relatively compact subset U of G containing $\operatorname{supp} h$, $|h|$ is majorized by $|V|^{-1} \sup \{|h(x)| : x \in U\} 1_V * 1_{V^{-1}U}$ where V is an arbitrary compact neighborhood of e because, for any $x \in U$,

$$1_V * 1_{V^{-1}U}(x) = \int_V 1_{V^{-1}U}(y^{-1}x) \, dy = |V|.$$

But obviously there exist $f, g \in \mathscr{K}_+(G)$ such that $|V|^{-1}\|h\| 1_V \leq f$ and $1_{V^{-1}U} \leq g$; hence $f * g$ majorizes $|h|$ on U.

Now for any $i \in I$, by (4) and Hölder's inequality, we have, for f, $g \in \mathcal{K}(G)$,

$$|\langle f * g, \mu_i \rangle| = |\langle f, \mu_i * \check{g} \rangle| = |\langle f, T(u_i) * \check{g} \rangle| = |\langle f, T(u_i * \check{g}) \rangle|$$

$$\leq \|f\|_{p'} \|T\|_{Cv^p} \|\check{g}\|_p < \infty.$$

(c) By (b) the set $\{\mu_i : i \in I\}$ is vaguely relatively compact in $\mathcal{M}(G)$ ([B INT] Chap. III, Section 1, No. 9, proposition 15); hence there exists $\mu \in \mathcal{M}(G)$ such that a subnet (μ_{i_k}) of (μ_i) converges vaguely to μ. Let $g \in \mathcal{K}(G)$; $\lim_k \|u_{i_k} * g - g\|_p = 0$ and then $\lim_k \|T(u_{i_k} * g - g)\|_p = 0$. For every $f \in \mathcal{K}(G)$, by (4) we have

$$\langle f, T(g) \rangle = \lim_k \langle f, T(u_{i_k} * g) \rangle = \lim_k \langle f, T(u_{i_k}) * g \rangle$$

$$= \lim_k \langle f, \mu_{i_k} * g \rangle = \lim_k \langle f * \check{g}, \mu_{i_k} \rangle = \langle f * \check{g}, \mu \rangle$$

$$= \langle f, L_\mu g \rangle,$$

that is, $T(g) = L_\mu g$.

(d) If $T \geq 0$, we have $\mu \geq 0$ and, for all $f, g \in \mathcal{K}(G)$, by (4) again

$$|\langle f, L_\mu g \rangle| = |\langle f * \check{g}, \mu \rangle| \leq \langle |f * \check{g}|, \mu \rangle \leq \langle |f| * |\check{g}|, \mu \rangle \tag{5}$$

hence

$$\|T\|_{Cv^p} = \sup\{\langle f * g, \mu \rangle : f, g \in \mathcal{K}_+(G), \|f\|_{p'} \leq 1, \|g\|_p \leq 1\}.$$

If T is hermitian, we have

$$\|T\|_{Cv^2} = \sup\{|\langle f, L_\mu f \rangle| : f \in L^2(G), \|f\|_2 = 1\}$$

([HS] 16.55). Density of $\mathcal{K}(G)$ in $L^2(G)$ and (5) imply that

$$\|T\|_{Cv^2} = \sup\{\langle f * \check{f}, \mu \rangle : f \in \mathcal{K}_+(G), \|f\|_2 = 1\}. \qquad \square$$

Proposition 9.3. *If G is a locally compact group and $p \in]1, \infty[$, $Cv^p(G)$ consists of all $T \in \mathcal{L}(L^p(G))$ commuting with right translations on $\mathcal{K}(G)$, that is, $(Tf)_a = T(f_a)$ whenever $f \in \mathcal{K}(G)$ and $a \in G$.*

Proof. (a) Let $T \in Cv^p(G)$; by Proposition 9.2 there exists $\mu \in \mathcal{M}(G)$ such that $Tf = L_\mu f$ whenever $f \in \mathcal{K}(G)$. For any $a \in G$, we have

$$(Tf)_a = (L_\mu f)_a = L_\mu(f_a) = T(f_a).$$

(b) Assume that $T \in \mathscr{L}(L^p(G))$ commutes with right translations on $\mathscr{K}(G)$. Let $f, g \in \mathscr{K}(G)$; for every $h \in L^{p'}(G)$,

$$\langle T(f * g), h \rangle = \langle f * g, \tilde{T} h \rangle = \iint f(xy) g(y^{-1}) \tilde{T} h(x) \, dx \, dy$$

$$= \int \langle f_y, \tilde{T} h \rangle g(y^{-1}) \, dy = \int \langle T(f_y), h \rangle g(y^{-1}) \, dy$$

$$= \int \langle (Tf)_y, h \rangle g(y^{-1}) \, dy$$

$$= \iint (Tf)(xy) h(x) g(y^{-1}) \, dx \, dy = \langle (Tf) * g, h \rangle,$$

that is, $T(f * g) = (Tf) * g$. □

B. Main Characterizations

Definition 9.4. *If G is a locally compact group and $p \in {]}1, \infty{[}$, consider the following properties*:

(C_p) *For every $\mu \in \mathscr{M}_+(G)$ such that $L_\mu \in Cv_+^p(G)$, one has $\mu \in \mathscr{M}_+^1(G)$, that is, μ is bounded.*
(C_p') *($\forall T \in Cv_+^p(G)$) ($\exists \mu \in \dot{\mathscr{M}}_+^1(G)) T = L_\mu$.*
(Cv_p) [*resp.* (Cv_p^*)] *($\forall \mu \in \mathscr{M}_+^1(G)$) [resp. ($\forall \mu \in \mathscr{M}_{c, +}^1(G)$)] $\|L_\mu\|_{Cv^p} = \|\mu\|$.*
(Cv_p') [*resp.* ($Cv_p'^*$)] *($\forall f \in L_+^1(G)$) [resp. ($\forall f \in \mathscr{K}_+(G)$)] $\|L_f\|_{Cv^p} = \|f\|_1$].*

Lemma 9.5. *Let G be a locally compact group and let $\mu \in \mathscr{M}_+^1(G)$. If there exists $p_0 \in {]}1, \infty{[}$ such that $\|L_\mu\|_{Cv^{p_0}} = \|\mu\|$, then for every $p \in {]}1, \infty{[}$, $\|L_\mu\|_{Cv^p} = \|\mu\|$.*

Proof. We consider the mapping $\theta: [0, 1] \to \mathbf{R}$ defined by

$$\theta(0) = \log \|L_\mu\|_{Cv^\infty}$$

$$\theta(t) = \log \|L_\mu\|_{Cv^{1/t}}; \quad 0 < t \le 1.$$

By the Riesz–Thorin convexity theorem, θ is a convex function ([DS] VI. 10.11). For any $t \in {]}0, 1{[}$, (1) implies that $\theta(t) \le \log \|\mu\|$; (2) and (3) imply that $\theta(0) = \theta(1) = \log \|\mu\|$. Then the property follows from convexity and the fact that $\|L_\mu\|_{Cv^{p_0}} = \|\mu\|$ for some $p_0 \in {]}1, \infty{[}$. □

Theorem 9.6. *The locally compact group G is amenable if and only if, for every $p \in {]}1, \infty{[}$ (for one $p \in {]}1, \infty{[}$) any one of the properties $(C_p), (C_p')$, $(Cv_p), (Cv_p^*), (Cv_p'), (Cv_p'^*)$ holds.*

Proof.

1. We begin by showing that if G is amenable, then (C_p) holds for every $p \in]1, \infty[$.

Let $\mu \in \mathcal{M}_+(G)$ such that $L_\mu \in Cv_+^p$. Choose $f_0 \in \mathcal{K}_+(G)$ such that $f_0 \leq 1_G$ and let $K = \operatorname{supp} f_0$. We put $\mu_0 = f_0 \mu \in \mathcal{M}_+^1(G)$. By Corollary 6.15, for any $\varepsilon \in]0,1[$, there exists $\varphi \in P^p(G) \cap \mathcal{K}(G)$ satisfying $\|_{y^{-1}}\varphi - \varphi\|_p < \varepsilon$ whenever $y \in K$. Let $\psi = \varphi^{p-1}$, then $\psi^{p'} = \varphi^p$ with $1/p + 1/p' = 1$. Therefore $\psi \in P^{p'}(G)$ and $\langle \varphi, \psi \rangle = 1$. For any $y \in K$, by Hölder's inequality,

$$|\langle _{y^{-1}}\varphi, \psi\rangle - 1| = |\langle _{y^{-1}}\varphi, \psi\rangle - \langle \varphi, \psi\rangle| = |\langle _{y^{-1}}\varphi - \varphi, \psi\rangle|$$

$$\leq \|_{y^{-1}}\varphi - \varphi\|_p \|\psi\|_{p'} < \varepsilon$$

hence

$$1 - \varepsilon < \int \varphi(y^{-1}x)\psi(x)\,dx$$

and also

$$(1-\varepsilon)\|\mu_0\| < \iint \varphi(y^{-1}x)\psi(x)\,dx\,d\mu_0(y) = \langle \mu_0 * \varphi, \psi\rangle \leq \|\mu_0 * \varphi\|_p \|\psi\|_{p'}$$

$$\leq \|L_{\mu_0}\|_{Cv^p} \|\varphi\|_p \|\psi\|_{p'} = \|L_{\mu_0}\|_{Cv^p}.$$

As $\varepsilon \in]0,1[$ may be chosen arbitrarily, we obtain $\|\mu_0\| \leq \|L_{\mu_0}\|_{Cv^p}$; so by (1) we have

$$\|L_{\mu_0}\|_{Cv^p} = \|\mu_0\|. \tag{6}$$

For any $f \in \mathcal{K}(G)$,

$$\|\mu_0 * f\|_p \leq \|\mu_0 * |f|\|_p \leq \|L_\mu(|f|)\|_p \leq \|L_\mu\|_{Cv^p} \|f\|_p;$$

therefore

$$\|L_{\mu_0}\|_{Cv^p} \leq \|L_\mu\|_{Cv^p}. \tag{7}$$

Notice that $\|\mu_0\| = \langle f_0, \mu\rangle$ and $\|\mu\| = \sup\{\langle f, \mu\rangle : f \in \mathcal{K}_+(G), f \leq 1_G\}$. Hence by (6) and (7) we have $\|\mu\| \leq \|L_\mu\|_{Cv^p}$; μ is bounded. With respect to (1) we obtain $\|L_\mu\|_{Cv^p} = \|\mu\|$.

2. For every $p \in]1, \infty[$, we establish the following diagram

$$\begin{array}{ccc} (Cv_p^*) & \Rightarrow & (Cv_p'^*) \\ \Updownarrow & & \Downarrow \\ (C_p) \Rightarrow (Cv_p) & & (Cv_p') \Rightarrow (Cv_2') \Rightarrow (Cv_2^*) \\ \Updownarrow & & \\ (C_p') & & \end{array}$$

$$(C_p) \Rightarrow (Cv_p); (C_p') \Rightarrow (Cv_p)$$

We assume a contrario that there exists $\mu \in M^1(G)$ such that $\alpha = \|L_\mu\|_{Cv^p} < 1$. Let $\varphi \in P^1(G) \cap \mathscr{K}(G)$; for $n \in \mathbf{N}^*$, we put $\varphi_n = \sum_{i=1}^n \mu^{*i} * \varphi$, hence

$$\|\varphi_n\|_1 = n. \tag{8}$$

For any $n \in \mathbf{N}^*$, $\|L_{\mu^{*n}}\|_{Cv^p} \leq \|L_\mu\|_{Cv^p}^n = \alpha^n$.

(a) Let $\psi \in \mathscr{K}(G)$ with $K = \operatorname{supp} \psi$. For any $n \in \mathbf{N}^*$, $\varphi_n \in L^p(G)$; if $1/p + 1/p' = 1$, we obtain via Hölder's inequality

$$|\langle \psi, \varphi_n \rangle| \leq \|\varphi_n\|_p \|\psi\|_{p'} \leq \sum_{i=1}^n \alpha^i \|\varphi\|_p \|\psi\| |K|^{1/p'} \leq \frac{1}{1-\alpha} \|\varphi\|_p \|\psi\| |K|^{1/p'}.$$

Hence $\nu = \sum_{i=1}^\infty \mu^{*i} * \varphi \in \mathscr{M}_+(G)$ and ν is unbounded by (8). On the other hand, for every $f \in \mathscr{K}(G)$ and every $n \in \mathbf{N}^*$,

$$\left\| \sum_{i=1}^n \mu^{*i} * \varphi * f \right\|_p \leq \left\| \sum_{i=1}^n \mu^{*i} * \varphi * |f| \right\|_p \leq \sum_{i=1}^n \alpha^i \|\varphi * |f|\|_p \leq \frac{1}{1-\alpha} \|f\|_p$$

hence $\|L_\nu f\|_p \leq 1/(1-\alpha) \|f\|_p$ and $L_\nu \in Cv_+^p(G)$. We have contradicted (C_p).

(b) The series $\sum_{n=1}^\infty L_{\mu^{*n}}$ is absolutely convergent in $Cv_+^p(G)$. If (C_p') holds, there exists $\sigma \in \mathscr{M}_+^1(G)$ such that $\sigma * \varphi = L_\sigma(\varphi) = \sum_{n=1}^\infty L_{\mu^{*n}}(\varphi) = \lim_{n \to \infty} \varphi_n$ in $L^p(G)$ and hence also in measure. By Riesz's theorem ([HS] 11.26), there exists a subsequence (φ_{n_k}) of (φ_n) converging λ-almost everywhere to $\sigma * \varphi$. We have $\sigma * \varphi \in L^1(G)$ and, by Lebesgue's dominated convergence theorem, $\lim_{k \to \infty} \|\varphi_{n_k} - \sigma * \varphi\|_1 = 0$. So we come to a contradiction as $\lim_{k \to \infty} \|\varphi_{n_k}\|_1 = \infty$ by (8).

$$(Cv_p) \Rightarrow (C_p')$$

Let $\mu \in \mathscr{M}_+(G)$ be associated to $T \in Cv_+^p(G)$ via Proposition 9.2. If K is a compact subset of G, we denote by μ_K the restriction of μ to K. By hypothesis and Proposition 9.2, we obtain

$$\|\mu_K\| = \|L_{\mu_K}\|_{Cv^p} = \sup \{ \langle f * g, \mu_K \rangle : f, g \in \mathscr{K}_+(G), \|f\|_{p'} \leq 1,$$

$$\|g\|_p \leq 1 \}$$

$$\leq \sup \{ \langle f * g, \mu \rangle : f, g \in \mathscr{K}_+(G), \|f\|_{p'} \leq 1, \|g\|_p \leq 1 \}$$

$$= \|T\|.$$

Therefore also $\|\mu\| \leq \|T\|$ and $\mu \in \mathscr{M}_+^1(G)$.

$$(Cv_p) \Rightarrow (Cv_p^*) \Rightarrow (Cv_p'^*) \quad \text{Trivial}$$

$$\left(Cv_p^*\right) \Rightarrow \left(Cv_p\right), \left(Cv_p'^*\right) \Rightarrow \left(Cv_p'\right)$$

Given $\mu \in M_+^1(G)$ [resp. $\mu \in L_+^1(G)$] and $\varepsilon \in \mathbf{R}_+^*$, there exists $\nu \in M_{c,+}^1(G)$ [resp. $\nu \in \mathcal{K}_+(G)$] such that $\|\nu - \mu\| < \varepsilon/2$. Then

$$\|L_\mu - L_\nu\|_{Cv^p} = \|L_{\mu-\nu}\|_{Cv^p} \leq \|\mu - \nu\| < \frac{\varepsilon}{2}$$

and

$$\|L_\mu\|_{Cv^p} \geq \|L_\nu\|_{Cv^p} - \frac{\varepsilon}{2} = \|\nu\| - \frac{\varepsilon}{2} > \|\mu\| - \varepsilon.$$

As $\varepsilon \in \mathbf{R}_+^*$ is chosen arbitrarily, $\|L_\mu\|_{Cv^p} = \|\mu\|$.

$$\left(Cv_p'\right) \Rightarrow \left(Cv_2'\right)$$

This implication is a consequence of Lemma 9.5.

$$\left(Cv_2'\right) \Rightarrow \left(Cv_2^*\right)$$

Let $\mu \in M_{c,+}^1(G)$. Choose $f_0 \in P^1(G) \cap \mathcal{K}(G)$ such that $\check{f}_0 = f_0$. For every $f \in \mathcal{K}_+(G)$ with $\|f\|_2 \leq 1$, we have $f_0 * f \in \mathcal{K}_+(G)$ and $\|f_0 * f\|_2 \leq \|f_0\|_1 \|f\|_2 \leq 1$. If also $f' \in \mathcal{K}_+(G)$, then by (4) we have

$$\langle f' * (f_0 * f)^{\vee}, \mu \rangle = \langle f' * \check{f} * \check{f}_0, \mu \rangle = \langle f' * \check{f}, \mu * f_0 \rangle.$$

We deduce from Proposition 9.2 that

$$\|L_\mu\|_{Cv^2} \geq \sup\left\{\langle f' * (f_0 * f)^{\vee}, \mu\rangle : f, f' \in \mathcal{K}_+(G), \|f\|_2 \leq 1, \|f'\|_2 \leq 1\right\}$$

$$= \sup\left\{\langle f' * \check{f}, \mu * f_0\rangle : f, f' \in \mathcal{K}_+(G), \|f\|_2 \leq 1, \|f'\|_2 \leq 1\right\}$$

$$= \|L_{\mu * f_0}\|_{Cv^2}.$$

By (Cv_2'), $\|L_{\mu * f_0}\|_{Cv^2} = \|\mu * f_0\|_1 = \|\mu\|$. Hence $\|L_\mu\|_{Cv^2} = \|\mu\|$.

3. Finally it suffices to show that (Cv_2) implies (ii') of Theorem 8.9.

Assume that there exists a positive-definite measure μ in $M^1(G)$ such that $\int d\mu \geq 0$ does not hold. As μ is hermitian, we must have $\int d\mu < 0$. Let $\nu = \mathcal{Re}\,\mu$ and $\nu = \nu^+ - \nu^-$ with $\nu^+, \nu^- \in M_+^1(G)$. As μ is positive-definite, for every $f \in \mathcal{K}_\mathbf{R}(G)$, $\langle f * \check{f}, \nu^+ \rangle \geq \langle f * \check{f}, \nu^- \rangle$. Hence by Proposition 9.2 we have $\|L_{\nu^+}\|_{Cv^2} \geq \|L_{\nu^-}\|_{Cv^2}$ and then by (Cv_2) also $\|\nu^+\| \geq \|\nu^-\|$. We contradict the assumption. □

C. Consequences

We state some more characterizations of amenability that are consequences of Theorem 9.6.

Corollary 9.7. *A locally compact group G is amenable if and only if, for one $p \in {]}1, \infty[$ (for every $p \in {]}1, \infty[$), the following property holds:*

$$\left(\forall \mu \in M^1(G)\right)\|L_\mu\|_{Cv^p} = 1$$

$$\left[resp. \left(\forall \varphi \in P^1(G)\right)\|L_\varphi\|_{Cv^p} = 1\right].$$

Proof. Property (Cv_p) [resp. (Cv'_p)] is implied by this special case as, for every nonzero $\mu \in \mathcal{M}^1_+(G)$ [resp. every nonzero $\mu = \varphi \in L^1_+(G)$], we have $\|L_{\mu/\|\mu\|}\|_{Cv^p} = 1$ and then

$$\|L_\mu\|_{Cv^p} = \|\mu\|\, \|L_{\mu/\|\mu\|}\|_{Cv^p} = \|\mu\|. \qquad \square$$

Proposition 9.8. *A locally compact group G is amenable if and only if, for every $\mu \in M^1_f(G)$ and one $p \in {]}1, \infty[$ (all $p \in {]}1, \infty[$), $\|L_\mu\|_{Cv^p} = 1$ holds.*

Proof. By Theorem 9.6 the condition is necessary. We now assume that it holds for some $p \in {]}1, \infty[$ and establish (G^*) which characterizes amenability by Proposition 8.5.

Let $F = \{a_1, \ldots, a_n\}$ and $\varepsilon \in \mathbf{R}^*_+$. We consider $\mu \in M^1_f(G)$ defined by

$$\mu(\{x\}) = \frac{1}{n} \quad \text{if } x \in F$$

$$\mu(\{x\}) = 0 \quad \text{if } x \in G \setminus F.$$

By hypothesis $\|L_\mu\|_{Cv^p} = 1$ and then by Lemma 9.5 also $\|L_\mu\|_{Cv^2} = 1$. Proposition 9.2 implies the existence of $\psi \in P^2(G) \cap \mathcal{K}(G)$ such that $0 < 1 - \langle \psi * \check\psi, \mu \rangle < \varepsilon$, that is,

$$0 < 1 - \frac{1}{n}\sum_{i=1}^{n} \psi * \check\psi(a_i) < \varepsilon.$$

For $\varphi = n^{-1/2}\psi \in L^2_+(G)$ we have $0 < 1 - \varphi * \check\varphi(a_i) < \varepsilon$ whenever $i = 1, \ldots, n$. $\qquad \square$

NOTES

The convolution problem was raised by Dieudonné [136], who proved $(P_p) \Rightarrow (Cv_p)$. The main characterization was demonstrated in the discrete case by Day [118]. Leptin proved the implications $(L) \Rightarrow (C_p)$ [349], $(Cv'_2) \Rightarrow (G)$ [353]. Gilbert [199] established $(G) \Leftrightarrow (Cv_p)$; see also that author's work for related properties and our proofs of 9.2, 9.6, and 9.8. Furthermore reference should be made to Eymard [172].

10. EXISTENCE OF APPROXIMATE UNITS

In Section 6 we characterized amenability by quasi-invariance properties; in this section we establish two more equivalent asymptotical invariance properties expressing the existence of approximate units.

A. Approximate Units in the Algebra $L^0(G)$

Theorem 10.1. *For a locally compact group G, the following properties characterize amenability:*
  (i) $L^0(G)$ *admits bounded approximate right [resp. left] units.*
  (ii) $L^0(G)$ *admits bounded multiple approximate right [resp. left] units.*
  (iii) $L^0(G)$ *admits multiple approximate right [resp. left] units.*

Proof. As $L^0(G)$ is invariant under involution, left-hand versions and right-hand versions are equivalent. We restrict ourselves to right-hand versions. Recall that, by Proposition 6.12 and Theorem 6.21, (P_1^*) and (GR_r) characterize amenability.

$$(GR_r) \Rightarrow \text{(i)}$$

Let $f \in L^0(G)$; by hypothesis, given $\varepsilon > 0$, there exist $\alpha_1, \ldots, \alpha_n \in \mathbf{R}_+^*$ and $x_1, \ldots, x_n \in G$ such that $\sum_{i=1}^n \alpha_i = 1$ and $\|\sum_{i=1}^n \alpha_i R_{x_i} f\|_1 < \varepsilon/3$. By Proposition 2.2 choose $\varphi \in P^1(G)$ such that $\|f * \varphi - f\|_1 < \varepsilon/3$. Let $\psi = \varphi - \sum_{i=1}^n \alpha_i R_{x_i} \varphi \in L^0(G)$; $\|\psi\|_1 \leq 2$. Then by (2.5) we obtain

$$\|f * \psi - f\|_1 \leq \|f * \varphi - f\|_1 + \left\| \sum_{i=1}^n \alpha_i (f * R_{x_i} \varphi) \right\|_1$$

$$= \|f * \varphi - f\|_1 + \left\| \sum_{i=1}^n \alpha_i R_{x_i}(f * \varphi) \right\|_1$$

$$\leq \|f * \varphi - f\|_1 + \left\| \sum_{i=1}^n \alpha_i R_{x_i}(f * \varphi - f) \right\|_1 + \left\| \sum_{i=1}^n \alpha_i R_{x_i} f \right\|_1$$

$$\leq 2\|f * \varphi - f\|_1 + \left\| \sum_{i=1}^n \alpha_i R_{x_i} f \right\|_1 < 2\frac{\varepsilon}{3} + \frac{\varepsilon}{3} = \varepsilon.$$

$$\text{(i)} \Rightarrow \text{(ii)}$$

The implication follows from Proposition 2.1.

$$\text{(ii)} \Rightarrow \text{(iii)} \quad \text{Trivial}$$

$$\text{(iii)} \Rightarrow (P_1^*)$$

EXISTENCE OF APPROXIMATE UNITS

Let F be a finite subset of G and $\varepsilon \in \mathbf{R}_+^*$. Consider a fixed element f in $L^1(G)$ such that $\int f = 1$; for every $a \in G$, $_a f - f \in L^0(G)$. By hypothesis, there exists $\psi \in L^0(G)$ such that $\|(_a f - f) * \psi - (_a f - f)\|_1 < \varepsilon$ whenever $a \in F$. Let $g = f - f * \psi$. As $f * \psi \in L^0(G)$, we have $\int g = \int f = 1$. For every $a \in F$,

$$\|_a g - g\|_1 = \|(_a f - f) + (_a f - f) * \psi\|_1 < \varepsilon.$$

We put $\varphi = |g|/\|g\|_1 \in P^1(G)$. Then for every $a \in F$,

$$\|_a \varphi - \varphi\|_1 = \int |_a \varphi(x) - \varphi(x)| \, dx = \frac{1}{\|g\|_1} \int \big| |_a g(x)| - |g(x)| \big| \, dx$$

$$\leq \frac{1}{\|g\|_1} \int |_a g(x) - g(x)| \, dx < \frac{\varepsilon}{\|g\|_1} \leq \frac{\varepsilon}{\left|\int g\right|} = \varepsilon. \qquad \Box$$

B. Approximate Units in the Algebras $A_p(G)$

Let G be a locally compact group and let $p \in]1, \infty[$. We summarize the fundamental properties of the Banach algebras $A_p(G)$ considered by Eymard [171] in case $p = 2$, by Herz [259] in the general case. A reference for tensor products is [HR] (Section 42).

If f and g are complex-valued functions defined on G, let

$$f \otimes g(x, y) = f(x)g(y)$$

for $x, y \in G$. If $1/p + 1/p' = 1$, we consider the Banach space $A_p(G)$ [resp. $L^p(G) \hat{\otimes} L^{p'}(G)$] generated by all complex-valued functions defined in a nonnecessarily unique way by $u = \sum_{n=1}^\infty f_n * \check{g}_n$ [resp. $u = \sum_{n=1}^\infty f_n \otimes g_n$] where $f_n \in \mathcal{K}(G)$, $g_n \in \mathcal{K}(G)$ $(n \in \mathbf{N}^*)$, $\sum_{n=1}^\infty \|f_n\|_p \|g_n\|_{p'} < \infty$ and $\|u\|_{A_p}$ [resp. $\|u\|_{\hat{\otimes}} = \|u\|_{L^p(G) \hat{\otimes} L^{p'}(G)}$] is the greatest lower bound of the set formed by the numbers $\sum_{n=1}^\infty \|f_n\|_p \|g_n\|_{p'}$ for all possible expressions of u. If $f \in L^p(G)$ and $g \in L^{p'}(G)$, $f * \check{g} \in \mathscr{C}_0(G)$ and $\|f * \check{g}\| \leq \|f\|_p \|g\|_{p'}$ (Section 2D.3); hence $A_p(G) \subset \mathscr{C}_0(G)$ and, for every $u \in A_p(G)$, $\|u\| \leq \|u\|_{A_p}$.

To every complex-valued function f defined on G we associate

$$\Gamma(f) : (x, y) \mapsto f(xy^{-1})$$

$$G \times G \to \mathbf{C}.$$

If $f \in L^p(G)$, $g \in L^{p'}(G)$, $x \in G$, we have $f * \check{g}(x) = \int f(xy) g(y) \, dy$. We then consider the linear mapping Γ_1 of $L^p(G) \hat{\otimes} L^{p'}(G)$ onto $A_p(G)$ defined by

$$\Gamma_1(\varphi)(x) = \int_G \varphi(xy, y) \, dy$$

for $\varphi \in L^p(G) \hat{\otimes} L^{p'}(G)$, $x \in G$. In particular, if $f \in L^p(G)$, $g \in L^{p'}(G)$, then $\Gamma_1(f \otimes g) = f * \check{g}$. For $\psi \in A_p(G)$,

$$\|\psi\|_{A_p} = \inf\{\|\varphi\|_{\hat{\otimes}} : \varphi \in L^p(G) \hat{\otimes} L^{p'}(G), \Gamma_1(\varphi) = \psi\};$$

$\|\Gamma_1\| = 1$ and $A_p(G) \simeq (L^p(G) \hat{\otimes} L^{p'}(G))/\mathcal{K}er\,\Gamma_1$. Let $u \in \mathscr{C}(G)$, $\varphi \in L^p(G) \hat{\otimes} L^{p'}(G)$, $x \in G$; we have

$$\Gamma_1(\Gamma(u)\varphi)(x) = \int_G (\Gamma(u)\varphi)(xy, y)\,dy$$

$$= u(x)\int_G \varphi(xy, y)\,dy = u(x)\Gamma_1(\varphi)(x),$$

that is,

$$\Gamma_1(\Gamma(u)\varphi) = u\Gamma_1(\varphi). \qquad (1)$$

If $f, g \in \mathscr{K}(G)$, $x, y \in G$, then

$$\Gamma(f * \check{g})(x, y) = f * \check{g}(xy^{-1}) = \int f(z)\check{g}(z^{-1}xy^{-1})\,dz$$

$$= \int f(xz)\check{g}(z^{-1}y^{-1})\,dz,$$

that is,

$$\Gamma(f * \check{g})(x, y) = \int f_z(x)g_z(y)\,dz. \qquad (2)$$

Let $p \in]1, \infty[$, $1/p + 1/p' = 1$ and $f, g, h, k \in \mathscr{K}(G)$. For all $x, y \in G$, by (2),

$$\Gamma(f * \check{g})h \otimes k(x, y) = \int (f_z h)(x)(g_z k)(y)\,dz.$$

Hölder's inequality implies that

$$\int \|(f_z h) \otimes (g_z k)\|_{\hat{\otimes}}\,dz \le \int \|f_z h\|_p \|g_z k\|_{p'}\,dz$$

$$\le \left(\int \|f_z h\|_p^p\,dz\right)^{1/p} \left(\int \|g_z k\|_{p'}^{p'}\,dz\right)^{1/p'}$$

$$= \left(\iint |f(xz)h(x)|^p\,dx\,dz\right)^{1/p}$$

$$\times \left(\iint |g(yz)k(y)|^{p'}\,dy\,dz\right)^{1/p'}$$

$$= \|f\|_p \|h\|_p \|g\|_{p'} \|k\|_{p'};$$

$\Gamma(f * \check{g})h \otimes k = \int (f_z h)(g_z k) \, dz \in L^p(G) \hat{\otimes} L^{p'}(G)$ ([B INT] Chap. VI, Section 2, No. 2) and

$$\|\Gamma(f * \check{g})h \otimes k\|_{\hat{\otimes}} \leq \|f\|_p \|g\|_{p'} \|h\|_p \|k\|_{p'}. \tag{3}$$

Let $\varphi \in L^p(G) \hat{\otimes} L^{p'}(G)$. If $\varepsilon > 0$, there exist sequences $(h_n), (k_n)$ formed by elements of $\mathcal{K}(G)$ such that $\varphi = \sum_{n=1}^{\infty} h_n \otimes k_n$ and

$$\sum_{n=1}^{\infty} \|h_n\|_p \|k_n\|_{p'} < \|\varphi\|_{\hat{\otimes}} + \varepsilon.$$

So by (3) the series $\psi = \sum_{n=1}^{\infty} \Gamma(f * \check{g}) h_n \otimes k_n$ converges absolutely in $L^p(G) \hat{\otimes} L^{p'}(G)$; we have $\psi = \Gamma(f * \check{g})\varphi$ and

$$\|\psi\|_{\hat{\otimes}} \leq \|f\|_p \|g\|_{p'} (\|\varphi\|_{\hat{\otimes}} + \varepsilon).$$

Therefore $\Gamma(f * \check{g})\varphi \in L^p(G) \hat{\otimes} L^{p'}(G)$ and

$$\|\Gamma(f * \check{g})\varphi\|_{\hat{\otimes}} \leq \|f\|_p \|g\|_{p'} \|\varphi\|_{\hat{\otimes}}. \tag{4}$$

Let A be any Banach space of complex-valued functions defined on a set X; we call a complex-valued function u defined on X a *multiplier function* of A if (a) $u\varphi \in A$ whenever $\varphi \in A$ and (b) the mapping $a \mapsto ua$ belongs to $\mathscr{L}(A)$. We define $B_p(G)$ [resp. $C_p(G)$] to be the Banach algebra, for pointwise addition and multiplication, formed by all elements u of $\mathscr{C}(G)$ that are multiplier functions of $A_p(G)$ [resp. that admit $\Gamma(u)$ as multiplier functions of $L^p(G) \hat{\otimes} L^{p'}(G)$], with $\|u\|_{B_p}$ [resp. $\|u\|_{C_p}$] being the operator norm of u [resp. $\Gamma(u)$]. As Γ_1 is surjective and $\|\Gamma_1\| = 1$, we deduce from (1) that $C_p(G) \subset B_p(G)$, $\|\cdot\|_{B_p} \leq \|\cdot\|_{C_p}$. We show briefly that $A_p(G)$ may be embedded continuously into $C_p(G)$ via Γ.

Proposition 10.2. *If G is a locally compact group and $p \in]1, \infty[$, the mapping Γ induces a continuous linear mapping of $A_p(G)$ into $C_p(G)$; $A_p(G) \subset C_p(G)$ and $\|\cdot\|_{C_p} \leq \|\cdot\|_{A_p}$.*

Proof. Let $1/p + 1/p' = 1$. If $u \in A_p(G)$ and $\varepsilon \in \mathbf{R}_+^*$, there exist sequences (f_n) and (g_n) formed by elements of $\mathcal{K}(G)$ such that $u = \sum_{n=1}^{\infty} f_n * \check{g}_n$ and $\sum_{n=1}^{\infty} \|f_n\|_p \|g_n\|_{p'} < \|u\|_{A_p} + \varepsilon$. So by (4), for every $\varphi \in L^p(G) \hat{\otimes} L^{p'}(G)$, the series $\sum_{n=1}^{\infty} \Gamma(f_n * \check{g}_n)\varphi$ is absolutely convergent to $\Gamma(u)\varphi$ in $L^p(G) \hat{\otimes} L^{p'}(G)$; we have

$$\|\Gamma(u)\varphi\|_{\hat{\otimes}} = \left\| \sum_{n=1}^{\infty} \Gamma(f_n * \check{g}_n)\varphi \right\|_{\hat{\otimes}} \leq \sum_{n=1}^{\infty} \|f_n\|_p \|g_n\|_{p'} \|\varphi\|_{\hat{\otimes}}$$

$$\leq (\|u\|_{A_p} + \varepsilon) \|\varphi\|_{\hat{\otimes}}.$$

Hence $\|\Gamma(u)\varphi\|_{\hat{\otimes}} \leq \|u\|_{A_p} \|\varphi\|_{\hat{\otimes}}$, that is, $u \in C_p(G)$ and $\|u\|_{C_p} \leq \|u\|_{A_p}$. $\square$

We may now prove that $A_p(G)$ is a Banach algebra. Let $u, v \in A_p(G)$ and $\varepsilon \in \mathbf{R}_+^*$. There exists $\varphi \in L^p(G) \hat{\otimes} L^{p'}(G)$ such that $v = \Gamma_1(\varphi)$ and $\|\varphi\|_{\hat{\otimes}} \leq \|v\|_{A_p} + \varepsilon$. Then by (1) we have $uv = u\Gamma_1(\varphi) = \Gamma_1(\Gamma(u)\varphi) \in A_p(G)$. Via the canonical embedding of $A_p(G)$ into $C_p(G)$ we obtain

$$\|uv\|_{A_p} \leq \|\Gamma(u)\varphi\|_{\hat{\otimes}} \leq \|u\|_{C_p} \|\varphi\|_{\hat{\otimes}} \leq \|u\|_{A_p} \|\varphi\|_{\hat{\otimes}} \leq \|u\|_{A_p}(\|v\|_{A_p} + \varepsilon).$$

Thus $\|uv\|_{A_p} \leq \|u\|_{A_p} \|v\|_{A_p}$ as $\varepsilon \in \mathbf{R}_+^*$ is arbitrary. The Banach algebra $A(G) = A_2(G)$ is the *Fourier algebra* of G. It consists of the functions $f * \check{g}$, where $f, g \in L^2(G)$; it is generated by the functions $f * g$, where $f, g \in \mathcal{K}(G)$, as well as the functions $f * \check{f}$, where $f \in \mathcal{K}(G)$ or $f \in L^2(G)$, and also by $P(G) \cap \mathcal{K}(G)$ ([171]).

The dual space of $L^p(G) \hat{\otimes} L^{p'}(G)$ ($1 < p < \infty, 1/p + 1/p' = 1$) may be identified with $\mathcal{L}(L^p(G))$ via the formula

$$\langle \varphi, T \rangle = \sum_{n=1}^{\infty} \langle Tf_n, g_n \rangle$$

where $T \in \mathcal{L}(L^p(G))$, $\varphi = \sum_{n=1}^{\infty} f_n \otimes g_n$, $f_n \in L^p(G)$, $g_n \in L^{p'}(G)$, $n \in \mathbf{N}^*$ and $\sum_{n=1}^{\infty} \|f_n\|_p \|g_n\|_{p'} < \infty$. The w*-topology on $(L^p(G) \hat{\otimes} L^{p'}(G))^*$ is identified with the ultraweak topology on $\mathcal{L}(L^p(G))$, that is, the weakest topology insuring the continuity of the linear functionals

$$T \mapsto \sum_{n=1}^{\infty} \langle Tf_n, g_n \rangle$$

on $\mathcal{L}(L^p(G))$ where $f_n \in L^p(G)$, $g_n \in L^{p'}(G)$, $n \in \mathbf{N}^*$ and $\sum_{n=1}^{\infty} \|f_n\|_p \|g_n\|_{p'} < \infty$.

If G is a locally compact group and $p \in]1, \infty[$, we consider the Banach space $PM^p(G)$ of *p-pseudomeasures* on G, that is, the smallest ultraweakly closed subspace of $Cv^p(G)$ containing $\{L_f : f \in L^1(G)\}$. We show that $A_p(G)^*$ may be identified with $PM^{p'}(G)$ where $1/p + 1/p' = 1$. We term 2-pseudomeasures simply pseudomeasures and write $PM(G)$ for $PM^2(G)$.

Proposition 10.3. *Let G be a locally compact group and let $p, p' \in]1, \infty[$ with $1/p + 1/p' = 1$. For any $F \in A_p(G)^*$ there exists a unique $F' \in PM^{p'}(G)$ such that, for all $f \in L^p(G)$, $g \in L^{p'}(G)$,*

$$\langle F'(g), f \rangle = \langle f * \check{g}, F \rangle. \tag{5}$$

The mapping

$$\theta : F \mapsto F'$$

$$A_p(G)^* \to PM^{p'}(G)$$

is a surjective isometry; it carries the w*-topology of $A_p(G)^*$ over to the ultraweak topology of $PM^{p'}(G)$. If $u = \sum_{n=1}^{\infty} f_n * \check{g}_n$ with $f_n \in L^p(G)$, $g_n \in L^{p'}(G)$ ($n \in \mathbf{N}^*$), $\sum_{n=1}^{\infty} \|f_n\|_p \|g_n\|_{p'} < \infty$, then

$$F(u) = \sum_{n=1}^{\infty} \langle F'(g_n), f_n \rangle.$$

If $\mu \in \mathcal{M}^1(G)$ and F_μ is the element of $A_p(G)^*$ defined by $F_\mu(v) = \langle v, \mu \rangle$ for $v \in A_p(G)$, then $L_\mu = (F_\mu)'$.

Proof. (a) If $F \in A_p(G)^*$ and $f \in L^p(G)$, $g \in L^{p'}(G)$, we have

$$|\langle f * \check{g}, F \rangle| \le \|F\|_{A_p(G)^*} \|f * \check{g}\|_{A_p} \le \|F\|_{A_p(G)^*} \|f\|_p \|g\|_{p'}.$$

For $g \in L^{p'}(G)$, $\theta_g : f \to \langle f * \check{g}, F \rangle$ is an element of $L^p(G) = L^p(G)^*$ with $\|\theta_g\| \le \|F\|_{A_p(G)^*} \|g\|_{p'}$ and therefore there exists $F' : g \mapsto \theta_g$ satisfying (5) and

$$\|F'\| \le \|F\|_{A_p(G)^*}. \tag{6}$$

As the series $u = \sum_{n=1}^{\infty} f_n * \check{g}_n$ is absolutely convergent, we have

$$F(u) = \sum_{n=1}^{\infty} F(f_n * \check{g}_n) = \sum_{n=1}^{\infty} \langle F'(g_n), f_n \rangle \tag{7}$$

and $|F(u)| \le \|F'\| \sum_{n=1}^{\infty} \|g_n\|_{p'} \|f_n\|_p$, $|F(u)| \le \|F'\| \|u\|_{A_p}$, hence

$$\|F\|_{A_p(G)^*} \le \|F'\|. \tag{8}$$

Now (6) and (8) imply that θ is an isometry and (7) proves that θ carries the w*-topology of $A_p(G)^*$ over to the ultraweak topology of $\mathcal{L}(L^p(G))$.

(b) Let $\mu \in \mathcal{M}^1(G)$. For all $f \in L^p(G)$, $g \in L^{p'}(G)$, we have

$$\langle (F_\mu)'(g), f \rangle = F_\mu(f * \check{g}) = \iint f(y) g(x^{-1}y) \, dy \, d\mu(x)$$

$$= \int f(y) \left(\int g(x^{-1}y) \, d\mu(x) \right) dy = \langle L_\mu(g), f \rangle,$$

that is, $L_\mu = (F_\mu)'$.

(c) It remains to show that θ maps $A_p(G)^*$ onto $PM^{p'}(G)$. For any $\varphi \in L^1(G)$, $(F_\varphi)' = L_\varphi$; so we have $\{ L_\varphi : \varphi \in L^1(G) \} \subset \theta(A_p(G)^*)$. Let f, $g \in \mathscr{C}_0(G)$; if $\langle f - g, \varphi \rangle = 0$ whenever $\varphi \in L^1(G)$, we have $\|f - g\|_\infty = 0$, hence $f = g \in \mathscr{C}_0(G)$, that is, $L^1(G)$ separates $\mathscr{C}_0(G)$. Therefore also $\{ F_\varphi : \varphi \in L^1(G) \}$ separates $A_p(G)$; it is then dense in $A_p(G)^*$. Hence $\{ L_\varphi : \varphi \in L^1(G) \}$ is ultraweakly dense in $\theta(A_p(G)^*)$ and $\theta(A_p(G)^*) \subset PM^{p'}(G)$.

We now consider $T \in PM^p(G)$ and a net (φ_i) in $L^1(G)$ such that (L_{φ_i}) converges ultraweakly to T. Let $u = \sum_{n=1}^{\infty} f_n * \check{g}_n \in A_p(G)$; for every $\varphi \in L^1(G)$,

$$F^L\varphi(u) = \sum_{n=1}^{\infty} \langle L_\varphi(g_n), f_n \rangle$$

is independent of the expression of u. Therefore we may define $F^T \in A_p(G)^*$ by putting

$$F^T(u) = \sum_{n=1}^{\infty} \langle T(g_n), f_n \rangle.$$

Obviously $(F^T)' = T$. We have $PM^p(G) \subset \theta(A_p(G)^*)$. □

Theorem 10.4. *A locally compact group G is amenable if and only if any one of the following conditions holds*:

(i) *For every $p \in]1, \infty[$ (for one $p \in]1, \infty[$), $A_p(G)$ admits approximate units belonging to $\mathcal{K}(G)$ that are bounded (by 1).*

(ii) *For every $p \in]1, \infty[$ (for one $p \in]1, \infty[$), $A_p(G)$ admits approximate units that are bounded (by 1).*

(iii) *For every $p \in]1, \infty[$ (for one $p \in]1, \infty[$), $A_p(G)$ admits multiple approximate units that are bounded (by 1).*

Proof. (a) Assume G to be amenable and let $p \in]1, \infty[$. If K is any compact subset of G and $\varepsilon \in \mathbf{R}^*_+$, by Proposition 7.11 there exists a compact subset $U = U_{K,\varepsilon}$ in G such that $0 < |U|$ and

$$|KU| < (1+\varepsilon)^p |U|. \tag{9}$$

We put

$$u_{K,\varepsilon} = \frac{1}{(1+\varepsilon)|U|} 1_{KU} * \check{1}_U \in A_p(G) \subset \mathscr{C}_0(G).$$

We have $\operatorname{supp} u_{K,\varepsilon} \subset KUU^{-1}$, hence $u_{K,\varepsilon} \in \mathcal{K}(G)$. Also, if $1/p + 1/p' = 1$, (9) implies that

$$\|u_{K,\varepsilon}\|_{A_p} \leq \frac{1}{(1+\varepsilon)|U|} |KU|^{1/p} |U|^{1/p'} < 1.$$

We consider the net $(u_{K,\varepsilon})$ in $A_p(G)$ where $(K, \varepsilon) \prec (K', \varepsilon')$ whenever $K, K' \in \Re(G)$, $\varepsilon, \varepsilon' \in \mathbf{R}^*_+$ with $K \subset K'$ and $\varepsilon' < \varepsilon$.

Let $f \in A_p(G) \cap \mathcal{K}(G)$ with $K = \operatorname{supp} f$ and let $\varepsilon \in \mathbf{R}^*_+$. Then

$$(u_{K,\varepsilon} f)(x) = \frac{f(x)}{(1+\varepsilon)|U|} \int 1_{KU}(xy) 1_U(y)\, dy = \begin{cases} 0 = \dfrac{f(x)}{1+\varepsilon} & \text{if } x \notin K \\ \dfrac{f(x)}{1+\varepsilon} & \text{if } x \in K. \end{cases}$$

Therefore

$$\|u_{K,\varepsilon} f - f\|_{A_p} = \left|\frac{1}{1+\varepsilon} - 1\right| \|f\|_{A_p} = \frac{\varepsilon}{1+\varepsilon} \|f\|_{A_p} \le \varepsilon \|f\|_{A_p}.$$

As the vector space generated by linear combinations of functions $g * \check{h}$, where $g, h \in \mathcal{K}(G)$, is dense in $A_p(G)$, we have proved that $(u_{K,\varepsilon})$ is a net of approximate units in $A_p(G)$.

(b) Assume that, for some $p \in \,]1, \infty[$, there exists a net (u_i) of approximate units bounded by c in $A_p(G)$. We establish $(Cv_p'^*)$ $(1/p + 1/p' = 1)$ which implies amenability by Theorem 9.6.

If K is any compact subset of G, we choose an arbitrary compact neighborhood V of e in G and put

$$f = \xi_V * 1_{V^{-1}K} \in A_{p'}(G);$$

if $x \in K$,

$$f(x) = \frac{1}{|V|} \int 1_V(y) 1_{V^{-1}K}(y^{-1}x)\, dy = 1.$$

As $\lim_i \|u_i f - f\|_{A_p} = 0$, a fortiori $\lim_i \|u_i f - f\| = 0$. Therefore if $\varepsilon > 0$, there exists $i_0 \in I$ such that $\inf \{\mathcal{R}e\, u_{i_0}(x) : x \in K\} \ge 1 - \varepsilon$.

Let $\varphi \in \mathcal{K}_+(G)$ with $K = \operatorname{supp} \varphi$. Proposition 10.3 implies that

$$|\langle u_{i_0}, \varphi \rangle| \le \|L_\varphi\|_{Cv^{p'}} \|u_{i_0}\|_{A_p} \le c \|L_\varphi\|_{Cv^{p'}}.$$

On the other hand,

$$\mathcal{R}e\, \langle u_{i_0}, \varphi \rangle = \int \mathcal{R}e\, u_{i_0}(x) \varphi(x)\, dx \ge (1 - \varepsilon) \|\varphi\|_1.$$

As $\varepsilon > 0$ may be chosen arbitrarily, $\|\varphi\|_1 \le c \|L_\varphi\|_{Cv^{p'}}$.

Let $\psi \in \mathcal{K}_+(G)$. For every $n \in \mathbf{N}^*$, we have now

$$\|\psi\|_1^n = \|\psi^{*n}\|_1 \le c \|L_{\psi^{*n}}\|_{Cv^{p'}} \le c \|L_\psi\|_{Cv^{p'}}^n.$$

Hence $\|\psi\|_1 \le \|L_\psi\|_{Cv^{p'}}$ and $\|\psi\|_1 = \|L_\psi\|_{Cv^{p'}}$.

(c) The equivalence (ii) $\Leftrightarrow$ (iii) follows from Proposition 2.1. $\square$

NOTES

For the proof of 10.1 we refer to Reiter [454]; also see Johnson [291] and Rindler [468, 470].

The Fourier algebra $A_2(G)$ was introduced by Eymard [171]. The general definition of the algebra $A_p(G)$ was given by Herz [259]; he had studied the case $p = 2$ in [255].

The proof of 10.4 in case $p = 2$ is due to Leptin [354], in the general case essentially to Herz [259], but Herz claims it to be folklore. See also Eymard [172]. For an alternative proof in case $p = 2$ based on (G) we refer to Derighetti [125].

11. COHOMOLOGICAL PROPERTIES

In the present section we state the characterization of amenability in the language of cohomology as given by Johnson [291]. The property is shown to be equivalent to the existence of a left invariant mean.

A. General Considerations

Let A be a complex Banach algebra and let X be a Banach A-module. We define a continuous linear mapping $d_1: X \to \mathscr{L}(A, X)$ by putting

$$d_1 \xi(a) = a\xi - \xi a$$

for $\xi \in X$ and $a \in A$. We also consider a continuous linear mapping d_2 of $\mathscr{L}(A, X)$ into the space $\mathscr{L}^2(A, X)$ of continuous bilinear mappings carrying A^2 into X; we put

$$d_2 F(a, b) = aF(b) - F(ab) + F(a)b$$

for $F \in \mathscr{L}(A, X)$ and $(a, b) \in A^2$. For all $(a, b) \in A^2$ and $\xi \in X$, we have $d_2 d_1 \xi(a, b) = 0$, that is, $\mathscr{I}m\, d_1 \subset \mathscr{K}er\, d_2$. The vector space $H_1(A, X) = \mathscr{K}er\, d_2 / \mathscr{I}m\, d_1$ is the *first cohomology group* of A over X.

A *derivation* from A into X is a linear mapping $D: A \to X$ such that

$$D(ab) = aD(b) + D(a)b$$

whenever $a, b \in A$; so $\mathscr{K}er\, d_2$ consists of all continuous derivations from A into X. For every $\xi \in X$, the mapping

$$d_1 \xi: a \mapsto a\xi - \xi a$$

$$A \to X$$

COHOMOLOGICAL PROPERTIES   **99**

is called *inner derivation* form A into X. Hence the condition $H_1(A, X) = \{0\}$ expresses that every continuous derivation from A into X is inner.

If G is a locally compact group and X is a Banach G-module, we consider similarly the space $Z(G, X^*)$ of all continuous *derivations* of G into X^*, that is, all norm-bounded mappings D of G into X^* that are continuous with respect to the w*-topology of X^* and satisfy

$$D(xy) = xD(y) + D(x)y$$

whenever $x, y \in G$. We denote by $N(G, X^*)$ the subspace of all *inner derivations*, that is, all mappings $d_1\xi$ ($\xi \in X^*$) defined by

$$d_1\xi(x) = x\xi - \xi x,$$

$x \in G$. Let $H_1(G, X^*) = Z(G, X^*)/N(G, X^*)$.

If A is a Banach algebra with unit e [resp. G is a locally compact group] and X is a Banach A-module [resp. a Banach G-module], we say that A [resp. G] acts as a *left (right) zero* on X if $a\xi = 0$ ($\xi a = 0$) whenever $\xi \in X$ and $a \in A$ [resp. $a \in G$].

Lemma 11.1. *If A is a Banach algebra with unit e [resp. G is a locally compact group] and A [resp. G] acts as a left (or right) zero on a Banach A-module X [resp. Banach G-module X], then $H_1(A, X) = \{0\}$ [resp. $H_1(G, X^*) = \{0\}$].*

Proof (for the left case). Let D be a continuous derivation. Let $\eta_0 = -D(e) \in X$ [resp. $\eta_0 = D(e) \in X^*$]. For every $a \in A$ [resp. $a \in G$], we have

$$D(a) = D(ea) = D(e)a = a\eta_0 - \eta_0 a$$

$$\bigl[\text{resp. } D(a) = D(ae) = aD(e) = a\eta_0 - \eta_0 a\bigr],$$

that is, D is inner. □

Lemma 11.2. *If A is a Banach algebra with unit e [resp. G is a locally compact group] and X is a Banach A-module [resp. a Banach G-module], then $H_1(A, X)$ [resp. $H_1(G, X^*)$] is isomorphic to $H_1(A, eXe)$ [resp. $H_1(G, eX^*e)$].*

Proof. Let $Y = X$ [resp. X^*] and consider $l: \xi \mapsto e\xi, r: \xi \mapsto \xi e$. One readily checks that Y is the direct sum of the closed A-submodules [resp. G-submodules] $eYe = l \circ r(Y)$, $Y_1 = (\text{id}_Y - r) \circ l(Y)$, $Y_2 = (\text{id}_Y - l) \circ r(Y)$, $Y_3 = (\text{id}_Y - l) \circ (\text{id}_Y - r)(Y)$, which are images of Y by pairwise commuting projections; $H_1(A, X)$ [resp. $H_1(G, X^*)$] is the direct sum of the cohomology groups over eYe, Y_1, Y_2, Y_3. The algebra A [resp. the group G] acts as a left zero on Y_2, Y_3, and as a right zero on Y_1. Therefore Lemma 11.1 implies the statement. □

Let A be a Banach algebra with unit e [resp. G be a locally compact group] and X a Banach A-module [resp. a Banach G-module]; X is called *unital* if $e\xi e = \xi$ whenever $\xi \in X$. Notice that X is unital if and only if $e\xi = \xi = \xi e$ whenever $\xi \in X$. For every Banach A-module X [resp. G-module X], eXe is unital. If D is a derivation into a unital Banach module, $D(e) = D(e^2) = eD(e) + D(e)e = D(e) + D(e)$, hence $D(e) = 0$.

Lemma 11.3. *Let A be a Banach algebra with unit [resp. G be a locally compact group]. If $H_1(A, Y) = \{0\}$ for every unital Banach A-module Y [resp. $H_1(G, Y^*) = \{0\}$ for every unital Banach G-module Y], then $H_1(A, X) = \{0\}$ for every Banach A-module X [resp. $H_1(G, X^*) = \{0\}$ for every Banach G-module X].*

Proof. The statement follows readily from Lemma 11.2. □

Recall that any Banach algebra A may be embedded canonically into a Banach algebra $\tilde{A}$ with unit u (Section 2B.2). If X is a Banach A-module, it may be equipped with a unital Banach $\tilde{A}$-module structure by putting $u\xi = \xi = \xi u$ for every $\xi \in X$. The induced Banach $\tilde{A}$-module structure on X^* is the same as the $\tilde{A}$-module structure associated to the A-module structure of X^*.

Lemma 11.4. *If A is a Banach algebra and X is a Banach A-module, then $H_1(A, X) \simeq H_1(\tilde{A}, X)$.*

Proof. If u is the unit of $\tilde{A}$, and D is a continuous derivation into the $\tilde{A}$-module X, then $D(u) = 0$. Therefore the canonical embedding of A into $\tilde{A}$ induces an isomorphism of $H_1(\tilde{A}, X)$ onto $H_1(A, X)$. □

Let A be a Banach algebra and X a Banach A-module. Then X is said to be an *essential* Banach A-module if X equals $AX = \{a\xi : a \in A, \xi \in X\}$ and $XA = \{\xi a : \xi \in X, a \in A\}$. Obviously any unital Banach A-module is essential.

Lemma 11.5. *If A is a Banach algebra with bounded two-sided approximate units $(u_i)_{i \in I}$ and X is a Banach A-module, then the following properties are equivalent:*

(i) *X is an essential Banach A-module.*

(ii) *$(\forall \xi \in X) \lim_i \|u_i \xi - \xi\| = 0 = \lim_i \|\xi u_i - \xi\|$.*

Proof. (i) $\Rightarrow$ (ii)

Given $\xi \in X$, by hypothesis there exist $a_1, a_2 \in A$ and $\xi_1, \xi_2 \in X$ such that $a_1\xi_1 = \xi = \xi_2 a_2$. For every $i \in I$, $\|u_i \xi - \xi\| \leq \|u_i a_1 - a_1\| \|\xi_1\|$ and $\|\xi u_i - \xi\| \leq \|a_2 u_i - a_2\| \|\xi_2\|$. As (u_i) constitutes a net of approximate units, we obtain (ii).

(ii) $\Rightarrow$ (i)

By hypothesis $\overline{XA} = X = \overline{AX}$. The statement follows from Cohen's factorization theorem ([HR] 32.22). □

Note that in the proof of the implication (i) ⇒ (ii), boundedness of the approximate units is not needed.

We now show that if G is a locally compact group, there exists a one-to-one correspondence between unital Banach G-modules and essential Banach $L^1(G)$-modules.

(a) Let X be a unital Banach G-module. We may define for X a Banach $\mathcal{M}^1(G)$-module structure via a vector-valued integral. We put

$$\mu\xi = \int_G x\xi \, d\mu(x), \quad \xi\mu = \int_G \xi x \, d\mu(x) \tag{1}$$

for $\mu \in \mathcal{M}^1(G)$, $\xi \in X$. It suffices to check the associativity of the law. By Fubini's theorem, if $\mu, \nu \in \mathcal{M}^1(G)$ and $\xi \in X$,

$$\mu(\nu\xi) = \int_G x \left(\int_G y\xi \, d\nu(y) \right) d\mu(x) = \int_G \int_G (xy)\xi \, d\mu(x) \, d\nu(y)$$

$$= \int_G z\xi \, d\mu * \nu(z) = \mu * \nu(\xi).$$

By restriction we may consider X to be a Banach $L^1(G)$-module. Notice also that if $x \in G$ and $\xi \in X$, then $\delta_x \xi = x\xi$, $\xi\delta_x = \xi x$.

If U runs over a basis of open relatively compact neighborhoods of e in G, we put

$$\varphi_U \xi = \frac{1}{|U|} \int_U x\xi \, dx = \frac{1}{|U|} \int_G x\xi 1_U(x) \, dx,$$

$\xi \in X$. By the continuity of the mapping $x \mapsto x\xi$ for any $\xi \in X$, we obtain $\lim_U \varphi_U \xi = e\xi = \xi$. Hence $\overline{L^1(G)X} = X$ and similarly $\overline{XL^1(G)} = X$. From Proposition 2.2 and Lemma 11.5 we conclude that X may be considered to be an essential Banach $L^1(G)$-module.

(b) Let X be an essential Banach $L^1(G)$-module and let (φ_i) be a net of approximate units in $L^1(G)$ obtained by Proposition 2.2. If $\xi \in X$, there exist $f_1, f_2 \in L^1(G)$ and $\xi_1, \xi_2 \in X$ such that $f_1 \xi_1 = \xi = \xi_2 f_2$. For any $\mu \in \mathcal{M}^1(G)$, we have

$$\lim_i (\mu * \varphi_i)\xi = \lim_i (\mu * \varphi_i) f_1 \xi_1$$

$$= \lim_i (\mu * \varphi_i * f_1)\xi_1 = (\mu * f_1)\xi_1$$

and

$$\lim_i \xi(\varphi_i * \mu) = \lim_i \xi_2 f_2(\varphi_i * \mu)$$
$$= \lim_i \xi_2(f_2 * \varphi_i * \mu) = \xi_2(f_2 * \mu).$$

Hence we may define

$$\mu\xi = \lim_i (\mu * \varphi_i)\xi, \qquad \xi\mu = \lim_i \xi(\varphi_i * \mu)$$

extending the Banach $L^1(G)$-module structure on X to a Banach $\mathscr{M}^1(G)$-module structure. By restriction to Dirac measures we obtain a unital Banach G-module structure on X. As a matter of fact, for $\xi = f_1\xi_1 \in X$ with $f_1 \in L^1(G)$, $\xi_1 \in X$, we have

$$e\xi = \delta_e\xi = \lim_i \varphi_i\xi = \lim_i (\varphi_i * f_1)\xi_1 = f_1\xi_1 = \xi$$

and similarly $\xi e = \xi$. We check the continuity of the mapping

$$(x, y) \mapsto x\xi y$$

$$G \times G \to X$$

for any $\xi \in X$. If $\lim_\alpha x_\alpha = x$ and $\lim_\beta y_\beta = y$ in G, then for every $g \in L^1(G)$, $\lim_\alpha \|\delta_{x_\alpha} * g - \delta_x * g\|_1 = \lim_\alpha \|L_{x_\alpha}g - L_x g\|_1 = 0$ and $\lim_\beta \|g * \delta_{y_\beta} - g * \delta_y\|_1 = \lim_\beta \|R_{y_\beta}g - R_y g\|_1 = 0$ (Section 2D.3). Given $\xi \in X$, there exist $\eta \in X$ and $h_1, h_2 \in L^1(G)$ such that $\xi = h_1\eta h_2$; hence

$$\lim_\alpha \lim_\beta x_\alpha \xi y_\beta = \lim_\alpha \lim_\beta \delta_{x_\alpha} \xi \delta_{y_\beta} = \lim_\alpha \lim_\beta (\delta_{x_\alpha} * h_1)\eta(h_2 * \delta_{y_\beta})$$
$$= (\delta_x * h_1)\eta(h_2 * \delta_y) = \delta_x \xi \delta_y = x\xi y.$$

B. Cohomological Characterizations of Amenability

Lemma 11.6. *Let G be an amenable group and let X be a Banach G-module such that $x\xi = \xi$ for all $x \in G$, $\xi \in X$. If D is any continuous derivation of G into X^*, there exists $\Phi \in X^*$ such that $D = -d_1\Phi$. For any $\xi \in X$,*

$$\inf\{\mathscr{R}e\langle\xi, D(x)\rangle : x \in G\} \leq \mathscr{R}e\langle\xi, \Phi\rangle.$$

COHOMOLOGICAL PROPERTIES

Proof. If $\xi \in X$ and F is any norm-bounded continuous mapping of G into X^*, the mapping

$$F'_\xi : x \mapsto \langle \xi, F(x) \rangle$$

belongs to $\mathscr{C}(G)$. For all $\xi \in X$, $a \in G$, and $x \in G$, we have $\langle \xi a, D(x) \rangle = \langle \xi, aD(x) \rangle$, that is, $D'_{\xi a} = (aD)'_\xi$; moreover

$$(aD)'_\xi(x) = \langle \xi, aD(x) \rangle = \langle \xi, D(ax) \rangle - \langle \xi, D(a)x \rangle$$
$$= \langle \xi, D(ax) \rangle - \langle \xi, D(a) \rangle =_a(D'_\xi)(x) - \langle \xi, D(a) \rangle.$$

If $M \in \text{LIM}(\mathscr{C}(G))$, we then have

$$M\big((aD)'_\xi\big) = M\big(D'_\xi\big) - \langle \xi, D(a) \rangle.$$

The mapping $\Phi : \xi \mapsto M(D'_\xi)$ belongs to X^*. For all $\xi \in X$, $a \in G$,

$$\langle \xi, a\Phi \rangle = \langle \xi a, \Phi \rangle = M\big(D'_{\xi a}\big) = M\big((aD)'_\xi\big) = \langle \xi, \Phi \rangle - \langle \xi, D(a) \rangle,$$

that is, $D(a) = \Phi - a\Phi = \Phi a - a\Phi$; $D = -d_1\Phi$.

Finally for $\xi \in X$,

$$\inf \{ \mathscr{Re} \langle \xi, D(x) \rangle : x \in G \} \leq M\big(\mathscr{Re} D'_\xi\big) = \mathscr{Re} M\big(D'_\xi\big) = \mathscr{Re} \langle \xi, \Phi \rangle. \quad \square$$

Lemma 11.7. *Let G be a locally compact group and let X be a unital Banach G-module. A new unital Banach G-module structure may be defined on X by $\xi \cdot x = x^{-1}\xi x$, $x \cdot \xi = \xi$ for $x \in G$, $\xi \in X$. If D is a continuous derivation into X^* for the given structure, then $E : x \mapsto D(x)x^{-1}$ is a continuous derivation into X^* for the new structure.*

Proof. For all $a, b \in G$,

$$E(ab) = D(ab)b^{-1}a^{-1} = (aD(b) + D(a)b)b^{-1}a^{-1}$$
$$= a\big(D(b)b^{-1}\big)a^{-1} + D(a)a^{-1}$$
$$= a \cdot E(b) + E(a) = a \cdot E(b) + E(a) \cdot b. \quad \square$$

Theorem 11.8. *If G is a locally compact group, the following conditions characterize amenability:*
  (i) *For every unital Banach G-module X, $H_1(G, X^*) = \{0\}$.*
  (ii) *For every Banach G-module X, $H_1(G, X^*) = \{0\}$.*

(iii) *For every essential Banach $L^1(G)$-module X, $H_1(L^1(G), X^*) = \{0\}$.*
(iv) *For every Banach $L^1(G)$-module X, $H_1(L^1(G), X^*) = \{0\}$.*
(v) $H_1(G, (\mathcal{RUC}(G)/\mathbf{Cl}_G)^*) = \{0\}$, where $xf = f$ and $fx = {}_x f$, for $f \in \mathcal{RUC}(G)$, $x \in G$.

Proof. Assume G to be amenable and let X be a unital Banach G-module. We consider the new unital Banach G-module structure given by Lemma 11.7. If D is a continuous derivation and E is the associated continuous derivation obtained in Lemma 11.7, then by Lemma 11.6 there exists $\Phi \in X^*$ such that, for every $a \in G$,

$$E(a) = -a \cdot \Phi + \Phi \cdot a = -a\Phi a^{-1} + \Phi.$$

As X is unital,

$$D(a) = E(a)a = -a\Phi + \Phi a,$$

that is, D is inner. Therefore (i) holds.

(i) $\Rightarrow$ (ii) follows from Lemma 11.3.

(ii) $\Rightarrow$ (iii)

Consider an essential Banach $L^1(G)$-module X and let D be a continuous derivation of $L^1(G)$ into X^*. We extend X to a Banach $\mathcal{M}^1(G)$-module.

(a) Let (φ_i) be a net of approximate units in $L^1(G)$ bounded by 1 obtained via Proposition 2.2. If $\xi \in X$, there exist $f_1 \in L^1(G)$, $\xi_1 \in X$ such that $\xi = f_1 \xi_1$. By the continuity of D and Lemma 11.5, for every $\mu \in \mathcal{M}^1(G)$,

$$\lim_i \langle \xi, D(\mu * \varphi_i) \rangle = \lim_i \langle f_1 \xi_1, D(\mu * \varphi_i) \rangle$$

$$= \lim_i \langle \xi_1, D(\mu * \varphi_i) f_1 \rangle$$

$$= \lim_i \langle \xi_1, D(\mu * (\varphi_i * f_1)) \rangle - \lim_i \langle \xi_1, (\mu * \varphi_i) D(f_1) \rangle$$

$$= \lim_i \langle \xi_1, D(\mu * (\varphi_i * f_1)) \rangle - \lim_i \langle (\xi_1 \mu) \varphi_i, D(f_1) \rangle$$

$$= \langle \xi_1, D(\mu * f_1) \rangle - \langle \xi_1 \mu, D(f_1) \rangle.$$

We may therefore define, for $\mu \in \mathcal{M}^1(G)$, an element $E(\mu) \in X^*$ by putting

$$\langle \xi, E(\mu) \rangle = \lim_i \langle \xi, D(\mu * \varphi_i) \rangle,$$

$\xi \in X$. We show that E is a derivation for the $\mathcal{M}^1(G)$-module structure.

Given $\xi \in X$, there exist $\xi_2 \in X$ and $f_2 \in L^1(G)$ such that $\xi = \xi_2 f_2$. For all $\mu, \nu \in \mathcal{M}^1(G)$, with respect to Lemma 11.5 we have

$$\langle \xi, E(\mu * \nu) \rangle = \lim_i \langle \xi_2 f_2, D(\mu * \nu * \varphi_i) \rangle$$

$$= \lim_i \langle \xi_2, f_2 D(\mu * \nu * \varphi_i) \rangle$$

$$= \lim_i \langle \xi_2, D(f_2 * \mu * \nu * \varphi_i) \rangle - \lim_i \langle \xi_2, D(f_2)(\mu * \nu * \varphi_i) \rangle$$

$$= \lim_i \langle \xi_2, (f_2 * \mu) D(\nu * \varphi_i) \rangle + \lim_i \langle \xi_2, D(f_2 * \mu)(\nu * \varphi_i) \rangle$$

$$- \lim_i \langle \xi_2, D(f_2)(\mu * \nu * \varphi_i) \rangle$$

$$= \lim_i \langle (\xi_2 f_2) \mu, D(\nu * \varphi_i) \rangle + \lim_i \langle \varphi_i \xi_2, D(f_2 * \mu) \nu \rangle$$

$$- \lim_i \langle \varphi_i \xi_2, D(f_2)(\mu * \nu) \rangle$$

$$= \lim_i \langle \xi \mu, D(\nu * \varphi_i) \rangle + \langle \xi_2, D(f_2 * \mu) \nu \rangle$$

$$- \langle \xi_2, D(f_2)(\mu * \nu) \rangle$$

$$= \langle \xi \mu, E(\nu) \rangle + \lim_i \langle \nu \xi_2, D(f_2 * \mu * \varphi_i) \rangle$$

$$- \langle \nu \xi_2, D(f_2) \mu \rangle$$

$$= \langle \xi, \mu E(\nu) \rangle + \lim_i \langle \nu \xi_2, f_2 D(\mu * \varphi_i) \rangle$$

$$+ \lim_i \langle \nu \xi_2, D(f_2)(\mu * \varphi_i) \rangle - \langle \nu \xi_2, D(f_2) \mu \rangle$$

$$= \langle \xi, \mu E(\nu) \rangle + \lim_i \langle \nu \xi, D(\mu * \varphi_i) \rangle$$

$$+ \lim_i \langle \varphi_i(\nu \xi_2), D(f_2) \mu \rangle - \langle \nu \xi_2, D(f_2) \mu \rangle$$

$$= \langle \xi, \mu E(\nu) \rangle + \langle \nu \xi, E(\mu) \rangle = \langle \xi, \mu E(\nu) + E(\mu) \nu \rangle.$$

Hence E is a derivation; its restriction to $L^1(G)$ coincides with D.

(b) We put

$$D_1(x) = E(\delta_x)$$

for $x \in G$. For every $x \in G$, we have

$$\|D_1(x)\| = \sup\{|\langle \xi, D_1(x)\rangle| : \|\xi\|_X \le 1\}$$
$$= \sup\left\{\left|\lim_i \langle \xi, D(\delta_x * \varphi_i)\rangle\right| : \|\xi\|_X \le 1\right\} \le \|D\|,$$

that is, D_1 is norm-bounded. We now show that D_1 is continuous for the w*-topology of X^*. Let $\xi \in X$. There exist $f', f'' \in L^1(G)$ and $\xi', \xi'' \in X$ such that $\xi = f'\xi'$ and $\xi' = \xi''f''$. If $\lim_j x_j = x$, we obtain

$$\lim_j \langle \xi, D_1(x_j) - D_1(x)\rangle = \lim_j \langle \xi', E(\delta_{x_j})f' - E(\delta_x)f'\rangle$$
$$= \lim_j \langle \xi', E(\delta_{x_j} * f' - \delta_x * f')\rangle$$
$$- \lim_j \langle \xi'\delta_{x_j} - \xi'\delta_x, E(f')\rangle$$
$$= \lim_j \langle \xi', D(\delta_{x_j} * f' - \delta_x * f')\rangle$$
$$- \lim_j \langle \xi''(f'' * \delta_{x_j} - f'' * \delta_x), D(f')\rangle = 0.$$

We conclude that D_1 is a continuous derivation of G into X^*.

(c) We show that, for all $\xi \in X$ and $f \in L^1(G)$,

$$\langle \xi, D(f)\rangle = \int f(x)\langle \xi, D_1(x)\rangle\, dx.$$

There exist $f' \in L^1(G)$, $\xi' \in X$ such that $\xi = f'\xi'$; then by (1) and Fubini's theorem we have

$$\int f(x)\langle \xi, D_1(x)\rangle\, dx = \int f(x)\langle f'\xi', D_1(x)\rangle\, dx$$
$$= \int f(x)\left\langle \int y\xi f'(y)\, dy, D_1(x)\right\rangle dx$$
$$= \int f'(y)\left(\int f(x)\langle y\xi', D_1(x)\rangle\, dx\right) dy$$
$$= \int f'(y)\left(\int f(x)\langle \xi', D_1(x)y\rangle\, dx\right) dy.$$

Let $y \in G$. Notice that, for every $g \in \mathscr{C}_0(G)$,

$$\int g(z)\left(\int f(x)\delta_{xy}(z)\,dx\right)dz = \int g(z)\left(\int f * \delta_y(x)\delta_x(z)\,dx\right)dz$$

$$= \int g(z) f * \delta_y(z)\,dz$$

hence $\int f(x)\delta_{xy}\,dx = f * \delta_y$ and then

$$\int f(x)\langle \xi', D_1(x)y\rangle\,dx = \int f(x)\langle \xi', D_1(xy)\rangle\,dx - \int f(x)\langle \xi', xD_1(y)\rangle\,dx$$

$$= \left\langle \xi', \int f(x)E(\delta_{xy})\,dx\right\rangle - \int f(x)\langle \xi'x, E(\delta_y)\rangle\,dx$$

$$= \left\langle \xi', E\left(\int f(x)\delta_{xy}\,dx\right)\right\rangle - \langle \xi'f, E(\delta_y)\rangle$$

$$= \langle \xi', E(f * \delta_y)\rangle - \langle \xi', E(f * \delta_y)\rangle + \langle \xi', E(f)\delta_y\rangle = \langle y\xi', D(f)\rangle.$$

Therefore

$$\int f(x)\langle \xi, D_1(x)\rangle\,dx = \int f'(y)\langle y\xi', D(f)\rangle\,dy = \left\langle \int f'(y)y\xi'dy, D(f)\right\rangle$$

$$= \langle f'\xi', D(f)\rangle = \langle \xi, D(f)\rangle.$$

(d) By hypothesis there exists $\Psi \in X^*$ such that $D_1(x) = x\Psi - \Psi x$ whenever $x \in G$. Taking (c) into account we obtain, for every $f \in L^1(G)$ and every $\xi \in X$,

$$\langle \xi, D(f)\rangle = \int f(x)\langle \xi, D_1(x)\rangle\,dx = \int f(x)\langle \xi, x\Psi - \Psi x\rangle\,dx$$

$$= \int f(x)\langle \xi x - x\xi, \Psi\rangle\,dx = \left\langle \int f(x)(\xi x - x\xi)\,dx, \Psi\right\rangle$$

$$= \langle \xi f - f\xi, \Psi\rangle = \langle \xi, f\Psi - \Psi f\rangle.$$

Thus $D(f) = f\Psi - \Psi f$; D is inner.

$$(\text{iii}) \Rightarrow (\text{iv})$$

The implication follows from Lemmas 11.4 and 11.2.

$$(\text{iv}) \Rightarrow (\text{i})$$

Let X be a unital Banach G-module. We may consider it to be an essential Banach $L^1(G)$-module. Let D be a continuous derivation of G into X^*. We put

$$D'(f) = \int D(x)f(x)\,dx,$$

$f \in L^1(G)$. Since D is norm-bounded, D' is a continuous derivation of $L^1(G)$ into X^*. Notice that, if $f, g \in L^1(G)$ and $\xi \in X$,

$$\langle \xi, D'(f*g) \rangle = \int \langle \xi, D(x) f*g(x) \rangle\,dx$$

$$= \iint \langle \xi, D(x) \rangle f(y) g(y^{-1}x)\,dy\,dx$$

$$= \iint \langle \xi, D(yx) \rangle f(y) g(x)\,dx\,dy$$

$$= \int \left(\int \langle x\xi, D(y) \rangle f(y)\,dy \right) g(x)\,dx + \int \left(\int \langle \xi y, D(x) \rangle g(x)\,dx \right) f(y)\,dy$$

$$= \int \langle x\xi, D'(f) \rangle g(x)\,dx + \int \langle \xi y, D'(g) \rangle f(y)\,dy$$

$$= \langle g\xi, D'(f) \rangle + \langle \xi f, D'(g) \rangle = \langle \xi, D'(f)g + fD'(g) \rangle.$$

By hypothesis there exists $\Psi \in X^*$ such that, for all $f \in L^1(G)$, $D'(f) = f\Psi - \Psi f$, that is,

$$\int \langle \xi, D(x) \rangle f(x)\,dx = \int \langle \xi, x\Psi - \Psi x \rangle f(x)\,dx$$

whenever $\xi \in X$; hence $D(x) = x\Psi - \Psi x$ for every $x \in G$. Therefore $H_1(G, X^*) = \{0\}$.

(ii) $\Rightarrow$ (v) Trivial

Finally, assume that (v) holds. Let $X = \mathscr{RUC}(G)/\mathbb{C}1_G$ and $Y = \widetilde{\mathscr{RUC}(G)}$. We have $\mathscr{RUC}(G) \simeq \mathscr{C}(Y)$ (Section 2B.3) and $\mathscr{RUC}(G)^* \simeq \mathscr{M}^1(Y)$. As X^* is isomorphic to $\{\Theta \in \mathscr{RUC}(G)^* : \langle 1_G, \Theta \rangle = 0\}$, X^* is isomorphic to the subspace $\{\theta \in \mathscr{M}^1(Y) : \theta(Y) = 0\}$ of $\mathscr{M}^1(Y)$.

Choose $\Phi \in \mathscr{RUC}(G)^*$ such that $\langle 1_G, \Phi \rangle \neq 0$ and, for $x \in G$, put

$$\Phi'(x) = x\Phi - \Phi x = x\Phi - \Phi; \qquad (2)$$

$\Phi'(x) \in \mathscr{RUC}(G)^*$ and $\langle 1_G, \Phi'(x) \rangle = 0$. We may identify Φ' with a continuous derivation Φ_1 of G into X^*. By hypothesis there exists $\Psi \in X^*$ such that

$$\Phi_1(x) = x\Psi - \Psi x = x\Psi - \Psi \tag{3}$$

for every $x \in G$.

Let μ [resp. ν] be the image of Φ [resp. Ψ] in $\mathscr{M}^1(Y)$. We have $(\mu - \nu)(Y) = \mu(Y) \neq 0$. Considering the action of G induced on $\mathscr{M}^1(Y)$, by (2) and (3), for every $x \in G$,

$$|\mu - \nu| = |x(\mu - \nu)| = x|\mu - \nu|.$$

Therefore $(|\mu - \nu|(Y))^{-1}|\mu - \nu| \in M^1(Y)$ is the image of a mean M on $\mathscr{RUC}(G)$ such that

$$\langle {}_x f, M \rangle = \langle fx, M \rangle = \langle f, xM \rangle = \langle f, M \rangle$$

whenever $f \in \mathscr{RUC}(G)$, $x \in G$. Amenability of G follows from Theorem 4.19. □

C. Derivations into $L^\infty(G)$

We transcribe a particular cohomological property characterizing amenability.

Proposition 11.9. *A locally compact group G is amenable if and only if, for any Banach G-module structure on $L^1(G)$ such that $xf = f$ whenever $f \in L^1(G)$, $x \in G$, and any continuous derivation D of G into $L^\infty(G)$, there exists $\Phi \in \overline{\mathrm{co}} D(G)$ such that $D = -d_1\Phi$.*

Proof. (a) Assume G to be amenable. Choose $\Phi \in L^\infty(G)$ given by Lemma 11.6 such that $D = -d_1\Phi$. If we had $\Phi \notin \overline{\mathrm{co}} D(G)$, then by the Hahn–Banach theorem there would exist $f_0 \in L^1(G)$ verifying

$$\mathscr{Re} \langle f_0, \Phi \rangle < \inf \{ \mathscr{Re} \langle f_0, \Psi \rangle : \Psi \in \overline{\mathrm{co}} D(G) \}.$$

Lemma 11.6 would be violated.

(b) Assume that the condition holds. We establish (i) of Theorem 11.8. Let X be a unital Banach G-module and D a continuous derivation of G into X^*. We consider the new unital Banach G-module structure on X given by Lemma 11.7 and also the continuous derivation E associated to D. We carry the latter structure over to the associated essential Banach $L^1(G)$-module structure on X. For $\xi \in X$, we define

$$\alpha_\xi : f \mapsto \xi \cdot f$$

$$L^1(G) \to X;$$

α_ξ is a continuous $L^1(G)$-module homomorphism of the right $L^1(G)$-module $L^1(G)$ into the right $L^1(G)$-module X and induces a w*-continuous $L^1(G)$-module homomorphism $\alpha_\xi^\sim$ of the left $L^1(G)$-module X^* into the left $L^1(G)$-module $L^\infty(G)$.

For $x \in G$, we consider the affine mapping

$$A_x : \Theta \mapsto x \cdot \Theta + E(x)$$

$$X^* \to X^*.$$

The set $\overline{\mathrm{co}}\, E(G)$ is nonvoid and w*-compact in X^*. For all $a, x \in G$,

$$A_x(E(a)) = x \cdot E(a) + E(x) = x \cdot E(a) + E(x) \cdot a = E(xa).$$

Hence $\overline{\mathrm{co}}\, E(G)$ is A_G-invariant, that is, invariant by all the transformations A_x ($x \in G$). Via Zorn's lemma there exists a minimal nonvoid w*-compact convex A_G-invariant subset K_0 in $\overline{\mathrm{co}}\, E(G)$; by minimality of K_0, we must have $\overline{\mathrm{co}}\,\{A_x(\Phi) : x \in G\} = K_0$ whenever $\Phi \in K_0$. Choose $\Phi_0 \in -K_0$, that is, $-\Phi_0 \in K_0$ and consider $\Psi = E - d_1\Phi_0$ for the new structure. If $x \in G$, $\Psi(x) = E(x) - x \cdot \Phi_0 + \Phi_0 = A_x(-\Phi_0) + \Phi_0$. Hence $\overline{\mathrm{co}}\,\Psi(G) = K_0 + \{\Phi_0\}$ and, by w*-compactness, $\overline{\mathrm{co}}\,\alpha_\xi^\sim(\Psi(G)) = \alpha_\xi^\sim(\overline{\mathrm{co}}\,\Psi(G))$. For $x \in G$, we also consider the affine mapping

$$B_x : \Theta \mapsto x \cdot \Theta + \Psi(x)$$

$$X^* \to X^*.$$

If $x \in G$, $\Theta \in X^*$, we have

$$B_x(\Theta) = x \cdot \Theta + E(x) - x \cdot \Phi_0 + \Phi_0 = A_x(\Theta - \Phi_0) + \Phi_0.$$

For $\Theta \in \overline{\mathrm{co}}\,\Psi(G)$, $\Theta - \Phi_0 \in K_0$; hence $\overline{\mathrm{co}}\,\{B_x(\Theta) : x \in G\} = K_0 + \{\Phi_0\}$, that is, $\overline{\mathrm{co}}\,\{x \cdot \Theta + \Psi(x) : x \in G\} = \overline{\mathrm{co}}\,\Psi(G)$.

If $\Phi \in \overline{\mathrm{co}}\,\alpha_\xi^\sim(\Psi(G))$, choose $\Phi_1 \in \overline{\mathrm{co}}\,\Psi(G)$ such that $\Phi = \alpha_\xi^\sim(\Phi_1)$; then we have also

$$\overline{\mathrm{co}}\,\{x \cdot \Phi + \alpha_\xi^\sim(\Psi(x)) : x \in G\} = \alpha_\xi^\sim\left(\overline{\mathrm{co}}\,\{x \cdot \Phi_1 + \Psi(x) : x \in G\}\right)$$

$$= \overline{\mathrm{co}}\,\alpha_\xi^\sim(\Psi(G)). \qquad (4)$$

For every $\xi \in X$, $\alpha_\xi^\sim \circ \Psi$ is a continuous derivation of G into $L^\infty(G)$ for the new structure; hence by hypothesis there must exist $\Phi_0 \in \overline{\mathrm{co}}(\alpha_\xi^\sim(\Psi(G)))$ such that $\alpha_\xi^\sim \circ \Psi = -d_1\Phi_0$, that is, for all $x \in G$,

$$\alpha_\xi^\sim \circ \Psi(x) = -x \cdot \Phi_0 + \Phi_0 \cdot x = -x \cdot \Phi_0 + \Phi_0.$$

By (4) we obtain

$$\overline{\mathrm{co}}\{\alpha_\xi^\sim \circ \Psi(x) : x \in G\} = \overline{\mathrm{co}}\{\alpha_\xi^\sim \circ \Psi(x) + x \cdot \Phi_0 : x \in G\} = \{\Phi_0\},$$

that is, $\alpha_\xi^\sim \circ \Psi(x) = \Phi_0$ for every $x \in G$. As $\alpha_\xi^\sim \circ \Psi$ is a derivation, $\alpha_\xi^\sim \circ \Psi(e) = 0$ and then $\alpha_\xi^\sim \circ \Psi = 0$. Thus for all $f \in L^1(G)$, $\xi \in X$, $x \in G$,

$$\langle \xi \cdot f, \Psi(x) \rangle = \langle \alpha_\xi(f), \Psi(x) \rangle = \langle f, \alpha_\xi^\sim(\Psi(x)) \rangle = 0.$$

As X is essential, we have $\langle \eta, \Psi(x) \rangle = 0$ for every $x \in G$ and every $\eta \in X$. Hence $\Psi = 0$, that is, $E = d_1\Phi_0$. As the given structure on X is unital, we conclude that, for all $a \in G$,

$$D(a) = E(a)a = (a \cdot \Phi_0 - \Phi_0 \cdot a)a$$
$$= (a\Phi_0 a^{-1} - \Phi_0)a = a\Phi_0 - \Phi_0 a,$$

that is, D is inner. □

NOTES

The results of this section are due to Johnson [291], who inaugurated the subject. Ringrose had shown the sufficiency of the condition 11.8 (iv) for amenability. Alternative proofs of the main characterization are due to Khelemskiĭ and Sheinberg [314]. A demonstration of 11.8 applying to discrete groups is given by Bonsall and Duncan [46]; also see Johnson [293].

3
THE CLASS OF AMENABLE GROUPS

In this chapter we list basic amenable groups. Stability properties of the class of amenable groups provide constructions of new amenable groups. We give indications on the localization of amenable groups in the class of all locally compact groups.

12. EXAMPLES OF AMENABLE GROUPS

We begin by stating for compact groups and abelian groups direct proofs of the amenability based on the fundamental characterizations. We then show that more generally all exponentially bounded groups are amenable. Other examples of amenable groups appear in Sections 13, 14, and 23.

A. Compact Groups and Abelian Groups

Proposition 12.1. *Any compact group G is amenable. The sets* $\text{LIM}(\mathscr{C}(G))$, $\text{RIM}(\mathscr{C}(G))$, $\text{IM}(\mathscr{C}(G))$, $\text{TLIM}(L^\infty(G))$, $\text{TRIM}(L^\infty(G))$, *and* $\text{TIM}(L^\infty(G))$ *are singletons reduced to the normalized Haar measure on G.*

Proofs. We indicate some alternative demonstrations.
(a) The normalized Haar measure on G is an invariant mean on $\mathscr{C}(G)$ and this (left) invariant mean is unique ([HR] 15.9).
(b) Property (P_p) is satisfied trivially by $1_G \in P^p(G)$ for any $p \in [1, \infty[$.
(c) Property (F) holds trivially too; for any $a \in G$, $|aG \triangle G|/|G| = |\phi|/|G| = 0$.
(d) As every measure on G is bounded, one readily verifies (C_p) for $p \in]1, \infty[$.

EXAMPLES OF AMENABLE GROUPS

Finally, note that if $M \in \text{TLIM}(L^\infty(G))$ [resp. $M \in \text{TRIM}(L^\infty(G))$], for all $f \in L^\infty(G)$, $\varphi \in P^1(G)$, we have $\varphi * f \in \mathscr{C}(G)$ [resp. $f * \check{\varphi} \in \mathscr{C}(G)$] and

$$M(f) = M(\varphi * f) = \int \varphi * f(x)\, dx = \int f$$

$$\left[\text{resp. } M(f) = M(f * \check{\varphi}) = \int f * \check{\varphi}(x)\, dx = \int f\right]. \qquad \square$$

Proposition 12.2. *Any locally compact abelian group G is amenable.*

Proofs. We also indicate some alternative demonstrations.

(a) We verify Proposition 4.29 for G_d; then G is amenable by Proposition 4.21.

Assume that there exist $f_1, \ldots, f_n \in l_{\mathbf{R}}^\infty(G)$, $a_1, \ldots, a_n \in G$ and $c \in \mathbf{R}_+^*$ such that $\sup h < -c$ for $h = \sum_{i=1}^n (f_i - {}_{a_i}f_i)$. Let $p \in \mathbf{N}^*$ and consider the set Θ of all mappings from $\{1, \ldots, n\}$ into $\{1, \ldots, p\}$; card $\Theta = p^n$. Define

$$\alpha: \theta \mapsto a_1^{\theta(1)} \cdots a_n^{\theta(n)}$$

$$\Theta \to G.$$

For a fixed $i \in \{1, \ldots, n\}$, in the sum $\sum_{\theta \in \Theta} (f_i(\alpha(\theta)) - f_i(a_i \alpha(\theta)))$ by commutativity all terms cancel except at most $2p^{n-1}$ terms for which $\theta(i) = 1$ or p. Therefore

$$\sum_{\theta \in \Theta} (f_i(\alpha(\theta)) - f_i(a_i \alpha(\theta))) \geq -2p^{n-1}\|f_i\|;$$

then we have

$$-cp^n \geq \sum_{\theta \in \Theta} h(\alpha(\theta)) = \sum_{\theta \in \Theta} \sum_{i=1}^n (f_i(\alpha(\theta)) - f_i(a_i \alpha(\theta)))$$

$$= \sum_{i=1}^n \sum_{\theta \in \Theta} (f_i(\alpha(\theta)) - f_i(a_i \alpha(\theta))) \geq -2p^{n-1} \sum_{i=1}^n \|f_i\|,$$

that is, $cp \leq 2\sum_{i=1}^n \|f_i\|$. As $p \in \mathbf{N}^*$ may be chosen arbitrarily, we come to a contradiction.

(b) In order to apply Theorem 5.4 it suffices to invoke the Markov-Kakutani fixed point theorem ([B EVT] Chap. IV, App. No. 1, théorème 1): Let E be a locally convex Hausdorff topological real vector space, Z a nonvoid compact convex subset of E, Γ a family of affine mappings of E into itself that are pairwise commuting and admit continuous restrictions to Z with ranges in Z; then there exists $\zeta_0 \in Z$ such that $u(\zeta_0) = \zeta_0$ whenever $u \in \Gamma$.

(c) As G is abelian, $L^1(G)$ is commutative. For all $\varphi_1, \varphi_2 \in P^1(G)$, we have $\|\varphi_1 * \varphi_2 - \varphi_2 * \varphi_1\|_1 = 0$ and (iii) of Proposition 6.17 is satisfied. $\square$

B. Growth Conditions

We show to what extent mild growth of powers of compact neighborhoods of e in a locally compact group G insures amenability.

Definition 12.3. *Let G be a locally compact group. We say that G has polynomial growth if the following condition holds*:
  (*PG*) *For every compact neighborhood V of e in G, there exists $d \in \mathbf{N}^*$ such that*

$$\limsup_{n \to \infty} \frac{|V^n|}{n^d} < \infty.$$

We say that G is exponentially bounded if the following condition holds:
  (*EB*) *For every compact neighborhood V of e in G and every $t \in \,]1, \infty[$,*

$$\limsup_{n \to \infty} \frac{|V^n|}{t^n} < \infty.$$

Consideration has been given to two related conditions.

Definition 12.4. *If G is a locally compact group, we define the properties*:
  (*A*) *For every compact neighborhood V of e in G and every $t \in \,]1, \infty[$,*

$$\lim_{n \to \infty} \frac{|V^n|}{t^n} = 0.$$

  (*B*) *For every compact neighborhood V of e in G, $\liminf_{n \to \infty} |V^{n+1}|/|V^n| = 1$.*

Obviously $(PG) \Rightarrow (A) \Rightarrow (EB)$.

Proposition 12.5. *Every exponentially bounded group G has property (B).*

Proof. If the statement did not hold, there would exist $\alpha \in \mathbf{R}^*_+$ and $n_0 \in \mathbf{N}^*$ such that, for any $n \geq n_0$, $|V^{n+1}|/|V^n| \geq 1 + \alpha$, hence for any $m \in \mathbf{N}^*$, $|V^{n_0+m}| \geq (1+\alpha)^m |V^{n_0}|$ and

$$\frac{|V^{n_0+m}|}{(1+\alpha/2)^{n_0+m}} \geq \left(\frac{1+\alpha}{1+\alpha/2}\right)^m \frac{|V^{n_0}|}{(1+\alpha/2)^{n_0}},$$

that is, (*EB*) would not hold. $\square$

EXAMPLES OF AMENABLE GROUPS

Any locally compact group G satisfying (B) is unimodular. As a matter of fact, if there existed $a \in G$ with $\Delta(a) > 1$, then for every compact neighborhood V of e and a, one should have

$$\liminf_{n \to \infty} \frac{|V^{n+1}|}{|V^n|} \geq \liminf_{n \to \infty} \frac{|V^n a|}{|V^n|} = \Delta(a) > 1.$$

Proposition 12.6. *A locally compact group G having property (B) is amenable.*

Proof. For every compact neighborhood V of e in G, we have

$$\liminf_{n \to \infty} \frac{|VV^n \triangle V^n|}{|V^n|} = \liminf_{n \to \infty} \frac{|V^{n+1} \triangle V^n|}{|V^n|} = \liminf_{n \to \infty} \left(\frac{|V^{n+1}|}{|V^n|} - 1 \right) = 0.$$

Therefore $(SF_{e,0})$ holds and G is amenable by Theorem 7.9. □

We show that all compact groups and all locally compact abelian groups have polynomial growth and thus are amenable.

If G is a *compact* group, for every compact neighborhood V of e in G and every $m \in \mathbb{N}^*$,

$$\limsup_{n \to \infty} \frac{|V^n|}{n^m} \leq \lim_{n \to \infty} \frac{1}{n^m} = 0.$$

Let G be a locally compact *abelian* group. Given a compact neighborhood V of e in G there exists a finite subset $F = \{a_1, \ldots, a_m\}$ in V such that $V^2 \subset FV$. By induction we obtain $V^{n+1} \subset F^n V$ for every $n \in \mathbb{N}^*$ as, if $V^{n+1} \subset F^n V$, then also $V^{n+2} = V^{n+1} V \subset F^n VV \subset F^{n+1} V$. But by commutativity every element of F^n is of the form $a_1^{r_1} \cdots a_m^{r_m}$ where $r_1, \ldots, r_m \in \{0, 1, \ldots, n\}$. Hence

$$\text{card } F^n \leq (n+1)^m$$

and $|V^{n+1}| \leq (n+1)^m |V|$; (PG) holds.

We add two more examples of (amenable) groups having polynomial growth.

Proposition 12.7. *The discrete free group G generated by two elements a and b of order 2 is of polynomial growth and hence amenable.*

Proof. If $n \in \mathbb{N}^*$, let E_n be the set of elements in G which equal e or, when written as reduced words, are products of at most n factors $a = a^{-1}$, $b = b^{-1}$; we have card $E_n = 1 + 2n$. If F is a finite subset of G containing e, there exists $n_0 \in \mathbb{N}^*$ such that $F \subset E_{n_0}$; for any $n \in \mathbb{N}^*$, $F^n \subset E_{n_0 n}$ and card $F^n \leq 1 + 2n_0 n$. For any $m \in \mathbb{N}^*$,

$$\limsup_{n \to \infty} \frac{\text{card } F^n}{n^m} \leq \limsup_{n \to \infty} \frac{1 + 2n_0 n}{n^m} < \infty. \qquad \square$$

Definition 12.8. *If G is a locally compact group and $a \in G$, the conjugacy class of a is the set $K_a = \{xax^{-1} : x \in G\}$. The class of all locally compact groups having all their conjugacy classes relatively compact is denoted by $[FC]^-$.*

Compact groups and abelian groups belong to $[FC]^-$.

Proposition 12.9. *Any locally compact totally disconnected group G belonging to $[FC]^-$ has polynomial growth and is amenable.*

Proof. As G is totally disconnected, there exists an open compact subgroup H in G ([HR] 7.7). Let V be a compact neighborhood of e in G; $U = HV$ is also a compact neighborhood of e. There exists a finite subset E of U such that $e \in E$ and $U \subset EH$. Then by induction, for every $n \in \mathbf{N}^*$,

$$U^n \subset E^n H. \tag{1}$$

As a matter of fact, if $U^n \subset E^n H$, then

$$U^{n+1} = U^n U \subset (E^n H)U = E^n(HU) = E^n U \subset E^{n+1}H.$$

Let $L = \cup_{c \in E} K_c$. We have $E \subset L$ and therefore $E \subset EH \subset LH$. As LH is relatively compact and H is open, there exists then a finite subset F such that $E \subset F \subset L$ and

$$LH = FH. \tag{2}$$

For every $n \in \mathbf{N}^*$, by (1) we have

$$U^n \subset F^n H. \tag{3}$$

For every $y \in H$, by (2), we obtain

$$yFH = yLH = yLy^{-1}H = LH = FH$$

and also $HFH = FH$. Then by induction, for every $n \in \mathbf{N}^*$,

$$HF^n H = F^n H \tag{4}$$

as, if (4) holds for $n \in \mathbf{N}^*$, then also

$$HF^{n+1}H = HF^n FH = HF^n HFH = (F^n H)FH = F^n(HFH) = F^{n+1}H.$$

Let $F = \{a_1, \ldots, a_m\}$. We establish inductively that every element a of $F^n H$ ($n \in \mathbf{N}^*$) may be written in the form

$$a = a_{j_1} \cdots a_{j_n} y$$

with $j_1, \ldots, j_n \in \{1, \ldots, m\}$, $j_1 \leq \cdots \leq j_n$ and $y \in H$. For
holds trivially. If the property holds for $F^{n-1}H(n \geq 2)$, it h/
Assume now that it holds for $a_k F^{n-1}H$ with $k \in \{1, \ldots,$
$a \in a_l F^{n-1}H$. By the induction hypothesis, we have c
with $l, k_1, \ldots, k_{n-1} \in \{1, \ldots, m\}$, $k_1 \leq \cdots \leq k_{n-1}$ and
$l \leq k_1$, we are done. In case $k_1 < l$, as $a_{k_1}^{-1} a_l a_{k_1} \in K_{a_l} \subset L$, (2) imp..

$$a_l a_{k_1} = a_{k_1} a_{k_1}^{-1} a_l a_{k_1} = a_{k_1} a_{l'} z$$

with $l' \in \{1, \ldots, m\}$ and $z \in H$. Now by (4) we have

$$a \in a_{k_1} a_{l'} z F^{n-2} H \subset a_{k_1} a_{l'} F^{n-2} H \subset a_{k_1} F^{n-1} H.$$

As $k_1 < l$, the induction hypothesis yields the desired decomposition.

Considering all these decompositions, we see that any coset cH ($c \in F^n$) is determined by the first position of each $a_1, \ldots, a_m$; hence we have $|F^n H| \leq (n+1)^m |H|$. Then by (3) we obtain $|V^n| \leq |U^n| \leq (n+1)^m |H|$. □

NOTES

Existence of invariant means on abelian groups was noticed by von Neumann [414]. Day [116] used the Markov–Kakutani fixed point theorem to verify (FP) in the commutative case. The proof given in 12.2(a) is due to Dixmier [138]; see also [HR]. Reiter proved (P_1) for locally compact abelian groups via Fourier transform arguments [450, 453]. Bonsall and Duncan [46] proved property 11.8(iv) for every discrete abelian group.

Growth conditions were studied systematically by Guivarc'h [241] and Jenkins [283, 285]; they established the basic characterizations of the different types of growth. An early reference is Dixmier [139]. Property (A) was introduced by Adel'son-Velskiĭ and Šreider [1] in the discrete case; they proved these groups to be amenable. Property (A) was studied extensively by Hulanicki in the discrete case [271] and in the general case [272]. Amenability of the discrete groups satisfying (B) was established by Avez [23] via Dixmier's criterion. For the general situation one should consult Jenkins [285, 286]. Dixmier [138] proved amenability of the group freely generated by two elements of order 2. Leptin [355] established 12.9.

13. STABILITY PROPERTIES OF AMENABILITY

A. Main Combinatorial Properties

To give a unified exposition we prove stability properties via existence conditions of invariant means making use of the interplay authorized by Theorem 4.19.

THE CLASS OF AMENABLE GROUPS

Proposition 13.1. *If π is a continuous homomorphism of an amenable group G onto a locally compact group H, then H is amenable.*

Proof. If $M \in \text{LIM}(\mathscr{C}(G))$, we define a mean M' on $\mathscr{C}(H)$ by putting

$$M'(f) = M(f \circ \pi),$$

$f \in \mathscr{C}(H)$. If $f \in \mathscr{C}(H)$ and $b \in H$, consider $a \in G$ such that $\pi(a) = b$;

$$M'(_b f) = M'(_{\pi(a)} f) = M(_a(f \circ \pi)) = M(f \circ \pi) = M'(f).$$

We have $M' \in \text{LIM}(\mathscr{C}(H))$. □

Corollary 13.2. *The quotient group of an amenable group G by a closed normal subgroup H is amenable.*

Proof. Apply Proposition 13.1 to the canonical continuous homomorphism of G onto G/H. □

If G is a locally compact group admitting a closed subgroup H, there exists a Bruhat function, that is, a continuous function $h: G \to \mathbf{R}_+$ such that, for every $x \in G$, $\int_H h(xz)\, d\lambda_H(z) = 1$ and, for every compact subset K in G, $(\text{supp}\, h) \cap KH$ is compact ([B INT] Chap. VII, Section 2, No. 5, théorème 2d).

Proposition 13.3. *Any closed subgroup H of an amenable group G is amenable.*

Proof. We establish the existence of a left invariant mean on $\mathscr{C}(H)$.

Let h be a Bruhat function attached to the pair (G, H). If $f \in \mathscr{C}(H)$, we may define

$$f'(x) = \int_H h(x^{-1}z) f(z)\, d\lambda_H(z),$$

$x \in G$. In case $f = 0$, $f' = 0$. If $f \neq 0$, let V be a fixed compact neighborhood of e in G. By Urysohn's lemma ([HS] 6.80), if $x \in G$, there exists $k \in \mathscr{K}_+(G)$ which coincides with h on $(\text{supp}\, h) \cap Vx^{-1}H$. Let C be a compact subset of H such that $(xV \text{supp}\, k) \cap H \subset C$ and $\mathring{C} \neq \emptyset$. Given $\varepsilon > 0$, there exists a symmetric neighborhood U of e in G such that $U \subset V$ and $|k(y) - k(y')| < \varepsilon(\|f\|\, |C|_H)^{-1}$ whenever $y, y' \in G$ with $yy'^{-1} \in U$. Hence if $y \in xU$ and $z \in H$,

$$|k(x^{-1}z) - k(y^{-1}z)| < \varepsilon(\|f\|\, |C|_H)^{-1}.$$

As $y^{-1}z \in Vx^{-1}H$, we have

$$|f'(x) - f'(y)| = \left| \int_H (h(x^{-1}z) - h(y^{-1}z)) f(z) \, d\lambda_H(z) \right|$$

$$= \left| \int_{(xV \operatorname{supp} h) \cap H} (h(x^{-1}z) - h(y^{-1}z)) f(z) \, d\lambda_H(z) \right|$$

$$= \left| \int_{(xV \operatorname{supp} h) \cap H} (k(x^{-1}z) - k(y^{-1}z)) f(z) \, d\lambda_H(z) \right|$$

$$\leq \int_{(xV \operatorname{supp} k) \cap H} |k(x^{-1}z) - k(y^{-1}z)| |f(z)| \, d\lambda_H(z) < \varepsilon,$$

that is, $f' \in \mathscr{C}(G)$.

If $M \in \mathrm{LIM}(\mathscr{C}(G))$, we may define a mean M' on $\mathscr{C}(H)$ by putting

$$M'(f) = M(f'),$$

$f \in \mathscr{C}(H)$. Obviously $M' \geq 0$; for every $x \in G$, $1'_H(x) = \int_H h(x^{-1}z) \, d\lambda_H(z) = 1$, that is, $1'_H = 1_G$ and therefore $M'(1_H) = 1$. Moreover if $f \in \mathscr{C}(G)$, $a \in H$ and $x \in G$, we have

$$({}_a f)'(x) = \int_H h(x^{-1}z) f(az) \, d\lambda_H(z) = \int_H h(x^{-1}a^{-1}z) f(z) \, d\lambda_H(z)$$

$$= \int_H h((ax)^{-1}z) f(z) \, d\lambda_H(z) = {}_a(f')(x);$$

$$M'({}_a f) = M(({}_a f)') = M({}_a(f')) = M(f') = M'(f).$$

Hence $M' \in \mathrm{LIM}(\mathscr{C}(H))$. □

Proposition 13.4. *Let G be a locally compact group admitting a closed normal subgroup H. If H and G/H are amenable, then G is amenable.*

Proof. Let $M_1 \in \mathrm{LIM}(\mathscr{C}(H))$ and $M_2 \in \mathrm{LIM}(\mathscr{C}(G/H))$. If $f \in \mathscr{RUC}(G)$, for every $a \in G$, the mapping $f_1^{(a)} = ({}_a f)|_H$ belongs to $\mathscr{C}(H)$ and $x \mapsto f_1^{(x)}$ is a right uniformly continuous mapping. We define $G \to \mathscr{C}(H)$

$$f' : x \mapsto M_1(f_1^{(x)})$$

$$G \to \mathbf{C}.$$

As $\|M_1\| \le 1$, $f' \in \mathscr{C}(G)$. Moreover for all $a \in G$ and $x \in H$, $M_1(f_1^{(ax)}) = M_1(f_1^{(a)})$; hence f' is constant on the coset aH and f' may be identified with an element $\dot{f}$ of $\mathscr{C}(G/H)$.

We finally put $M(f) = M_2(\dot{f})$ for $f \in \mathscr{RUC}(G)$; M is a mean on $\mathscr{RUC}(G)$. For all $f \in \mathscr{RUC}(G)$, $a \in G$, we have $(_af)' = _a\dot{f}$ and then

$$M(_af) = M_2(_{\dot a}\dot{f}) = M_2(\dot{f}) = M(f).$$

Therefore $M \in \text{LIM}(\mathscr{RUC}(G))$. □

Proposition 13.4 stipulates that the cartesian product of two (and hence any finite number of) amenable groups is amenable.

Corollary 13.5. *Every locally compact solvable group G is amenable.*

Proof. If A is a subset of G, let $[A, A]$ denote the closed subgroup of G generated by the commutators $xyx^{-1}y^{-1}$, where $x, y \in A$. We consider the derived group $D^{(1)}(G) = [G, G]$ and then define inductively $D^{(n+1)}(G) = [D^{(n)}(G), D^{(n)}(G)]$, $n \in \mathbf{N}^*$. We have $D^{(0)}(G) = G \supset D^{(1)}(G) \supset \cdots \supset D^{(n)}(G) \supset D^{(n+1)}(G) \cdots$ and, for every $n \in \mathbf{N}^*$, $D^{(n+1)}(G)$ is a closed normal subgroup of $D^{(n)}(G)$ such that $D^{(n)}(G)/D^{(n+1)}(G)$ is abelian. The solvability of G signifies that there exist $n_0 \in \mathbf{N}^*$ such that $D^{(n_0)}(G) = \{e\}$ ([B A] Chap. I, Section 6, No. 4). As the abelian quotient groups are amenable by Proposition 12.2 and the trivial group $D^{(n_0)}(G)$ is amenable too, one establishes stepwise, via Proposition 13.4, the amenability of $D^{(n_0-1)}(G), \ldots, D^{(1)}(G), D^{(0)}(G) = G$. □

Proposition 13.6. *If $(H_i)_{i \in I}$ is a net of closed amenable subgroups of a locally compact group G, directed by inclusion, then $H = \overline{\bigcup_{i \in I} H_i}$ is an amenable subgroup of G.*

Proof. If $M_i \in \text{LIM}(\mathscr{UC}(H_i))$ for $i \in I$, we put

$$M_i'(f) = M_i(f|_{H_i})$$

for $f \in \mathscr{UC}(H)$. For $i \in I$, let $\mathfrak{M}_i$ be the w*-closed set of all means on $\mathscr{UC}(H)$ that are invariant for the left translations by elements of H_i; we have $M_i' \in \mathfrak{M}_i$ and $\mathfrak{M}_i \ne \varnothing$. As $(H_i)_{i \in I}$ is directed by inclusion and $\mathfrak{M}(\mathscr{UC}(H))$ is w*-compact by Proposition 3.3, necessarily $\cap_{i \in I} \mathfrak{M}_i \ne \varnothing$. Each element of this intersection is a mean on $\mathscr{UC}(H)$ such that $M(_af) = M(f)$ whenever $f \in \mathscr{UC}(H)$ and $a \in \bigcup_{i \in I} H_i$. The statement now follows from Proposition 4.23. □

Definition 13.7. *We denote by EG the class of all discrete elementary groups, that is, all discrete groups obtained from abelian or finite groups by taking (I)*

subgroups, (*II*) *quotient groups* (*or equivalently homomorphic images*), (*III*) *group extensions*, (*IV*) *direct unions*.

Propositions 13.3, 13.2, 13.4, 13.6 ensure that all discrete elementary groups are amenable.

B. Complementary Properties

Recall that the *weak direct product* of the discrete groups G_i ($i \in I$) is the subgroup of the cartesian product $\prod_{i \in I} G_i$ formed by all elements $(x_i)_{i \in I}$ such that $x_i = e_i$ for all but a finite number of indices ([HR] 2.3).

Proposition 13.8. *A weak direct product of discrete amenable groups is amenable.*

Proof. The assertion is a consequence of Proposition 13.6. □

Proposition 13.9. *A locally compact group is amenable if and only if the closure of every finitely generated subgroup is amenable.*

Proof. The statement follows from Propositions 13.3 and 13.6. □

Corollary 13.10. *A discrete group is amenable if and only if every finitely generated subgroup is amenable.*

Proof. The corollary is a special case of Proposition 13.9. □

In particular amenability holds for every locally finite group, that is, every discrete group for which any finite subset generates a finite subgroup.

Proposition 13.11. *The locally compact group G is amenable if and only if the family $\mathfrak{A}(G)$ of all closed amenable subgroups of G is directed by inclusion.*

Proof. Assume that the condition holds; by Proposition 13.6 we have $K = \overline{\cup_{H \in \mathfrak{A}(G)} H} \in \mathfrak{A}(G)$. Proposition 13.1 asserts that, if $H \in \mathfrak{A}(G)$ and $a \in G$, the image aHa^{-1} of H by inner automorphism is amenable, hence $\cup_{H \in \mathfrak{A}(G)} H$ and K are normal subgroups of G. If we had $K \neq G$, then for $a \in G \setminus K$, the closed subgroup G'/K of G/K generated by $\dot{a}$ would be finite or isomorphic to $\mathbf{Z}$, hence amenable. As K is also amenable, the closed subgroup G' of G would be amenable by Proposition 13.4. But then we should have $G' \subset K$ and we would come to a contradiction. □

NOTES

All characterizations of amenability have been used to establish stability properties. Surveys studying the exploitation of invariant means conditions in order to prove combinatorial properties appear in Hewitt and Ross [HR] for the discrete case and in Greenleaf [226] for the general case.

In his study on discrete groups von Neumann [414] established 13.4, and 13.5 as well as a first version of 13.6. For discrete groups also Day enounced 13.4 and 13.6 [112] as well as 13.2 and 13.3 [113]; explicit proofs are given in [114, 115]. General demonstrations of 13.2, 13.6 and a version of 13.4 were obtained by Dixmier [138]. For discrete groups Følner [187] demonstrated 13.2, 13.3, and 13.4. Hulanicki [267] studied 13.4 and obtained a version of 13.3; he established 13.2, and also 13.4 considering topologically left invariant means on essentially bounded functions [268]. Using powerful structure theorems Rickert [462] gave a proof of 13.3 which applies to means on continuous bounded functions; Greenleaf [226] performed a simplified version producing a transversal for cosets and making use of the equivalence properties established in 4.19. For our proof of 13.3 we refer to Reiter [453]. See Eymard [174] for our proof of 13.6. The fixed point characterization allowed Rickert [462] to establish 13.1, 13.4, 13.6. Also see Bourbaki [B EVT], especially for a demonstration of 13.5. Dieudonné [136] showed 13.2, via (P_1), and 13.4, in case the subgroup is compact, via (P_p). More observations on the relevance of (P_1) for the study of combinatorial properties were made by Reiter [453]; in particular he proved 13.4. Also using (P_1) Derighetti [131] established 13.3. Various versions of 13.4 were obtained by Greenleaf [228] via (SF). By making use of (G), Takenouchi [527] proved 13.2 and, for open subgroups, also 13.3. Hulanicki [268] established the general property 13.3 via (G). Relying on von Neumann's fundamental work [414], Day [115, 120] was the first to consider elementary groups systematically.

Day proved 13.10 [114, 115]. Rickert [462] gave 13.9. Schochetman [504] established 13.11 and obtained interesting related statements.

14. THE STRUCTURE OF AMENABLE GROUPS

We begin by considering standard examples of nonamenable groups. The simplest example is the discrete free group on two generators. We then give the full characterization of amenable groups among all almost connected groups, that is, groups that are compact extensions of connected groups.

A. Standard Nonamenable Groups

We establish the nonamenability of the free group F_2 on two generators by proving that there does not exist a left invariant mean on $l^\infty(F_2)$.

Proposition 14.1. *The discrete free group F_2 on two generators a and b is not amenable.*

Proof. Assume that $M \in \text{LIM}(l^\infty(F_2))$. Let A be the subset of all elements in F_2 that, written as reduced words, admit first factors of the form

THE STRUCTURE OF AMENABLE GROUPS

$a^n (n \in \mathbf{Z}^*)$. Then A, bA, and $b^2 A$ are disjoint subsets and

$$3M(1_A) = M(1_A) + M(1_{bA}) + M(1_{b^2 A}) \le M(1_G) = 1$$

hence $M(1_A) \le \frac{1}{3}$. On the other hand, $A \cup aA = G$ and therefore

$$1 = M(1_G) \le M(1_A) + M(1_{aA}) = 2M(1_A)$$

hence $M(1_A) \ge \frac{1}{2}$. We come to a contradiction. □

We now show that the group examined in Proposition 12.7 is the only example of a free amenable group of at least two generators with bounded order.

Proposition 14.2. *Let G be the discrete free group that is freely generated by the elements $a_1, \ldots, a_n \in G (n \ge 2)$ of orders p_i ($i = 1, \ldots, n$). Then G is nonamenable unless $n = 2$ and $p_1 = p_2 = 2$.*

Proof. If $i = 1, \ldots, n$, let A_i be the subset of G consisting of e and all elements that written as reduced words do not begin with a_i^k ($k = 1, \ldots, p_i - 1$).

(a) Assume that $n \ge 3$ and let $h = \sum_{i=1}^{n}(-1_{A_i} + {}_{a_i^{-1}}1_{A_i})$. Then

$$\sum_{i=1}^{n} 1_{A_i}(e) = n, \quad \sum_{i=1}^{n} {}_{a_i^{-1}}1_{A_i}(e) = 0$$

and

$$\sum_{i=1}^{n}(-1_{A_i} + {}_{a_i^{-1}}1_{A_i})(e) = -n.$$

If $x \ne e$,

$$\sum_{i=1}^{n} 1_{A_i}(x) = n - 1 \quad \text{and} \quad \sum_{i=1}^{n} {}_{a_i^{-1}}1_{A_i}(x) \le 1$$

hence

$$\sum_{i=1}^{n}(-1_{A_i} + {}_{a_i^{-1}}1_{A_i})(x) \le -n + 2 \le -1.$$

(b) Assume that $n = 2$ and $a_1^2 \ne e$. Let

$$h = 3\big(-1_{A_1} + {}_{a_1^{-1}}1_{A_1}\big) + 2\big(-1_{A_1} + {}_{a_1^{-2}}1_{A_1}\big) + 4\big(-1_{A_2} + {}_{a_2^{-1}}1_{A_2}\big).$$

We have $h(e) = -3 - 2 - 4 = -9$, $h(a_1) = 3 - 4 = -1$, $h(a_1^2) = 2 - 4 = -2$, $h(a_2) = -3 - 2 + 4 = -1$. If a reduced word x in G begins with $a_1^{k_1}(2 < k_1 < p_1)$, we have $h(x) = -4$; if it begins with $a_2^{k_2}(1 < k_2 < p_2)$, we have $h(x) = -3 - 2 = -5$. Hence $h \le -1_G$.

In both cases nonamenability follows from Proposition 4.29. □

The next corollary shows that the converse of Proposition 4.21 does not hold.

Corollary 14.3. *Let G be the compact (amenable) rotation group $SO(3, \mathbf{R})$ of the unit sphere in $\mathbf{R}^3$. Then G_d is nonamenable.*

Proof. Consider the rotations

$$a = \begin{pmatrix} 0 & 1 & 0 \\ 1 & 0 & 0 \\ 0 & 0 & -1 \end{pmatrix}, \quad b = \begin{pmatrix} 1 & 0 & 0 \\ 0 & -1/2 & \sqrt{3}/2 \\ 0 & -\sqrt{3}/2 & -1/2 \end{pmatrix}$$

and let H be the subgroup of G generated by a and b. We have $a^2 = b^3 = e$ and also

$$ba = \frac{1}{2}\begin{pmatrix} 0 & 2 & 0 \\ -1 & 0 & -\sqrt{3} \\ -\sqrt{3} & 0 & 1 \end{pmatrix}, \quad b^2 a = \frac{1}{2^2}\begin{pmatrix} 0 & 2 & 0 \\ -1 & 0 & \sqrt{3} \\ \sqrt{3} & 0 & 1 \end{pmatrix}.$$

One checks inductively that, for all $k \in \mathbf{N}^*$ and $n_1, \ldots, n_k \in \{1, 2\}$, the element $b^{n_1}ab^{n_2} \cdots ab^{n_k}a$ in H has the form

$$\frac{1}{2^k}\begin{pmatrix} p_1 & p_2 & p_3\sqrt{3} \\ q_1 & p_4 & q_2\sqrt{3} \\ q_3\sqrt{3} & p_5\sqrt{3} & q_4 \end{pmatrix}$$

where p_1, p_2, p_3, p_4, and p_5 are even integers and q_1, q_2, q_3, and q_4 are odd integers. Hence this element differs from e and a; there cannot exist a relation of the form $a^{\varepsilon_1}b^{n_1}ab^{n_2} \cdots b^{n_k}a^{\varepsilon_2} = e$ with $k \in \mathbf{N}^*$, $n_1, \ldots, n_k \in \{1, 2\}$, $\varepsilon_1, \varepsilon_2 \in \{0, 1\}$. So H is nonamenable by Proposition 14.2. Then also G_d is nonamenable by Proposition 13.3. □

Next we establish the nonamenability of some classical groups by direct computations.

Lemma 14.4. *Let $\alpha = \begin{pmatrix} 0 & 1 \\ -1 & 0 \end{pmatrix}$, $\beta = \begin{pmatrix} 1 & 2 \\ 0 & 1 \end{pmatrix} \in G = SL(2, \mathbf{R})$. There does not exist a relation of the form*

$$\alpha^{\varepsilon_1}\beta^{p_1} \cdots \alpha^{\varepsilon_m}\beta^{p_m} = e, \tag{1}$$

where $m \in \mathbf{N}^*$, $\varepsilon_1, \ldots, \varepsilon_m \in \{1, 2, 3\}$ and $p_1, \ldots, p_m \in \mathbf{Z}$ with $p_k \neq 0$ for any $k = 1, \ldots, m$ or, in case $m \ge 2$, $p_k \neq 0$ for any $k = 1, \ldots, m - 1$.

Proof. No nontrivial power of α is a nontrivial power of β. So (1) cannot hold in case $m = 1$. As α^2 belongs to the center of G, it suffices to show that we cannot have $\alpha^{\varepsilon_1}\beta^{p_1} \cdots \alpha^{\varepsilon_m}\beta^{p_m} = \alpha^2$ or e with $\varepsilon_1, \ldots, \varepsilon_m \in \{1, 3\}$ and $p_1, \ldots, p_{m-1} \in \mathbf{Z}^*$, $m \geq 2$.

One readily checks that $SL(2, \mathbf{R})$ may be realized by fractional transformations of $\mathbf{R}$ associating

$$x \mapsto \frac{ax + b}{cx + d}$$

to $\begin{pmatrix} a & b \\ c & d \end{pmatrix} \in SL(2, \mathbf{R})$. Notice that α, $\alpha^{-1} = \alpha^3$ correspond to $x \mapsto -1/x$, α^2 corresponds to $x \mapsto x$, and β^p corresponds to $x \mapsto x + 2p$ for $p \in \mathbf{Z}$. Assume that, for any $x \in \mathbf{R}^*$,

$$-\cfrac{1}{-\cfrac{1}{-\cfrac{1}{x + 2p_m} + 2p_{m-1}} + \cdots + 2p_1} = x. \qquad (2)$$

We let $a_1 = 2p_{m-1}$; so $|a_1| \geq 2$. If $m \geq 3$, we define inductively $a_k = -1/a_{k-1} + 2p_{m-k}$ for $k = 2, \ldots, m-1$; then $|a_{k-1}| \geq k/(k-1)$ implies that $|a_k| \geq 2|p_{m-k}| - 1/|a_{k-1}| \geq 2 - (k-1)/k = (k+1)/k$. Hence we have

$$|a_h| \geq \frac{h+1}{h} \qquad (3)$$

for $h = 1, \ldots, m-1$. By (3) the absolute value of the left-hand side of (2) should remain bounded when x tends to infinity and we come to a contradiction. We conclude that (1) cannot hold. $\square$

Proposition 14.5. *The group $G = SL(2, \mathbf{R})$ contains F_2.*

Proof. If G did not contain F_2, then for all $\gamma, \delta \in G$ we could find $m \in \mathbf{N}^*$ and $i_1, \ldots, i_m; j_1, \ldots, j_m \in \mathbf{Z}$ with $i_1, \ldots, i_m; j_1, \ldots, j_m \neq 0$ or, in case $m \geq 2$, $i_1, \ldots, i_m; j_1, \ldots, j_{m-1} \neq 0$ such that

$$\gamma^{i_1}\delta^{j_1} \cdots \gamma^{i_m}\delta^{j_m} = e. \qquad (4)$$

Choose in particular $\gamma = \alpha\beta\alpha$, $\delta = \beta$ with the notations of Lemma 14.4. Then (4) would give a relation of the form (1). $\square$

Recall that the discrete free group F_2 is closed in any locally compact group into which it is embedded ([HR] 5.10).

Corollary 14.6. *The groups $GL(n, \mathbf{R}), GL(n, \mathbf{C}), SL(n, \mathbf{R}), SL(n, \mathbf{C})$ ($n \geq 2$) admit the discrete (closed) free subgroup F_2 and are nonamenable.*

Proof. For any $n \geq 2$, $H = SL(2, \mathbf{R})$ [resp. $H = SL(2, \mathbf{C})$] may be considered to be a closed subgroup of $G = SL(n, \mathbf{R})$ [resp. $G = SL(n, \mathbf{C})$] by identification of the element $\begin{pmatrix} a & b \\ c & d \end{pmatrix}$ of H with the element

$$\begin{pmatrix} a & b & & \bigcirc & \\ c & d & & & \\ & & 1 & & \\ & \bigcirc & & \ddots & \\ & & & & 1 \end{pmatrix}$$

of G. Moreover $SL(2, \mathbf{R})$ is a closed subgroup of $SL(2, \mathbf{C})$. Hence all the groups considered here admit F_2 as a discrete (closed) subgroup by Proposition 14.5. As F_2 is nonamenable by Proposition 14.1, we conclude from Proposition 13.3 that all these groups are nonamenable. □

We now summarize some general results on Haar measures of a locally compact group G and a closed subgroup H ([B INT] Chap. VII, Section 2, No. 5, théorème 2; No. 6, théorème 3, corollaire 2).

(a) There exists a continuous mapping $\rho: G \to \mathbf{R}_+^*$ such that

$$\rho(xy) = \rho(x) \frac{\Delta_H(y)}{\Delta_G(y)}$$

whenever $x \in G$, $y \in H$.

(b) For every function $\rho: G \to \mathbf{R}_+^*$ satisfying (a), there exists $\mu_\rho = \mu \in \mathcal{M}_+(G/H)$, $\mu \neq 0$ such that

$$\int_G f(x) \rho(x) \, d\lambda_G(x) = \int_{G/H} \left(\int_H f(xy) \, d\lambda_H(y) \right) d\mu_\rho(\dot{x}) \tag{5}$$

whenever $f \in \mathcal{K}(G)$. Moreover μ is G-quasi-invariant: For every $a \in G$, the measure $\mu' \in \mathcal{M}_+(G/H)$ defined by

$$\langle g, \mu' \rangle = \int_{G/H} g(a\dot{x}) \, d\mu_\rho(\dot{x}),$$

$g \in \mathcal{K}(G/H)$, is equivalent to μ, that is, μ' and μ have the same locally null sets. For every $g \in \mathcal{K}(G/H)$ and every $a \in G$,

$$\int_{G/H} g(a\dot{x}) \, d\mu(\dot{x}) = \int_{G/H} \frac{\rho(a^{-1}x)}{\rho(x)} g(\dot{x}) \, d\mu(\dot{x}).$$

(c) The equality $\Delta_G|_H = \Delta_H$ is a necessary and sufficient condition for the existence of $\mu \in \mathcal{M}_+(G/H)$, $\mu \neq 0$ satisfying

$$\int_{G/H} g(a\dot{x}) \, d\mu(\dot{x}) = \int_{G/H} g(\dot{x}) \, d\mu(\dot{x})$$

whenever $g \in \mathcal{K}(G/H)$ and $a \in G$; the measure μ is said to be G-invariant.

THE STRUCTURE OF AMENABLE GROUPS

In particular, if H is a closed normal subgroup of G, then $\Delta_G|_H = \Delta_H$; we may normalize the Haar measures on G, H, and G/H in such a way that $\rho = 1_G$ and *Weil's formula*

$$\int_G f(x)\,d\lambda_G(x) = \int_{G/H}\left(\int_H f(xy)\,d\lambda_H(y)\right) d\lambda_{G/H}(\dot{x})$$

holds for every $f \in \mathcal{K}(G)$. ([B INT] Chap. VII, Section 2, No. 7, proposition 10).

We come back to the general case and consider

$$J_{H,\rho} : f \mapsto \dot{f}$$

$$L^1(G) \to L^1(G/H, \mu_\rho)$$

where $\dot{f}(\dot{x}) = \int_H f(xy)/\rho(xy)\,d\lambda_H(y)$, $\dot{x} \in G/H$. We write J_H for $J_{H,\rho}$ if $\rho = 1_G$. From (5) we deduce the *Mackey–Bruhat formula*

$$\int_G f(x)\,dx = \int_{G/H} \dot{f}(\dot{x})\,d\mu_\rho(\dot{x}),$$

$f \in L^1(G)$. Moreover for every $f \in L^1(G)$,

$$\|\dot{f}\|_{L^1(G/H)} = \int_{G/H} \left|\int_H \frac{f(xy)}{\rho(xy)}\,d\lambda_H(y)\right| d\mu_\rho(\dot{x})$$

$$\leq \int_{G/H} \int_H \frac{|f(xy)|}{\rho(xy)}\,d\lambda_H(y)\,d\mu_\rho(\dot{x}) = \|f\|_{L^1(G)}$$

hence $\|J_{H,\rho}\| \leq 1$. If $F \in \mathcal{K}(G/H)$, $F \neq 0$, then $K = \operatorname{supp} F$ has nonvoid interior, and there exists a compact subset L in G such that K is the canonical image of L ([HR] 5.24.b) with $\mathring{L} \neq \varnothing$. Let $\varphi \in \mathcal{K}_+(G)$ such that $\varphi(z) > 0$ whenever $z \in L$; then also $\varphi(z)/\rho(z) > 0$ whenever $z \in L$. For any $\dot{z} \in K$,

$$J_{H,\rho}\varphi(\dot{z}) > 0$$

and we may put

$$f(x) = F(\dot{x})\frac{\varphi(x)}{J_{H,\rho}\varphi(\dot{x})} \quad \text{if } x \in L,$$

$$f(x) = 0 \quad \text{if } x \in G \setminus L.$$

Then f is continuous, and, more precisely, $f \in \mathcal{K}(G)$ and $J_{H,\rho}(f) = F$. By density we conclude that $J_{H,\rho}$ is surjective. If $f \in L^1(G)$, $\dot{x} \in G/H$, and

$a \in H$, then

$$\int_H \frac{f(xya^{-1})}{\rho(xy)} \Delta_G(a^{-1}) \, dy = \int_H \frac{f(xy)}{\rho(xya)} \frac{\Delta_H(a)}{\Delta_G(a)} \, dy = \int_H \frac{f(xy)}{\rho(xy)} \, dy,$$

that is,

$$J_{H,\rho}(R_a f) = J_{H,\rho}(f). \tag{6}$$

If in particular, H is a closed normal subgroup, let π be the canonical homomorphism of G onto G/H. For any $F \in \mathscr{C}_0(G/H)$ and $f \in \mathscr{K}(G)$,

$$\int_{G/H} F(\dot{x}) J_H(f^*)(\dot{x}) \, d\dot{x} = \int_G F \circ \pi(x) f^*(x) \, dx$$

$$= \int_G F \circ \pi(x) \overline{f(x^{-1})} \Delta(x^{-1}) \, dx$$

$$= \int_G \overline{(F \circ \pi)^{\sim}(x) f(x)} \, dx = \int_{G/H} \overline{F^{\sim}(\dot{x}) J_H f(\dot{x})} \, d\dot{x}$$

$$= \int_{G/H} F(\dot{x}) \overline{J_H f(\dot{x}^{-1})} \Delta_{G/H}(\dot{x}^{-1}) \, d\dot{x} = \int_{G/H} F(\dot{x}) (J_H f)^*(\dot{x}) \, d\dot{x}.$$

Hence we proved that, for every $f \in \mathscr{K}(G)$,

$$J_H(f^*) = (J_H f)^* \in L^1(G/H). \tag{7}$$

The equality then holds for every $f \in L^1(G)$ by density. Also J_H is an algebraic homomorphism. For density reasons, it suffices to consider $f, g \in \mathscr{K}(G)$. By Weil's formula we obtain, for $\dot{x} \in G/H$,

$$J_H(f *_G g)(\dot{x}) = \int_H \int_G f(y) g(y^{-1} xz) \, d\lambda_G(y) \, d\lambda_H(z)$$

$$= \int_H \int_{G/H} \int_H f(yt) g(t^{-1} y^{-1} xz) \, d\lambda_{G/H}(\dot{y}) \, d\lambda_H(t) \, d\lambda_H(z)$$

$$= \int_H \int_{G/H} f(yt) \left(\int_H g(t^{-1} y^{-1} xz) \, d\lambda_H(z) \right) d\lambda_{G/H}(\dot{y}) \, d\lambda_H(t)$$

$$= \int_H \int_{G/H} f(yt) \left(\int_H g(y^{-1} xz) \, d\lambda_H(z) \right) d\lambda_{G/H}(\dot{y}) \, d\lambda_H(t)$$

$$= \int_H \int_{G/H} \int_H f(yt) g(y^{-1} xz) \, d\lambda_{G/H}(\dot{y}) \, d\lambda_H(t) \, d\lambda_H(z)$$

$$= J_H f *_{G/H} J_H g(\dot{x}).$$

If π is the canonical mapping of G onto G/H and $f: G/H \to \mathbf{C}$, then $f \circ \pi \in L^\infty(G)$ if and only if $f \in L^\infty(G/H)$ ([B INT] Chap. VII, Section 2, No. 3, proposition 6).

Lemma 14.7. *Let G be a unimodular locally compact group admitting a nonunimodular closed subgroup H and a compact subgroup K such that $G = KH$ and the mapping $(z, y) \mapsto zy$ of $K \times H$ onto G is open. Then there exist $\alpha \in \mathbf{R}_+^*$ and $a_0 \in H$ such that the following property holds: For every $f \in \mathcal{K}(G)$ such that $\int f = 1$ and $f(bx) = f(x)$ whenever $x \in G$, $b \in K$, one has $\|_{a_0} f - f\|_1 \geq \alpha$.*

Proof. Since G is unimodular and the compact group K is also unimodular we may assume that $\rho = 1_G$ and deduce from (5) the existence of a G-invariant measure $\mu \in \mathcal{M}_+(G/K)$ such that $\mu \neq 0$ and, for every $f \in \mathcal{K}(G)$,

$$\int_G f(x)\,dx = \int_{G/K} \left(\int_K f(xz)\,d\lambda_K(z) \right) d\mu(\dot{x}). \tag{8}$$

The canonical mapping of $H/H \cap K$ onto HK/K is open ([HR] 5.32, 5.33), hence by hypothesis there exists a canonical open mapping of $H/H \cap K$ onto G/K. Therefore, μ induces on $H/H \cap K$ a measure ν that is H-invariant. As $H \cap K$ is a compact subgroup of H, we may choose the Haar measure λ_H on H in such a way that the following relation is satisfied for any $g \in \mathcal{K}(H)$ for which $g(ab) = g(a)$ whenever $a \in H$ and $b \in H \cap K$:

$$\int_H g(y)\,d\lambda_H(y) = \int_{H/H \cap K} \left(\int_{H \cap K} g(yz)\,d\lambda_{H \cap K}(z) \right) d\nu(\dot{y})$$

$$= \int_{H/H \cap K} g(y)\,d\nu(\dot{y}). \tag{9}$$

Let $g_0(y) = \int_K f(yz)\,d\lambda_K(z)$, $y \in H$. Since $f \in \mathcal{K}(G)$, we have $g_0 \in \mathcal{K}(H)$ and, for all $a \in H$, $b \in H \cap K$,

$$g_0(ab) = \int_K f(abz)\,d\lambda_K(z) = \int_K f(az)\,d\lambda_K(z) = g_0(a).$$

The relations (8) and (9) imply that

$$\int_G f(x)\,dx = \int_H \left(\int_K f(yz)\,d\lambda_K(z) \right) d\lambda_H(y)$$

for every $f \in \mathscr{K}(G)$. As G and K are unimodular, we obtain

$$\int_G f(x)\,dx = \int_G \check{f}(x)\,dx = \int_H \left(\int_K f(z^{-1}y^{-1})\,d\lambda_K(z)\right) d\lambda_H(y)$$

$$= \int_H \left(\int_K f(zy^{-1})\,d\lambda_K(z)\right) d\lambda_H(y)$$

$$= \int_K \left(\int_H f(zy)\Delta_H(y^{-1})\,d\lambda_H(y)\right) d\lambda_K(z). \tag{10}$$

Suppose now that $f \in \mathscr{K}(G)$, $\int_G f = 1$, and $f(bx) = f(x)$ whenever $x \in G$, $b \in K$. By (10) we have

$$1 = \int_G f(x)\,dx = \int_H f(y)\Delta_H(y^{-1})\,d\lambda_H(y) = \int_H f(y^{-1})\,d\lambda_H(y) \tag{11}$$

as λ_K is normalized on the compact group K. By nonunimodularity of H, there exists $a_0 \in H$ such that $\Delta_H(a_0) \neq 1$. For every $x \in G$, there exist $p(x) \in H$ and $q(x) \in K$ such that $x = q(x)p(x)$. If $t \in K$, we determine $y_1 = p(a_0 t) \in H$ and $z_1 = q(a_0 t) \in K$ such that $z_1 y_1 = a_0 t \in G$. If also $z_1 y_1 = z_1' y_1'$ for $y_1' \in H$, $z_1' \in K$, then $y_1 y_1'^{-1} = z_1^{-1} z_1' \in H \cap K$. Because $H/H \cap K$ admits an H-invariant measure ν and $H \cap K$ is a compact group, with respect to (c) we have $\Delta_H(y_1 y_1'^{-1}) = 1$, that is, $\Delta_H(y_1) = \Delta_H(y_1')$. In particular, $\Delta_H(p(a_0)) = \Delta_H(a_0) \neq 1$. Since $_{a_0}f - f \in \mathscr{K}(G)$, by (10) we obtain

$$\int_G |f(a_0 x) - f(x)|\,dx = \int_K \int_H |f(a_0 zy) - f(zy)|\Delta_H(y^{-1})\,d\lambda_H(y)\,d\lambda_K(z).$$

$$\tag{12}$$

The mapping

$$\varphi: z \mapsto \left|\int_H (f(a_0 zy) - f(zy))\Delta_H(y^{-1})\,d\lambda_H(y)\right|$$
$$G \to \mathbf{C}$$

is continuous and with respect to the definition of f and (11), for every $z \in K$, we have

$$\int_H (f(a_0 zy) - f(zy))\Delta_H(y^{-1})\, d\lambda_H(y)$$
$$= \int_H (f(q(a_0 z)p(a_0 z)y) - f(zy))\Delta_H(y^{-1})\, d\lambda_H(y)$$
$$= \int_H (f(p(a_0 z)y) - f(y))\Delta_H(y^{-1})\, d\lambda_H(y)$$
$$= \int_H (f(p(a_0 z)y^{-1}) - f(y^{-1}))\, d\lambda_H(y)$$
$$= \int_H f(y^{-1})(\Delta_H(p(a_0 z)) - 1)\, d\lambda_H(y) = \Delta_H(p(a_0 z)) - 1.$$

As $\varphi(e) = |\Delta_H(p(a_0)) - 1| = |\Delta_H(a_0) - 1| > 0$, by continuity we have

$$\alpha = \int_K |\Delta_H(p(a_0 z)) - 1|\, d\lambda_K(z) > 0.$$

So finally (12) implies that

$$\|_{a_0}f - f\|_1 \geq \int_K \left|\int_H (f(a_0 zy) - f(zy))\Delta_H(y^{-1})\, d\lambda_H(y)\right| d\lambda_K(z) = \alpha. \quad \square$$

Proposition 14.8. *Let G be a unimodular locally compact group admitting a nonunimodular closed subgroup H and a compact subgroup K such that $G = KH$ and the mapping $(z, y) \mapsto zy$ from $K \times H$ into G is open. Then G is nonamenable.*

Proof. Given $\varepsilon > 0$ and the compact subset C of G, we consider the compact subset $C' = (KC) \cup K$ of G. If G were amenable, by Corollary 6.15 there would exist $\varphi \in P^1(G) \cap \mathcal{K}(G)$ such that $\|_a\varphi - \varphi\|_1 < \varepsilon/2$ for every $a \in C'$; so for all $c \in C$ and $z \in K$, we would have $\|_{zc}\varphi - \varphi\|_1 < \varepsilon/2$ and $\|_z\varphi - \varphi\|_1 < \varepsilon/2$, hence $\|_{zc}\varphi - _z\varphi\|_1 < \varepsilon$.

Let $g(x) = \int_K \varphi(zx)\, d\lambda_K(z)$, $x \in G$. Since $\varphi \in \mathcal{K}(G)$ and K is compact, $g \in \mathcal{K}(G)$. By unimodularity of K, for every $x \in G$ and every $b \in K$,

$$g(bx) = \int_K \varphi(zbx)\, d\lambda_K(z) = \int_K \varphi(zx)\, d\lambda_K(z) = g(x).$$

We have

$$\int_G g(x)\, dx = \int_G \left(\int_K \varphi(zx)\, d\lambda_K(z)\right) dx = \int_K \left(\int_G \varphi(zx)\, dx\right) d\lambda_K(z)$$
$$= \int_K d\lambda_K(z) = 1.$$

For every $c \in C$,

$$\|_c g - g\|_1 = \int_G |g(cx) - g(x)| d\lambda_G(x)$$

$$\leq \int_K \int_G |\varphi(zcx) - \varphi(zx)| d\lambda_K(z) d\lambda_G(x)$$

$$= \int_K \|_{zc}\varphi - {}_z\varphi\|_1 d\lambda_K(z) < \varepsilon.$$

Since the compact subset C of G and $\varepsilon > 0$ may be chosen arbitrarily, Lemma 14.7 would be violated. $\square$

We state an example to which Proposition 14.8 applies. If p is a prime number, we consider the p-adic field $\mathbf{Q}_p$ ([HR] 10.10, [B TG] Chap. IX, Section 3, No. 2).

Proposition 14.9. *The groups $GL(n, \mathbf{Q}_p), SL(n, \mathbf{Q}_p)$ ($n \geq 2$) are nonamenable.*

Proof. Let $G = GL(2, \mathbf{Q}_p)$. We consider the closed subgroup G_1 of all $\begin{pmatrix} a & b \\ bp & a \end{pmatrix} \in G$ such that $a, b \in \mathbf{Q}_p$ and $a = 0, b = 0$ do not hold simultaneously. We also consider the closed affine subgroup H of G formed by all $\begin{pmatrix} c & d \\ 0 & 1 \end{pmatrix} \in G$ with $c \in \mathbf{Q}_p^*, d \in \mathbf{Q}_p$. Obviously $G_1 \cap H = \{e\}$. Any element $x = \begin{pmatrix} s & t \\ u & v \end{pmatrix} \in G$ may be decomposed uniquely in the form $\begin{pmatrix} a & b \\ bp & a \end{pmatrix}\begin{pmatrix} c & d \\ 0 & 1 \end{pmatrix}$ with

$$a = \frac{sp(sv - tu)}{s^2 p - u^2}, \quad b = \frac{u(sv - tu)}{s^2 p - u^2}, \quad c = \frac{s^2 p - u^2}{p(sv - tu)},$$

$$d = \frac{stp - uv}{p(sv - tu)} \in \mathbf{Q}_p,$$

where $sv - tu \neq 0$ and also $s^2 p - u^2 \neq 0$ since p is a prime number.

In G_1 we consider the center $Z = \left\{\begin{pmatrix} a & 0 \\ 0 & a \end{pmatrix} : a \in \mathbf{Q}_p^*\right\}$ of G. Because $Z \cap H = \{e\}$, we may identify $G' = G/Z = (G_1 H)/Z$ with $(G_1/Z)H$. Let π be the canonical homomorphism of G_1 onto $K = G_1/Z$. We consider the nonvoid compact subset

$$C = \left\{\begin{pmatrix} a & b \\ bp & a \end{pmatrix} : \sup\{|a|_p, |b|_p\} = 1\right\} \text{ of } G_1.$$

For every $x \in G_1$, $xZ \cap C \neq \emptyset$. In view of establishing compactness of K, let $\{U_i : i \in I\}$ be an open covering of K and put $V_i = \pi^{-1}(U_i)$ for $i \in I$;

$\{V_i : i \in I\}$ is an open covering of G_1 and there exist $i_1,\ldots,i_n \in I$ such that $C \subset \cup_{k=1}^n V_{i_k}$. For every $x \in G_1$, we may determine $z \in Z$ such that $xz \in C$ and $k = 1,\ldots,n$ such that $xz \in V_{i_k}$, $x \in V_{i_k} z^{-1}$. Therefore $G_1 \subset \cup_{k=1}^n V_{i_k} Z$ and $K \subset \cup_{k=1}^n U_{i_k}$.

The group H is nonunimodular ([B INT] Chap. VII, Section 3, No. 3, exemple 2) and the group G is unimodular (ibid. Chap. VII, Section 3, No. 3, exemple 1); then also $G' = G/Z$ is unimodular (ibid. Chap. VII, Section 2, No. 7, corollaire). Proposition 14.8 applied to G', H, and K yields nonamenability of G'. Then G is also nonamenable by Proposition 13.2.

The quotient group $GL(2,\mathbf{Q}_p)/SL(2,\mathbf{Q}_p)$ is abelian, hence amenable. We deduce from Proposition 13.4 that $SL(2,\mathbf{Q}_p)$ is nonamenable. Moreover as $SL(n,\mathbf{Q}_p)$ [resp. $GL(n,\mathbf{Q}_p)$] admits a closed subgroup isomorphic to $SL(2,\mathbf{Q}_p)$ [resp. $GL(2,\mathbf{Q}_p)$], it is nonamenable by Proposition 13.3. □

B. Amenable Almost Connected Groups

We begin by quoting some general results about the structures of topological groups and locally compact groups.

Let G be a topological group. If A is a closed subset and B is a compact subset, then AB is closed ([HR] 4.4). If H and K are normal subgroups and $K \subset H$, then G/H is topologically isomorphic to $(G/K)/(H/K)$ ([HR] 5.35). For a closed [resp. compact] subgroup H of G, the natural mapping of G onto G/H is open [resp. open and closed] ([HR] 5.17–18). If H is a compact subgroup of G and G/H is compact, then G is compact ([HR] 5.25).

In a topological group G the connected component G_0 of e is a closed normal subgroup ([HR] 7.1). As G_0 is the intersection of all open subgroups of G ([HR] 7.8), one has $H_0 = G_0$ for every open subgroup H of G. If G is a topological group, H is a closed normal subgroup of G and π is the canonical homomorphism of G onto G/H, then $\pi(G_0)$, that is, $G_0 H/H$ ($\simeq G_0/G_0 \cap H$) equals $(G/H)_0$ ([HR] 5.32, 7.13). The locally compact group G is termed *almost connected* if G/G_0 is compact.

We recall the classical result due to Montgomery and Zippin ([397] 4.6) concerning the approximation of locally compact groups by Lie groups: If G is a locally compact almost connected group, then any neighborhood of e contains a compact normal subgroup K of G such that G/K is a Lie group. We state some more classical results due to Iwasawa ([276] Theorem 2, Theorem 15): If G is a connected topological group and K is a compact normal subgroup of G, we have $G = HK$, where H is the centralizer of K in G. Every connected locally compact group G contains a unique (closed) maximal solvable normal subgroup S; S_0 is the (closed) maximal connected solvable normal subgroup of G and is called the *radical* $\operatorname{rad} G$ of G. If G is any locally compact group, we define the radical $\operatorname{rad} G$ of G by $\operatorname{rad} G = \operatorname{rad} G_0$. In case G is a connected Lie group and H is a closed normal (Lie) subgroup of G, the canonical homomorphism of G onto G/H maps $\operatorname{rad} G$ onto $\operatorname{rad}(G/H)$

([B LIE] Chap. III, Section 9, No. 7, proposition 24). The locally compact group G is called *semisimple* if its radical is trivial, that is, reduced to $\{e\}$; G is semisimple if and only if G_0 is semisimple. Notice that any quotient group of a semisimple connected Lie group is also semisimple. We say that a locally compact group G is *almost solvable* if $G/\operatorname{rad} G$ is compact.

We now recall another classical fact for which references are Helgason [254], Warner [550], Wallach [548], and Dieudonné [137]. Every semisimple connected Lie group G admits an *Iwasawa decomposition*: There exists a diffeomorphism of a product of three connected Lie subgroups onto G; $G = KAN$, where K is a maximal compact subgroup, A is abelian and N is nilpotent. Also $G = KH$ where H is a solvable Lie subgroup of G ([254] Chap. VI, Section 5; [550] p. 27; [137] 21.21.10). If $\mathfrak{n}$ is the Lie algebra of N, we have

$$\Delta_H(an) = e^{-\operatorname{Trace}((ad \log a)|_{\mathfrak{n}})},$$

$a \in A$, $n \in N$ ([254] Ch. X, Theorem 1.15; [548] 7.6.3), and we may assume that $ad \log a$ $(a \in A)$ admits positive roots ([137] 21.21.10); hence if G is noncompact, the solvable group H is nontrivial and nonunimodular. But the semisimple Lie group G is unimodular ([254] Ch. X, Proposition 1.4; [137] 21.6.6). Therefore a noncompact semisimple connected Lie group satisfies the conditions of Proposition 14.8 and is nonamenable. A more precise statement is at hand.

Proposition 14.10. *A noncompact semisimple connected Lie group G admits a discrete free subgroup F_2 on two generators and is nonamenable.*

Proof. The Iwasawa decomposition of G insures the existence of a nontrivial nilpotent Lie subgroup of G; its nilpotent Lie algebra is nontrivial. Therefore the semisimple Lie algebra of G admits a subalgebra that is isomorphic to $sl(2, \mathbf{R})$ ([B LIE] Chap. VIII, Section 11, No. 2, proposition 2). We conclude that the connected Lie group G admits a subgroup that is isomorphic to $SL(2, \mathbf{R})$. Then, by Proposition 14.5, G admits the discrete (closed) free subgroup F_2. Therefore Propositions 14.1 and 13.3 yield nonamenability of G. □

Proposition 14.11. *If G is a semisimple locally compact group and K is a compact normal subgroup of G, then G/K is also semisimple.*

Proof. (a) Assume G to be connected. We have $G = HK$, where H is the centralizer of K in G. If π is the canonical homomorphism of H onto the quotient group $H/H \cap K$ ($\simeq HK/K = G/K$) and R is the radical of $H/H \cap K$, we consider the closed subgroup $T = \pi^{-1}(R)$ of H. As H is the centralizer of K, $H \cap K$ belongs to the center of H; hence T is solvable ([B A] I, Section 6, No. 4, proposition 10). As R is normal in G/K, T is normal in G. Thus T is contained in the maximal (closed) normal solvable subgroup S of G.

THE STRUCTURE OF AMENABLE GROUPS   135

We have $T_0 \subset S_0 = \mathrm{rad}\, G = \{e\}$, that is, T is totally disconnected. Therefore $R = \{\dot{e}\}$, that is, G/K is semisimple.

(b) Consider now the general case. The group G_0 is semisimple and admits the normal compact subgroup $G_0 \cap K$. By (a), $G_0/G_0 \cap K$ is semisimple. Since $(G/K)_0 \simeq G_0/G_0 \cap K$, $(G/K)_0$ is semisimple, hence G/K is semisimple. □

Proposition 14.12. *Let G be a locally compact group that is almost connected and semisimple. Then G is amenable if and only if it is compact.*

Proof. If G is compact, it is amenable by Proposition 12.1.

If G is amenable, the closed subgroup G_0 is amenable by Proposition 13.3. There exists a compact normal subgroup K of G_0 such that G_0/K is a Lie group; G_0/K is also amenable by Corollary 13.2. As G is semisimple, G_0 is semisimple, and then G_0/K is semisimple by Proposition 14.11. The connected semisimple amenable Lie group G_0/K is compact by Proposition 14.10. As K is also compact, we conclude that G_0 is compact. In the same way, G/G_0 being compact, we prove compactness of G. □

Theorem 14.13. *Let G be an almost connected locally compact group. Then G is amenable if and only if it is almost solvable.*

Proof. Assume that $G/\mathrm{rad}\, G$ is compact. Since the solvable group $\mathrm{rad}\, G$ is amenable, G is then amenable by Proposition 13.4.

If G is amenable, $G/\mathrm{rad}\, G$ is amenable by Corollary 13.2. As $(G/\mathrm{rad}\, G)_0 = G_0/\mathrm{rad}\, G$ and $(G/\mathrm{rad}\, G)/(G_0/\mathrm{rad}\, G) \simeq G/G_0$, $G/\mathrm{rad}\, G$ is almost connected. We may apply Proposition 14.12 to the semisimple group $G/\mathrm{rad}\, G$; $G/\mathrm{rad}\, G$ is compact. □

Corollary 14.14. *A connected locally compact group G is amenable if and only if it is an extension of a closed connected normal solvable subgroup by a compact group.*

Proof. Assume G to be amenable; $G/\mathrm{rad}\, G$ is connected and by Corollary 13.2 also amenable. Therefore Proposition 14.12 implies that $G/\mathrm{rad}\, G$ is compact and the condition is satisfied.

Conversely, if the condition holds, let H be a closed connected normal solvable subgroup of G such that G/H is compact. Since H and G/H are amenable, G is also amenable by Proposition 13.4. □

Lemma 14.15. *Let G be a locally compact group that is not semisimple. There then exists a neighborhood U of e in G such that, for every closed normal subgroup H contained in U, G/H is not semisimple.*

Proof. Since G is not semisimple, it admits a nontrivial closed connected normal solvable subgroup S. Let $x \in S$, $x \neq e$ and $U = G \setminus \{x\}$. If H is any closed normal subgroup contained in U and π is the canonical homomorphism of G onto G/H, $\pi(S)$ is a connected normal solvable subgroup of G/H, and $\pi(S)$ is not trivial. □

Proposition 14.16. *Let G be a semisimple locally compact group and H a closed normal subgroup of G. Then G/H is semisimple.*

Proof. Let π be the canonical homomorphism of G onto G/H.

(a) We first consider the special case in which G is almost connected.

Assume G/H to be nonsemisimple, and consider the neighborhood U of $\dot{e}$ in G/H given by Lemma 14.15. There exists a compact normal subgroup K of G such that K is contained in the neighborhood $\pi^{-1}(U)$ of e and G/K is a Lie group. Then $KH/H \subset U$ and, by Lemma 14.15, $(G/K)/(KH/K) \simeq G/KH \simeq (G/H)/(KH/H)$ is not semisimple. By Proposition 14.11, G/K is semisimple and the quotient group $(G/K)/(KH/K)$ of this semisimple Lie group should also be semisimple. A contradiction would arise.

(b) We now consider the general case.

Since G/G_0 is a totally disconnected group, there exists a compact open subgroup L' in G/G_0 ([HR] 7.7). Hence there exists an open subgroup L of G such that $L/G_0 = L'$ is compact. We have $L_0 = G_0$ and, since G is semisimple, L is semisimple. By (a), $\pi(L) = LH/H \simeq L/L \cap H$ is semisimple, so $\pi(L)_0$ is semisimple. Since L is open and π is an open mapping, $\pi(L)$ is an open subgroup of G/H and $\pi(L)_0 = (G/H)_0$; therefore G/H is semisimple. □

Corollary 14.17. *Let G be a locally compact group with radical R and let H be a closed normal subgroup of G. If π is the canonical homomorphism of G onto G/H, then $\overline{\pi(R)}$ is the radical of G/H.*

Proof. Since $\pi(R)$ is a connected normal solvable subgroup of G/H, so is $S = \overline{\pi(R)}$. We have $(G/H)/S \simeq G/\pi^{-1}(S) \simeq (G/R)/(\pi^{-1}(S)/R)$. Since G/R is semisimple, by Proposition 14.16, $(G/H)/S$ is semisimple. Therefore S must be the radical of G/H. □

We adopt a definition introduced by Iwasawa ([276]).

Definition 14.18. (a) *A Lie group G is called (C)-group if G_0 is almost solvable.*

(b) *A locally compact group G is called (C)-group if every neighborhood U of e contains a compact normal subgroup K such that G/K is a Lie group that is a (C)-group.*

Lemma 14.19. *Let G be an almost connected locally compact group with radical R. If G is a (C)-group, then G/R is compact.*

Proof. Let K be a compact normal subgroup of G such that G/K is a Lie group and $(G/K)_0$ is almost solvable. Denote by π the canonical homomor-

THE STRUCTURE OF AMENABLE GROUPS

phism of G onto G/K. Since R is closed and K is compact, RK is closed in G and, by Corollary 14.17, $\overline{\pi(R)} = \pi(R) = RK/K$ is the radical of G/K; it is also the radical of $(G/K)_0$. By hypothesis $((G/K)_0)/(RK/K)$ is compact, that is, $(G_0K/K)/(RK/K)$ is compact. Since G is almost connected, G/G_0 is compact and then the quotient of $(G/K)/(RK/K)$, by its closed normal subgroup $(G_0K/K)/(RK/K)$, is compact. We obtain compactness of $(G/K)/(RK/K)$, G/RK, and $(G/R)/(RK/R)$. But the image RK/R of the compact subset K in the canonical homomorphism of G onto G/R is compact; we may thus conclude that G/R is compact. □

Lemma 14.20. *Let G be a locally compact group that is almost solvable. Then for every closed normal subgroup H of G, $(G/H)_0$ is almost solvable.*

Proof. Let R be the radical of G and let π be the canonical homomorphism of G onto G/H. By Corollary 14.17, for $S = \overline{RH}$, $\pi(S)$ is the radical of G/H and $(G/H)_0$. Since G/R is compact, $(G/H)/(\pi(S)) \simeq G/S \simeq (G/R)/(S/R)$ is compact. The closed subgroup $(G/H)_0/\pi(S)$ of $(G/H)/\pi(S)$ is therefore also compact. □

Proposition 14.21. *If G is an almost connected locally compact group, the following properties are equivalent:*
  (i) *G is almost solvable.*
  (ii) *G is a (C)-group.*
  (iii) *Every Lie group G/H, where H is a closed normal subgroup of G, is a (C)-group.*

Proof. (i) ⇒ (iii)
The statement follows from Lemma 14.20.

$$(iii) \Rightarrow (ii)$$

As G is almost connected, every neighborhood U of e in G contains a compact normal subgroup K such that G/K is a Lie group; by (iii), G/K is a (C)-group.

$$(ii) \Rightarrow (i)$$

The statement is Lemma 14.19. □

Proposition 14.22. *An almost solvable locally compact group G is amenable and does not admit the discrete subgroup F_2.*

Proof. Let R be the radical of G. Since the solvable group R and the compact group G/R are amenable, G is amenable by Proposition 13.4. Therefore every closed subgroup of G must be amenable by Proposition 13.3, and the nonamenable discrete group F_2 cannot be a closed subgroup of G. □

Lemma 14.23. *Let G be a locally compact group. If G has a closed normal subgroup H such that G/H admits a discrete free subgroup on two generators, then also G admits a discrete free subgroup on two generators.*

Proof. Let $\dot{a}$ and $\dot{b}$ be the generators of F_2 in G/H. If π is the canonical homomorphism of G onto G/H, choose $a' \in \pi^{-1}(\dot{a})$ and $b' \in \pi^{-1}(\dot{b})$; a' and b' generate a free subgroup F_2' on two generators in G. Since the restriction to F_2' of the continuous mapping π is a bijection from F_2' onto F_2, F_2' is discrete.
□

Lemma 14.24. *A noncompact, semisimple, almost connected locally compact group G admits a discrete free subgroup on two generators.*

Proof. There exists a compact normal subgroup K in G_0 such that G_0/K is a Lie group. If G_0/K were compact, G_0 would be compact and then, as G/G_0 is compact, G would be compact; hence G_0/K is noncompact. Since G is semisimple, G_0 is semisimple. By Proposition 14.11, G_0/K is also semisimple; moreover G_0/K is connected. Proposition 14.10 implies that G_0/K admits the discrete subgroup F_2, then G_0 does so by Lemma 14.23. Finally G also admits the discrete subgroup F_2.
□

Theorem 14.25. *Let G be an almost connected locally compact group. Then G is amenable if and only if it does not admit a discrete free subgroup on two generators.*

Proof. Since F_2 is nonamenable, by Proposition 13.3 it cannot appear as a closed subgroup of an amenable group.

Assume G to be nonamenable and let R be the (amenable) radical of G. The quotient group G/R is nonamenable by Proposition 13.4. We have $(G/R)/(G/R)_0 = (G/R)/(G_0/R) \simeq G/G_0$; hence the semisimple group G/R is almost connected. Proposition 14.12 implies that G/R is noncompact. Then G/R admits the discrete subgroup F_2 by Lemma 14.24, so the statement follows from Lemma 14.23.
□

We round up this section by giving a particular example illustrating the general results.

Corollary 14.26. *The class $[FC]^-$ belongs to the class of amenable groups.*

Proof. Let G be a locally compact group of the class $[FC]^-$. The groups G_0 and G/G_0 have all their conjugacy classes relatively compact. The group G_0 cannot admit the discrete free subgroup on two generators a and b because the subset $\{b^n a b^{-n} : n \in \mathbf{N}\}$ of the conjugacy class K_a of a would then be infinite. Thus G_0 is amenable by Theorem 14.25. Since the totally disconnected group G/G_0 is also amenable by Proposition 12.9, Proposition 13.4 implies that G is amenable.
□

NOTES

The first formulation of 14.1 is due to von Neumann [414]. Other characterizations of amenability may be used to establish the nonamenability of F_2. Hulanicki [268] noticed that (F) does not hold for this group, and Leptin [349] showed failure of (L_e). Yoshizawa [576] proved that (G) is not satisfied for F_2 and mentioned having been told later that his demonstration had already been given in a lecture by Kakutani. Dieudonné [136] noticed that F_2 does not satisfy (Cv_2). Dixmier [138] established 14.2; Darsow [109] showed that the group freely generated by at least three elements each of order 2 does not satisfy (G). Hausdorff's [253] original proof of 14.3 led to the Hausdorff–Banach–Tarski paradox. Elementary demonstrations of the fact were given by Locher–Ernst [366], Dubins [148], and Stromberg [522]. The considerations made in 14.4–5 go back to von Neumann [414], and are exposed in Greenleaf [226]. Reiter [453] proved 14.7 in case $H \cap K = \{e\}$; the generalization is due to Van Dijk [544]. Reiter [453] established the proof of 14.8; he noticed that it applies to any noncompact connected semisimple Lie group with finite center via the Iwasawa decomposition. He also gave 14.9. Van Dijk [544] proved that the group of rational points of a connected semisimple linear algebraic group over a p-adic field is amenable if and only if it is compact.

Furstenberg [191] showed that a connected Lie group has property (FP) if and only if it is almost solvable; for related results see Guivarc'h [241]. Takenouchi [527] demonstrated that an almost connected locally compact group has property (G) if and only if it is a (C)-group. The results given in 14.10–25 were established by Rickert [462, 463]. For 14.26 we refer to Leptin [355]. In fact, the groups belonging to $[FC]^-$ have polynomial growth; see Palmer [425].

4

COMPLEMENTS ON THE CHARACTERISTIC PROPERTIES OF AMENABLE GROUPS

In this chapter we give supplementary information on the characteristic properties of amenable groups stated in Sections 6–10. Considerations on invariant means are postponed to the next chapter.

15. QUASI-INVARIANCE PROPERTIES

We indicate refinements concerning the properties of Day, Reiter, and Glicksberg–Reiter and existence of approximate units in the algebras $A_p(G)$.

A. Dichotomy Properties

We come back to the notations of Section 6B.

Lemma 15.1. *Let G be a locally compact group.*
  (a) *If* $\gamma_1 = 2$ [*resp.* $\gamma_1^* = 2$], *then* $\gamma_2 = 2^{1/2}$ [*resp.* $\gamma_2^* = 2^{1/2}$].
  (b) *If* $\gamma_2 = 2^{1/2}$ [*resp.* $\gamma_2^* = 2^{1/2}$], *then* $\gamma_p = 2^{1/p}$ [*resp.* $\gamma_p^* = 2^{1/p}$] *whenever* $p \in [1, \infty[$.

Proof. By (6.3) we have $\gamma_p \leq 2^{1/p}$ and $\gamma_p^* \leq 2^{1/p}$ whenever $p \in [1, \infty[$. Note also that $P^2(G) = \{\varphi^{1/2} : \varphi \in P^1(G)\}$ and $P^p(G) = \{\varphi^{2/p} : \varphi \in P^2(G)\}$ for $p \in [1, \infty[$.
  (a) Let $\varepsilon \in]0, 1[$. By hypothesis there exists a compact subset K [resp. a finite subset F] in G such that, for every $\varphi \in P^1(G)$, there exists $a \in K$ [resp.

$a \in F]$ satisfying $\|_a\varphi - \varphi\|_1 > 2 - \varepsilon$. Let $\psi = \varphi^{1/2} \in P^2(G)$; by the Cauchy–Schwarz inequality and (8.2) we have

$$2 - \varepsilon < \|_a\psi^2 - \psi^2\|_1 = \|(_a\psi + \psi)(_a\psi - \psi)\|_1$$

$$\leq \|_a\psi + \psi\|_2 \|_a\psi - \psi\|_2 \leq (2(1 + \psi * \tilde{\psi}(a))2(1 - \psi * \tilde{\psi}(a)))^{1/2}$$

$$= 2(1 - \psi * \tilde{\psi}(a)^2)^{1/2},$$

hence

$$\psi * \tilde{\psi}(a)^2 < \varepsilon - \frac{\varepsilon^2}{4} < \varepsilon, \qquad \psi * \tilde{\psi}(a) < \varepsilon^{1/2}$$

and then by (8.2) also

$$\|_a\psi - \psi\|_2^2 = 2(1 - \psi * \tilde{\psi}(a)) > 2(1 - \varepsilon^{1/2}).$$

Since $\varepsilon \in\,]0,1[$ may be chosen arbitrarily, we conclude that $\gamma_2 = 2^{1/2}$ [resp. $\gamma_2^* = 2^{1/2}$].

(b) Let $\varepsilon \in\,]0,1[$. By hypothesis there exists a compact subset K [resp. a finite subset F] in G such that, for every $\psi \in P^2(G)$, there exists $a \in K$ [resp. $a \in F$] satisfying $\|_a\psi - \psi\|_2 > 2^{1/2}(1 - \varepsilon^2)^{1/2}$. Then, by (8.2), we have $\psi * \tilde{\psi}(a) < \varepsilon^2$. In G we consider the measurable subsets $A_a' = \{x \in G : \psi(a^{-1}x) \leq \varepsilon\psi(x)\}$ and $A_a'' = \{x \in G : \psi(x) \leq \varepsilon\psi(a^{-1}x)\}$, so

$$\int_{G\setminus A_a'} \psi(x)^2 \, dx \leq \varepsilon^{-1} \int_{G\setminus A_a'} \psi(x)\psi(a^{-1}x) \, dx \leq \varepsilon^{-1}\psi * \tilde{\psi}(a) < \varepsilon,$$

$\int_{A_a'} \psi(x)^2 \, dx > 1 - \varepsilon$, and similarly $\int_{G\setminus A_a''} \psi(a^{-1}x)^2 \, dx < \varepsilon$, $\int_{A_a''} \psi(a^{-1}x)^2 \, dx > 1 - \varepsilon$. Moreover in the normed space $L^p(G)$ we have

$$\left(\int_{A_a'} |\psi(x)^{2/p} - \psi(a^{-1}x)^{2/p}|^p \, dx\right)^{1/p}$$

$$\geq \left|\left(\int_{A_a'} \psi(x)^2 \, dx\right)^{1/p} - \left(\int_{A_a'} \psi(a^{-1}x)^2 \, dx\right)^{1/p}\right|$$

$$= \left(\int_{A_a'} \psi(x)^2 \, dx\right)^{1/p} - \left(\int_{A_a'} \psi(a^{-1}x)^2 \, dx\right)^{1/p}$$

$$\geq \left(\int_{A_a'} \psi(x)^2 \, dx\right)^{1/p} - \left(\int_{G\setminus A_a''} \psi(a^{-1}x)^2 \, dx\right)^{1/p}$$

$$> (1 - \varepsilon)^{1/p} - \varepsilon^{1/p}$$

and similarly

$$\left(\int_{A''_a} \left|\psi(a^{-1}x)^{2/p} - \psi(x)^{2/p}\right|^p dx\right)^{1/p} > (1-\varepsilon)^{1/p} - \varepsilon^{1/p}.$$

Since $|A'_a \cap A''_a| = 0$, for $\varphi = \psi^{2/p} \in P^p(G)$ and $a \in K$ [resp. $a \in F$], we have

$$\|_a\varphi - \varphi\|_p^p = \|\varphi - _{a^{-1}}\varphi\|_p^p > 2\left((1-\varepsilon)^{1/p} - \varepsilon^{1/p}\right)^p.$$

As $\varepsilon \in\,]0,1[$ may be chosen arbitrarily, we conclude that $\gamma_p = 2^{1/p}$ [resp. $\gamma_p^* = 2^{1/p}$]. □

Lemma 15.2. *Let G be a locally compact group. If $\gamma_1^* < 2$, then G is amenable.*

Proof. We establish the amenability of G via Proposition 9.8. Let us assume that there exists $\mu \in M_f^1(G)$ with $\|L_\mu\|_{Cv^2} < 1$.

By hypothesis there exists $\varepsilon \in\,]0,1[$ such that, for any finite subset F in G, there exists $\varphi \in P^1(G)$ satisfying $\|_{a^{-1}}\varphi - \varphi\|_1 < 2(1-\varepsilon)$ whenever $a \in F$. By density of $\mathscr{S}(G)$ in $L^1(G)$ we may determine $\theta \in P^1(G) \cap \mathscr{S}(G)$ such that $\|\theta - \varphi\|_1 < \varepsilon/2$ (Section 2B.2). Then

$$\|_{a^{-1}}\theta - \theta\|_1 \leq \|_{a^{-1}}(\theta - \varphi)\|_1 + \|_{a^{-1}}\varphi - \varphi\|_1 + \|\varphi - \theta\|_1 < 2 - \varepsilon.$$

Via Lemma 7.2 we may choose measurable subsets $A_1,\ldots,A_n$ in G such that $0 < |A_i| < \infty$ whenever $i = 1,\ldots,n$ and $\alpha_1,\ldots,\alpha_n \in \mathbf{R}^*_+$ with $\sum_{i=1}^n \alpha_i = 1$ such that

$$\sum_{i=1}^n \alpha_i \frac{|aA_i \triangle A_i|}{|A_i|} < 2 - \varepsilon.$$

Hence

$$2\sum_{i=1}^n \alpha_i \frac{|aA_i \cap A_i|}{|A_i|} = \sum_{i=1}^n \alpha_i \frac{|aA_i| + |A_i| - |aA_i \triangle A_i|}{|A_i|} > \varepsilon. \tag{1}$$

Let $f_i = |A_i|^{-1/2} 1_{A_i} \in P^2(G) \cap \mathscr{S}(G)$ for $i = 1,\ldots,n$.
If $p \in \mathbf{N}^*$, by (1) and (9.4) we have

$$\frac{\varepsilon}{2} \sum_{k=1}^p \mu^{*k}(F) \leq \sum_{i=1}^n \sum_{k=1}^p \alpha_i \int_G \frac{|xA_i \cap A_i|}{|A_i|} d\mu^{*k}(x)$$

$$= \sum_{i=1}^n \sum_{k=1}^p \alpha_i \langle f_i * \check{f}_i, \mu^{*k}\rangle = \sum_{i=1}^n \sum_{k=1}^p \alpha_i \langle f_i, \mu^{*k} * f_i\rangle$$

$$\leq \sum_{i=1}^n \sum_{k=1}^p \alpha_i \|L_\mu\|_{Cv^2}^k \|f_i\|_2^2 \leq \|L_\mu\|_{Cv^2} \left(1 - \|L_\mu\|_{Cv^2}\right)^{-1}.$$

So for every $p \in \mathbf{N}^*$ and every finite subset F in G, we have

$$\sum_{k=1}^{p} \mu^{*k}(F) < \left(\frac{\varepsilon}{2}\right)^{-1} \|L_\mu\|_{Cv^2}(1 - \|L_\mu\|_{Cv^2})^{-1}. \quad (2)$$

As the probability measure μ has finite support, given $n_0 \in \mathbf{N}^*$, there exist finite subsets $F_1, \ldots, F_{n_0}$ in G such that $\mu^{*i}(F_i) = 1$ whenever $i = 1, \ldots, n_0$. So for the finite subset $F'_{n_0} = \cup_{i=1}^{n_0} F_i$ in G,

$$\sum_{i=1}^{n_0} \mu^{*i}(F'_{n_0}) = \sum_{i=1}^{n_0} \mu^{*i}(F_i) = n_0.$$

We contradict (2). Thus necessarily $\|L_\mu\|_{Cv^2} = 1$ and G is amenable by Proposition 9.8. □

Theorem 15.3. *In a locally compact group G the following dichotomy holds*:
  (a) *G is amenable, $\sigma = \sigma^* = 0$ and, for every $p \in [1, \infty[$, $\gamma_p = \gamma_p^* = 0$.*
  (b) *G is nonamenable, $\sigma = \sigma^* = 1$ and, for every $p \in [1, \infty[$, $\gamma_p = \gamma_p^* = 2^{1/p}$.*

Proof. The case (a) is ruled by Theorem 6.14 and (8.2).

If G is nonamenable, by Lemma 15.2, $\gamma_1^* = 2$, and then also $\gamma_1 = 2$ by (6.3). Lemma 15.1 yields that $\gamma_2 = 2^{1/2} = \gamma_2^*$ and then $\gamma_p = 2^{1/p} = \gamma_p^*$ for every $p \in [1, \infty[$. Finally (8.2) implies that $\sigma = \sigma^* = 1$. □

B. Various Characterizations of Amenability

We begin by establishing a characterization that complements Theorem 6.14 and Corollary 6.15.

Proposition 15.4. *Let G be a locally compact group. For $p \in [1, \infty[$, we consider the following properties*:

  (i_p) *[resp. (i_p^*)] If K is any compact subset [resp. F is any finite subset] of G and $\varepsilon > 0$, there exists $\varphi \in P^p(G) \cap \mathscr{K}(G)$ such that $\|_a\varphi_b \Delta(b)^{1/p} - \varphi\|_p < \varepsilon$ whenever $a, b \in K$ [resp. $a, b \in F$].*

  (ii_p) *[resp. (ii_p^*)] If K is any compact subset [resp. F is any finite subset] of G and $\varepsilon > 0$, there exists $\varphi \in P^p(G)$ such that $\|_a\varphi_b \Delta(b)^{1/p} - \varphi\|_p < \varepsilon$ whenever $a, b \in K$ [resp. $a, b \in F$].*

The group G is amenable if and only if, for one $p \in [1, \infty[$ (for all $p \in [1, \infty[$), any one of the conditions (i_p), (i_p^*), (ii_p), (ii_p^*) holds.

Proof. For every $p \in [1, \infty[$, the following diagram is trivial:

$$\begin{array}{ccc} (i_p) & \Rightarrow & (i_p^*) \\ \Downarrow & & \Downarrow \\ (ii_p) & \Rightarrow & (ii_p^*) \end{array}$$

Applying (ii_p^*) to finite subsets of G containing e we verify that (ii_p^*) implies amenability of the group by Theorem 6.14. We now show that if G is amenable, then (i_p) holds for every $p \in [1, \infty[$.

Let $p \in [1, \infty[$, $\varepsilon \in \mathbf{R}_+^*$, and K a compact subset of G. By Corollary 6.15 there exists $\psi \in P^1(G) \cap \mathcal{K}(G)$ such that $\|L_{a^{-1}}\psi - \psi\|_1 < \varepsilon^p/2$ whenever $a \in K \cup K^{-1}$. Let $\varphi = \psi * \psi^* \in P^1(G) \cap \mathcal{K}(G)$. For all $a, b \in K$, by (2.1) and (2.4) we have

$$\|_a\varphi_b \Delta(b) - \varphi\|_1 = \|L_{a^{-1}}R_{b^{-1}}(\psi * \psi^*) - \psi * \psi^*\|_1$$

$$= \|R_{b^{-1}}((L_{a^{-1}}\psi - \psi) * \psi^*)\|_1 + \|R_{b^{-1}}(\psi * \psi^*) - \psi * \psi^*\|_1$$

$$\leq \|L_{a^{-1}}\psi - \psi\|_1 + \|L_b(\psi * \psi^*) - \psi * \psi^*\|_1$$

$$\leq \|L_{a^{-1}}\psi - \psi\|_1 + \|L_b\psi - \psi\|_1 < \varepsilon^p.$$

Let $\varphi' = \varphi^{1/p} \in P^p(G) \cap \mathcal{K}(G)$. For all $a, b \in K$, by (6.2) we have

$$\|_a\varphi'_b \Delta(b)^{1/p} - \varphi'\|_p = \left(\int |_a\varphi'_b(x) \Delta(b)^{1/p} - \varphi'(x)|^p dx\right)^{1/p}$$

$$\leq \left(\int |_a\varphi_b(x) \Delta(b) - \varphi(x)| dx\right)^{1/p} < \varepsilon. \quad \square$$

Properties (P_1) and (P_1^*), which characterize amenability by Proposition 6.12, lead to the next result.

Proposition 15.5. *A locally compact group G is amenable if and only if the following condition holds: For all $n \in \mathbf{N}^*$ and all $\mu_1, \ldots, \mu_n \in \mathcal{M}^1(G)$,*

$$\inf\{\sup\{\|\mu_i * \varphi\|_1 : i = 1, \ldots, n\} : \varphi \in P^1(G)\} \leq \sup\{|\mu_i(G)| : i = 1, \ldots, n\}.$$

Proof. (a) Assume that (P_1) holds.

Let $\mu_1, \ldots, \mu_n \in \mathcal{M}^1(G)$, $\varepsilon \in \mathbf{R}_+^*$ and put $\delta = \varepsilon(2 + \sup\{\|\mu_i\| : i = 1, \ldots, n\})^{-1}$. There exists a compact subset K in G such that $|\mu_i|(G \setminus K) < \delta$ whenever $i = 1, \ldots, n$. By hypothesis there exists $\varphi \in P^1(G)$ such that $\|_{y^{-1}}\varphi - \varphi\|_1 < \delta$ whenever $y \in K$. Thus for every $i = 1, \ldots, n$,

$$\|\mu_i * \varphi - \mu_i(G)\varphi\|_1 = \int_G \left| \int_G \varphi(y^{-1}x) d\mu_i(y) - \left(\int_G d\mu_i(y)\right)\varphi(x)\right| dx$$

$$\leq \int_G \int_G |\varphi(y^{-1}x) - \varphi(x)| dx \, d|\mu_i|(y)$$

$$\leq \int_K \left(\int_G |\varphi(y^{-1}x) - \varphi(x)| dx\right) d|\mu_i|(y) + 2\|\varphi\|_1 |\mu_i|(G \setminus K)$$

$$< \delta\|\mu_i\| + 2\delta \leq \varepsilon;$$

hence

$$\sup \{\|\mu_i * \varphi\|_1 : i = 1, \ldots, n\} < \varepsilon + \sup \{|\mu_i(G)| : i = 1, \ldots, n\}.$$

Since $\varepsilon > 0$ may be chosen arbitrarily, the property holds.

(b) We assume that the condition holds and establish (P_1^*).

Let $F = \{a_1, \ldots, a_n\}$ be a finite subset of G. For any $i = 1, \ldots, n$, consider $\mu_i = \delta_{a_i^{-1}} - \delta_e \in \mathcal{M}^1(G)$; $\mu_i(G) = 0$. By assumption, given $\varepsilon > 0$, there exists $\varphi \in P^1(G)$ such that, for every $i = 1, \ldots, n$, $\|\mu_i * \varphi\|_1 < \varepsilon$, that is, $\|_{a_i}\varphi - \varphi\|_1 < \varepsilon$. $\square$

We add several characterizations of amenability in terms of asymptotical invariance.

Proposition 15.6. *A locally compact group G is amenable if and only if there exists a net $(\mu_i)_{i \in I}$ in $M^1(G)$ satisfying any one of the following conditions:*
  (i) $(\forall \mu \in M^1(G))\, (\forall \varphi \in P^1(G)) \lim_i \|\mu * \mu_i * \varphi - \mu_i * \varphi\|_1 = 0.$
  (ii) $(\forall \mu \in M^1(G))\, (\forall f \in \mathcal{UC}(G)) \lim_i \langle f, \mu * \mu_i - \mu_i \rangle = 0.$
  (iii) *For every compact subset K of G and every $f \in \mathcal{UC}(G)$* $\lim_i \sup \{|\langle f, \delta_x * \mu_i - \mu_i \rangle| : x \in K\} = 0.$

A net (μ_i) in $M^1(G)$ satisfies one of these conditions if and only if, for every $\psi \in P^1(G)$, the net $(\mu_i * \psi)$ does.

Proof. (a) If G is amenable, Proposition 6.7 implies the existence of a net (μ_i) in $M^1(G)$ satisfying (i).

$$\text{(i)} \Rightarrow \text{(ii)}$$

Let $f \in \mathcal{UC}(G)$ and $\varepsilon \in \mathbf{R}_+^*$. There exists a compact neighborhood U of e in G such that, for every $x \in G$ and every $y \in U$, $|f(xy) - f(x)| < \varepsilon/2$. For any $\nu \in M^1(G)$,

$$|\langle f, \nu * \xi_U - \nu \rangle| = \left| \iint f(xy)\, d\nu(x) \xi_U(y)\, dy - \int f(x)\, d\nu(x) \int \xi_U(y)\, dy \right|$$

$$\leq \iint |f(xy) - f(x)| \xi_U(y)\, dy\, d\nu(x) < \frac{\varepsilon}{2}.$$

In particular, for every $i \in I$ and every $\mu \in M^1(G)$,

$$|\langle f, \mu_i * \xi_U - \mu_i \rangle| < \frac{\varepsilon}{2}$$

and

$$|\langle f, \mu * \mu_i * \xi_U - \mu * \mu_i \rangle| < \frac{\varepsilon}{2}.$$

Hence

$$|\langle f, \mu * \mu_i - \mu_i \rangle| \le |\langle f, \mu * \mu_i - \mu * \mu_i * \xi_U \rangle| + |\langle f, \mu * \mu_i * \xi_U - \mu_i * \xi_U \rangle|$$
$$+ |\langle f, \mu_i * \xi_U - \mu_i \rangle| < \varepsilon + \|f\| \, \|\mu * \mu_i * \xi_U - \mu_i * \xi_U\|_1.$$

Since $\varepsilon \in \mathbf{R}_+^*$ is arbitrary, the conclusion now follows from (i).

$$\text{(ii)} \Rightarrow \text{(iii)}$$

If (iii) did not hold, there would exist a compact subset K of G, $f \in \mathcal{UC}(G)$, $\varepsilon > 0$, a subnet $(\mu_{i'})_{i' \in I'}$ of $(\mu_i)_{i \in I}$ and a net $(a_{i'})_{i' \in I'}$ in K such that

$$|\langle f, \delta_{a_{i'}} * \mu_{i'} - \mu_{i'} \rangle| > 2\varepsilon \qquad (3)$$

whenever $i' \in I'$. There exists a subnet $(a_j)_{j \in J}$ of $(a_{i'})_{i' \in I'}$ converging to an element a of the compact subset K. Since $f \in \mathcal{UC}(G)$, there exists $j_0 \in J$ such that

$$\|_a f - _{a_j} f\| < \varepsilon$$

whenever $j \succ j_0$; hence

$$|\langle f, \delta_a * \mu_j - \delta_{a_j} * \mu_j \rangle| = |\langle _a f - _{a_j} f, \mu_j \rangle| \le \|_a f - _{a_j} f\| < \varepsilon \qquad (4)$$

and then by (3) and (4)

$$|\langle f, \delta_a * \mu_j - \mu_j \rangle| > \varepsilon.$$

So (ii) would be contradicted.

$$\text{(iii)} \Rightarrow \text{(ii)}$$

By density we may suppose that μ has compact support. Let $f \in \mathcal{UC}(G)$. By Fatou's lemma and (iii) we have

$$\lim |\langle f, \mu * \mu_i - \mu_i \rangle|$$

$$= \lim \left| \iint f(xy) \, d\mu(x) \, d\mu_i(y) - \iint f(y) \, d\mu_i(y) \, d\mu(x) \right|$$

$$\le \int \left(\lim \left| \int (f(xy) - f(y)) \, d\mu_i(y) \right| \right) d\mu(x) = 0.$$

If (ii) holds, then for every $f \in \mathcal{UC}(G)$ and every $a \in G$,

$$\lim \langle _a f - f, \mu_i \rangle = \lim \langle f, \delta_a * \mu_i - \mu_i \rangle = 0.$$

Every w*-cluster point of (μ_i) belongs to $\mathrm{LIM}(\mathscr{UC}(G))$ and G is amenable by Theorem 4.19.

(b) If (μ_i) satisfies (i), then trivially $(\mu_i * \psi)$ satisfies (i) for every $\psi \in P^1(G)$.

Assume now that, for all $\psi \in P^1(G)$, $(\mu_i * \psi)$ satisfies (i). Consider $\mu \in M^1(G)$, $\varphi \in P^1(G)$ and let $\varepsilon > 0$. By Proposition 2.2 there exists $\varphi_1 \in P^1(G)$ such that $\|\varphi * \varphi_1 - \varphi\|_1 < \varepsilon$. Then for every $i \in I$,

$$\|\mu * \mu_i * \varphi - \mu_i * \varphi\|_1 \leq \|\mu * \mu_i * (\varphi - \varphi * \varphi_1)\|_1$$
$$+ \|\mu * (\mu_i * \varphi) * \varphi_1 - (\mu_i * \varphi) * \varphi_1\|_1$$
$$+ \|\mu_i * (\varphi * \varphi_1 - \varphi)\|_1 < 2\varepsilon$$
$$+ \|\mu * (\mu_i * \varphi) * \varphi_1 - (\mu_i * \varphi) * \varphi_1\|_1,$$

so

$$\lim_i \|\mu * \mu_i * \varphi - \mu_i * \varphi\|_1 < 2\varepsilon$$

and, since $\varepsilon > 0$ is arbitrary, (μ_i) satisfies (i). □

C. Asymptotical Invariance Properties of Amenable Groups

Proposition 15.7. *Let G be an amenable group and let $(\mu_i)_{i \in I}$ be a net in $M^1(G)$. The following properties are equivalent:*

(i) (μ_i) *is asymptotically $M^1(G)$-invariant.*

(ii) $(\forall \sigma \in \mathscr{M}^1(G)) \lim_i \|\sigma * \mu_i\| = |\sigma(G)|.$

(iii) $(\forall \nu \in M^1(G)) (\forall \nu' \in M^1(G)) \lim_i \|(\nu - \nu') * \mu_i\| = 0.$

Proof. (i) $\Rightarrow$ (ii)

Let $\sigma \in \mathscr{M}^1(G)$, $\alpha = \inf\{\|\sigma * \mu\| : \mu \in M^1(G)\}$ and $\beta = \inf\{\|\sigma * \mu_i\| : i \in I\}$; $\alpha \leq \beta$. If $\varepsilon > 0$, there exists $\mu_0 \in M^1(G)$ such that

$$\alpha \leq \|\sigma * \mu_0\| < \alpha + \frac{\varepsilon}{2}.$$

Thus for every $i \in I$, $\|\sigma * \mu_0 * \mu_i\| < \alpha + \varepsilon/2$. We have

$$\lim_i \|\sigma * \mu_0 * \mu_i - \sigma * \mu_i\| \leq \|\sigma\| \lim_i \|\mu_0 * \mu_i - \mu_i\| = 0,$$

that is, there exists $i_0 \in I$ such that, for every $i \succ i_0$,

$$\|\sigma * \mu_0 * \mu_i - \sigma * \mu_i\| < \frac{\varepsilon}{2}$$

and

$$\alpha \leq \beta \leq \|\sigma * \mu_i\| \leq \|\sigma * \mu_0 * \mu_i\| + \frac{\varepsilon}{2} < \alpha + \varepsilon.$$

Hence $\alpha = \beta = \lim_i \|\sigma * \mu_i\|$ and it remains to show that $\alpha = |\sigma(G)|$.

Note that, for every $\mu \in M^1(G)$,

$$\|\sigma * \mu\| = |\sigma * \mu|(G) \geq |\sigma * \mu(G)| = |\sigma(G)|$$

hence $\alpha \geq |\sigma(G)|$. By Proposition 15.5 we have

$$\inf\{\|\sigma * \varphi\|_1 : \varphi \in P^1(G)\} \leq |\sigma(G)|$$

and thus a fortiori $\alpha \leq |\sigma(G)|$. Therefore $\alpha = |\sigma(G)|$.

(ii) $\Rightarrow$ (iii)

By assumption $\lim_i \|(\nu - \nu') * \mu_i\| = |(\nu - \nu')(G)| = |1 - 1| = 0$.

(iii) $\Rightarrow$ (i)

If (iii) holds, then for every $\mu \in M^1(G)$,

$$\lim_i \|\mu * \mu_i - \mu_i\| = \lim_i \|(\mu - \delta_e) * \mu_i\| = 0. \qquad \square$$

The next statement complements Proposition 6.7.

Proposition 15.8. *Let G be an amenable group. If $(\varphi_i)_{i \in I}$ is an asymptotically $P^1(G)$-invariant net in $P^1(G)$ and $\varphi \in P^1(G)$, then for every compact subset K in G,*

$$\lim_i \|_a(\varphi * \varphi_i) - \varphi * \varphi_i\|_1 = 0$$

uniformly for every $a \in K$.

Proof. Note that by hypothesis, for every $\varphi \in P^1(G)$ and every $a \in G$,

$$\lim_i \|_a(\varphi * \varphi_i) - \varphi * \varphi_i\|_1 \leq \lim_i \|(_a\varphi) * \varphi_i - \varphi_i\|_1 + \lim_i \|\varphi * \varphi_i - \varphi_i\|_1 = 0. \tag{5}$$

If this convergence were not uniform on compacta, there would exist $\varepsilon \in \mathbf{R}_+^*$, a compact subset K of G, a subnet $(\varphi_{i'})_{i' \in I'}$ of $(\varphi_i)_{i \in I}$, and a net $(a_{i'})_{i' \in I'}$ in K such that, for every $i' \in I'$,

$$\|_{a_{i'}}(\varphi * \varphi_{i'}) - \varphi * \varphi_{i'}\|_1 > 2\varepsilon. \tag{6}$$

There exists a neighborhood U of e in G such that $\|_a\varphi - \varphi\|_1 < \varepsilon$ whenever $a \in U$. A subnet $(a_j)_{j \in J}$ of $(a_{i'})_{i' \in I'}$ would converge to an element c in the compact subset K. For $ca_j^{-1} \in U$, $j \in J$, we would have

$$\|_c(\varphi * \varphi_j) - _{a_j}(\varphi * \varphi_j)\|_1 = \|(_{ca_j^{-1}}\varphi) * \varphi_j - \varphi * \varphi_j\|_1 < \varepsilon; \tag{7}$$

(6) and (7) would imply that

$$\left\|_c(\varphi * \varphi_j) - \varphi * \varphi_j\right\|_1 > \varepsilon$$

contradicting (5). □

Remark 15.9. In (ii) of Proposition 6.7 one must require uniform convergence for all compact subsets.

Let us consider the abelian group $(\mathbf{R}, +)$, which is amenable. If $i = (F, n)$, $i' = (F', n') \in \mathfrak{F}(\mathbf{R}) \times \mathbf{N}^*$, we put $i \prec i'$ in case $F \subset F'$ and $n < n'$. To every $i = (F, n) \in \mathfrak{F}(\mathbf{R}) \times \mathbf{N}^*$ with $F = \{a_1, \ldots, a_m\}$ we associate

$$E_i = \left\{ \sum_{j=1}^m k_j a_j : (k_1, \ldots, k_m) \in \{1, \ldots, n\}^m \right\}$$

and

$$\mu_i = \frac{1}{n^m} \sum_{x \in E_i} m_x \delta_x \in M_f^1(\mathbf{R}),$$

where

$$m_x = \operatorname{card}\left\{ (k_1, \ldots, k_m) \in \{1, \ldots, n\}^m : x = \sum_{j=1}^m k_j a_j \right\} \quad \text{for } x \in E_i.$$

For a fixed $j = 1, \ldots, m$, consider

$$\delta_{a_j} * \mu_i - \mu_i = \frac{1}{n^m} \sum_{x \in E_i} m_x \left(\delta_{a_j + x} - \delta_x \right).$$

Canceling out opposite terms we see that this sum has at most $2n^{m-1}$ terms. Therefore

$$\|\delta_{a_j} * \mu_i - \mu_i\| \leq \frac{2n^{m-1}}{n^m} = \frac{2}{n}.$$

If $\mu = \sum_{j=1}^m \alpha_j \delta_{a_j} \in M^1(\mathbf{R})$ with $\alpha_j \geq 0$ whenever $j = 1, \ldots, m$, then also

$$\|\mu * \mu_i - \mu_i\| = \left\| \sum_{j=1}^m \alpha_j \delta_{a_j} * \mu_i - \mu_i \right\| \leq \sum_{j=1}^m \alpha_j \|\delta_{a_j} * \mu_i - \mu_i\| \leq \frac{2}{n}.$$

Thus $\lim_i \|\mu * \mu_i - \mu_i\| = 0$ uniformly for all $\mu \in M^1(\mathbf{R})$ having their supports contained in F. On the other hand, for every $i \in I$, $1_{[0,1]} * \mu_i \in P^1(\mathbf{R})$ is absolutely continuous with respect to the Lebesgue measure whereas $\mu_i \in M^1_f(\mathbf{R})$ is singular with respect to that measure, hence $\|1_{[0,1]} * \mu_i - \mu_i\| = \|1_{[0,1]} * \mu_i\|_1 + \|\mu_i\| = 2$ (Section 2D.3).

D. Quasi-Invariance Properties for Specific Amenable Groups

We first give formulations of the properties (P_1) and (P_1^*) for unimodular groups.

Proposition 15.10. *A unimodular locally compact group G is amenable if and only if, for every compact subset K [resp. every finite subset F] in G and every $\varepsilon > 0$, there exists $\varphi = \check{\varphi} \in P^1(G)$ such that $\|_a\varphi - \varphi\|_1 < \varepsilon$ and $\|\varphi_a - \varphi\|_1 < \varepsilon$ whenever $a \in K$ [resp. $a \in F$].*

Proof. By Proposition 6.12 the condition is obviously sufficient. Conversely if G is amenable, by (P_1) [resp. (P_1^*)] there exists $\psi \in P^1(G)$ such that $\|_a\psi - \psi\|_1 < \varepsilon$ for every $a \in K \cup K^{-1}$ [resp. $a \in F \cup F^{-1}$]. As G is unimodular, $\check{\psi} \in P^1(G)$; also $\varphi = \psi * \check{\psi} \in P^1(G)$ and $\varphi = \check{\varphi}$. For every $a \in K$ [resp. $a \in F$], by (2.4) and (2.1) we have

$$\|_a\varphi - \varphi\|_1 = \|(_a\psi - \psi) * \check{\psi}\|_1 \leq \|_a\psi - \psi\|_1 < \varepsilon$$

and

$$\|\varphi_a - \varphi\|_1 = \|(\varphi_a - \varphi)\check{}\,\|_1 = \|_{a^{-1}}\check{\varphi} - \check{\varphi}\|_1 = \|_{a^{-1}}(\psi * \check{\psi}) - \psi * \check{\psi}\|_1$$
$$\leq \|_{a^{-1}}\psi - \psi\|_1 < \varepsilon. \qquad \square$$

We show that, for any compact group G, every asymptotically $M^1(G)$-invariant net in $M^1(G)$ converges to the normalized Haar measure.

Proposition 15.11. *Let G be a compact group. A net $(\mu_i)_{i \in I}$ in $M^1(G)$ is asymptotically $M^1(G)$-invariant if and only if $\lim_i \|\mu_i - \lambda\| = 0$.*

Proof. Since G is unimodular, for every $\mu \in M^1(G)$ and every $f \in \mathscr{C}(G)$,

$$\langle f, \mu * \lambda - \lambda \rangle = \iint f(yx)\, d\mu(y)\, dx - \int f(x)\, dx$$
$$= \int \left(\int f(x)\, dx \right) d\mu(y) - \int f(x)\, dx = 0,$$

$$\langle f, \lambda * \mu - \mu \rangle = \iint f(xy)\, dx\, d\mu(y) - \int f(x)\, dx$$
$$= \int \left(\int f(x)\, dx \right) d\mu(y) - \int f(x)\, dx = 0;$$

hence $\mu * \lambda = \lambda = \lambda * \mu$.

QUASI-INVARIANCE PROPERTIES

If (μ_i) is asymptotically $M^1(G)$-invariant,

$$\lim_i \|\mu_i - \lambda\| = \lim_i \|\mu_i - \lambda * \mu_i\| = 0.$$

On the other hand, for every $\mu \in M^1(G)$ and every $i \in I$,

$$\|\mu * \mu_i - \mu_i\| \leq \|\mu * \mu_i - \mu * \lambda\| + \|\mu * \lambda - \lambda\| + \|\lambda - \mu_i\| \leq 2\|\lambda - \mu_i\|.$$

Therefore if the condition holds,

$$\lim_i \|\mu * \mu_i - \mu_i\| = 0$$

and (μ_i) is asymptotically $M^1(G)$-invariant. □

Proposition 15.12. *Let G be an amenable group. Then G is discrete if and only if, for any net (μ_i) in $M^1(G)$, the following conditions are equivalent*:
  (i) (μ_i) *is asymptotically $M^1(G)$-invariant.*
  (ii) $\lim \|(\mu * \mu_i - \mu_i) * f\|_1 = 0$ *whenever $\mu \in M^1(G)$ and $f \in L^1(G)$.*

Proof. (a) If G is discrete, in particular $\delta_e \in l^1(G)$ and the two conditions are equivalent.

(b) If $(\mu_i)_{i \in I}$ is an asymptotically $M^1(G)$-invariant net in $M^1(G)$, for every $i \in I$, μ_i admits a Lebesgue decomposition $\mu_i = \mu_i^a + \mu_i'$, where $\mu_i^a \in \mathcal{M}_a^1(G)$; for a fixed $\varphi \in P^1(G)$ and every $i \in I$,

$$\|\mu_i'\| + \|\mu_i^a - \varphi * \mu_i\| = \|\mu_i - \varphi * \mu_i\|$$

(Section 2D.3). So we must have $\lim_i \|\mu_i'\| = 0$, and we conclude that there cannot exist an asymptotically $M^1(G)$-invariant net formed by probability measures having finite support in case G is nondiscrete.

(c) Now assume that the two conditions are equivalent and start from a net $(\mu_i)_{i \in I}$ that is asymptotically $M^1(G)$-invariant. We choose $\varphi_0 \in P^1(G)$ and let $\varphi_i = \mu_i * \varphi_0 \in P^1(G)$, $i \in I$; (φ_i) is asymptotically $M^1(G)$-invariant. Let $\varepsilon \in {]0,1[}$; as $\mathcal{K}(G)$ is dense in $L^1(G)$, for every $i \in I$, there exists $\varphi_{i,\varepsilon} \in \mathcal{K}_+(G) \cap P^1(G)$ such that $\|\varphi_{i,\varepsilon} - \varphi_i\|_1 < \varepsilon/4$. Via proposition 2.2 we may determine $\psi_{i,\varepsilon} \in P^1(G)$ such that

$$\|(\varphi_{i,\varepsilon} - \varphi_i) * \psi_{i,\varepsilon}\|_1 < \frac{\varepsilon}{2}.$$

There exist open neighborhoods $U_1, \ldots, U_m$ of $z_1, \ldots, z_m \in G$ such that

$$\operatorname{supp} \varphi_{i,\varepsilon} \subset \bigcup_{k=1}^m U_k \quad \text{and} \quad \|_{x^{-1}}\psi_{i,\varepsilon} - {}_{z_k^{-1}}\psi_{i,\varepsilon}\|_1 < \varepsilon/2$$

whenever $x \in U_k$, $k = 1, \ldots, m$. So we may determine a partition $\{A_l : l = 1, \ldots, n\}$ of $\operatorname{supp} \varphi_{i,\varepsilon}$ formed by Borel subsets and $x_1, \ldots, x_n \in G$ such that $x_l \in A_l$ whenever $l = 1, \ldots, n$ and $\|_{x^{-1}}\psi_{i,\varepsilon} - _{x_l^{-1}}\psi_{i,\varepsilon}\|_1 < \varepsilon/2$ whenever $x \in A_l$, $l = 1, \ldots, n$. Let $\nu_{i,\varepsilon} = \sum_{l=1}^{n} (\int_{A_l} \varphi_{i,\varepsilon}(x)\, dx)\, \delta_{x_l} \in M_f^1(G)$. We have

$$\|\varphi_{i,\varepsilon} * \psi_{i,\varepsilon} - \nu_{i,\varepsilon} * \psi_{i,\varepsilon}\|_1$$

$$= \int_G \left| \int_G \varphi_{i,\varepsilon}(x) \psi_{i,\varepsilon}(x^{-1}y)\, dx - \sum_{l=1}^{n} \left(\int_{A_l} \varphi_{i,\varepsilon}(x)\, dx \right) \psi_{i,\varepsilon}(x_l^{-1}y) \right| dy$$

$$\leq \sum_{l=1}^{n} \int_{A_l} \left(\int_G |\psi_{i,\varepsilon}(x^{-1}y) - \psi_{i,\varepsilon}(x_l^{-1}y)|\, dy \right) \varphi_{i,\varepsilon}(x)\, dx < \frac{\varepsilon}{2}$$

and

$$\|\varphi_i * \psi_{i,\varepsilon} - \nu_{i,\varepsilon} * \psi_{i,\varepsilon}\|_1 < \varepsilon.$$

As (φ_i) is asymptotically $M^1(G)$-invariant, we may therefore determine a net $(\nu_{i'})$ in $M_f^1(G)$ and a net $(\psi_{i'})$ of approximate units in $P^1(G)$ such that

$$\lim_{i'} \|(\mu * \nu_{i'} - \nu_{i'}) * \psi_{i'}\|_1 = 0$$

whenever $\mu \in M^1(G)$. Then for every $f \in L^1(G)$,

$$\lim_{i'} \|(\mu * \nu_{i'} - \nu_{i'}) * \psi_{i'} * f\|_1 = 0.$$

As $(\psi_{i'})$ is a net of approximate units, we have also

$$\lim_{i'} \|(\mu * \nu_{i'} - \nu_{i'}) * f\|_1 = 0.$$

Thus the net $(\nu_{i'})$ satisfies (ii) and by assumption it is asymptotically $M^1(G)$-invariant. Since it is formed by measures with finite support, we deduce from (b) that the group must be discrete. □

We next study a complex version of the Glicksberg–Reiter property. It characterizes amenability for almost connected locally compact groups. We consider only the left-hand version.

Definition 15.13. *Let G be a locally compact group. For every $f \in L^1(G)$, we define*

$$d_l^c(f) = \inf\{\|g\|_1 : g \in L_G(f)\}.$$

Definition 15.14. *If G is a locally compact group, we consider the property*

$$(GR_l^c) \ (\forall f \in L^0(G)) \, d_l^c(f) = 0.$$

Obviously for every $f \in L^1(G)$,

$$\left| \int f \right| \leq d_l^c(f) \leq d_l(f).$$

We deduce from Theorem 6.21 that amenability implies (GR_l^c).

Lemma 15.15. *If G is a locally compact group, for any $f \in L^1(G)$, $d_l^c(f) = \inf \{ \|h * f - f\|_1 : h \in L^0(G) \}$.*

Proof. By Proposition 2.2, given $\varepsilon > 0$, there exists $\psi \in P^1(G)$ such that $\|\psi * f - f\|_1 < \varepsilon$.

(a) Let $\varphi \in L^1(G)$ such that $\int \varphi = 1$. We put $h = \psi - \varphi \in L^0(G)$; then

$$\|h * f - f\|_1 = \|(\psi - \varphi) * f - f\|_1 \leq \|\psi * f - f\|_1 + \|\varphi * f\|_1 < \varepsilon + \|\varphi * f\|_1. \tag{8}$$

(b) Let $h \in L^0(G)$. We put $\varphi = \psi - h \in L^1(G)$; then $\int \varphi = 1$ and

$$\|\varphi * f\|_1 = \|(\psi * f - f) - (h * f - f)\|_1 \leq \|\psi * f - f\|_1 + \|h * f - f\|_1$$

$$< \varepsilon + \|h * f - f\|_1. \tag{9}$$

Proposition 3.6 and relations (8) and (9) establish the statement. □

Proposition 15.16. *A locally compact group G has property (GR_l^c) if and only if $L^0(G)$ admits approximate left units.*

Proof. The statement follows directly from Lemma 15.15. □

Proposition 15.17. *If G is a locally compact group having property (GR_l^c), then every open subgroup H has property (GR_l^c).*

Proof. As H is open in G, λ_H may be considered to be induced on H by λ_G. Let $f \in L^0(H)$, and extend f to $f' \in L^0(G)$ by putting $f'(x) = 0$ for $x \in G \setminus H$. By hypothesis, given $\varepsilon > 0$, there exist $\alpha_1, \ldots, \alpha_n \in \mathbf{C}$ with $\sum_{i=1}^n \alpha_i = 1$ and $x_1, \ldots, x_n \in G$ such that $\|\sum_{i=1}^n \alpha_i L_{x_i} f'\|_1 < \varepsilon$. There exist $a_1, \ldots, a_p \in G$ such that $x_1, \ldots, x_n \in \cup_{j=1}^p a_j H$ and the subsets $a_j H$ ($j = 1, \ldots, p$) are pairwise disjoint. Given $i \in \{1, \ldots, n\}$, let $j(i) \in \{1, \ldots, p\}$ and $y_i \in H$ such that $x_i = a_{j(i)} y_i$. If $z = a_j h$ for $j \in \{1, \ldots, p\}$ with $j \neq j(i)$ and

$h \in H$, then $x_i^{-1}z = y_i^{-1}a_{j(i)}^{-1}a_j h \notin H$ and $f'(x_i^{-1}z) = 0$. Therefore we have

$$\left\| \sum_{i=1}^n \alpha_i L_{y_i} f \right\|_{L^1(H)} = \int_H \left| \sum_{i=1}^n \alpha_i f(y_i^{-1}x) \right| dx = \int_H \left| \sum_{i=1}^n \alpha_i f(x_i^{-1}a_{j(i)}x) \right| dx$$

$$= \int_G \left| \sum_{i=1}^n \alpha_i f'(x_i^{-1}a_{j(i)}x) \right| 1_H(x) \, dx$$

$$= \int_G \left| \sum_{i=1}^n \alpha_i f'(x_i^{-1}y) \right| 1_{\bigcup_{j=1}^p a_j H}(y) \, dy$$

$$\leq \left\| \sum_{i=1}^n \alpha_i L_{x_i} f' \right\|_1 < \varepsilon.$$

We conclude that $d_l^c(f) = 0$. $\square$

Proposition 15.18. *If G is a locally compact group having property (GR_l^c) and H is a closed normal subgroup of G, then G/H has property (GR_l^c).*

Proof. Let $F \in L^1(G/H)$. There then exists $f \in L^1(G)$ such that $J_H f = F$ (Section 14A). If also $F \in L^0(G/H)$, then by Weil's formula we have $f \in L^0(G)$ and therefore $d_{G,l}^c(f) = 0$. As $\|J_H\| \leq 1$ (Section 14A), we must have then $d_{G/H,l}^c(F) = 0$. $\square$

Lemma 15.19. *Let G be a unimodular locally compact group admitting a nonunimodular closed subgroup H and a compact subgroup K such that $G = KH$ and the mapping $(z, y) \mapsto zy$ of $K \times H$ onto G is open. Then G does not satisfy (GR_l^c).*

Proof. By Lemma 14.7 there exist $\alpha \in \mathbf{R}_+^*$ and $a_0 \in H$ such that $\|_{a_0}f - f\|_1 \geq \alpha$ whenever $f \in \mathcal{K}(G)$ with $\int f = 1$ and $f(bx) = f(x)$ for all $x \in G$, $b \in K$.

Let $\psi \in \mathcal{K}_+(G) \cap P^1(G)$ and let

$$\varphi(x) = \int_K \psi(yx) \, dy,$$

$x \in G$; then $\varphi \in \mathcal{K}_+(G)$ and

$$\|\varphi\|_1 = \int_G \left(\int_K \psi(yx) \, dy \right) dx = \int_K \left(\int_G \psi(yx) \, dx \right) dy = \|\psi\|_1 = 1.$$

We have $_b\varphi = \varphi$ whenever $b \in K$. Let $g = {_{a_0}\varphi} - \varphi \in L^0(G)$. If $\alpha_1, \ldots, \alpha_n \in \mathbf{C}$ with $\sum_{i=1}^n \alpha_i = 1$ and $x_1, \ldots, x_n \in G$, then $h = \sum_{i=1}^n \alpha_i R_{x_i} \varphi \in \mathcal{K}(G)$; also

$\int h = 1$ and $h(bx) = h(x)$ for all $x \in G$, $b \in K$. Therefore $\|_{a_0}h - h\|_1 \geq \alpha$, that is, $\|\sum_{i=1}^n \alpha_i R_{x_i} g\|_1 \geq \alpha$, so $d_l^c(g^*) \geq \alpha$ and (GR_l^c) cannot hold. □

Proposition 15.20. *For an almost connected locally compact group G amenability is characterized by (GR_l^c).*

Proof. As we noted already, amenability implies (GR_l^c) by Theorem 6.21.

Assume now G to be a nonamenable group. There exists a compact normal subgroup K in G such that $G' = G/K$ is a Lie group (Section 14B); $G'' = G_0'/\mathrm{rad}\, G_0'$ is a semisimple connected Lie group. If G'' were compact, by Proposition 13.4 the group G_0' would be amenable. Since $G_0' \simeq G_0/(G_0 \cap K)$ and $G_0 \cap K$ is a compact subgroup of G_0, similarly G_0 would be amenable and also the almost connected group G. Thus G'' is a noncompact semisimple connected Lie group and verifies the conditions of Lemma 15.19 (Section 14 B); (GR_l^c) does not hold for G''. Proposition 15.18 insures then that G_0' does not satisfy (GR_l^c). As G_0' is the identity component of the Lie group G', it is an open subgroup of G'. Proposition 15.17 implies that G' has not property (GR_l^c). Finally by Proposition 15.18 we conclude that G does not satisfy (GR_l^c). □

E. Approximate Units in $A_p(G)$

We first establish a general technical lemma.

Lemma 15.21. *Let G be a locally compact group, K a compact subset of G, and $p \in {]}1, \infty[$.*

(a) *There exists $u \in A_p(G) \cap \mathcal{K}_+(G)$ such that $u(x) = 1$ whenever $x \in K$.*

(b) *If G is amenable, for every $\alpha > 1$, there exist $f \in P^p(G)$, $g \in L_+^{p'}(G)$ with $1/p + 1/p' = 1$ such that $\|g\|_{p'} < \alpha$, $u = f * \check{g} \in A_p(G) \cap \mathcal{K}_+(G)$, $\|u\|_{A_p} < \alpha$ and $u(x) = 1$ whenever $x \in K$.*

Proof. Let U be a compact subset of G such that $0 < |U|$ and let $f = (1/|U|^{1/p})1_U \in P^p(G)$, $g = (1/|U|^{1/p'})1_{K^{-1}U} \in L_+^{p'}(G)$; then $u = f * \check{g} \in A_p(G)$. For every $x \in G$, we have $u(x) = (1/|U|)\int 1_U(y) 1_{K^{-1}U}(x^{-1}y)\, dy$. For every $x \in K$, $u(x) = 1$. Moreover $u \in \mathcal{K}_+(G)$ with $\mathrm{supp}\, u \subset UU^{-1}K$.

If G is amenable, by (SF) and Proposition 7.11 we may choose U in such a way that the inequality $|K^{-1}U| < \alpha^{p'}|U|$ holds. Then $\|g\|_{p'} = (|K^{-1}U|/|U|)^{1/p'} < \alpha$ and also $\|u\|_{A_p} < \alpha$. □

We consider in particular the Fourier algebra (Section 10B).

Proposition 15.22. *Let G be a locally compact group. If $A(G)$ admits (nonnecessarily bounded) approximate units belonging to $A(G) \cap P(G)$, then G is amenable.*

Proof. We establish (G) which characterizes amenability by Proposition 8.5. Let K be a compact subset of G and $\varepsilon > 0$. By Lemma 15.21 there exists $u \in A(G) \cap \mathscr{K}_+(G)$ such that $u(x) = 1$ whenever $x \in K$. By hypothesis there exists $v \in A(G) \cap P(G)$ such that $\|u - uv\|_{A(G)} < \varepsilon/2$. Since $\|\cdot\| \leq \|\cdot\|_{A(G)}$ on $A(G)$, for every $x \in K$, $|1 - v(x)| < \varepsilon/2$. By density of $P(G) \cap \mathscr{K}(G)$ in $P(G) \cap A(G)$ we may choose $v' \in P(G) \cap \mathscr{K}(G)$ such that $\|v - v'\|_{A(G)} < \varepsilon/2$. Also there exists $\varphi \in L^2(G)$ such that $v' = \varphi * \tilde{\varphi}$. Then for every $x \in K$ we have $|1 - \varphi * \tilde{\varphi}(x)| < \varepsilon$. □

NOTES

The dichotomy properties 15.1–3 are essentially due to Losert [371], who considered a more general setting involving a representation of the group on $L^2(X, \mu)$, where X is a locally compact space and μ is a quasi-invariant measure. Derighetti [126] had shown that $[\gamma_1^* < 1] \Rightarrow [\gamma_1^* = 0]$ giving general versions mainly concerning groups acting on homogeneous spaces.

For 15.4 we refer to Skudlarek [513]. The proof of 15.5 is due to Anker [10], who emphasized the analogy of the condition with the Glicksberg–Reiter property. Gerl [197] gave 15.6 and 15.7 and exhibited a collection of interesting related conditions. General asymptotical properties of these types applying to semigroups had been established previously by Wong [562]. Maxones and Rindler [383] made further investigations on these properties. Granirer [215] proved 15.8. Maxones and Rindler [382] formulated 15.9. Johnson [291] showed 10.1(i) to characterize amenability via cohomological arguments. Rindler [470, 474] noticed that a locally compact group G is amenable if and only if $M^0(G)$ has bounded left approximate units. More on the subject is to be found in Kotzmann and Rindler [322] and Kotzmann, Losert, and Rindler [321].

Gerl [197] established 15.11. For 15.12 we refer to Maxones and Rindler [382]; see also [381] for special results concerning Heisenberg-type groups. The complex version of the Glicksberg–Reiter property developed in 15.15–20 is due to Rindler [468]. Rosenblatt [488] proved that if G is an amenable, unimodular, σ-compact group, there exists an element φ in $P^1(G)$ such that $\lim_{n \to \infty} \|\varphi^{*n} * f\|_1 = 0$ for all $f \in L^0(G)$.

The useful property 15.21 was considered, for instance, by Herz and Rivière [262]. Derighetti [125] gave 15.22. Stegmeir [518] made interesting comments on approximate units in Fourier algebras. Granirer and Leinert [225] proved that if G is an amenable group, $p \in]1, \infty[$, and E is a relatively compact subset of $A_p(G)$, then for every $\varepsilon > 0$ there exists $v \in A_p(G)$ such that $\|uv - v\|_{A_p} < \varepsilon$ whenever $u \in E$. From general considerations and 10.4 Lau [343] deduced that a locally compact group G is amenable if and only if $\{u \in A(G) : u(e) = 0\}$ admits bounded approximate units. Haagerup [245] proved that if G is a free group on at least two generators, there exists an (unbounded) net $(u_i)_{i \in I}$ of functions with finite support in $A(G)$ such that, for every $v \in A(G)$ and every $i \in I$, $\|u_i v\|_A \leq \|v\|_A$ and $\lim_i \|u_i v - v\|_A = 0$.

By adapting the direct proofs of the equivalence of (P_1) [resp. (P_1^*)] with the existence of a left invariant mean which were given in Section 6, Eymard [174] obtained the following extension: Let G be a locally compact group and assume $L^1(G)$ to be equipped with a seminorm N such that $|\int f| \leq N(f) \leq \|f\|_1$, $N(f) = N(\check{f}) = N(L_a f) = N(R_a f)$ whenever $f \in L^1(G)$ and $a \in G$. Let X be the subspace of $L^\infty(G)$ consisting of all $f \in L^\infty(G)$ such that $\sup\{|\int \varphi f| : \varphi \in L^1(G), N(\varphi) \leq 1\} < \infty$; then $1_G \in X$ and X is left invariant. There exists a left invariant mean on X if and only if, for every compact subset K [resp. every finite subset F] in G and every $\varepsilon > 0$, there exists $\varphi \in P^1(G)$ such that $N(_a\varphi - \varphi) < \varepsilon$ for any $a \in K$ [resp. $a \in F$]. If a locally compact group G is amenable, by 6.7 there exists a net (μ_i) of probability measures on G such that $\lim_i |\mu_i(xA) - \mu_i(A)| = 0$ whenever $A \in \mathfrak{B}(G)$ and $a \in G$. The converse holds since any w*-limit point of (μ_i) may be considered to be a left invariant mean on $L^\infty_{\mathbb{R}}(G)$. For comments on these topics in relation to statistical applications see Bondar and Milnes [44]. Such asymptotically invariant nets of probability measures had been considered in an early work of Lehmann [345] concerning mathematical statistics. Paterson [432] studied translation experiments on amenable groups.

16. SUPPLEMENTARY STRUCTURAL PROPERTIES

We give complements on Følner and Leptin properties. The conditions admit particularly convenient formulations for amenable σ-compact groups.

A. Dichotomy Properties

Definition 16.1. *Let G be a locally compact group. We define*

$$\vartheta = \sup\left\{\inf\left\{\sup\left\{\frac{|aU \triangle U|}{|U|} : a \in K\right\} : U \in \mathfrak{K}_0(G)\right\} : K \in \mathfrak{K}(G)\right\},$$

$$\vartheta^* = \sup\left\{\inf\left\{\sup\left\{\frac{|aU \triangle U|}{|U|} : a \in F\right\} : U \in \mathfrak{K}_0(G)\right\} : F \in \mathfrak{F}(G)\right\},$$

$$\kappa = \sup\left\{\inf\left\{\frac{|KU|}{|U|} : U \in \mathfrak{K}_0(G)\right\} : K \in \mathfrak{K}(G)\right\},$$

$$\kappa^* = \sup\left\{\inf\left\{\frac{|FU|}{|U|} : U \in \mathfrak{K}_0(G)\right\} : F \in \mathfrak{F}(G)\right\}.$$

We have $\vartheta, \vartheta^* \in [0, 2]$ and $\kappa, \kappa^* \in [1, \infty]$.

Theorem 16.2. *In a locally compact group G the following dichotomy holds:*
(a) *G is amenable and $\vartheta = \vartheta^* = 0$, $\kappa = \kappa^* = 1$.*
(b) *G is nonamenable and $\vartheta = \vartheta^* = 2$, $\kappa = \kappa^* = \infty$.*

Proof. (a) Theorems 7.3, 7.9, and Propositions 7.4, 7.11 imply that amenability of G is characterized by each of the conditions $\vartheta = 0$, $\vartheta^* = 0$, $\kappa = 1$, and $\kappa^* = 1$.

(b) If G is nonamenable, by Theorem 15.3, given $\varepsilon \in [0,2]$, there exists $K \in \mathfrak{K}(G)$ [resp. $F \in \mathfrak{F}(G)$] such that for every $U \in \mathfrak{K}_0(G)$ there exists $a \in K$ [resp. $a \in F$] with $\|{}_a\xi_U - \xi_U\|_1 > \varepsilon$; hence $\|\xi_U - {}_{a^{-1}}\xi_U\|_1 > \varepsilon$ and $|aU \triangle U|/|U| > \varepsilon$, that is, $\vartheta = 2$ [resp. $\vartheta^* = 2$].

If $\kappa^* > 1$, there exist $F \in \mathfrak{F}(G)$ and $\delta > 1$ such that

$$\inf\left\{\frac{|FU|}{|U|} : U \in \mathfrak{K}_0(G)\right\} > \delta.$$

For all $n \in \mathbf{N}^*$ and $U \in \mathfrak{K}_0(G)$,

$$\frac{|F^n U|}{|U|} = \frac{|F^n U|}{|F^{n-1}U|}\frac{|F^{n-1}U|}{|F^{n-2}U|}\cdots\frac{|FU|}{|U|} = \frac{|FF^{n-1}U|}{|F^{n-1}U|}\frac{|FF^{n-2}U|}{|F^{n-2}U|}\cdots\frac{|FU|}{|U|}$$

and $F^{n-1}U, F^{n-2}U, \ldots, FU \in \mathfrak{K}_0(G)$; so we have

$$\inf\left\{\frac{|F^n U|}{|U|} : U \in \mathfrak{K}_0(G)\right\} > \delta^n$$

with $F^n \in \mathfrak{F}(G)$. Necessarily $\kappa^* = \infty$. Then a fortiori $\kappa = \infty$. $\square$

B. Structural Characterizations of Amenability

The first proposition states a general Følner-type characterization.

Proposition 16.3. *The locally compact group G is amenable if and only if the following condition holds for one (for every) $p \in]1, \infty[$:*

$$\sup\left\{\inf\left\{\sup\left\{\frac{|U|^{1/p}|V|^{1/p'}}{|aU \cap V|} : a \in K\right\} : U, V \in \mathfrak{K}_0(G)\right\} : K \in \mathfrak{K}(G)\right\} = 1$$

with $1/p + 1/p' = 1$.

Proof. (a) Assume G to be amenable. By Proposition 7.11, for every $p \in]1, \infty[$, every $K \in \mathfrak{K}(G)$, and every $\varepsilon \in \mathbf{R}_+^*$ there exists $U \in \mathfrak{K}_0(G)$ such that $|KU|/|U| < (1+\varepsilon)^{p'}$. Let $V = KU \in \mathfrak{K}_0(G)$. For every $a \in K$, $aU \cap V = aU$ and

$$1 \leq \frac{|U|^{1/p}|V|^{1/p'}}{|aU \cap V|} = \frac{|U|^{1/p}|KU|^{1/p'}}{|aU|} = \frac{|KU|^{1/p'}}{|U|^{1-1/p}} = \left(\frac{|KU|}{|U|}\right)^{1/p'} < 1 + \varepsilon.$$

As $\varepsilon \in \mathbf{R}_+^*$ is arbitrary, the condition holds.

SUPPLEMENTARY STRUCTURAL PROPERTIES

(b) We show that, for every $p \in]1, \infty[$, the condition implies (Cv_p^*), hence amenability by Theorem 9.6. Let $\mu \in \mathscr{M}_{c,+}^1(G)$, $\mu \neq 0$ with $K = \operatorname{supp}\mu$. We have $\|L_\mu\|_{Cv^p} \leq \|\mu\|$ and prove the reverse inequality.

Given $\varepsilon > 0$, by assumption there exist $U, V \in \Re_0(G)$ such that

$$\sup\left\{\frac{|U|^{1/p}|V|^{1/p'}}{|aU \cap V|} : a \in K\right\} < 1 + \varepsilon.$$

Let $f = (1/|U|^{1/p})1_U \in P^p(G)$ and $g = (1/|V|^{1/p'})1_V \in P^{p'}(G)$. We have

$$0 < \|\mu\| \inf\{|aU \cap V| : a \in K\} \leq \int_K |xU \cap V| \, d\mu(x)$$

$$= |U|^{1/p}|V|^{1/p'} \int_G \frac{|xU \cap V|}{|U|^{1/p}|V|^{1/p'}} d\mu(x)$$

$$= |U|^{1/p}|V|^{1/p'} \int_G \int_G f(x^{-1}y)g(y) \, dy \, d\mu(x)$$

$$= |U|^{1/p}|V|^{1/p'} \int_G \mu * f(y)g(y) \, dy = |U|^{1/p}|V|^{1/p'} \langle \mu * f, g \rangle$$

$$\leq |U|^{1/p}|V|^{1/p'} \|\mu * f\|_p \|g\|_{p'} \leq |U|^{1/p}|V|^{1/p'} \|L_\mu\|_{Cv^p}.$$

Hence

$$\|\mu\| \leq \sup\left\{\frac{|U|^{1/p}|V|^{1/p'}}{|aU \cap V|} : a \in K\right\}\|L_\mu\|_{Cv^p} \leq (1 + \varepsilon)\|L_\mu\|_{Cv^p}.$$

As $\varepsilon > 0$ is arbitrary, $\|\mu\| \leq \|L_\mu\|_{Cv^p}$. □

We add some refinements for Følner and Leptin conditions.

Lemma 16.4. *Let G be a noncompact amenable group. If $\varepsilon > 0$ and K is a compact subset of G such that $e \in K$ [resp. K is a compact subset of G], there exist $c > 0$ and a sequence $(U_n)_{n \in \mathbb{N}}$ of compact subsets of G such that, for every $n \in \mathbb{N}$, $|U_n| \geq 2^n c$ and $|KU_n \triangle U_n|/|U_n| < \varepsilon$ [resp. $|aU_n \triangle U_n|/|U_n| < \varepsilon$ whenever $a \in K$].*

Proof. As G is amenable, by Proposition 7.11 [resp. 7.4] there exists $U \in \Re(G)$ such that $|U| > 0$ and $|KU \setminus U|/|U| = |KU \triangle U|/|U| < \varepsilon$ [resp. $|aU \triangle U|/|U| < \varepsilon$ whenever $a \in K$]. Let $c = |U| > 0$. We construct the sequence inductively. Let $U_0 = U$ and suppose U_n to be defined. There exists $y_n \in G$ such that $U_n y_n \cap K^{-1}KU_n = \emptyset$ because otherwise $G = U_n^{-1}K^{-1}KU_n$

and G would be compact. If G is unimodular, then $\Delta(y_n) = 1$. For given compact subsets A and B in $\mathbf{R}^*_+$ we may find $\alpha_0 \neq 1$ such that $A\alpha \cap B = \emptyset$ whenever $\alpha \geq \alpha_0$. So if G is nonunimodular, there exists $y'_n \in G$ such that $\Delta(y'_n) > 1$ and $\Delta(U_n y'_n) \cap \Delta(K^{-1}KU_n) = \Delta(U_n)\Delta(y'_n) \cap \Delta(K^{-1}KU_n) = \emptyset$; hence $U_n y'_n \cap K^{-1}KU_n = \emptyset$. Therefore, in any case, there exists $x_n \in G$ such that $\Delta(x_n) \geq 1$ and $U_n x_n \cap K^{-1}KU_n = \emptyset$. We have

$$U_n x_n \cap U_n = \emptyset \tag{1}$$

and

$$KU_n x_n \cap KU_n = \emptyset. \tag{2}$$

Let $U_{n+1} = U_n x_n \cup U_n$. The induction hypothesis now implies that by (1)

$$|U_{n+1}| = |U_n|\Delta(x_n) + |U_n| \geq 2|U_n| \geq 2^{n+1}c.$$

By (1) and (2) we obtain

$$|KU_{n+1} \triangle U_{n+1}| = |KU_{n+1} \setminus U_{n+1}| = |KU_n x_n \setminus U_n x_n| + |KU_n \setminus U_n|$$
$$= (\Delta(x_n) + 1)|KU_n \setminus U_n| < \varepsilon(\Delta(x_n) + 1)|U_n|$$
$$= \varepsilon|U_n x_n \cup U_n| = \varepsilon|U_{n+1}|$$

[resp. by (1) for every $a \in K$,

$$|aU_{n+1} \triangle U_{n+1}| \leq |aU_n x_n \triangle U_n x_n| + |aU_n \triangle U_n|$$
$$= (\Delta(x_n) + 1)|aU_n \triangle U_n| < \varepsilon(\Delta(x_n) + 1)|U_n|$$
$$= \varepsilon|U_n x_n \cup U_n| = \varepsilon|U_{n+1}|. \qquad \square$$

Proposition 16.5. *A locally compact group G is amenable if and only if, for every compact subset C in G, every compact subset K of G such that $e \in K$ [resp. every compact subset K of G] and every $\varepsilon > 0$, there exists a compact subset U of G such that $0 < |U|$, $C \subset U$, and $|KU \triangle U|/|U| < \varepsilon$ [resp. $|aU \triangle U|/|U| < \varepsilon$ whenever $a \in K$].*

Proof. The condition obviously implies (SF_e) [resp. (F)] hence amenability by Theorem 7.9 [resp. Theorem 7.3].

Conversely, if G is compact, it suffices to choose $U = G$. If G is noncompact, we consider the sequence (U_n) constructed in Lemma 16.4. Let $n \in \mathbf{N}^*$ such that $|KC|/|U_n| < \varepsilon/4$ and $|KU_n \triangle U_n|/|U_n| < \varepsilon/2$ [resp.

$|aU_n \triangle U_n|/|U_n| < \varepsilon/2$ whenever $a \in K$]. Let $U = U_n \cup C$. Then

$$\frac{|KU \triangle U|}{|U|} \leq \frac{|KU \triangle KU_n| + |KU_n \triangle U_n| + |U_n \triangle U|}{|U_n|}$$

$$\leq \frac{|KC|}{|U_n|} + \frac{|KU_n \triangle U_n|}{|U_n|} + \frac{|C|}{|U_n|} < \frac{\varepsilon}{4} + \frac{\varepsilon}{2} + \frac{\varepsilon}{4} = \varepsilon$$

[resp. $\frac{|aU \triangle U|}{|U|} \leq \frac{|aU \triangle aU_n| + |aU_n \triangle U_n| + |U_n \triangle U|}{|U_n|}$

$$\leq \frac{|aC|}{|U_n|} + \frac{|aU_n \triangle U_n|}{|U_n|} + \frac{|C|}{|U_n|} < \frac{\varepsilon}{4} + \frac{\varepsilon}{2} + \frac{\varepsilon}{4} = \varepsilon$$

whenever $a \in K$]. □

Proposition 16.6. *A locally compact group G is amenable if and only if, for every compact subset K in G and every $\varepsilon > 0$, there exists a compact symmetric subset U in G such that $|U| > 0$ and $|aU \triangle U|/|U| < \varepsilon$ whenever $a \in K$.*

Proof. The condition is obviously sufficient by Theorem 7.3. Now assume G to be amenable.

CASE 1: G IS UNIMODULAR.

(a) Let $\varepsilon \in \,]0, 1[$. By Proposition 15.10 there exists $\varphi = \check{\varphi} \in P^1(G)$ such that $\|{}_a\varphi - \varphi\|_1 < \varepsilon/3$ whenever $a \in K$. There also exists $\theta_0 \in \mathscr{S}_+(G)$ such that $\|\varphi - \theta_0\|_1 < \varepsilon/6$. Let

$$\theta = \frac{\theta_0 + \check{\theta}_0}{2} \in \mathscr{S}_+(G);$$

$$\theta = \check{\theta}, \|\varphi - \theta\|_1 = \frac{1}{2}\|(\varphi - \theta_0) + (\varphi - \check{\theta}_0)\|_1 < \frac{\varepsilon}{6},$$

and $\|\theta\|_1 > 0$. Then let $\theta' = \theta/\|\theta\|_1 \in \mathscr{S}(G) \cap P^1(G)$. We have

$$\|\theta - \theta'\|_1 = |\|\theta\|_1 - 1| = |\|\theta\|_1 - \|\varphi\|_1| \leq \|\theta - \varphi\|_1$$

and

$$\|\varphi - \theta'\|_1 \leq \|\varphi - \theta\|_1 + \|\theta - \theta'\|_1 \leq 2\|\theta - \varphi\|_1 < \frac{\varepsilon}{3}.$$

Moreover for every $a \in K$,

$$\|{}_a\theta' - \theta'\|_1 \leq \|{}_a(\theta' - \varphi)\|_1 + \|{}_a\varphi - \varphi\|_1 + \|\varphi - \theta'\|_1 < \frac{\varepsilon}{3} + \frac{\varepsilon}{3} + \frac{\varepsilon}{3} = \varepsilon.$$

(b) Now if $\varepsilon > 0$, by Lemma 7.2, the proof of $(P_1) \Rightarrow (WF) \Rightarrow (F)$ given in Theorem 7.3 implies the existence of a symmetric measurable subset V in G

such that $|V| > 0$ and $|aV \triangle V|/|V| < \varepsilon/6$ whenever $a \in K$. By regularity of the Haar measure there exists a compact subset V_1 of G such that $V_1 \subset V$, $|V_1| > 0$ and $|V \triangle V_1| < \gamma = \inf\{|V|/2, \varepsilon/6 \, |V|\}$. Let $U = V_1 \cup V_1^{-1}$; then $|V \triangle U| \leq |V \triangle V_1| < \gamma$. Moreover

$$|V| = |U| + |V \setminus U| < |U| + \frac{|V|}{2}$$

hence $|V|/|U| < 2$. For every $a \in K$,

$$\frac{|aU \triangle U|}{|U|} \leq \frac{|V|}{|U|} \frac{|a(U \triangle V)| + |aV \triangle V| + |V \triangle U|}{|V|} < 2\left(\frac{\varepsilon}{6} + \frac{\varepsilon}{6} + \frac{\varepsilon}{6}\right) = \varepsilon.$$

CASE 2: G IS NONUNIMODULAR

By Proposition 7.4, given a compact subset K in G and $\varepsilon > 0$, there exists a compact subset V in G such that $|V| > 0$ and $|aV \triangle V|/|V| < \varepsilon/2$ whenever $a \in K$. Consider $s \in G$ such that $\Delta(s) > 1$. There exists $n \in \mathbf{N}^*$ such that $\Delta(s)^n > 4|V^{-1}|/\varepsilon|V|$. For $t = s^n$, we have $\Delta(t) = \Delta(s)^n$ and $2|V^{-1}|/|V|\Delta(t) < \varepsilon/2$. Let $U = (Vt) \cup (Vt)^{-1}$; $|U| > 0$ and U is compact. For every $a \in K$,

$$\frac{|aU \triangle U|}{|U|} \leq \frac{|aVt \triangle Vt| + |a(Vt)^{-1} \triangle (Vt)^{-1}|}{|Vt|}$$

$$= \frac{1}{|V|\Delta(t)}\left(|aV \triangle V|\Delta(t) + |at^{-1}V^{-1} \triangle t^{-1}V^{-1}|\right)$$

$$\leq \frac{|aV \triangle V|}{|V|} + \frac{2|V^{-1}|}{|V|\Delta(t)} < \frac{\varepsilon}{2} + \frac{\varepsilon}{2} = \varepsilon. \qquad \square$$

Proposition 16.7. *Let G be a locally compact group. Then G is amenable if and only if there exists a net $(U_i)_{i \in I}$ of compact subsets of G such that $|U_i| > 0$ whenever $i \in I$ and the following equivalent conditions hold:*

(i) *[resp. (i_0)] For every compact subset V of G [resp. every compact neighborhood V of e]*

$$\lim_i \frac{|\cap_{x \in V} x^{-1} U_i|}{|VU_i|} = 1.$$

(ii) *[resp. (ii_0)] For every compact subset V of G [resp. every compact neighborhood V of e],*

$$\lim_i \frac{|\cap_{x \in V} x^{-1} U_i|}{|U_i|} = 1.$$

SUPPLEMENTARY STRUCTURAL PROPERTIES

Proof. By Theorem 7.9 the properties (L) and $(L_{e,0})$ characterize amenability. Therefore the statement may be proved via the following diagram

$$\begin{array}{ccc} (L) \Rightarrow (\mathrm{i}) & \Rightarrow & (\mathrm{i}_0) \\ \Downarrow & & \Downarrow \\ (\mathrm{ii}) \Rightarrow (\mathrm{ii}_0) & \Rightarrow & (L_{e,0}) \end{array}$$

where only the first and last implications are nontrivial.

$$(L) \Rightarrow (\mathrm{i})$$

Let I be the directed set $\Re(G) \times \mathbf{R}_+^*$ with $(V, \varepsilon) = i \prec i' = (V', \varepsilon')$ for $i, i' \in I$ if and only if $V \subset V'$ and $\varepsilon' < \varepsilon$. By Proposition 7.11, to $i = (V, \varepsilon) \in I$ we may associate a compact subset W_i of G such that $0 < |W_i|$ and $|V^2 W_i|/|W_i| < 1 + \varepsilon$. Let $U_i = VW_i$, so $0 < |W_i| \leq |U_i| < \infty$. We have

$$1 \leq \frac{|VU_i|}{|\cap_{x \in V} x^{-1} U_i|} = \frac{|V^2 W_i|}{|\cap_{x \in V} x^{-1} VW_i|} \leq \frac{|V^2 W_i|}{|W_i|} < 1 + \varepsilon.$$

Therefore (i) holds.

$$(\mathrm{ii}_0) \Rightarrow (L_{e,0})$$

Let V be a compact neighborhood of e in G. By hypothesis there exists a net $(U_i)_{i \in I}$ of compact subsets in G such that $|U_i| > 0$ whenever $i \in I$ and $\lim_i |\cap_{x \in V} x^{-1} U_i|/|U_i| = 1$. There exists a subnet $(U_j)_{j \in J}$ of $(U_i)_{i \in I}$ such that $|\cap_{x \in V} x^{-1} U_j| > 0$ whenever $j \in J$. For every $j \in J$, we consider the measurable subset $V_j = \cap_{x \in V} x^{-1} U_j$. Since $e \in V$, we have $V_j \subset VV_j \subset U_j$, $0 < |V_j| < \infty$ and

$$\frac{|\cap_{x \in V} x^{-1} U_j|}{|U_j|} \leq \frac{|V_j|}{|VV_j|} \leq 1.$$

Therefore also $\lim_j |VV_j|/|V_j| = 1$, that is, $(L_{e,0})$ holds. □

We state a characterization of amenability on discrete groups in terms of an averaging Følner relation.

Proposition 16.8. *A discrete group G is amenable if and only if there exists $\alpha \in \,]0, 1[$ such that to every finite subset $F = \{a_1, \ldots, a_n\}$ of G there corresponds a nonvoid finite subset U of G satisfying the relation*

$$\frac{1}{n} \sum_{i=1}^n \operatorname{card}(U \cap a_i U) \geq \alpha \operatorname{card} U.$$

Proof. The necessity of the condition follows from (F). Assume now that the property holds. In $\mathcal{N}(l_\mathbf{R}^\infty(G))$ we consider reduced sums $h = f_1 - {}_{a_1}f_1 + \cdots + f_n - {}_{a_n}f_n$, where $f_1, \ldots, f_n \in l_\mathbf{R}^\infty(G)$, $a_1, \ldots, a_n \in G$, $n \in \mathbf{N}^*$. Let then

$$m(h) = 2n \sup \{\|f_i\| : i = 1, \ldots, n\}$$

so $\|h\| \leq m(h)$.

(a) To every $h \in \mathcal{N}(l_\mathbf{R}^\infty(G))$ we associate $h_1 \in \mathcal{N}(l_\mathbf{R}^\infty(G))$ such that $m(h_1) \leq (1 - \alpha)m(h)$.

If $h = \sum_{i=1}^{n}(f_i - {}_{a_i}f_i)$ with $f_i \in l_\mathbf{R}^\infty(G)$, $a_i \in G$, $i = 1, \ldots, n$, let U be the finite subset associated to $F = \{a_1, \ldots, a_n\}$ by hypothesis; we define

$$h_1 = \frac{1}{\operatorname{card} U} \sum_{y \in U} {}_y h = \frac{1}{\operatorname{card} U} \sum_{i=1}^{n} \sum_{y \in U} \left({}_y f_i - {}_{a_i y} f_i\right).$$

For a fixed $i \in \{1, \ldots, n\}$, $2 \operatorname{card}(U \cap a_i U)$ terms in the sum $\sum_{y \in U}({}_y f_i - {}_{a_i y} f_i)$ cancel, therefore

$$m(h_1) \leq \frac{2}{\operatorname{card} U} \sum_{i=1}^{n} (\operatorname{card} U - \operatorname{card}(U \cap a_i U)) \sup \{\|f_i\| : i = 1, \ldots, n\}$$

$$= 2 \sum_{i=1}^{n} \left(1 - \frac{\operatorname{card}(U \cap a_i U)}{\operatorname{card} U}\right) \sup \{\|f_i\| : i = 1, \ldots, n\}$$

$$\leq 2n(1 - \alpha) \sup \{\|f_i\| : i = 1, \ldots, n\} = (1 - \alpha)m(h).$$

Moreover $\sup h_1 \leq \sup h$.

(b) If G were not amenable, by Proposition 4.29 there would exist $h_0 \in \mathcal{N}(l_\mathbf{R}^\infty(G))$ and $\beta \in \mathbf{R}_-^*$ such that $\sup h_0 = \beta$. Applying (a) p times to h_0 we determine $h_p \in \mathcal{N}(l_\mathbf{R}^\infty(G))$ such that $m(h_p) \leq (1 - \alpha)^p m(h_0)$ and $\sup h_p \leq \beta$. Since $\|h_p\| \leq m(h_p)$, for every $x \in G$, we have

$$-(1 - \alpha)^p m(h_0) \leq -m(h_p) \leq h_p(x) \leq \beta < 0.$$

As $\lim_{p \to \infty}(1 - \alpha)^p = 0$, a contradiction would arise. We conclude that G is amenable. □

C. Amenable σ-Compact Groups

Proposition 16.7 characterizes amenability by the existence of nets of compact subsets satisfying an asymptotical invariance property. In case the amenable group is σ-compact, one may consider asymptotical invariance properties for *sequences* of compact subsets obtained via Følner conditions. We establish a collection of precise statements.

SUPPLEMENTARY STRUCTURAL PROPERTIES

Lemma 16.9. *Let G be a locally compact group admitting a measurable subset U and a sequence $(V_n)_{n \in \mathbf{N}^*}$ of measurable subsets such that, for every $n \in \mathbf{N}^*$, $0 < |V_n| < \infty$ and, for every $x \in U$, $\lim_{n \to \infty} |xV_n \triangle V_n|/|V_n| = 0$. Then $\liminf_{n \to \infty} |V_n^{-1}| \geq |U|$.*

Proof. (a) Assume that $|U| < \infty$. By hypothesis, for every $x \in U$, $\lim_{n \to \infty} |xV_n \cap V_n|/|V_n| = 1$. We deduce from Fatou's lemma that

$$|U| = \int_U dx = \int_U \lim_{n \to \infty} \frac{|xV_n \cap V_n|}{|V_n|} dx \leq \liminf_{n \to \infty} \int_U \frac{|xV_n \cap V_n|}{|V_n|} dx.$$

For every $n \in \mathbf{N}^*$,

$$\int_G \frac{|xV_n \cap V_n|}{|V_n|} dx = \frac{1}{|V_n|} \int_G \left(\int_{V_n} 1_{V_n}(x^{-1}y) \, dy \right) dx$$

$$= \frac{1}{|V_n|} \int_{V_n} \left(\int_G 1_{V_n}(x^{-1}y) \, dx \right) dy$$

$$= \frac{1}{|V_n|} \int_{V_n} |yV_n^{-1}| \, dy \leq |V_n^{-1}|.$$

Thus the statement holds.

(b) Assume that $|U| = \infty$. If $\alpha \in \mathbf{R}^*_+$, there exists a measurable subset U_1 of U such that $\alpha < |U_1| < \infty$. We deduce from (a) that

$$\liminf_{n \to \infty} |V_n^{-1}| \geq |U_1| > \alpha.$$

As $\alpha \in \mathbf{R}^*_+$ may be chosen arbitrarily, we have $\liminf_{n \to \infty} |V_n^{-1}| = \infty$. □

Proposition 16.10. *Let G be a locally compact σ-compact group. Then G is amenable if and only if there exists a sequence $(U_n)_{n \in \mathbf{N}^*}$ in $\Re_0(G)$ such that $G = \cup_{n \in \mathbf{N}^*} U_n$ and, for every $x \in G$, $\lim_{n \to \infty} |xU_n \triangle U_n|/|U_n| = 0$, that is, $\lim_{n \to \infty} \|_x\xi_{U_n} - \xi_{U_n}\|_1 = 0$ whenever $x \in G$. The limit is uniform on compacta. One may assume that each $U_n (n \in \mathbf{N}^*)$ is symmetric. The sequence may be chosen to be nondecreasing.*

Proof. The condition is obviously sufficient by Theorem 7.3.
Conversely assume G to be amenable. If G is compact, it suffices to choose $U_n = G$ for every $n \in \mathbf{N}^*$. Now let G be noncompact. Since G is σ-compact, there exists a nondecreasing sequence (K_n) of relatively compact symmetric neighborhoods of e in G such that $\cup_{n \in \mathbf{N}^*} K_n = G$. By Proposition 16.6, for any $n \in \mathbf{N}^*$, there exists a compact symmetric subset V_n of G such that $|V_n| > 0$ and $|aV_n \triangle V_n|/|V_n| < 1/3n$ whenever $a \in K_n$. We construct (U_n) inductively.

Let $U_1 = V_1$ and assume that, for $p = 1, \ldots, m$, we have defined a compact symmetric subset U_p of G such that $|U_p| > 0$ and $|aU_p \triangle U_p|/|U_p| < 1/p$ whenever $a \in K_p$. As G is noncompact, $\lim_{n \to \infty} |K_n| = \infty$. Thus by Lemma 16.9 we have $\lim_{n \to \infty} |V_n| = \lim_{n \to \infty} |V_n^{-1}| = \infty$. There exists $n_m \in \mathbf{N}^*$ such that $n_m > m$ and

$$\frac{|U_m| + |K_m|}{|V_{n_m}|} < \frac{1}{3(m+1)}.$$

We put $U_{m+1} = V_{n_m} \cup U_m \cup K_m$. Then U_{m+1} is symmetric and, for every $a \in K_m$,

$$\frac{|aU_{m+1} \triangle U_{m+1}|}{|U_{m+1}|} \leq \frac{|aV_{n_m} \triangle V_{n_m}|}{|V_{n_m}|} + 2\frac{|U_m| + |K_m|}{|V_{n_m}|}$$

$$< \frac{1}{3n_m} + \frac{2}{3(m+1)} < \frac{1}{m+1}.$$

Thus $G = \bigcup_{m \in \mathbf{N}^*} U_m$ and $\lim_{m \to \infty} |xU_m \triangle U_m|/|U_m| = 0$ for every $x \in G$. More precisely, if K is any compact subset of G and $\varepsilon > 0$, there exists $q \in \mathbf{N}^*$ such that $K \subset K_q$, $1/q < \varepsilon$ and

$$\frac{|xU_q \triangle U_q|}{|U_q|} < \frac{1}{q}$$

whenever $x \in K_q$. Convergence is uniform on compacta. □

Proposition 16.11. *Let G be a locally compact σ-compact group. Then G is amenable if and only if there exists a sequence $(U_n)_{n \in \mathbf{N}^*}$ in $\mathfrak{K}_0(G)$ such that $G = \bigcup_{n \in \mathbf{N}^*} U_n$ and, for every nonvoid compact subset K in G, $\lim_{n \to \infty} |KU_n \triangle U_n|/|U_n| = 0$. The sequence may be chosen to be nondecreasing.*

Proof. If the condition holds, G is amenable by Theorem 7.9.

Conversely assume G to be amenable. There exists a nondecreasing sequence $(V_n)_{n \in \mathbf{N}^*}$ of compact neighborhoods of e such that $G = \bigcup_{n \in \mathbf{N}^*} V_n$. Let $U_0 = V_1$. We define $(U_n)_{n \in \mathbf{N}^*}$ inductively. For $n \in \mathbf{N}^*$, via Proposition 16.5 one may determine a compact subset U_n of G such that $V_n \cup U_{n-1} \subset U_n$ and $|V_n U_n \triangle U_n|/|U_n| < 1/n$. Then $G = \bigcup_{n \in \mathbf{N}^*} U_n$. If K is a compact subset in G, there exists $n_0 \in \mathbf{N}^*$ such that $K \subset V_{n_0}$. For every $n \geq n_0$, by (7.2) we have

$$|KU_n \triangle U_n| \leq 2||KU_n \setminus U_n| \leq 2|V_n U_n \setminus U_n| = 2|V_n U_n \triangle U_n| < \frac{2}{n}|U_n|. \quad \square$$

Definition 16.12. *If G is a locally compact group, let $\mathscr{A}(G)$ [resp. $\mathscr{SA}(G)$] be the class of all averaging sequences [resp. strongly averaging sequences] in G,*

SUPPLEMENTARY STRUCTURAL PROPERTIES

that is, all sequences $(U_n)_{n \in \mathbf{N}^*}$ consisting of measurable subsets in G such that $0 < |U_n| < \infty$ whenever $n \in \mathbf{N}^*$ and

$$\lim_{n \to \infty} \frac{|xU_n \triangle U_n|}{|U_n|} = 0 \quad \text{whenever } x \in G$$

$$\left[\text{resp. } \lim_{n \to \infty} \frac{|KU_n \triangle U_n|}{|U_n|} = 0 \quad \text{whenever } K \text{ is a compact subset of } G \right].$$

We denote by $\mathscr{A}_c(G)$ [resp. $\mathscr{SA}_c(G)$] the subclass of $\mathscr{A}(G)$ [resp. $\mathscr{SA}(G)$] formed by the sequences admitting at most a finite number of members that are not relatively compact.

Obviously $\mathscr{SA}(G) \subset \mathscr{A}(G)$ and $\mathscr{SA}_c(G) \subset \mathscr{A}_c(G)$.

Note that if $(U_n) \in \mathscr{SA}(G)$, for every $n \in \mathbf{N}^*$ and every compact subset K of G, $KU_n \triangle U_n$ is measurable. Thus we have also measurability of the sets

$$KU_n \cup U_n = (KU_n \triangle U_n) \cup U_n,$$

$$KU_n \setminus U_n = (KU_n \triangle U_n) \setminus U_n,$$

$$U_n \setminus KU_n = (KU_n \triangle U_n) \setminus (KU_n \setminus U_n),$$

$$KU_n = (KU_n \cup U_n) \setminus (U_n \setminus KU_n).$$

The following proposition emphasizes the equivalence $(SF) \Leftrightarrow (L)$.

Proposition 16.13. *Let (U_n) be a sequence of measurable subsets of the locally compact group G such that $0 < |U_n| < \infty$ whenever $n \in \mathbf{N}^*$. Then $(U_n) \in \mathscr{SA}(G)$ if and only if $\lim_{n \to \infty} |KU_n|/|U_n| = 1$ whenever K is a compact subset of G.*

Proof. (a) Assume that $(U_n) \in \mathscr{SA}(G)$. Let K be a compact subset and consider $K' = K \cup \{e\}$. Then

$$1 \leq \liminf_{n \to \infty} \frac{|KU_n|}{|U_n|} \leq \limsup_{n \to \infty} \frac{|KU_n|}{|U_n|} \leq 1 + \limsup_{n \to \infty} \frac{|KU_n \setminus U_n|}{|U_n|}$$

$$= 1 + \lim_{n \to \infty} \frac{|K'U_n \triangle U_n|}{|U_n|} = 1.$$

Hence $\lim_{n \to \infty} |KU_n|/|U_n| = 1$.

(b) Assume that the condition holds. If K is a compact subset of G, consider again $K' = K \cup \{e\}$. For any $\in \mathbf{N}^*$,

$$\frac{|K'U_n|}{|U_n|} = 1 + \frac{|KU_n \setminus U_n|}{|U_n|}$$

and

$$\frac{|K'U_n|}{|U_n|} = \frac{|KU_n|}{|U_n|} + \frac{|U_n \setminus KU_n|}{|U_n|}.$$

Therefore $\lim_{n \to \infty} |KU_n \triangle U_n|/|U_n| = 0$, and we have $(U_n) \in \mathscr{SA}(G)$. □

Proposition 16.14. *Let G be a locally compact group.*
(a) *We have $\mathscr{A}(G) \neq \varnothing$ if and only if $\mathscr{A}_c(G) \neq \varnothing$ and $\mathscr{SA}(G) \neq \varnothing$ if and only if $\mathscr{SA}_c(G) \neq \varnothing$. More precisely, given $(U_n) \in \mathscr{A}(G)$ [resp. $(U_n) \in \mathscr{SA}(G)$] there exists $(V_n) \in \mathscr{A}_c(G)$ [resp. $(V_n) \in \mathscr{SA}_c(G)$] formed by compact subsets of G such that $V_n \subset U_n$ and $|U_n \setminus V_n| \leq |U_n|/(n+1)$ whenever $n \in \mathbf{N}^*$. If, for $n \in \mathbf{N}^*$, U_n is symmetric, we may choose V_n to be symmetric.*
(b) *The members of each averaging sequence generate G.*

Necessary and sufficient conditions for a locally compact group G to be amenable σ-compact are: $\mathscr{A}(G) \neq \varnothing$, $\mathscr{A}_c(G) \neq \varnothing$, $\mathscr{SA}(G) \neq \varnothing$, *and* $\mathscr{SA}_c(G) \neq \varnothing$.

Proof. (a) Let $(U_n) \in \mathscr{A}(G)$ [resp. $(U_n) \in \mathscr{SA}(G)$]. By the regularity of the Haar measure, for every $n \in \mathbf{N}^*$, there exists a compact subset W_n in G such that $W_n \subset U_n$, $0 < |W_n|$ and $|U_n \setminus W_n| \leq |U_n|/(n+1)$. Let $n \in \mathbf{N}^*$. We put $V_n = W_n$ and, in case U_n is symmetric, $V_n = W_n \cup W_n^{-1}$. We have $V_n \subset U_n$, $0 < |V_n|$, $|U_n \setminus V_n| \leq |U_n|/(n+1)$ and

$$|U_n| \leq |V_n| + |U_n \setminus V_n| \leq |V_n| + \frac{|U_n|}{n+1}.$$

Hence $|U_n|/|V_n| \leq (n+1)/n$.

Let K be any singleton [resp. any compact subset] of G. For every $n \in \mathbf{N}^*$,

$$KV_n \triangle V_n \subset (KV_n \triangle KU_n) \cup (KU_n \triangle U_n) \cup (U_n \setminus V_n)$$
$$= (KU_n \setminus KV_n) \cup (KU_n \triangle U_n) \cup (U_n \setminus V_n),$$

but

$$|KU_n \setminus KV_n| = |KU_n| - |KV_n| \leq |KU_n| - |V_n|$$
$$\leq |KU_n \triangle U_n| + |U_n| - |V_n| = |KU_n \triangle U_n| + |U_n \setminus V_n|.$$

Hence

$$|KV_n \triangle V_n| \leq 2(|KU_n \triangle U_n| + |U_n \setminus V_n|)$$

and

$$\frac{|KV_n \triangle V_n|}{|V_n|} \leq 2\left(\frac{|KU_n \triangle U_n|}{|U_n|} + \frac{|U_n \setminus V_n|}{|U_n|}\right)\frac{|U_n|}{|V_n|}$$

$$\leq 2\left(\frac{|KU_n \triangle U_n|}{|U_n|} + \frac{1}{n+1}\right)\frac{n+1}{n}.$$

Therefore $(V_n) \in \mathscr{A}_c(G)$ [resp. $(V_n) \in \mathscr{SA}_c(G)$].

(b) The members of the averaging sequence (V_n) formed by compact subsets generate a σ-compact subgroup H of G. If we had $H \neq G$, there would exist $a \in G$ such that, for every $n \in \mathbf{N}^*$, $a \notin V_n V_n^{-1}$, hence $aV_n \cap V_n = \emptyset$, $|aV_n \triangle V_n| = 2|V_n|$, and (V_n) could not be averaging. We conclude that $G = H$. Therefore G is σ-compact and generated by (U_n).

The necessity and sufficiency of the conditions are now immediate if we take Propositions 16.10 or 16.11 into account. □

In Proposition 16.16 we give the precise version of the equivalence $(F^*) \Leftrightarrow (L^*)$ for σ-compact groups.

Lemma 16.15. *Let V be a compact neighborhood of e in a locally compact group G. Suppose that there exists a uniformly bounded sequence (f_n) of measurable functions in $l_+^\infty(G)$ satisfying the following properties:*
  (i) *For every $x \in V^2$, $\lim_{n \to \infty} f_n(x) = 0$.*
  (ii) *There exists $\alpha > 0$ such that $f_n(xy) \geq \alpha|f_n(x) - f_n(y)|$ for all $x, y \in V$ and $n \in \mathbf{N}^*$.*

Then (f_n) converges uniformly on V to the null function.

Proof. For every $n \in \mathbf{N}^*$ and every $\varepsilon \in \mathbf{R}_+^*$, let

$$A_{n,\varepsilon} = \{x \in V^2 : f_n(x) \geq \varepsilon\}.$$

As the uniformly bounded sequence (f_n) converges pointwise to the null function on the compact subset V^2, it converges in measure by Lebesgue's theorem, that is, for every $\varepsilon \in \mathbf{R}_+^*$, $\lim_{n \to \infty} |A_{n,\varepsilon}| = 0$.

Assume now that the convergence is not uniform on V. Then there exist $\delta > 0$, a subsequence (f_{n_m}) of (f_n) and a sequence (x_m) in V such that, for every $m \in \mathbf{N}^*$, $f_{n_m}(x_m) > \delta$. For $m \in \mathbf{N}^*$, we put

$$B_{m,\delta} = A_{n_m, \delta/2} \cap V$$

and

$$U_{m,\delta} = B_{m,\delta} \cup (x_m(V \setminus B_{m,\delta})).$$

Thus $U_{m,\delta} \subset V^2$. If $|V \setminus B_{m,\delta}| < |V|/2$, we have $|B_{m,\delta}| \geq |V|/2$ and $|U_{m,\delta}| \geq |V|/2$. If $|V \setminus B_{m,\delta}| \geq |V|/2$, we have $x_m(V \setminus B_{m,\delta}) \geq |V|/2$ and $|U_{m,\delta}| \geq |V|/2$. Hence in all cases

$$|U_{m,\delta}| \geq \frac{|V|}{2} > 0. \tag{3}$$

Let $x \in U_{m,\delta}$. If $x \in B_{m,\delta}$, $f_{n_m}(x) \geq \delta/2$. If $x \in x_m(V \setminus B_{m,\delta})$, there exists $y_m \in V \setminus B_{m,\delta}$ such that $x = x_m y_m \in V^2$, $f_{n_m}(x_m) > \delta$, $f_{n_m}(y_m) < \delta/2$ and by (ii)

$$f_{n_m}(x) \geq \alpha\big(f_{n_m}(x_m) - f_{n_m}(y_m)\big) > \alpha\frac{\delta}{2}.$$

Thus for $\gamma = \inf\{\delta/2, \alpha\delta/2\}$ we have $U_{m,\delta} \subset A_{n_m,\gamma}$. As $\lim_{m \to \infty} |A_{n_m,\gamma}| = 0$, we would contradict (3). So we must have uniform convergence on V. □

Proposition 16.16. *Let G be a locally compact group and let $(U_n)_{n \in \mathbf{N}^*}$ be a sequence of measurable subsets in G such that $0 < |U_n| < \infty$ whenever $n \in \mathbf{N}^*$. Then the following properties are equivalent:*
  (i) $(U_n) \in \mathcal{A}(G)$.
  (ii) *If K is a compact subset of G, $\lim_{n \to \infty} |aU_n \triangle U_n|/|U_n| = 0$ uniformly for $a \in K$.*
  (iii) *If K is a compact subset of G and $k \in \mathbf{N}^*$, $\lim_{n \to \infty} |\cap_{i=1}^{k} a_i U_n|/|U_n| = 1$ uniformly for $\{a_1, \ldots, a_k\} \subset K$.*
  (iii') *If F is a finite subset of G, $\lim_{n \to \infty} |\cap_{a \in F} aU_n|/|U_n| = 1$.*
  (iv) *If K is a compact subset of G and $k \in \mathbf{N}^*$, then $\lim_{n \to \infty} |\{a_1, \ldots, a_k\}U_n|/|U_n| = 1$ uniformly for $\{a_1, \ldots, a_k\} \subset K$.*
  (iv') *If F is a finite subset of G, $\lim_{n \to \infty} |FU_n|/|U_n| = 1$.*

Proof. (i) ⇒ (ii)
For $n \in \mathbf{N}^*$, let

$$f_n(x) = \frac{|xU_n \triangle U_n|}{|U_n|},$$

$x \in G$; $\lim_{n \to \infty} f_n(x) = 0$ for every $x \in G$ and $\|f_n\| \leq 2$ for all $n \in \mathbf{N}^*$. If $n \in \mathbf{N}^*$, $x, y \in G$, we have

$$|xU_n \triangle U_n| \leq |xU_n \triangle xyU_n| + |xyU_n \triangle U_n| = |U_n \triangle yU_n| + |xyU_n \triangle U_n|.$$

Hence

$$\big||xU_n \triangle U_n| - |yU_n \triangle U_n|\big| \leq |xyU_n \triangle U_n|.$$

SUPPLEMENTARY STRUCTURAL PROPERTIES

Therefore the sequence (f_n) satisfies the conditions of Lemma 16.15 on any compact neighborhood of $K \cup \{e\}$ and (ii) holds.

$$(\text{ii}) \Rightarrow (\text{iii})$$

Let $\varepsilon > 0$. By hypothesis there exists $n_0 \in \mathbf{N}^*$ such that, for every element x of the compact subset $K^{-1}K$ and $n \geq n_0$, $|xU_n \triangle U_n|/|U_n| < \varepsilon/k$. Because

$$\bigcap_{i=1}^{k} a_i U_n = (\cdots(a_1 U_n \setminus (a_1 U_n \triangle a_2 U_n)) \setminus \cdots) \setminus (a_{k-1} U_n \triangle a_k U_n),$$

we have

$$|U_n| \geq \left|\bigcap_{i=1}^{k} a_i U_n\right| \geq |a_1 U_n| - \sum_{i=2}^{k} |a_{i-1} U_n \triangle a_i U_n|$$

$$= |U_n| - \sum_{i=2}^{k} |a_i^{-1} a_{i-1} U_n \triangle U_n| > (1 - (k-1)\varepsilon/k)|U_n| > (1 - \varepsilon)|U_n|$$

and then

$$1 - \varepsilon < \frac{|\bigcap_{i=1}^{k} a_i U_n|}{|U_n|} \leq 1.$$

(iii) $\Rightarrow$ (iii'), (iv) $\Rightarrow$ (iv')   Trivial

(iii') $\Rightarrow$ (i)

For any $n \in \mathbf{N}^*$ and any $a \in G$,

$$|aU_n \triangle U_n| = 2(|U_n| - |aU_n \cap U_n|).$$

Applying (iii') to $F = \{e, a\}$ we obtain $\lim_{n \to \infty} |aU_n \cap U_n|/|U_n| = 1$, hence $\lim_{n \to \infty} |aU_n \triangle U_n|/|U_n| = 0$.

$$(\text{ii}) \Rightarrow (\text{iv})$$

Given $\varepsilon > 0$, there exists $n_0 \in \mathbf{N}^*$ such that $|aU_n \triangle U_n|/|U_n| < \varepsilon/k$ whenever $a \in K$ and $n \geq n_0$. For any $n \in \mathbf{N}^*$,

$$\{a_1, \ldots, a_k\} U_n \subset U_n \cup \left(\bigcup_{i=1}^{k} (a_i U_n \triangle U_n)\right),$$

therefore

$$1 \le \frac{|\{a_1,\ldots,a_k\}U_n|}{|U_n|} \le 1 + \sum_{i=1}^{k} \frac{|a_i U_n \triangle U_n|}{|U_n|} < 1 + \varepsilon.$$

(iv') ⇒ (i)

Let $a \in G$, $a \ne e$ and consider $F = \{a, a^{-1}, e\}$. For any $n \in \mathbf{N}^*$, we have

$$|aU_n \triangle U_n| = |aU_n \setminus U_n| + |U_n \setminus aU_n| = |aU_n \setminus U_n| + |a^{-1}U_n \setminus U_n|$$

$$\le 2|FU_n \setminus U_n| = 2(|FU_n| - |U_n|).$$

Hence

$$\limsup_{n \to \infty} \frac{|aU_n \triangle U_n|}{|U_n|} \le 2 \lim_{n \to \infty} \left(\frac{|FU_n|}{|U_n|} - 1 \right) = 0. \qquad \square$$

Proposition 16.17. *If G is an amenable σ-compact group, let $\mathcal{A} = \mathcal{A}(G)$, $\mathcal{A}_c(G)$, $\mathcal{SA}(G)$, and $\mathcal{SA}_c(G)$.*
 (a) *If $(U_n) \in \mathcal{A}$ and $(V_n) \in \mathcal{A}$, then $(U_n \cup V_n) \in \mathcal{A}$.*
 (b) *If $(U_n) \in \mathcal{A}$ and (a_n) is a sequence in G, then $(U_n a_n) \in \mathcal{A}$.*

Proof. (a) Obviously for every $n \in \mathbf{N}^*$, $0 < |U_n \cup V_n| < \infty$ and $U_n \cup V_n$ is relatively compact if U_n and V_n are. Notice that if K is a compact subset of G and $n \in \mathbf{N}^*$,

$$\frac{|K(U_n \cup V_n) \triangle (U_n \cup V_n)|}{|U_n \cup V_n|} \le \frac{|KU_n \triangle U_n| + |KV_n \triangle V_n|}{|U_n \cup V_n|}$$

$$\le \frac{|KU_n \triangle U_n|}{|U_n|} + \frac{|KV_n \triangle V_n|}{|V_n|}.$$

Hence $(U_n \cup V_n) \in \mathcal{A}$.
 (b) It suffices to notice that, if K is any compact subset of G, for every $n \in \mathbf{N}^*$,

$$\frac{|KU_n a_n \triangle U_n a_n|}{|U_n a_n|} = \frac{|KU_n \triangle U_n|\Delta(a_n)}{|U_n|\Delta(a_n)} = \frac{|KU_n \triangle U_n|}{|U_n|}. \qquad \square$$

Proposition 16.18. *Let G be a noncompact amenable σ-compact group. If $(U_n) \in \mathcal{A}(G)$, one has $\lim_{n \to \infty} |U_n^{-1}| = \infty$ [and $\lim_{n \to \infty} |U_n| = \infty$ in case G is unimodular]. In case G is nonunimodular, for every $\varepsilon > 0$, there exists $(V_{\varepsilon,n})_{n \in \mathbf{N}^*} \in \mathcal{A}(G)$ such that $\sum_{n=1}^{\infty} |V_{\varepsilon,n}| < \varepsilon$.*

Proof. Applying Lemma 16.9 to $U = G$ we obtain $\lim_{n \to \infty} |U_n^{-1}| = \infty$.
If G is nonunimodular, given $\varepsilon > 0$, there exists a sequence $(a_{\varepsilon,n})_{n \in \mathbf{N}^*}$ in G such that $\Delta(a_{\varepsilon,n}) < \varepsilon/2^n |U_n|$ for every $n \in \mathbf{N}^*$. Let $V_{\varepsilon,n} = U_n a_{\varepsilon,n}$, $n \in \mathbf{N}^*$. By Proposition 16.17 we have $(V_{\varepsilon,n}) \in \mathscr{A}(G)$. Also

$$\sum_{n=1}^{\infty} |V_{\varepsilon,n}| = \sum_{n=1}^{\infty} |U_n| \Delta(a_{\varepsilon,n}) < \sum_{n=1}^{\infty} \frac{\varepsilon}{2^n} = \varepsilon. \qquad \square$$

Proposition 16.19. *Let G be an amenable σ-compact group. If $(U_n) \in \mathscr{A}(G)$, K is a compact subset of G and (a_n) is a sequence in K, then $(a_n U_n) \in \mathscr{A}(G)$.*

Proof. Let $\varepsilon > 0$ and $x \in G$. By Proposition 16.16 there exists $n_0 \in \mathbf{N}^*$ such that $|zU_n \triangle U_n|/|U_n| < \varepsilon$ whenever z belongs to the compact subset $K^{-1}xK$ and $n \geq n_0$. Hence, in particular we have

$$\frac{|xa_n U_n \triangle a_n U_n|}{|a_n U_n|} = \frac{|a_n^{-1} x a_n U_n \triangle U_n|}{|U_n|} < \varepsilon. \qquad \square$$

Proposition 16.20. *Let G be an amenable σ-compact group. Then $G \in [FC]^-$ if and only if, for every $(U_n) \in \mathscr{A}(G)$ and every sequence (a_n) in G, $(a_n U_n) \in \mathscr{A}(G)$.*

Proof. (a) Assume that $G \in [FC]^-$. For every $a \in G$ and every $n \in \mathbf{N}^*$, $a_n^{-1} a a_n \in \overline{K_a}$. By Proposition 16.16, given $n_0 \in \mathbf{N}^*$, there exists $n_1 \geq n_0$ such that

$$\frac{|aa_n U_n \triangle a_n U_n|}{|a_n U_n|} = \frac{|a_n^{-1} aa_n U_n \triangle U_n|}{|U_n|} < \frac{1}{n_0}$$

whenever $n \geq n_1$.

(b) Assume that there exists $a \in G$ such that $\overline{K_a}$ is noncompact. There exists then a net $(a_i)_{i \in I}$ in G such that $(a_i^{-1} aa_i)$ does not converge. By Proposition 16.14 we determine $(V_n) \in \mathscr{A}_c(G)$ such that, for every $n \in \mathbf{N}^*$, V_n is relatively compact. For every $n \in \mathbf{N}^*$, $\overline{V_n V_n^{-1}}$ is compact, so to every $i' \in I$ there corresponds $i'' \in I$ such that $i'' \succ i'$ and $a_{i''}^{-1} aa_{i''} \notin \overline{V_n V_n^{-1}}$. Hence we may determine a subsequence (a_{i_n}) of (a_i) such that, for every $n \in \mathbf{N}^*$, $a_{i_n}^{-1} aa_{i_n} \notin V_n V_n^{-1}$. Therefore

$$aa_{i_n} V_n \triangle a_{i_n} V_n = aa_{i_n} V_n \cup a_{i_n} V_n$$

and

$$\frac{|aa_{i_n} V_n \triangle a_{i_n} V_n|}{|a_{i_n} V_n|} = 2.$$

We conclude that $(a_{i_n} V_n) \notin \mathscr{A}(G)$. $\qquad \square$

Proposition 16.21. *Let G be a noncompact locally compact group. If A is a relatively compact subset of G, there exists an infinite sequence (a_n) in G such that the sets Aa_n $(n \in \mathbf{N}^*)$ are pairwise disjoint and $\Delta(a_n) \geq 1$ whenever $n \in \mathbf{N}^*$.*

Proof. Let $B = \{x \in G : \Delta(x) \geq 1\}$. The closed subset B is noncompact because otherwise $G = B \cup B^{-1}$ would be compact. Let C be the maximal subset of B such that the sets $\bar{A}x$ $(x \in C)$ are pairwise disjoint. For any $a \in B$, $\bar{A}a \cap \bar{A}C \neq \emptyset$, and $B \subset \bar{A}^{-1}\bar{A}C$. If C were finite, $\bar{A}^{-1}\bar{A}C$ would be compact, and then so would be its closed subset B. Therefore C must be infinite. □

Proposition 16.22. *Let G be an amenable σ-compact group. Then the following properties are equivalent:*

  (i)   *G is compact or discrete.*
  (ii) $\mathscr{A}(G) = \mathscr{A}_c(G)$.
  (iii) $\mathscr{S}\mathscr{A}(G) = \mathscr{A}(G)$.
  (iv) $\mathscr{S}\mathscr{A}_c(G) = \mathscr{A}_c(G)$.

Proof. (a) If G is discrete, we have trivially $\mathscr{A}(G) = \mathscr{A}_c(G)$ and $\mathscr{S}\mathscr{A}(G) = \mathscr{S}\mathscr{A}_c(G)$. Propositions 16.13 and 16.16 imply that $\mathscr{A}(G) = \mathscr{S}\mathscr{A}(G)$.

(b) Let G be compact. We have trivially $\mathscr{A}(G) = \mathscr{A}_c(G)$ and $\mathscr{S}\mathscr{A}(G) = \mathscr{S}\mathscr{A}_c(G)$.

If $(U_n) \in \mathscr{A}(G)$, for any $x \in G$, $\lim_{n \to \infty} |xU_n \cap U_n|/|U_n| = 1$. Also

$$\int_G |xU_n \cap U_n|\, dx = \iint 1_{xU_n}(y) 1_{U_n}(y)\, dy\, dx = \int \left(\int 1_{U_n}(x^{-1}y)\, dx\right) 1_{U_n}(y)\, dy$$

$$= \int |yU_n^{-1}| 1_{U_n}(y)\, dy = |U_n^{-1}|\, |U_n|.$$

As G is compact we deduce from Lebesgue's dominated convergence theorem that

$$1 = \lim_{n \to \infty} \int_G \frac{|xU_n \cap U_n|}{|U_n|}\, dx = \lim_{n \to \infty} |U_n^{-1}| = \lim_{n \to \infty} |U_n|.$$

As the Haar measure is normalized on G, we have a fortiori, for every compact subset K in G, $\lim_{n \to \infty} |KU_n|/|U_n| = 1$. We conclude from Proposition 16.13 that $(U_n) \in \mathscr{S}\mathscr{A}(G)$.

(c) Assume G to be neither compact nor discrete.

Let U be a compact neighborhood of e in G, and let (a_m) be an infinite sequence in G associated via Lemma 16.21. We consider $B = \{a_m : m \in \mathbf{N}^*\}$. If the sequence did converge to an element a in G, there would exist $m_0 \in \mathbf{N}^*$ such that $a_m \in U^{-1}a$, that is, $a \in Ua_m$ whenever $m \geq m_0$. The sets Ua_m

SUPPLEMENTARY STRUCTURAL PROPERTIES 175

($m \geq m_0$) could not be pairwise disjoint. Therefore (a_m) does not converge.
Let (U_n) $\in \mathscr{A}_c(G)$ formed by compacta.

We consider $V_n = U_n \cup B$, $n \in \mathbf{N}^*$. For every $n \in \mathbf{N}^*$ and every $x \in G$, $xV_n \triangle V_n \subset (xU_n \triangle U_n) \cup B \cup xB$. Since G is nondiscrete,

$$\frac{|xV_n \triangle V_n|}{|V_n|} = \frac{|xU_n \triangle U_n|}{|U_n|}.$$

Then also (V_n) $\in \mathscr{A}(G)$. Since $B \subset V_n$ and the sequence (a_n) does not converge, (V_n) $\notin \mathscr{A}_c(G)$.

There exists a nondecreasing sequence (k_n) in $\mathbf{N}^*$ such that, for every $n \in \mathbf{N}^*$, $|\cup_{i=1}^{k_n} Ua_i| = \sum_{i=1}^{k_n} |U|\Delta(a_i) \geq n|U_n|$. Let $W_n = U_n \cup \{a_i : i = 1, \ldots, k_n\}$. Since G is nondiscrete, for every $n \in \mathbf{N}^*$ and every $a \in G$,

$$\frac{|aW_n \triangle W_n|}{|W_n|} = \frac{|aU_n \triangle U_n|}{|U_n|}$$

and we have (W_n) $\in \mathscr{A}_c(G)$. But on the other hand, for every $n \in \mathbf{N}^*$,

$$\frac{|UW_n|}{|W_n|} \geq \frac{|\cup_{i=1}^{k_n} Ua_i|}{|U_n|} \geq n$$

hence $\lim_{n \to \infty} |UW_n|/|W_n| = \infty$ and, by Proposition 16.13, (W_n) $\notin \mathscr{S\!A}(G)$. □

Proposition 16.23. *Let G be an amenable σ-compact group. Then $\mathscr{S\!A}_c(G) = \mathscr{S\!A}(G)$ if and only if G is unimodular.*

Proof. (a) Let G be unimodular. Assume that there exists (U_n) $\in \mathscr{S\!A}(G)$ such that, for every $n \in \mathbf{N}^*$, $\overline{U_n}$ is noncompact.

Let K be any compact subset of G such that $|K| > 0$. For a fixed $n \in \mathbf{N}^*$, we may define a sequence (x_p) in U_n such that

$$Kx_{p+1} \cap \left(\bigcup_{k=1}^{p} Kx_p\right) = \varnothing$$

whenever $p \in \mathbf{N}^*$. As a matter of fact, we may choose $x_1 \in U_n$ arbitrarily. If $x_1, \ldots, x_p$ have been defined, because $K^{-1}(\cup_{k=1}^{p} Kx_k)$ is compact and $\overline{U_n}$ is noncompact, $U_n \not\subset K^{-1}(\cup_{k=1}^{p} Kx_k)$, and we may choose $x_{p+1} \in U_n \setminus K^{-1}(\cup_{k=1}^{p} Kx_k)$.

For every $p \in \mathbf{N}^*$, by unimodularity, $|KU_n| \geq |\cup_{k=1}^{p} Kx_k| = p|K|$. As $p \in \mathbf{N}^*$ may be chosen arbitrarily and $|K| > 0$, $|KU_n| = \infty$ whereas $0 < |U_n| < \infty$. We would then contradict Proposition 16.13.

(b) Let G be nonunimodular and consider (U_n) $\in \mathscr{S\!A}_c(G)$. There exists a sequence (a_n) in G such that, for every $n \in \mathbf{N}^*$,

$$\Delta(a_n) \leq \frac{1}{2^n n} \inf\{|U_1|, \ldots, |U_n|\};$$

then $\sum_{m \geq n} \Delta(a_m) \leq (1/n)|U_n|$. For every $n \in \mathbf{N}^*$, let $E_n = \{a_m : m \geq n\}$ and $U_n' = U_n \cup E_n$. Since G is nondiscrete, $|U_n'| = |U_n|$ and $0 < |U_n'| < \infty$. Moreover U_n' is noncompact because $\lim_{n \to \infty} \Delta(a_n) = 0$. It suffices now to show that however $(U_n') \in \mathcal{SA}(G)$.

Let K be any nonvoid compact subset of G. For every $n \in \mathbf{N}^*$,

$$\frac{|KU_n' \triangle KU_n|}{|U_n|} \leq \sum_{m \geq n} \frac{|Ka_m|}{|U_n|} = \frac{|K|}{|U_n|} \sum_{m \geq n} \Delta(a_m) \leq \frac{|K|}{n}$$

and

$$\frac{|KU_n' \triangle U_n'|}{|U_n'|} \leq \frac{|KU_n' \triangle KU_n|}{|U_n|} + \frac{|KU_n \triangle U_n|}{|U_n|} + \frac{|E_n|}{|U_n|}.$$

So we have

$$\lim_{n \to \infty} \frac{|KU_n' \triangle U_n'|}{|U_n'|} = \lim_{n \to \infty} \frac{|KU_n \triangle U_n|}{|U_n|} = 0. \qquad \square$$

Proposition 16.24. *Let G be a locally compact group and let U be an open relatively compact subset. If $(U^n) \in \mathcal{SA}(G)$, then $\lim_{n \to \infty} |U^{n+1}|/|U^n| = 1$. The converse statement holds in case $G = \bigcup_{n \in \mathbf{N}^*} U^n$.*

Proof. (a) Let $(U^n) \in \mathcal{SA}(G)$. For every $n \in \mathbf{N}^*$,

$$0 \leq \frac{|U^{n+1}|}{|U^n|} - 1 \leq \frac{|\overline{U}U^n|}{|U^n|} - 1 = \frac{|\overline{U}U^n| - |U^n|}{|U^n|} \leq \frac{|\overline{U}U^n \setminus U^n|}{|U^n|}$$

and by hypothesis

$$\lim_{n \to \infty} \frac{|\overline{U}U^n \setminus U^n|}{|U^n|} \leq \lim_{n \to \infty} \frac{|\overline{U}U^n \triangle U^n|}{|U^n|} = 0.$$

(b) Assume that the condition holds. If K is any compact subset of G, there exists $n_0 \in \mathbf{N}^*$ such that $K \cup \{e\} \subset U^{n_0}$. For every $n \geq n_0$, we have $KU^n \subset U^{n+n_0}$, so by (7.2) we obtain

$$\frac{|KU^n \triangle U^n|}{|U^n|} \leq 2\frac{|KU^n \setminus U^n|}{|U^n|} \leq 2\frac{|U^{n+n_0} \setminus U^n|}{|U^n|} = 2\left(\frac{|U^{n+n_0}|}{|U^n|} - 1\right)$$

$$= 2\left(\frac{|U^{n+n_0}|}{|U^{n+n_0-1}|} \cdots \frac{|U^{n+1}|}{|U^n|} - 1\right).$$

Hence

$$\lim_{n \to \infty} \frac{|KU^n \triangle U^n|}{|U^n|} = 0. \qquad \square$$

SUPPLEMENTARY STRUCTURAL PROPERTIES

We consider a particular type of averaging sequences: *right* [resp. *left*] *regular averaging sequences*.

Definition 16.25. *Let G be a locally compact group and $k \in \mathbf{N}^*$. We denote by $\mathcal{R}_r\mathcal{A}(G,k)$ [resp. $\mathcal{R}_l\mathcal{A}(G,k)$] the set of all sequences (U_n) belonging to $\mathcal{A}(G)$ such that $G = \bigcup_{n \in \mathbf{N}^*} U_n$ and $|U_n U_n^{-1}| \leq k|U_n|$ [resp. $|U_n^{-1} U_n| \leq k|U_n|$] whenever $n \in \mathbf{N}^*$.*

Proposition 16.26. *Let G be an amenable σ-compact group. If there exists $k \in \mathbf{N}^*$ such that $\mathcal{R}_r\mathcal{A}(G,k) \neq \emptyset$, then G is unimodular.*

Proof. Let $(U_n) \in \mathcal{R}_r\mathcal{A}(G,k)$. If a is any element in G, there exists $n_0 \in \mathbf{N}^*$ such that $a^{-1} \in U_{n_0}$, hence $U_{n_0} a \subset U_{n_0} U_{n_0}^{-1}$ and

$$|U_{n_0}|\Delta(a) \leq |U_{n_0} U_{n_0}^{-1}| \leq k|U_{n_0}|.$$

As $|U_{n_0}| > 0$, we have $\Delta(a) \leq k$.

If we had $\Delta(a) \neq 1$, then either $\lim_{n \to \infty} \Delta(a)^n = \infty$ or $\lim_{n \to \infty} \Delta(a^{-1})^n = \infty$. We would come to a contradiction. Therefore G must be unimodular. □

Proposition 16.27. *Let G be a locally compact group and let $k \in \mathbf{N}^*$. If $(U_n) \in \mathcal{R}_r\mathcal{A}(G,k)$, then $(U_n a) \in \mathcal{R}_r\mathcal{A}(G,k)$ and $(aU_n) \in \mathcal{R}_r\mathcal{A}(G,k)$ whenever $a \in G$.*

Proof. The first statement is trivial.
For all $n \in \mathbf{N}^*$ and $x \in G$,

$$\frac{|xaU_n \triangle aU_n|}{|aU_n|} \leq \frac{|xaU_n \triangle U_n|}{|U_n|} + \frac{|aU_n \triangle U_n|}{|U_n|}.$$

Hence $(aU_n) \in \mathcal{A}(G)$. Since G is unimodular by Proposition 16.26, for every $n \in \mathbf{N}^*$,

$$\left|(aU_n)(aU_n)^{-1}\right| = |aU_n U_n^{-1} a^{-1}| = |U_n U_n^{-1}| \leq k|U_n|. \quad \square$$

Proposition 16.28. *Let G be a locally compact group having polynomial growth. If U is a compact neighborhood of e, there exist $\alpha > 1$ and a sequence (p_m) in $\mathbf{N}^*$ such that $|U^{2p_m}| \leq \alpha |U^{p_m}|$ whenever $m \in \mathbf{N}^*$ and $\lim_{m \to \infty} |U^{p_m + k}|/|U^{p_m}| = 1$ whenever $k \in \mathbf{N}^*$.*

Proof. By hypothesis there exist $d \in \mathbf{N}^*$ and $n_0 \in \mathbf{N}^*$ such that $|U^n| \leq n^d$ whenever $n \geq n_0$; hence

$$|U^{2^n}|^{1/n} \leq \left((2^n)^d\right)^{1/n} = 2^d. \qquad (4)$$

If $\beta > 1$, we consider

$$E_\beta = \left\{ r \in \mathbf{N}^* : \frac{|U^{2^r}|}{|U^{2^{r-1}}|} \geq \beta \right\}.$$

If $n \in \mathbf{N}^*$, let $n' = \operatorname{card}(E_\beta \cap \{1,\ldots,n\})$ and $c_{\beta,n} = n'/n$. We have

$$|U^{2^n}| = \frac{|U^{2^n}|}{|U^{2^{n-1}}|} \cdots \frac{|U^2|}{|U|}|U| \geq \beta^{n'}|U|. \tag{5}$$

The relations (4) and (5) imply that

$$2^d \geq \beta^{c_{\beta,n}}|U|^{1/n},$$

$$c_{\beta,n} \leq \frac{d\log 2 - (1/n)\log|U|}{\log\beta} < \frac{d\log 2}{\log\beta}.$$

Therefore we may choose $\beta > 1$ such that, for every $n \in \mathbf{N}^*$, $c_{\beta,n} < \frac{1}{3}$, hence $n' < n/3$. Thus it is possible to determine an increasing sequence (n_m) in $\mathbf{N}^*$ such that, for every $m \in \mathbf{N}^*$, we have $n_m - 1, n_m, n_m + 1 \notin E_\beta$. For every $m \in \mathbf{N}^*$, let

$$I_{n_m} = \{2^{n_m-1} + 1, 2^{n_m-1} + 2, \ldots, 2^{n_m}\}.$$

Since $n_m - 1 \notin E_\beta$, we have

$$1 \leq \frac{|U^{2^{n_m-1}+1}|}{|U^{2^{n_m-1}}|} \frac{|U^{2^{n_m-1}+2}|}{|U^{2^{n_m-1}+1}|} \cdots \frac{|U^{2^{n_m}}|}{|U^{2^{n_m}-1}|} = \frac{|U^{2^{n_m}}|}{|U^{2^{n_m-1}}|} < \beta.$$

So there cannot exist $k \in \mathbf{N}^*$, $\varepsilon \in \mathbf{R}_+^*$ and an infinite subsequence $(n'_{m'})_{m' \in \mathbf{N}^*}$ of (n_m) such that, for every $q_{m'} \in I_{n'_{m'}}$ $(m' \in \mathbf{N}^*)$,

$$\frac{|U^{q_{m'}+k}|}{|U^{q_{m'}}|} \geq 1 + \varepsilon.$$

Thus we may determine a sequence (p_m) with $p_m \in I_{n_m}$ $(m \in \mathbf{N}^*)$ such that

$$1 \leq \frac{|U^{p_m+m}|}{|U^{p_m}|} < 1 + \frac{1}{m}$$

and therefore

$$\lim_{m \to \infty} \frac{|U^{p_m+k}|}{|U^{p_m}|} = 1$$

whenever $k \in \mathbf{N}^*$. Moreover for every $m \in \mathbf{N}^*$, we have $2^{n_m-1} < p_m \leq 2^{n_m}$, so $2p_m \leq 2^{n_m+1}$. Since $n_m + 1 \notin E_\beta$ and $n_m \notin E_\beta$, we obtain

$$|U^{2p_m}| \leq |U^{2^{n_m+1}}| < \beta|U^{2^{n_m}}| < \beta^2|U^{2^{n_m-1}}| \leq \beta^2|U^{p_m}|.$$

It suffices to put $\alpha = \beta^2$. $\square$

Corollary 16.29. *Let G be a locally compact connected (amenable) group having polynomial growth. Then G admits a nondecreasing, left and right regular averaging sequence belonging to $\mathscr{SA}(G)$.*

Proof. With the notations of Proposition 16.28 let $V_m = U^{p_m}(m \in \mathbf{N}^*)$, U being symmetric; (V_m) is a nondecreasing sequence of symmetric neighborhoods of e and, as G is connected, $G = \cup_{m \in \mathbf{N}^*} V_m$. By Propositions 16.13 and 16.16, $(V_m) \in \mathscr{SA}(G)$ and $(V_m) \in \mathscr{A}(G)$. Also for every $m \in \mathbf{N}^*$,

$$|V_m^{-1} V_m| = |V_m^2| = |U^{2p_m}| \le \alpha |V_m|. \qquad \square$$

The foregoing results on averaging sequences concerning amenable σ-compact groups may be carried over, up to a certain extent, to more formal properties concerning arbitrary amenable groups. We briefly indicate the principle.

Definition 16.30. *If G is a locally compact group, we call averaging net any net $(U_i)_{i \in I}$ of compact subsets in G such that $|U_i| > 0$ whenever $i \in I$, $U_i \subset U_{i'}$ whenever $i, i' \in I$ with $i \prec i'$, $\cup_{i \in I} U_i = G$ and*

$$\lim_i \left(\sup \left\{ \frac{|aU_i \triangle U_i|}{|U_i|} : a \in K \right\} \right) = 0$$

for every compact subset K of G. The class of all averaging nets is denoted by $\mathscr{AN}(G)$.

Proposition 16.31. *A locally compact group G is amenable if and only if $\mathscr{AN}(G) \ne \varnothing$. The amenable group G admits an averaging net formed by symmetric subsets. In particular, if G is an amenable σ-compact group, the sequence given by Proposition 16.10 provides an averaging net.*

Proof. The property implies (F) and hence amenability by Theorem 7.3.

If G is amenable, consider the net $(K_j)_{j \in J}$ of all open σ-compact subgroups ordered by inclusion. For every $j \in J$, there exists a sequence $(C_{j,n})_{n \in \mathbf{N}^*}$ of compact subsets in G such that $K_j = \cup_{n \in \mathbf{N}^*} C_{j,n}$. By proposition 13.3, the closed subgroup K_j is amenable and, by Proposition 16.10, there exists a nondecreasing sequence $(U_{j,n})_{n \in \mathbf{N}^*}$ of compact subsets in K_j such that $|U_{j,n}| > 0$ whenever $n \in \mathbf{N}^*$, $\cup_{n \in \mathbf{N}^*} U_{j,n} = K_j$ and $\sup\{|aU_{j,n} \triangle U_{j,n}|/|U_{j,n}| : a \in C_{j,n}\} < 1/n$ whenever $n \in \mathbf{N}^*$. The subsets may be chosen to be symmetric. We consider the net $\mathscr{U} = (U_{j,n})_{j \in J, n \in \mathbf{N}^*}$ where, for $(j,n), (j',n') \in J \times \mathbf{N}^*$, $(j,n) \prec (j',n')$ if and only if $j \prec j'$, $n \le n'$, $C_{j,n} \subset C_{j',n'}$, and $U_{j,n} \subset U_{j',n'}$; then $\mathscr{U} \in \mathscr{AN}(G)$. $\qquad \square$

Averaging nets of the amenable group G give rise to asymptotically $P^1(G)$-invariant nets in $P^1(G)$.

Proposition 16.32. *Let G be an amenable group and let $(U_i)_{i \in I} \in \mathscr{AN}(G)$. Then for every $\varphi \in P^1(G)$, $\lim_i \|\varphi * \xi_{U_i} - \xi_{U_i}\|_1 = 0$.*

Proof. Let $\varphi \in P^1(G)$ and $\varepsilon \in \,]0, 1[$. There exists $\varphi_0 \in \mathscr{K}_+(G)$ such that $\|\varphi - \varphi_0\|_1 < \varepsilon/3$, $\|\varphi_0\|_1 > 1 - \varepsilon/3 > 0$. Let $K = \operatorname{supp} \varphi_0$. There exists $i_0 \in I$ such that, for every $i \in I$ with $i \succ i_0$ and every $x \in K$,

$$\frac{|xU_i \triangle U_i|}{|U_i|} < \frac{\varepsilon}{3\|\varphi_0\|_1}.$$

Then

$$\|\varphi * \xi_{U_i} - \xi_{U_i}\|_1 = \int_G \left| \varphi * \xi_{U_i}(y) - \xi_{U_i}(y) \right| dy$$

$$= \int_G \left| \int_G \varphi(x) \xi_{U_i}(x^{-1}y) \, dx - \int_G \varphi(x) \xi_{U_i}(y) \, dx \right| dy$$

$$= \int_G \left(\int_G \left| \xi_{U_i}(x^{-1}y) - \xi_{U_i}(y) \right| dy \right) \varphi(x) \, dx$$

$$= \int_G \frac{|xU_i \triangle U_i|}{|U_i|} \varphi(x) \, dx$$

$$\leq \int_G \frac{|xU_i \triangle U_i|}{|U_i|} |\varphi(x) - \varphi_0(x)| \, dx + \int_K \frac{|xU_i \triangle U_i|}{|U_i|} \varphi_0(x) \, dx$$

$$< 2\frac{\varepsilon}{3} + \frac{\varepsilon}{3} = \varepsilon. \qquad \square$$

NOTES

On a general locally compact group G admitting a closed subgroup H and for an invariant nonnegative measure μ on G/H, Leptin [349, 350] investigated properties of the number $\sup \{\inf \{ \mu(KU)/\mu(U) : U \in \mathfrak{K}(G/H), \mu(U) > 0 \} : K \in \mathfrak{K}(G)\}$. Mocanu [395] generalized some of the results replacing the invariant measure by an arbitrary, relatively invariant measure. Leptin mentioned [353] that the implication $\kappa > 1 \Rightarrow \kappa = \infty$ was pointed out to him by Kneser. He established 16.3 in [349]. Emerson and Greenleaf [169] demonstrated 16.4–5, and Emerson [163] proved 16.6. For 16.7 we refer to Emerson [165]. Kieffer [317] and Emerson [165, 166] were interested in the study of a ratio limit problem which may be formulated in the following way. Let $S: \mathfrak{B}(G) \to \mathbf{R}_-$ such that $S(\varnothing) = 0$, $S(A \cup B) + S(A \cap B) \leq S(A) + S(B)$ whenever $A, B \in \mathfrak{B}(G)$, and $S(aA) = S(A)$ whenever $A \in \mathfrak{B}(G)$, $a \in G$. If

$A \in \mathcal{B}(G)$ with $|A| > 0$, consider the average $|A|^{-1}S(A)$. In case G is unimodular, limits of such averages may be determined for the nets (U_i) considered in 16.7. These results were carried over to the nonunimodular case by Moulin Ollagnier and Pinchon [404]. Kieffer's motivation had been the consideration of an entropy for a measurable partition of a probability space under the action of an amenable discrete group; see [315, 316].

Følner [187] proved 16.8. An attempt to generalize this characterization to an arbitrary amenable group was made by Hulanicki [267]. Further refinements of (F) concerning discrete groups and involving nonamenability of F_2 were obtained by Melven and Myren Krom [324]. Bonic [45] gave a criterion for nonamenability of a countable discrete group; it expresses that the group must be sufficiently free: $G = \{a_n : n \in \mathbb{N}^*\}$ fails to be amenable if it contains a finite subset F with $k = \operatorname{card} F \geq 3$ such that F generates an infinite subgroup of G and there exists $n_0 \geq 2$ such that $\operatorname{card}(Fa_n \cap (\cup_{i=1}^{n-1} Fa_i)) \leq k/2$ whenever $n \geq n_0$.

The subject studied throughout 16.9–23 was developed essentially by Emerson [162, 163]; see also [161] for related properties concerning integral operators. As an early survey on these topics we recommend Stone and von Randow [520]. The σ-compactness of a locally compact group G with nonvoid $\mathscr{A}(G)$ had been noticed by Davis [110]. Blum and Eisenberg [37] examined averaging sequences for locally compact abelian groups; see Blum, Eisenberg, and L. Hahn [38] for further comments, especially in relation to ergodic theory. Douglass [146] studied averaging sequences on amenable semigroups. He also considered the following simple example: Let G be an (amenable) countable locally finite group and let (F_n) be an increasing sequence of finite subsets in G with $F_1 = \{e\}$ and $G = \cup_{n=1}^{\infty} F_n$. For $n \in \mathbb{N}^*$, let U_n be the subgroup of G generated by F_n. Then $(U_n) \in \mathscr{A}(G)$. By combinatorial methods Greenleaf [228] performed constructions of averaging sequences in amenable connected groups. Emerson and Greenleaf [170] gave 16.24. Douglass [146] noticed that for the amenable group G on two generators a and b of order 2, if $U = \{e, a, b\}$, then $\operatorname{card}(U^{n+1} \setminus U^n) = 2$ whenever $n \in \mathbb{N}^*$; $(U^n) \in \mathscr{A}(G)$. Milnes [386] showed that for the amenable affine group G of the real line there exist a compact subset K and a compact symmetric neighborhood U of e such that $\lim_n \sup\{|xU^n \triangle U^n|/|U^n| : x \in K\} = 0$ fails. Studying the particular action of the discrete countable group G given by right translations on the compact metrizable space of total orders on G, Moulin Ollagnier and Pinchon [403] proved that $\mathscr{A}(G) \neq \emptyset$ if G has the fixed point property. Right regular averaging sequences were studied by Chatard [67], who gave 16.26–27. Jenkins [285] established 16.28–29; see also [283]. For 16.31 we refer to Milnes [388]. Brooks [53] interpreted amenability of the fundamental group of a compact manifold via the existence of averaging sequences.

Bożejko [50] studied a condition of uniform amenability for a discrete group G. To every $\varepsilon > 0$ there corresponds a mapping $d_\varepsilon : \mathbb{N}^* \to \mathbb{N}^*$ such that, for every finite subset F in G with $\operatorname{card} F = n$, one may determine a nonvoid finite subset U in G satisfying $\operatorname{card} U \leq d_\varepsilon(\operatorname{card} F)$ and $\operatorname{card} FU < (1 + \varepsilon)$

card U. He showed the condition to be implied by a uniform polynomial growth condition that is stronger than polynomial growth and is shared by all discrete nilpotent groups. Keller [310] had considered pointwise uniform amenability. Nonstandard analysis arguments using a full structure that includes the discrete group G and $\mathbf{R}$ had allowed him to show that G has the property if and only if every nonstandard model of G is amenable. Thus stability properties of the class of amenable groups may be carried over to the class of uniformly amenable groups.

We quote some results concerning discrete amenable semigroups, that is, semigroups S admitting left invariant means on $l^\infty(S)$. They satisfy the following Følner condition: For any finite subset F in S and any $\varepsilon > 0$ there exists a nonvoid finite subset U in S such that card $(xU \setminus U) < \varepsilon \operatorname{card} U$ whenever $x \in F$. The proof is due to Frey; see Namioka [411]. However, the condition is not sufficient for a semigroup S to be amenable because any finite semigroup satisfies the condition with $U = S$ and there exist nonamenable finite semigroups. A sufficient condition for amenability is the existence, for any finite subset F in S and any $\varepsilon > 0$, of a nonvoid finite subset U in S such that card $(U \setminus xU) < \varepsilon \operatorname{card} U$ whenever $x \in F$. Finite semigroups and commutative semigroups satisfy the latter condition. These facts were established by Argabright and Wilde [17]. The problem of the necessity of the second condition was investigated by Rajagopalan and Ramakrishnan [447] and solved negatively by Klawe [318]. Making use of Følner conditions, del Junco and Rosenblatt [297] showed that if (X, μ) is a probability space for a nonatomic measure and $\tau_1, \ldots, \tau_n$ are measurable measure-preserving transformations of X generating a discrete semigroup S that admits a right invariant mean on $l^\infty(S)$, then for every $\varepsilon > 0$ there exists a μ-measurable subset A of X such that $0 < \mu(A) < \varepsilon$ and $\mu(\tau_i^{-1}(A) \delta A) \leq \varepsilon \mu(A)$ whenever $i = 1, \ldots, n$.

17. SUPPLEMENTARY UNITARY GROUP REPRESENTATION PROPERTIES

We state amenability properties expressed in Fell topology language. The Sz. Nagy–Dixmier property shows how invariant means perform on amenable groups the role played by normalized Haar measures on compact groups.

A. Weak and Strong Containment Properties

Let G be a locally compact group admitting a closed subgroup H. We consider the associated continuous mapping $\rho: G \to \mathbf{R}^*_+$ defined in Section 14A as well as the corresponding G-quasi-invariant measure $\mu = \mu_\rho$ on the homogeneous space G/H. If $a \in G$, $x \in G$, and $s \in H$, we have

$$\frac{\rho(axs)}{\rho(xs)} = \frac{\rho(ax)(\Delta_H(s)/\Delta_G(s))}{\rho(x)(\Delta_H(s)/\Delta_G(s))} = \frac{\rho(ax)}{\rho(x)}.$$

Hence we may consider the mapping

$$\chi: G \times G/H \to \mathbf{R}_+^*$$

$$(a, \dot{x}) \mapsto \frac{\rho(ax)}{\rho(x)}.$$

By the statement made in Section 14A(b), for $F \in L^1(G/H, \mu)$ and $a \in G$, we have

$$\int_{G/H} F(a\dot{x}) \, d\mu(\dot{x}) = \int_{G/H} \chi(a^{-1}, \dot{x}) F(\dot{x}) \, d\mu(\dot{x}) \tag{1}$$

([B INT] Chap. VII, Section 2, No. 5, théorème 2). Let $p \in [1, \infty[$, $F \in L^p(G/H, \mu)$, $a \in G$, and $\dot{x} \in G/H$. We put

$$[J^p(a)F](\dot{x})\bigl(= [J^p_{H,\rho}(a)F](\dot{x})\bigr) = \chi(a^{-1}, \dot{x})^{1/p} F(a^{-1}\dot{x}).$$

If $F \in L^p(G/H, \mu)$ and $a, b \in G$, we have, for μ-almost every $\dot{x} \in G/H$,

$$[J^p(ab)F](\dot{x}) = \left(\frac{\rho(b^{-1}a^{-1}x)}{\rho(x)}\right)^{1/p} F(b^{-1}a^{-1}\dot{x})$$

$$= \left(\frac{\rho(a^{-1}x)}{\rho(x)}\right)^{1/p} \left(\frac{\rho(b^{-1}a^{-1}x)}{\rho(a^{-1}x)}\right)^{1/p} F(b^{-1}a^{-1}\dot{x})$$

$$= J^p(a)\left(\left(\frac{\rho(b^{-1}x)}{\rho(x)}\right)^{1/p} F(b^{-1}\dot{x})\right) = [J^p(a)J^p(b)F](\dot{x}).$$

Hence J^p is a continuous representation of G on $L^p(G/H, \mu)$. Moreover if $F \in L^p(G/H, \mu)$ and $a \in G$,

$$\int_{G/H} |[J^p(a)F](\dot{x})|^p \, d\mu(\dot{x}) = \int_{G/H} \chi(a^{-1}, \dot{x}) |F(a^{-1}\dot{x})|^p \, d\mu(\dot{x})$$

$$= \int_{G/H} |F(\dot{x})|^p \, d\mu(\dot{x}),$$

that is, $\|J^p(a)F\|_p = \|F\|_p$. In particular, J^2 is a continuous unitary representation of G on the Hilbert space $L^2(G/H, \mu)$.

The theory of *induced representations* for separable groups was inaugurated by Mackey [377, 378]. Blattner [36] developed a theory applying to general locally compact groups that is equivalent to Mackey's method in the separable case. Greenleaf adapted Mackey's construction to the nonseparable case [227]. We briefly outline the canonical performance of induced representations. Let G be a locally compact group admitting a closed subgroup H. Let T be a continuous unitary representation of H on the Hilbert space $\mathcal{H}$. We consider the set $\mathcal{X}_T$ of all functions $F: G \to \mathcal{H}$ having the following properties:

(a) For every $\xi \in \mathcal{H}$, $x \mapsto (f(x)|\xi)$ is a measurable function on G.
(b) For all $x \in G$ and $z \in H$, $f(xz) = T_{z^{-1}}f(x)$.
(c) $\int_{G/H} \|f(x)\|_{\mathcal{H}}^2 \, d\mu(\dot{x}) < \infty$.

By (a), (b) and the fact that T is unitary, (c) makes sense. Also, the relation

$$((f_1|f_2)) = \int_{G/H} (f_1(x)|f_2(x)) \, d\mu(\dot{x})$$

for $f_1, f_2 \in \mathcal{X}_T$, defines a Hilbert space structure on $\mathcal{X}_T$. Then a continuous unitary representation U^T of G with representation space $\mathcal{X}_T$ may be defined by

$$U_a^T f(x) = \chi(a^{-1}, \dot{x})^{1/2} f(a^{-1}x),$$

$f \in \mathcal{X}_T$, $a \in G$, $x \in G$. Notice that by (1) we have

$$((U_a^T f | U_a^T f)) = \int_{G/H} \chi(a^{-1}, \dot{x})(f(a^{-1}x)|f(a^{-1}x)) \, d\mu(\dot{x})$$

$$= \int_{G/H} (f(x)|f(x)) \, d\mu(\dot{x}) = ((f|f)).$$

The representation U^T is said to be induced from H to G by T and is denoted also by $\mathrm{ind}_{H \uparrow G} T$.

In particular, J^2 is induced from H to G by the trivial representation i_H on $L^2(G/H, \mu)$; J^2 is called the *quasi-regular representation* of G on $L^2(G/H, \mu)$. If $H = \{e\}$, let $\rho = 1_G$; so $\chi = 1_{G \times G/H}$ and the left regular representation of G on $L^2(G)$ may be considered to be induced from $\{e\}$ to G by $i_{\{e\}}$. Therefore, Theorem 8.9 characterizes amenability of the locally compact group G by saying that i_G or even more generally every continuous unitary representation of G is weakly contained in $\mathrm{ind}_{\{e\} \uparrow G} i_{\{e\}}$.

Let G be a locally compact group and let ν be a positive-definite measure defined on G. We may equip $\mathcal{K}(G)$ with a prehilbert structure by putting

$$(f|g)_\nu = \langle \tilde{g} * f, \nu \rangle$$

for $f, g \in \mathcal{K}(G)$ ([D] 13.7.9). The corresponding Hilbert space is denoted by

$\mathcal{H}_\nu$. Let now

$$\pi_a^\nu f = \Delta^{1/2}(a) L_a f = \Delta^{1/2}(a) \delta_a * f$$

for $f \in \mathcal{K}(G)$, $a \in G$; π^ν induces a continuous unitary representation of G on $\mathcal{H}_\nu$ also denoted by π^ν (ibid.). For every $f \in \mathcal{K}(G)$, $f' = f\Delta^{1/2} \in \mathcal{K}(G)$. We consider $p_f^\nu \in P(G)$ defined, for $x \in G$, by

$$p_f^\nu(x) = (\pi_x^\nu f' | f')_\nu = \langle \Delta^{1/2}(x) f'^\frown * \delta_x * f', \nu \rangle$$

$$= \iint \overline{f'(y^{-1})} f'(x^{-1} y^{-1} z) \Delta^{1/2}(x) \, dy \, d\nu(z)$$

$$= \iint \overline{f(y^{-1})} f(x^{-1} y^{-1} z) \Delta(y^{-1}) \Delta^{1/2}(z) \, dy \, d\nu(z)$$

$$= \iint \overline{f(y)} f(x^{-1} yz) \Delta^{1/2}(z) \, dy \, d\nu(z). \tag{2}$$

Proposition 17.1. *Let G be an amenable group and let ν be a positive-definite bounded measure on G such that $\nu(G) \neq 0$. Then i_G is weakly contained in π^ν.*

Proof. With respect to Theorem 8.9 the hypothesis implies that $\nu(G) > 0$. Let K be a compact subset of G and let $\varepsilon > 0$. There exists a compact subset C in G such that $K \subset C$ and $|\nu|(G \setminus C) < \varepsilon/3$. Given $\delta \in]0, \varepsilon(3|\nu|(G))^{-1}[$, by Proposition 15.4 there exists $\varphi \in P^2(G) \cap \mathcal{K}(G)$ such that

$$\left\| _{a^{-1}}\varphi_b \Delta(b)^{1/2} - \varphi \right\|_2 < \delta$$

whenever $a, b \in C$. For every $a \in C$, by (2) and the Cauchy–Schwarz inequality we have

$$|p_\varphi^\nu(a) - \nu(G)| \leq \left| \int_{G \setminus C} \left(\int_G \varphi(y) \varphi(a^{-1} yz) \Delta^{1/2}(z) \, dy \right) d\nu(z) \right|$$

$$+ \left| \int_C \left(\int_G (\varphi(a^{-1} yz) \Delta^{1/2}(z) - \varphi(y)) \varphi(y) \, dy \right) d\nu(z) \right|$$

$$+ \left| \int_C \left(\int_G \varphi(y)^2 \, dy \right) d\nu(z) - \nu(G) \right|$$

$$\leq \|\varphi\|_2^2 |\nu|(G \setminus C) + \|\varphi\|_2 \delta |\nu|(G) + |\nu|(G \setminus C)$$

$$< 2|\nu|(G \setminus C) + \frac{\varepsilon}{3} < \varepsilon.$$

We proved that $\nu(G)1_G$ may be approximated uniformly on compacta by positive-definite functions associated to π^ν, hence i_G is weakly contained in π^ν. □

If G is a locally compact group, let $P_L(G)$ denote the subset of $P(G)$ formed by the positive-definite functions that are associated to the left regular representation.

Proposition 17.2. *Let G be a locally compact group admitting a nonamenable closed proper subgroup H. Then $P_L(G) \subsetneq \mathrm{co}(P_L(G) \cup \{1_G\}) \subsetneq P(G)$, where $\mathrm{co}(P_L(G) \cup \{1_G\})$ denotes the convex hull of $P_L(G) \cup \{1_G\}$ in the positive cone $P(G)$.*

Proof. As G is nonamenable by Proposition 13.3, Theorem 8.9 implies that $1_G \notin P_L(G)$.

We have $1_H \in P(G)$ ([HR] 32.43). If we had also $1_H \in \mathrm{co}(P_L(G) \cup \{1_G\})$, there would exist $\alpha \in [0,1]$ and $f \in P_L(G)$ such that $1_H = \alpha f + (1-\alpha)1_G$. Since H is proper, necessarily $\alpha \neq 0$. Then
$$1_H|_H = \alpha f|_H + (1-\alpha)1_H|_H$$
would imply that $1_H = 1_H|_H = f|_H \in P(H)$ and, by Theorem 8.9, H should be amenable. □

We make a final remark on groups having the following property called *Kazhdan's property*.

Definition 17.3. *A locally compact group G is said to have property (T) if every $U \in \hat{G}$, that contains i_G weakly, also contains i_G strongly, that is, there exists $\xi \in \mathcal{H}_U$ such that $\xi \neq 0$ and $U_x\xi = \xi$ whenever $x \in G$.*

Proposition 17.4. *Any amenable group G admitting property (T) is compact.*

Proof. As i_G is weakly contained in the left regular representation by Theorem 8.9, the hypothesis implies that there exists $f \in L^2(G)$, $f \neq 0$ such that $_x f = f$ whenever $x \in G$; then f is a constant function in $L^2(G)$ and necessarily G is compact. □

B. The Sz. Nagy–Dixmier Property

We state a generalized form of the property that applies to topological (nonnecessarily locally compact) groups.

A *representation* of a group G on a Banach space E, that is, a homomorphism T of G into $\mathcal{L}(E)$, is said to be *strongly bounded* or *uniformly bounded* if $\sup\{\|T_x\| : x \in G\} < \infty$.

Proposition 17.5. *Let G be a topological group and let X be a subspace of $\mathscr{C}(G)$ that is closed under complex conjugation, right invariant with $1_G \in X$. Assume* that there exists a right invariant mean M on X. Let T be a strongly continuous, uniformly bounded representation of G on a Hilbert space $\mathscr{H}$ such that, for all ξ, $\eta \in \mathscr{H}$, the mapping $\psi_{\xi,\eta}: \begin{array}{c} x \mapsto (T_x\xi|T_x\eta) \\ G \to \mathbb{C} \end{array}$ belongs to X. Then T is equivalent to a strongly continuous unitary representation on $\mathscr{H}$ equipped with an appropriate norm.

Proof. By hypothesis, we have $\alpha = \sup\{\|T_x\| : x \in G\} < \infty$. We define a bilinear form on $\mathscr{H} \times \mathscr{H}$ by putting

$$((\xi|\eta)) = M(\psi_{\xi,\eta}),$$

$\xi, \eta \in \mathscr{H}$. Since M is a mean, for all $\xi, \eta \in \mathscr{H}$, we have

$$|((\xi|\eta))| \leq \|\psi_{\xi,\eta}\| \leq \alpha^2 \|\xi\| \|\eta\|.$$

There exists $A \in \mathscr{L}(\mathscr{H})$ such that $((\xi|\eta)) = (A\xi|\eta)$ whenever $\xi, \eta \in \mathscr{H}$ ([HR] B.60); A is positive-definite. In particular, for every $\xi \in \mathscr{H}$,

$$((\xi|\xi)) \leq \|\psi_{\xi,\xi}\| \leq \alpha^2 \|\xi\|^2.$$

Moreover for every $x \in G$ and every $\xi \in \mathscr{H}$,

$$\|\xi\|^2 = (T_{x^{-1}}T_x\xi|T_{x^{-1}}T_x\xi) \leq \alpha^2(T_x\xi|T_x\xi) = \alpha^2\psi_{\xi,\xi}(x).$$

Hence as M is a mean,

$$\|\xi\|^2 = M(\|\xi\|^2 1_G) \leq \alpha^2((\xi|\xi)).$$

Therefore $((\cdot|\cdot))$ defines an inner product for $\mathscr{H}$ and the associated norm is equivalent to $\|\cdot\|$. Let B be the hermitian square root $A^{1/2}$ of A. For every $\xi \in \mathscr{H}$, $(B\xi|B\xi) = (A\xi|\xi) = ((\xi|\xi))$. Hence B is an isometry, and B^{-1} exists.

As M is right invariant, for all $a \in G$ and $\xi, \eta \in \mathscr{H}$,

$$(T_a^*AT_a\xi|\eta) = (AT_a\xi|T_a\eta) = ((T_a\xi|T_a\eta)) = ((\xi|\eta)).$$

Hence $T_a^*AT_a = A$ for all $a \in G$. Let also $T_a' = BT_aB^{-1}$, $a \in G$, that is, T' is equivalent to T and T' is strongly continuous. For every $a \in G$,

$$T_a'^*T_a' + B^{-1}T_a^*BBT_aB^{-1} = B^{-1}T_a^*AT_aB^{-1} = B^{-1}AB^{-1} = \mathrm{id}_{\mathscr{H}}$$

and also $T_a'T_a'^* = \mathrm{id}_{\mathscr{H}}$, that is, T' is unitary. □

Corollary 17.6. *Every continuous uniformly bounded representation of an amenable group on a Hilbert space is equivalent to a continuous unitary representation.*

Proof. If G is an amenable group, Proposition 17.5 applies as $\text{RIM}(\mathscr{C}(G)) \neq \emptyset$ by Theorem 4.19. □

Comments

We introduce a definition in view of pointing out a partial converse of Corollary 17.6.

Definition 17.7. *Let G be a topological group and let X be a subspace of $\mathscr{C}(G)$ that is closed under complex conjugation, right invariant with $1_G \in X$. We say that G has property (U_X) if the following condition holds: Let E be any Banach space equipped with a prehilbert structure $((\cdot|\cdot))$ and let T be any uniformly bounded representation of G on E such that (a) the mapping*
$$\psi_{\xi,\eta} : \begin{matrix} x \mapsto \\ G \to \mathbb{C} \end{matrix} ((T_x\xi|T_x\eta))$$
belongs to X whenever $\xi, \eta \in E$ and (b) there exists $k \in \mathbb{R}_+^*$ such that $|((T_x\xi|T_x\eta))| \leq k\|\xi\|\|\eta\|$ whenever $x \in G$ and $\xi, \eta \in E$. Then there exist a unitary representation T' of G on a Hilbert space $\mathscr{H}$ and $A \in \mathscr{L}(E, \mathscr{H})$ such that

(i) $T_x'A = AT_x$ whenever $x \in G$,
(ii) $\inf\{\mathscr{R}e((T_x\xi|T_x\eta)) : x \in G\} \leq \mathscr{R}e(A\xi|A\eta)_{\mathscr{H}}$
$\leq \sup\{\mathscr{R}e((T_x\xi|T_x\eta)) : x \in G\}$ *whenever $\xi, \eta \in E$.*

Proposition 17.8. *Let G be a topological group and let X be a subspace of $\mathscr{C}(G)$ that is closed under complex conjugation and right invariant with $1_G \in X$. If there exists a right invariant mean on X, then G satisfies (U_X).*

Proof. Let E be a Banach space equipped with a prehilbert structure $((\cdot|\cdot))$, and let T be a uniformly bounded representation of G on E satisfying (a) and (b) of Definition 17.7. We consider a right invariant mean M on X and let

$$(\xi|\eta) = M(\psi_{\xi,\eta}),$$

$\xi, \eta \in E$. Since M is a mean, for all $\xi, \eta \in E$,

$$\inf\{\mathscr{R}e((T_x\xi|T_x\eta)) : x \in G\} \leq \mathscr{R}e(\xi|\eta) \leq \sup\{\mathscr{R}e((T_x\xi|T_x\eta)) : x \in G\}.$$

Since M is right invariant, for all $a \in G$ and $\xi, \eta \in E$, we have

$$(T_a\xi|T_a\eta) = (\xi|\eta). \tag{3}$$

Let $E_0 = \{\zeta \in E : (\zeta|\zeta) = 0\}$, $\mathscr{H}_0 = E/E_0$, and let Φ be the canonical homomorphism of E onto $\mathscr{H}_0$. We obtain an induced inner product on $\mathscr{H}_0$ defined by

$$(\dot{\xi}|\dot{\eta}) = (\xi|\eta)$$

for $\dot{\xi} = \Phi(\xi)$, $\dot{\eta} = \Phi(\eta)$. For every $x \in G$, $T_x(E_0) \subset E_0$ by (3), so we may

define
$$V_x\dot\xi = \Phi(T_x\xi),$$

$\xi \in E$. For every $\xi \in E$, $V_e\dot\xi = \Phi(\xi) = \dot\xi$ and $V_e = id_{\mathcal{H}_0}$. If $x, y \in G$, $V_{xy} = V_x V_y$ and also $V_{x^{-1}} = V_x^{-1}$.

From (3) we deduce that, for all $a \in G$ and $\dot\xi, \dot\eta \in \mathcal{H}_0$,
$$(V_a\dot\xi | V_a\dot\eta) = (\dot\xi | \dot\eta).$$

Let $\mathcal{H}$ be the completion of $\mathcal{H}_0$. We may define the unitary extension T' of V over $\mathcal{H}$. The statement now holds for $A = j \circ \Phi$, j being the canonical injection of $\mathcal{H}_0$ into $\mathcal{H}$. □

Proposition 17.9. *A locally compact group G is amenable if and only if $(U_{\mathscr{C}(G)})$ holds.*

Proof. If G is amenable, $\text{RIM}(\mathscr{C}(G)) \neq \varnothing$ by Theorem 4.19. Therefore Proposition 17.8 applied to $\mathscr{C}(G)$ yields property $(U_{\mathscr{C}(G)})$.

Assume now G to have property $(U_{\mathscr{C}(G)})$. We show that $\text{LIM}(\mathscr{C}(G)) \neq \varnothing$ establishing amenability of G via Theorem 4.19. We define $((f|g)) = f(e)g(e)$ for $f, g \in \mathscr{C}(G)$ and apply $(U_{\mathscr{C}(G)})$ to the left regular representation. Let L' be the corresponding unitary representation of G on a Hilbert space $\mathcal{H}$ and A the corresponding element in $\mathscr{L}(\mathscr{C}(G), \mathcal{H})$. We put
$$M(f) = (Af | A1_G)_{\mathcal{H}}$$
for $f \in \mathscr{C}(G)$.

Obviously M is linear. For every $f \in \mathscr{C}(G)$, by $(U_{\mathscr{C}(G)})$ we have
$$\inf \mathscr{Re} f = \inf \{ \mathscr{Re}((L_x f | L_x 1_G)) : x \in G \} \leq \mathscr{Re} M(f)$$
$$\leq \sup \{ \mathscr{Re}((L_x f | L_x 1_G)) : x \in G \} = \sup \mathscr{Re} f.$$

Hence M is a mean on $\mathscr{C}(G)$. Moreover for all $f \in \mathscr{C}(G)$, $a \in G$,
$$M(_a f) = (A(_a f)|A1_G)_{\mathcal{H}} = (AL_{a^{-1}} f | A1_G)_{\mathcal{H}} = (L'_{a^{-1}} Af | A1_G)_{\mathcal{H}}$$
$$= (Af | L'_a A1_G)_{\mathcal{H}} = (Af | AL_a 1_G)_{\mathcal{H}} = (Af | A1_G)_{\mathcal{H}} = M(f),$$
that is, M is a left invariant mean on $\mathscr{C}(G)$. □

If G is a locally compact group, the elements U, V of $\hat{G}$ are said to be *weakly equivalent* if U is weakly contained in V and V is weakly contained in U. If $U, V \in \hat{G}$, we denote by $U \otimes V$ the tensor product of the representations U and V ([HR] 27.33). If $T \in \hat{G}$, $\mathscr{S} \subset \hat{G}$ and T is weakly contained in every element S of $\mathscr{S}$, then $T \otimes U$ is weakly contained in $S \otimes U$, whenever $S \in \mathscr{S}$ and $U \in \hat{G}$.

Let G be a locally compact group admitting the closed subgroup H. If $T \in \hat{G}$ and $U \in \hat{H}$, $(\mathrm{ind}_{H \uparrow G} U) \otimes T$ is weakly equivalent to $\mathrm{ind}_{H \uparrow G} (U \otimes (T|_H))$; in particular, if $T \in \hat{G}$, $(\mathrm{ind}_{H \uparrow G} i_H) \otimes T$ is weakly equivalent to $\mathrm{ind}_{H \uparrow G} T|_H$ ([182] Lemma 4.2, Corollary 1). If H is a closed normal subgroup of G and π is the canonical homomorphism of G onto G/H, then $\mathrm{ind}_{H \uparrow G} i_H$ is weakly equivalent to $\mathrm{ind}_{\{\dot{e}\} \uparrow G/H} i_{\{\dot{e}\}} \circ \pi$ ([26] Proposition 1.2-E). We indicate a property of amenable quotient groups expressed in terms of induced representations.

Proposition 17.10. *Let G be a locally compact group admitting a closed normal subgroup H such that G/H is amenable. Then every $T \in \hat{G}$ is weakly contained in* $\mathrm{ind}_{H \uparrow G} T|_H$.

Proof. Let π be the canonical homomorphism of G onto G/H. As G/H is amenable, $i_{G/H}$ is weakly contained in $\mathrm{ind}_{\{\dot{e}\} \uparrow G/H} i_{\{\dot{e}\}}$. Hence $i_G \otimes T = (i_{G/H} \circ \pi) \otimes T$ is weakly contained in $(\mathrm{ind}_{H \uparrow G} i_H) \otimes T$, that is, T is weakly contained in $\mathrm{ind}_{H \uparrow G} T|_H$. $\square$

NOTES

Greenleaf [227] investigated in detail amenability properties in the framework of induced representations. For 17.1 and further comments see Skudlarek [513]. Paterson [428] established 17.2. Kazhdan's property was studied in relation to amenability by Eymard [174], Margulis [379], Furstenberg [193], and Wang [549].

Sz. Nagy [525] established 17.6 for $G = \mathbf{R}$, and Dixmier [138] gave the proof for an arbitrary amenable group. The extension 17.5 is taken from K. Sakai [498]. He proved the converse version 17.8–9 [499]. See also Vasilescu and Zsidó [545].

Baggett [26] gave 17.10. Related questions are dealt with by Schochetman [505].

Kaniuth [305, 306] studied representation properties for the (amenable) groups belonging to $[FC]^-$. Such properties were also examined extensively for amenable groups that are contained in various important classes of locally compact groups. References are Kaniuth [304] for the class $[IN]$ of groups admitting a neighborhood of the identity that is invariant by inner automorphisms, Kaniuth [303], Hauenschild [252], and Liukkonen [363] for the class $[SIN]$ of groups admitting a basis of neighborhoods of the identity that are invariant by inner automorphisms. Robert [475] studied particular amenable groups G that are also (a) Fell groups, that is, groups for which the left regular representation may be decomposed into a Hilbert sum of irreducible representations; (b) maximally almost periodic, that is, injectable into compact groups; and (c) CCR-groups, that is, for every $T \in \hat{G}$ and every $f \in L^1(G)$, T_f is compact. The characterization of amenability in terms of weak containment was used by Margulis in his investigations on discrete subgroups of Lie groups; see Tits [539].

Arsac [20] considered for a locally compact group G an arbitrary element π in $\hat{G}$ and defined $A_\pi(G)$ to be the closure of the subspace of $B(G)$ spanned by the positive-definite functions associated to π. He proved that, in case $\pi' \in \hat{G}$ admits a cyclic vector ζ and $u \in P(G)$ is associated to (π', ζ), π' is weakly contained in π if and only if there exists a bounded net (u_i) in $A_\pi(G)$ such that $\lim_i \|u_i v - uv\|_B = 0$ whenever $v \in A_\pi(G)$. This general result implies the characterization 10.4 ($p = 2$) for amenable groups.

Generalized weak containment properties were studied by Baggett [25]. If G is a separable locally compact group, one considers a cocycle mapping, that is, a Borel measurable mapping α from $G \times G$ into the torus $\mathbf{T}$ having the following properties: $(\forall x \in G)$ $\alpha(x, e) = \alpha(e, x) = 1$; $(\forall x \in G)$ $(\forall y \in G)$ $(\forall z \in G)$ $\alpha(xy, z)\alpha(x, y) = \alpha(x, yz)\alpha(y, z)$. Then $G_\alpha = G \times \mathbf{T}$ is a separable locally compact group for the group structure defined by $(x, s)(y, t) = (xy, \alpha(x, y)st)$, $(x, s), (y, t) \in G_\alpha$. An α-representation of G is a mapping $U: x \mapsto U_x$ of G into the group of unitary operators on a separable Hilbert space $\mathscr{H}$ verifying the following conditions: For all $\xi \in \mathscr{H}$ and $\eta \in \mathscr{H}$, $x \mapsto (U_x \xi | \eta)$ is a Borel measurable mapping. For all $x, y \in G$, $U_x U_y = \alpha(x, y)U_{xy}$. In particular, $L^{(\alpha)}$ is the left regular α-representation of $L^2(G)$ defined by $L_x^{(\alpha)} f(z) = \alpha(x, x^{-1}z) L_x f(z)$ for $f \in L^2(G)$, $x \in G$ and $z \in G$. If T is a unitary representation of G_α such that $T|_\mathbf{T}$ is a multiple of the trivial character, let T^α be the α-representation of G defined by $T_x^\alpha = T_{(x,1)}$, $x \in G$. This correspondence is one-to-one and an α-representation S^α is said to be weakly contained in the α-representation T^α in case S is weakly contained in T. For an amenable separable group G, $L^{(\alpha)}$ weakly contains all irreducible α-representations of G. Recent publications: Anker [589], Boidol [590], and Lance [593].

18. SUPPLEMENTARY CONVOLUTION PROPERTIES

W give refinements concerning the characteristic property (Cv_2) of amenable groups and consider a general property (G_p) $(1 < p < \infty)$ that coincides with Godement's property (G) in case $p = 2$.

A. Sufficient Conditions for Amenability

Let G be a locally compact group; $Cv^2(G)$ is a Banach algebra (Section 9A). For every $\mu \in M^1(G)$, by Proposition 9.3 we have $L_\mu \in Cv^2(G)$ and by (2.13) also $L_\mu^* = L_{\mu^*}$, hence $\|L_{\mu * \mu^*}\|_{Cv^2} = \|L_\mu L_\mu^*\|_{Cv^2} = \|L_\mu\|_{Cv^2}^2$. We consider the spectral radius $r(L_\mu)$ of L_μ; $r(L_\mu) \leq \|L_\mu\|_{Cv^2} \leq \|\mu\|$. (Section 2B.2). In Corollary 9.7 amenability of G is characterized by the fact that $\|L_\mu\|_{Cv^2} = 1$ whenever $\mu \in M^1(G)$. Hence G is amenable if $r(L_\mu) = 1$ for every $\mu \in M^1(G)$.

We show that if the latter condition holds for one $\mu \in M^1(G)$, then the group is amenable.

Definition 18.1. *Let G be a locally compact group. A probability measure μ in $M^1(G)$ is called aperiodic if the closed subgroup of G generated by $\operatorname{supp}\mu$ is G.*

Lemma 18.2. *Let $\mathcal{H}$ be a Hilbert space and let $T \in \mathcal{L}(\mathcal{H})$ with $\|T\| = 1$ and $1 \in \operatorname{sp} T$. Then there exists a sequence (ξ_n) in the unit sphere of $\mathcal{H}$ such that $\lim_{n \to \infty} \|T\xi_n - \xi_n\| = 0$.*

Proof. If 1 is an eigenvalue for T, the statement is trivial. Otherwise, let $\eta \in \mathcal{H}$ such that $(\xi - T\xi|\eta) = 0$ for every $\xi \in \mathcal{H}$; then $\eta = T^*\eta$ and

$$\|T\eta - \eta\|^2 = \|T\eta\|^2 + \|\eta\|^2 - 2\mathcal{R}e\,(\eta|T^*\eta) = \|T\eta\|^2 + \|\eta\|^2 - 2\|\eta\|^2 \leq 0.$$

We have $T\eta = \eta$. As 1 is not eigenvalue, $\eta = 0$, and $(T - \operatorname{id}_{\mathcal{H}})\mathcal{H}$ is dense in $\mathcal{H}$. □

Proposition 18.3. *Let G be a locally compact group admitting an aperiodic probability measure μ. If $r(L_\mu) = 1$, then G is amenable.*

Proof. By Theorem 6.14, it suffices to establish (D_2).

As $\operatorname{sp} L_\mu$ is compact (Section 2B.2), there exists $\alpha \in \operatorname{sp} L_\mu$ such that $|\alpha| = 1$; hence $1 \in \operatorname{sp}(\bar{\alpha} L_\mu)$ by the Hilbert–Dirac spectral theorem ([B TS] Chap. I, Section 1, No. 2, Remarque 4). Thus by Lemma 18.2 there exists a sequence (f_n) in $L^2(G)$ such that, for every $n \in \mathbf{N}^*$, $\|f_n\|_2 = 1$ and

$$\lim_{n \to \infty} \|\mu * f_n - \alpha f_n\|_2 = \lim_{n \to \infty} \|\bar{\alpha}\mu * f_n - f_n\|_2 = \lim_{n \to \infty} \|\bar{\alpha}L_\mu(f_n) - f_n\|_2 = 0.$$

As for every $n \in \mathbf{N}^*$, by the Cauchy–Schwarz inequality,

$$|(\mu * f_n|f_n) - \alpha| = |(\mu * f_n - \alpha f_n|f_n)| \leq \|\mu * f_n - \alpha f_n\|_2,$$

we have $\alpha = \lim_{n \to \infty}(\mu * f_n|f_n)$ and $1 = |\alpha| = \lim_{n \to \infty}|(\mu * f_n|f_n)|$. Also for every $n \in \mathbf{N}^*$, by the Cauchy–Schwarz inequality,

$$|(\mu * f_n|f_n)| = \left|\int (_{x^{-1}}f_n|f_n)\,d\mu(x)\right| \leq \int (_{x^{-1}}|f_n|\,|f_n|)\,d\mu(x) \leq \|f_n\|_2^2 = 1.$$

Therefore $\lim_{n \to \infty} \int (1 - (_{x^{-1}}|f_n|\,|\,|f_n|))\,d\mu(x) = 0$. Then (8.2) and (2.9) imply that

$$\lim_{n \to \infty} \int \||f_n| - {}_x|f_n|\|_2^2\,d\mu(x) = 2 \lim_{n \to \infty} \int (1 - |f_n| * |\check{f}_n|(x))\,d\mu(x)$$

$$= 2 \lim_{n \to \infty} \int \left(1 - (_{x^{-1}}|f_n|\,|\,|f_n|)\right) d\mu(x) = 0.$$

So there exists a subsequence (φ_m) of $(|f_n|)$ in $P^2(G)$ and a measurable subset A of G such that $\mu(A) = 1$ and $\lim_{m \to \infty} \|\varphi_m - {}_x\varphi_m\|_2 = 0$ whenever $x \in A$ ([HS] 11.26). For all $x, y \in A$, $\lim_{m \to \infty} \|{}_{x^{-1}}\varphi_m - \varphi_m\|_2 = 0$ and

$$\lim_{m \to \infty} \|\varphi_m - {}_{xy}\varphi_m\|_2 \le \lim_{m \to \infty} \|\varphi_m - {}_y\varphi_m\|_2 + \lim_{m \to \infty} \|{}_y(\varphi_m - {}_x\varphi_m)\|_2 = 0.$$

Hence A generates a closed subgroup H of G such that $\mu(H) = 1$ and $\lim_{m \to \infty} \|\varphi_m - {}_z\varphi_m\|_2 = 0$ whenever $z \in H$. As μ is aperiodic, $H = G$ and (D_2) is established. □

Remark 18.4. Proposition 18.3 does not hold if one weakens the hypothesis replacing it by the condition $\|L_\mu\|_{Cv^2} = 1$ for one $\mu \in M^1(G)$.

Consider the free group F_2 generated by a and b, and let $\mu = \frac{1}{2}(\delta_a + \delta_b)$; μ is aperiodic. Since F_2 is nonamenable by Proposition 14.1, Proposition 18.3 implies that $r(L_\mu) < 1$. On the other hand the subgroup H of F_2 generated by ab^{-1} is isomorphic to $\mathbf{Z}$ and hence amenable. Since $\text{supp}(\mu * \mu^*) = (\text{supp}\,\mu)(\text{supp}\,\mu)^{-1} = \{e, ab^{-1}, ba^{-1}\}$, $\mu * \mu^*$ induces an aperiodic measure ν on H. By Corollary 9.7 we have $\|L_\nu\|_{Cv^2(H)} = 1$; therefore a fortiori $\|L_{\mu*\mu^*}\|_{Cv^2(F_2)} = 1$. Then we have

$$\|L_\mu\|^2_{Cv^2(F_2)} = \|L_\mu L_\mu^*\|_{Cv^2(F_2)} = \|L_{\mu*\mu^*}\|_{Cv^2(F_2)} = 1.$$

Corollary 18.5. *Let G be a locally compact group admitting $\mu \in M^1(G)$ such that $\|L_\mu\|_{Cv^2} = 1$ and the subgroup of G generated by $(\text{supp}\,\mu)(\text{supp}\,\mu)^{-1}$ is dense in G. Then G is amenable.*

Proof. Since $(\text{supp}\,\mu)(\text{supp}\,\mu)^{-1} = \text{supp}(\mu * \mu^*)$, the probability measure $\mu * \mu^*$ is aperiodic on G. Since $L_{\mu*\mu^*}$ is hermitian, for every $n \in \mathbf{N}^*$,

$$\|L_{(\mu*\mu^*)^{2^n}}\|^{1/2^n}_{Cv^2} = \|L_{\mu*\mu^*}\|_{Cv^2} = \|L_\mu\|^2_{Cv^2}$$

and

$$r(L_{\mu*\mu^*}) = \lim_{n \to \infty} \|L_{(\mu*\mu^*)^{2^n}}\|^{1/2^n}_{Cv^2} = 1.$$

Thus the statement follows from Proposition 18.3 applied to $\mu * \mu^*$. □

Corollary 18.6. *Let G be a locally compact group admitting an aperiodic probability measure μ such that $\|L_\mu\|_{Cv^2} = 1$ and $e \in \text{supp}\,\mu$. Then G is amenable.*

Proof. In this case $\text{supp}\,\mu \subset (\text{supp}\,\mu)(\text{supp}\,\mu)^{-1}$ and the conditions of Corollary 18.5 are satisfied. □

B. Characterizations of Amenability

We first consider properties that generalize Godement's condition. If G is a locally compact group, $p \in]1, \infty[$, $1/p + 1/p' = 1$, $\varphi \in P^p(G)$, we have $\psi = \varphi^{p/p'} \in P^{p'}(G)$ and $\langle \psi, \varphi \rangle = \langle \varphi^{p/p'}, \varphi \rangle = \int \varphi^p = 1$. For every $a \in G$, $\psi * \check{\varphi}(a) = (_a\psi | \varphi)$ by (2.9). Moreover $\psi * \check{\varphi} \in \mathscr{C}_0(G)$ and $0 \leq \psi * \check{\varphi}(a) \leq \|\psi * \check{\varphi}\| \leq \|\varphi\|_p \|\psi\|_{p'} = 1$ (Section 2D.3).

Definition 18.7. *Let G be a locally compact group. If $p \in]1, \infty[$ and $1/p + 1/p' = 1$, we define*

$$\sigma_p = \sup \left\{ \inf \left\{ \sup \left\{ 1 - \varphi^{p/p'} * \check{\varphi}(a) : a \in K \right\} : \varphi \in P^p(G) \right\} : K \in \mathfrak{K}(G) \right\},$$

$$\sigma_p^* = \sup \left\{ \inf \left\{ \sup \left\{ 1 - \varphi^{p/p'} * \check{\varphi}(a) : a \in F \right\} : \varphi \in P^p(G) \right\} : F \in \mathfrak{F}(G) \right\},$$

$$\tau_p = \sup \left\{ \inf \left\{ 1 - \langle \varphi^{p/p'} * \check{\varphi}, \mu \rangle : \varphi \in P^p(G) \right\} : \mu \in M^1(G) \right\},$$

$$\tau_p' = \sup \left\{ \inf \left\{ 1 - \langle \varphi^{p/p'} * \check{\varphi}, \psi \rangle : \varphi \in P^p(G) \right\} : \psi \in P^1(G) \right\}.$$

Note that $\sigma_2 = \sigma$ and $\sigma_2^* = \sigma^*$.

Definition 18.8. *Let G be a locally compact group and let $p \in]1, \infty[$. We define the property*

$$(G_p)\left[\text{resp. } (G_p^*), (SG_p), (SG_p')\right] \sigma_p = 0 \left[\text{resp. } \sigma_p^* = 0, \tau_p = 0, \tau_p' = 0\right].$$

Property (G_2) [resp. (G_2^*)] is (G) [resp. (G^*)] by Proposition 8.7.

Proposition 18.9. *A locally compact group G is amenable if and only if, for every $p \in]1, \infty[$ (for one $p \in]1, \infty[$), (G_p) [resp. $(G_p^*), (SG_p), (SG_p')$] holds.*

Proof. Let $p \in]1, \infty[$ and $1/p + 1/p' = 1$. We show that

$$(P_p) \Rightarrow (G_p) \Rightarrow (G_p^*) \Rightarrow (P_{p'}^*)$$
$$\Downarrow$$
$$(SG_p) \Rightarrow (SG_p') \Rightarrow (Cv_p').$$

The statement then follows from Theorems 6.14 and 9.6. Notice that the implications $(G_p) \Rightarrow (G_p^*)$ and $(SG_p) \Rightarrow (SG_p')$ are trivially verified.

$$(P_p) \Rightarrow (G_p)$$

Let $\varepsilon \in \mathbf{R}_+^*$ and let K be a compact subset of G. There exists $\varphi \in P^p(G)$ such that $\|\varphi - {}_{a^{-1}}\varphi\|_p < \varepsilon$ whenever $a \in K$. By Hölder's inequality we have

$$1 - \varphi^{p/p'} * \check{\varphi}(a) = 1 - \langle {}_a\varphi^{p/p'}, \varphi \rangle = \langle \varphi^{p/p'}, \varphi - {}_{a^{-1}}\varphi \rangle$$

$$\leq \|\varphi^{p/p'}\|_{p'} \|\varphi - {}_{a^{-1}}\varphi\|_p = \|\varphi - {}_{a^{-1}}\varphi\|_p < \varepsilon.$$

$$(G_p^*) \Rightarrow (P_{p'}^*)$$

Let F be a finite subset of G. For every $\varepsilon \in]0,1[$, there exists $\varphi \in P^p(G)$ such that $\varphi^{p/p'} * \check{\varphi}(a) > 1 - \varepsilon$ whenever $a \in F$. Then

$$\langle {}_a\varphi^{p/p'} + \varphi^{p/p'}, \varphi \rangle = \langle {}_a\varphi^{p/p'}, \varphi \rangle + \langle \varphi^{p/p'}, \varphi \rangle = \varphi^{p/p'} * \check{\varphi}(a) + 1 > 2 - \varepsilon$$

and also

$$\|{}_a\varphi^{p/p'} + \varphi^{p/p'}\|_{p'} = \|{}_a\varphi^{p/p'} + \varphi^{p/p'}\|_{p'} \|\varphi\|_p \geq \langle {}_a\varphi^{p/p'} + \varphi^{p/p'}, \varphi \rangle > 2 - \varepsilon$$

whenever $a \in F$.

Now Clarkson's inequalities ([HS] 15.5, 15.8) say that, for every $a \in F$,

$$\left\| \frac{{}_a\varphi^{p/p'} + \varphi^{p/p'}}{2} \right\|_{p'}^{p'} + \left\| \frac{{}_a\varphi^{p/p'} - \varphi^{p/p'}}{2} \right\|_{p'}^{p'} \leq \frac{1}{2}\left(\|{}_a\varphi^{p/p'}\|_{p'}^{p'} + \|\varphi^{p/p'}\|_{p'}^{p'}\right) = 1$$

in Case $2 \leq p'$, and

$$\left\| \frac{{}_a\varphi^{p/p'} + \varphi^{p/p'}}{2} \right\|_{p'}^{p} + \left\| \frac{{}_a\varphi^{p/p'} - \varphi^{p/p'}}{2} \right\|_{p'}^{p}$$

$$\leq \left(\frac{1}{2}\left(\|{}_a\varphi^{p/p'}\|_{p'}^{p'} + \|\varphi^{p/p'}\|_{p'}^{p'}\right)\right)^{1/(p'-1)} = 1$$

in Case $1 < p' < 2$. Thus we have

$$\|{}_a\varphi^{p/p'} - \varphi^{p/p'}\|_{p'} < \left(2^{p'} - (2-\varepsilon)^{p'}\right)^{1/p'}$$

in Case $2 \leq p'$, and

$$\|{}_a\varphi^{p/p'} - \varphi^{p/p'}\|_{p'} < \left(2^{p} - (2-\varepsilon)^{p}\right)^{1/p}$$

in Case $1 < p' < 2$. Therefore $(P_{p'}^*)$ holds.

$$(G_p) \Rightarrow (SG_p)$$

Let $\mu \in M^1(G)$ and $\varepsilon \in]0,1[$. There exist $\mu_0 \in \mathcal{M}_c^1(G)$ such that $\|\mu - \mu_0\| < \varepsilon/3$, hence $\|\mu_0\| > 1 - \varepsilon/3 > 0$. By hypothesis there exists $\varphi \in P^p(G)$

such that $1 - \varphi^{p/p'} * \check{\varphi}(a) < \varepsilon/3\|\mu_0\|$ whenever $a \in \operatorname{supp}\mu_0$. Then

$$1 - \langle \varphi^{p/p'} * \check{\varphi}, \mu \rangle = \langle 1_G - \varphi^{p/p'} * \check{\varphi}, \mu \rangle \leq 2\|\mu - \mu_0\| + \langle 1_G - \varphi^{p/p'} * \check{\varphi}, \mu_0 \rangle$$

$$< 2\frac{\varepsilon}{3} + \frac{\varepsilon}{3} = \varepsilon.$$

$$(SG'_p) \Rightarrow (Cv'_p)$$

Let $\psi \in P^1(G)$. By assumption, given $\varepsilon \in]0,1[$, there exists $\varphi \in P^p(G)$ such that $1 - \langle \varphi^{p/p'} * \check{\varphi}, \psi \rangle < \varepsilon/4$. Choose $\varphi_1, \varphi_2 \in \mathcal{K}_+(G)$ such that $\|\varphi - \varphi_1\|_p < \varepsilon/4$ and $\|\varphi^{p/p'} - \varphi_2\|_{p'} < \varepsilon/4$. Then $\|\varphi_1\|_p < 1 + \varepsilon/4$, $\|\varphi_2\|_{p'} < 1 + \varepsilon/4$ and

$$1 - \langle \varphi_2 * \check{\varphi}_1, \psi \rangle \leq 1 - \langle \varphi^{p/p'} * \check{\varphi}, \psi \rangle + \left|\langle \varphi^{p/p'} * (\varphi - \varphi_1)^{\vee}, \psi \rangle\right|$$

$$+ \left|\langle (\varphi_2 - \varphi^{p/p'}) * \check{\varphi}_1, \psi \rangle\right|$$

$$< \frac{\varepsilon}{4} + \|\varphi^{p/p'}\|_{p'}\|\varphi - \varphi_1\|_p + \|\varphi_2 - \varphi^{p/p'}\|_{p'}\|\varphi_1\|_p$$

$$< \frac{\varepsilon}{4} + \frac{\varepsilon}{4} + \frac{\varepsilon}{4}\left(1 + \frac{\varepsilon}{4}\right) < \varepsilon.$$

For $i = 1, 2$, let $\varphi'_i = (1 + \varepsilon/4)^{-1}\varphi_i \in \mathcal{K}_+(G)$; $\|\varphi'_1\|_p < 1$ and $\|\varphi'_2\|_{p'} < 1$. We have

$$\langle \varphi'_2 * (\varphi'_1)^{\vee}, \psi \rangle > \left(1 + \frac{\varepsilon}{4}\right)^{-2}(1 - \varepsilon).$$

Therefore Proposition 9.2 implies that $\|L_\psi\|_{Cv^p} = 1$. □

The next lemmas are heading toward an evaluation of $\|L_\mu\|_{Cv^2}$ for $\mu \in \mathcal{M}^1_+(G)$ in an arbitrary locally compact group G.

Lemma 18.10. *Let G be a locally compact group. If V is a compact neighborhood of e, there exist $g, h \in \mathcal{K}_+(G)$, $g \neq 0$, $h \neq 0$ such that $g * \check{g} \leq 1_V \leq h * \check{h}$.*

Proof. (a) Let V_1 be a compact symmetric neighborhood of e in G such that $V_1^2 \subset V$. By Urysohn's lemma there exists $f_1 \in \mathcal{K}_+(G)$ such that $\|f_1\| \leq 1$ and $f_1(x) = 0$ whenever $x \notin V_1$. Let $g = |V_1|^{-1/2}f_1$. Then by (2.10) we have $\|g * \check{g}\| \leq \|g\|_2^2 \leq 1$. If $x \in G$,

$$g * \check{g}(x) = \int_G g(xy)g(y)\,dy = \int_{V_1} g(xy)g(y)\,dy.$$

Hence $g * \check{g}(x) \neq 0$ implies that $x \in V_1^2$. We conclude that $g * \check{g} \leq 1_V$.

(b) There exists $f_2 \in \mathcal{K}_+(G)$ such that $f_2(x) = 1$ whenever $x \in V^2$. Let $h = |V|^{-1/2} f_2$. If $x \in V$,

$$h * \check{h}(x) = \int_G h(xy) h(y) \, dy \geq \frac{1}{|V|} \int_V f_2(xy) f_2(y) \, dy = \frac{1}{|V|} \int_V f_2(y) \, dy = 1.$$

□

Let $\mathcal{H}$ be a Hilbert space and let T be a positive-definite operator on $\mathcal{H}$. For all $\xi \in \mathcal{H}$, $(T\xi|\xi) \geq 0$ and, as T is hermitian, $(T^2\xi|\xi) = (T\xi|T\xi) \geq 0$. Assume that, for $q \leq m$ and every $\xi \in \mathcal{H}$, $(T^q\xi|\xi) \geq 0$; then also $(T^{m+1}\xi|\xi) = (T^{m-1}T\xi|T\xi) \geq 0$. So we see inductively that every $T^n (n \in \mathbf{N}^*)$ is a positive-definite operator. As T admits a square root $T^{1/2}$ and $T^{1/2}$ is a positive-definite operator (Section 8A), for every $n \in \mathbf{N}^*$, there exists a positive-definite (hermitian) operator $T^{1/2^n}$ with $\|T^{1/2^n}\| = \|T\|^{1/2^n}$. Let also $T^0 = id_\mathcal{H}$. Then for every $s \in D = \{p2^{-n} : p \in \mathbf{N}, n \in \mathbf{N}\}$ we may define a positive-definite operator T^s on $\mathcal{H}$. Let $\xi \in \mathcal{H}$ with $\|\xi\| = 1$. We put

$$\theta(s) = \log(T^s \xi | \xi),$$

$s \in D$. By the Cauchy–Schwarz inequality, for all $s, t \in D$, we have

$$(T^{(s+t)/2}\xi|\xi) = (T^{s/2}\xi|T^{t/2}\xi) \leq (T^{s/2}\xi|T^{s/2}\xi)^{1/2} (T^{t/2}\xi|T^{t/2}\xi)^{1/2}$$

$$= (T^s\xi|\xi)^{1/2} (T^t\xi|\xi)^{1/2};$$

therefore θ is a convex mapping. In particular,

$$(T\xi|\xi) \leq (T^0\xi|\xi)^{1/2} (T^2\xi|\xi)^{1/2} = (T^2\xi|\xi)^{1/2}$$

and

$$\theta(1) \leq \frac{\theta(2)}{2}.$$

Assume now that $\theta(m)/m \leq \theta(m+1)/(m+1)$ for $m = 1, \ldots, 2n-1$; $n \in \mathbf{N}^*$. Then

$$\theta(2n) \leq \tfrac{1}{2}(\theta(2n+1) + \theta(2n-1)) \leq \tfrac{1}{2}\theta(2n+1) + \frac{2n-1}{4n}\theta(2n),$$

$$\frac{\theta(2n)}{2n} \leq \frac{\theta(2n+1)}{2n+1};$$

so also

$$\theta(2n+1) \leq \tfrac{1}{2}(\theta(2n+2)+\theta(2n)) \leq \tfrac{1}{2}\theta(2n+2) + \frac{1}{2}\frac{2n}{2n+1}\theta(2n+1),$$

$$\frac{\theta(2n+1)}{2n+1} \leq \frac{\theta(2n+2)}{2n+2}.$$

We proved by induction that the sequence $(\theta(n)/n)$ is nondecreasing, so also is the sequence $((T^n\xi|\xi)^{1/n})$. For every $n \in \mathbf{N}^*$, $(T^n\xi|\xi)^{1/n} \leq \|T\|$. Therefore the sequence $((T^n\xi|\xi)^{1/n})$ converges and

$$\lim_{n\to\infty} (T^n\xi|\xi)^{1/n} = \sup\left\{(T^n\xi|\xi)^{1/n} : n \in \mathbf{N}^*\right\} \in \mathbf{R}_+.$$

Let G be a locally compact group and let $\mu \in \mathcal{M}^1(G)$. By (9.4), for every $f \in \mathcal{K}(G)$ and every $n \in \mathbf{N}^*$, $(f|\mu^{*n} * f) = \langle f * \tilde{f}, \mu^{*n} \rangle$. If μ is a positive-definite element of $\mathcal{M}^1(G)$, by density of $\mathcal{K}(G)$ in $L^2(G)$, we have $(f|\mu * f) \geq 0$ whenever $f \in L^2(G)$, that is, L_μ is a positive-definite operator on $L^2(G)$. Then for every $n \in \mathbf{N}^*$, $L_{\mu^{*n}}$ is a positive-definite operator on $L^2(G)$. For all $f \in L^2(G)$, $n \in \mathbf{N}^*$, we have $(\mu^{*n} * f|f) = (f|\mu^{*n} * f) \in \mathbf{R}_+$ and

$$\lim_{n\to\infty} \langle f * \tilde{f}, \mu^{*n} \rangle^{1/n} = \lim_{n\to\infty} (\mu^{*n} * f|f)^{1/n}$$

$$= \sup\left\{(\mu^{*n} * f|f)^{1/n} : n \in \mathbf{N}^*\right\} \in \mathbf{R}_+.$$

Let G be a locally compact group, $\mu \in \mathcal{M}^1(G)$ and $g \in \mathcal{K}_\mathbf{R}(G)$. For every $a \in G$,

$$\langle g * \delta_a * \check{g}, \mu \rangle = \int g * \delta_a * \check{g}(x)\,d\mu(x) = \iint g(y)\delta_a * \check{g}(y^{-1}x)\,d\mu(x)\,dy$$

$$= \iint g(y)\check{g}(a^{-1}y^{-1}x)\,d\mu(x)\,dy = \iint g(y)g(x^{-1}ya)\,d\mu(x)\,dy$$

$$= \iint g^*(y)g(x^{-1}y^{-1}a)\,d\mu(x)\,dy$$

$$= \int g^*(y)\mu * g(y^{-1}a)\,dy = g^* * \mu * g(a). \tag{1}$$

Lemma 18.11. *Let G be a locally compact group and let μ be a positive-definite element in $\mathcal{M}^1_+(G)$. If $f \in \mathcal{K}_+(G)$ with $f \neq 0$ and V is a compact neighborhood of e, then*

$$\|L_\mu\|_{Cv^2} = \lim_{n\to\infty} \langle f * \check{f}, \mu^{*n} \rangle^{1/n} = \lim_{n\to\infty} \mu^{*n}(V)^{1/n}.$$

SUPPLEMENTARY CONVOLUTION PROPERTIES

Proof. (a) Consider $g, h \in \mathcal{K}_+(G)$ such that $g \neq 0$, $h \neq 0$. There exist $\alpha \in \mathbf{R}_+^*$, $x_0 \in G$, and a compact neighborhood U of x_0 in G such that $h(y) > \alpha$ whenever $y \in U$. For every $a \in G$, $U' = x_0 a^{-1} x_0^{-1} U a$ is a compact neighborhood of x_0 and

$$h * \delta_a * \check{h}(x_0 a x_0^{-1}) = \int_G h(y) h(x_0 a^{-1} x_0^{-1} y a) \, dy$$

$$> \alpha \int_{U \cap U'} h(z) \, dz > \alpha^2 |U \cap U'| > 0.$$

We put

$$h^{(a)} = h * \delta_a * \check{h},$$

$a \in G$; $h^{(a)} \in \mathcal{K}_+(G)$ and $x_0 a x_0^{-1} \in \overset{\circ}{\operatorname{supp}} h^{(a)}$. Since $g * \check{g} \in \mathcal{K}_+(G)$, there exist $\beta \in \mathbf{R}_+^*$ and $a_1, \ldots, a_m \in G$ such that

$$g * \check{g} \leq \beta \sum_{i=1}^{m} h^{(a_i)}.$$

For $n \in \mathbf{N}^*$, by (1) we have

$$\langle g * \check{g}, \mu^{*n} \rangle \leq \beta \sum_{i=1}^{m} \langle h^{(a_i)}, \mu^{*n} \rangle = \beta_i \sum_{i=1}^{m} \langle h * \delta_{a_i} * \check{h}, \mu^{*n} \rangle$$

$$= \beta \sum_{i=1}^{m} h^* * \mu^{*n} * h(a_i). \tag{2}$$

If $n \in \mathbf{N}^*$, for every $f_0 \in \mathcal{K}(G)$, (2.12) and (9.4) imply that

$$\langle f_0 * \tilde{f}_0, h^* * \mu^{*n} * h \rangle = \langle h * f_0 * \tilde{f}_0, \mu^{*n} * h \rangle$$

$$= \langle h * f_0 * \tilde{f}_0 * \check{h}, \mu^{*n} \rangle = \langle h * f_0 * (h * f_0)^{\sim}, \mu^{*n} \rangle \geq 0,$$

that is, $h^* * \mu^{*n} * h$ is a positive-definite measure. Then $\Delta^{1/2} h^* * \mu^{*n} * h \in P(G)$ ([D] 13.7.6). Thus for every $i = 1, \ldots, m$,

$$\Delta^{1/2}(a_i) h^* * \mu^{*n} * h(a_i) \leq \Delta^{1/2}(e) h^* * \mu^{*n} * h(e) = h^* * \mu^{*n} * h(e)$$

and, by (1) again,

$$\Delta^{1/2}(a_i) h^* * \mu^{*n} * h(a_i) \leq \langle h * \delta_e * \check{h}, \mu^{*n} \rangle = \langle h * \check{h}, \mu^{*n} \rangle.$$

Therefore by (2) we obtain

$$\langle g * \check{g}, \mu^{*n} \rangle \leq \beta m \sup \{ \Delta^{-1/2}(a_i) : i = 1, \ldots, m \} \langle h * \check{h}, \mu^{*n} \rangle$$

where $\sup \{ \Delta^{-1/2}(a_i) : i = 1, \ldots, m \} > 0$. Finally

$$\lim_{n \to \infty} \langle g * \check{g}, \mu^{*n} \rangle^{1/n} \leq \lim_{n \to \infty} \langle h * \check{h}, \mu^{*n} \rangle^{1/n}.$$

Since g and h play similar roles,

$$\lim_{n \to \infty} \langle g * \check{g}, \mu^{*n} \rangle^{1/n} = \lim_{n \to \infty} \langle h * \check{h}, \mu^{*n} \rangle^{1/n}.$$

(b) By Lemma 18.10 we may choose $g, h \in \mathcal{K}_+(G)$ such that $g \neq 0$, $h \neq 0$, and

$$g * \check{g} \leq 1_V \leq h * \check{h}.$$

Then (a) implies that

$$\lim_{n \to \infty} \mu^{*n}(V)^{1/n} = \lim_{n \to \infty} \langle f * \check{f}, \mu^{*n} \rangle^{1/n}. \tag{3}$$

(c) As μ is hermitian, we have $\|L_\mu\|_{Cv^2} = r(L_\mu)$. For every $n \in \mathbf{N}^*$, also $\|L_{\mu^{*n}}\|_{Cv^2} = r(L_{\mu^{*n}})$. The Hilbert–Dirac spectral theorem ([B TS] Chap. I, Section 1, No. 2, Remarque 4) and Proposition 9.2 imply that $\|L_\mu\|_{Cv^2} = \sup\{\langle k * \check{k}, \mu^{*n} \rangle^{1/n} : k \in \mathcal{K}_+(G), \|k\|_2 = 1\}$. So by (a) we obtain

$$\|L_\mu\|_{Cv^2} = \sup \{ \sup\{\langle k * \check{k}, \mu^{*n} \rangle^{1/n} : k \in \mathcal{K}_+(G), \|k\|_2 = 1\} : n \in \mathbf{N}^* \}$$

$$= \sup \{ \sup \{\langle k * \check{k}, \mu^{*n} \rangle^{1/n} : n \in \mathbf{N}^* \} : k \in \mathcal{K}_+(G), \|k\|_2 = 1 \}$$

$$= \sup \Big\{ \lim_{n \to \infty} \langle k * \check{k}, \mu^{*n} \rangle^{1/n} : k \in \mathcal{K}_+(G), \|k\|_2 = 1 \Big\}$$

$$= \lim_{n \to \infty} \{ f * \check{f}, \mu^{*n} \rangle^{1/n}. \tag{4}$$

The relations (3) and (4) establish the statement. □

Lemma 18.12. *Let G be a locally compact group, and let μ be a hermitian element in $M_+^1(G)$. If $f \in \mathcal{K}_+(G)$ with $f \neq 0$ and V is a compact neighborhood of e,*

$$\|L_\mu\|_{Cv^2} = \limsup_{n \to \infty} \langle f * \check{f}, \mu^{*n} \rangle^{1/n} = \limsup_{n \to \infty} \mu^{*n}(V)^{1/n}.$$

Proof. Let $\mu' = \mu * \mu$. As μ is hermitian, $\|L_{\mu'}\|_{Cv^2} = \|L_\mu\|_{Cv^2}^2$ and, for every $g \in \mathscr{K}(G)$,

$$\langle g * \check{g}, \mu * \mu \rangle = \iint g * \check{g}(tu)\, d\mu(t)\, d\mu(u)$$

$$= \iiint g(y)g(u^{-1}t^{-1}y)\, d\mu(t)\, d\mu(u)\, dy$$

$$= \iiint g(ty)g(u^{-1}y)\, d\mu(t)\, d\mu(u)\, dy$$

$$= \int \left(\int \left(\int g(u^{-1}y)\, d\mu(u) \right) g(t^{-1}y)\, d\mu(t) \right) dy$$

$$= \langle \mu * g, \mu * g \rangle \geq 0,$$

that is, μ' is a positive-definite measure. By Lemma 18.11 we have

$$\|L_{\mu'}\|_{Cv^2} = \lim_{m \to \infty} \langle f * \check{f}, \mu^{*2m} \rangle^{1/m} = \lim_{m \to \infty} \mu^{*2m}(V)^{1/m}$$

hence

$$\|L_\mu\|_{Cv^2} = \lim_{m \to \infty} \langle f * \check{f}, \mu^{*2m} \rangle^{1/2m} = \lim_{m \to \infty} \mu^{*2m}(V)^{1/2m}. \tag{5}$$

If $\nu_1, \nu_2 \in \mathscr{M}^1(G)$, we put

$$F(\nu_1, \nu_2) = \langle f * \check{f}, \nu_1 * \nu_2^* \rangle.$$

We have

$$F(\nu_2, \nu_1) = \langle f * \check{f}, \nu_2 * \nu_1^* \rangle = \overline{\langle (f * \check{f})^{\vee}, \nu_1 * \nu_2^* \rangle} = \overline{\langle f * \check{f}, \nu_1 * \nu_2^* \rangle}$$

and F defines a prehilbert structure on $\mathscr{M}^1(G)$. For $\nu_1, \nu_2 \in \mathscr{M}^1(G)$, the Cauchy–Schwarz inequality implies that

$$|\langle f * \check{f}, \nu_1 * \nu_2^* \rangle| \leq \langle f * \check{f}, \nu_1 * \nu_1^* \rangle^{1/2} \langle f * \check{f}, \nu_2 * \nu_2^* \rangle^{1/2}.$$

In particular, for every $m \in \mathbf{N}^*$,

$$\langle f * \check{f}, \mu^{*(2m+1)} \rangle^{1/(2m+1)} \leq \langle f * \check{f}, \mu^{*4m} \rangle^{1/(4m+2)} \langle f * \check{f}, \mu^{*2} \rangle^{1/(4m+2)}.$$

Therefore

$$\limsup_{m \to \infty} \langle f * \check{f}, \mu^{*(2m+1)} \rangle^{1/(2m+1)} \leq \lim_{m \to \infty} \langle f * \check{f}, \mu^{*4m} \rangle^{1/4m}. \tag{6}$$

From (5) and (6) we deduce that

$$\limsup_{m \to \infty} \langle f * \check{f}, \mu^{*(2m+1)} \rangle^{1/(2m+1)} \leq \|L_\mu\|_{Cv^2}. \tag{7}$$

By Lemma 18.10 there exists $h \in \mathcal{K}_+(G)$ such that $1_V \leq h * \check{h}$. Then also

$$\limsup_{m \to \infty} \mu^{*(2m+1)}(V)^{1/(2m+1)} \leq \limsup_{m \to \infty} \langle h * \check{h}, \mu^{*(2m+1)} \rangle^{1/(2m+1)}$$

$$\leq \|L_\mu\|_{Cv^2}. \tag{8}$$

The relations (5), (7), and (8) establish the statement. □

Proposition 18.13. *A necessary and sufficient condition for the amenability of a locally compact group G is given by each of the following properties*:
  (i) *For one (for every) aperiodic hermitian probability measure μ on G such that $e \in \operatorname{supp}\mu$ and for one (for every) $f \in \mathcal{K}_+(G)$ such that $f \neq 0$ one has*

$$\limsup_{n \to \infty} \langle f * \check{f}, \mu^{*n} \rangle = 1.$$

  (ii) *For one (for every) aperiodic hermitian probability measure μ on G such that $e \in \operatorname{supp}\mu$ and for one (for every) compact neighborhood V of e one has*

$$\limsup_{n \to \infty} \mu^{*n}(V)^{1/n} = 1.$$

Proof. The statement follows from Corollary 9.7, Lemma 18.12, and Corollary 18.6. □

Definition 18.14. *Let G be a discrete group admitting an aperiodic hermitian probability measure μ. For every $x \in G$, we define*

$$(K_{\mu, x}) \quad \limsup_{n \to \infty} \mu^{*n}(x)^{1/n} = 1.$$

We say that (K_μ) holds if $(K_{\mu, x})$ holds for every $x \in G$.

Lemma 18.15. *Let G be a discrete group admitting an aperiodic hermitian probability measure μ. If $(K_{\mu, x})$ holds for one $x \in G$, then (K_μ) holds.*

Proof. Let $x \in G$. As μ is aperiodic and hermitian, there exists $m_0 \in \mathbf{N}^*$ such that $\alpha = \mu^{*m_0}(x) = \mu^{*m_0}(x^{-1}) > 0$. We put $t = x$ [resp. e], $u = x^{-1}$ [resp. x], and $v = e$ [resp. x]; hence $v = tu$. We establish the implication $(K_{\mu, t}) \Rightarrow (K_{\mu, v})$.

Let $\varepsilon \in \,]0,1[$. By assumption there exists an increasing sequence (n_k) in $\mathbf{N}^*$ such that, for every $k \in \mathbf{N}^*$, $\mu^{*n_k}(t) > \varepsilon^{n_k}$. Then for every $k \in \mathbf{N}^*$,

$$1 \geq \mu^{*(m_0+n_k)}(v) \geq \mu^{*n_k}(t)\mu^{*m_0}(u) > \varepsilon^{n_k}\alpha,$$

$$1 \geq \mu^{*(m_0+n_k)}(v)^{1/(m_0+n_k)} > \varepsilon^{n_k/(m_0+n_k)}\alpha^{1/(m_0+n_k)}.$$

Hence

$$1 \geq \limsup_{k \to \infty} \mu^{*(m_0+n_k)}(v)^{1/(m_0+n_k)} \geq \varepsilon$$

and

$$1 \geq \limsup_{n \to \infty} \mu^{*n}(v)^{1/n} \geq \varepsilon.$$

As $\varepsilon \in \,]0,1[$ may be chosen arbitrarily,

$$\limsup_{n \to \infty} \mu^{*n}(v)^{1/n} = 1. \qquad \square$$

Proposition 18.16. *A discrete group G is amenable if and only if, for one (for every) aperiodic hermitian probability measure μ and for one (for every) element x of G, property $(K_{\mu,x})$ holds.*

Proof. The statement is a consequence of Corollary 9.7, Lemma 18.12, Lemma 18.15, and Proposition 18.13 as $(K_{\mu,e})$ can only hold if $e \in \operatorname{supp}\mu$. $\quad\square$

Comments

If G and H are locally compact groups and $1 < p, q < \infty$, we denote by $L^p(G, L^q(H))$ the Banach space of all (classes of) complex-valued measurable functions f defined on $G \times H$ such that

$$\|f\|_{p,q} = \left(\int_G \left(\int_H |f(x,y)|^q d\lambda_H(y) \right)^{p/q} d\lambda_G(x) \right)^{1/p} < \infty.$$

If $g \in L^p(G)$ and $h \in L^q(H)$, $\|g \otimes h\|_{p,q} = \|g\|_p \|h\|_q$. A general *Marcienkiewicz interpolation theorem-type* result established by Herz and Rivière ([258] Theorem 1, [262] Lemma 2) leads to the following statement. If $1 < p \leq q \leq 2$ or $2 \leq q \leq p < \infty$, then to every $T \in \mathscr{L}(L^p(G))$ there corresponds $T_G^H \in \mathscr{L}(L^p(G, L^q(H)))$ such that $\|T_G^H\| \leq \|T\|$ and, for all $f = g \otimes h$, $g \in L^p(G)$, $h \in L^q(H)$, the relation $T_G^H f = Tg \otimes h$ holds.

Lemma 18.17. *Let G be a locally compact group and $p \in \,]1,\infty[$. If $T \in Cv^p(G)$, there exists a net $(u_i)_{i \in I}$ in $P^1(G) \cap \mathscr{K}(G)$ such that T is the strong limit of (L_{φ_i}) for $\varphi_i = Tu_i \in L^p(G)$, $i \in I$, and $\|L\varphi_i\|_{Cv^p} \leq \|T\|_{Cv^p}$ whenever $i \in I$.*

Proof. By Proposition 2.2 we may determine a net $(u_i)_{i \in I}$ in $P^1(G) \cap \mathcal{K}(G)$ such that $\lim_i \|u_i * g - g\|_p = 0$ whenever $g \in L^p(G)$. Let $\varphi_i = Tu_i \in L^p(G)$, $i \in I$. For every $i \in I$ and every $f \in \mathcal{K}(G)$, $L_{\varphi_i} f = \varphi_i * f = T(u_i * f)$, $\|L_{\varphi_i} f\|_p \leq \|T\|_{Cv^p} \|f\|_p$, so $\|L_{\varphi_i}\|_{Cv^p} \leq \|T\|_{Cv^p}$. Furthermore

$$\lim_i \|(L_{\varphi_i} - T)f\|_p = \lim_i \|T(u_i * f - f)\|_p \leq \|T\|_{Cv^p} \lim_i \|u_i * f - f\|_p = 0.$$

□

Proposition 18.18. *Let G be an amenable group. If $1 < p \leq q \leq 2$ or $2 \leq q \leq p < \infty$, and $T \in Cv^p(G)$, then $T \in Cv^q(G)$ and $\|T\|_{Cv^q} \leq \|T\|_{Cv^p}$.*

Proof. By Lemma 18.17 we may suppose that there exists $k \in L^p(G)$ such that $Tf = k * f$ whenever $f \in \mathcal{K}(G)$. The mapping $\Theta : \mathcal{K}(G \times G) \to \mathcal{K}(G \times G)$ defined by

$$\Theta \varphi(x, z) = \varphi(x, xz)$$

for $\varphi \in \mathcal{K}(G \times G)$, $(x, z) \in G \times G$, may be extended to an isometric automorphism of $L^p(G, L^q(G))$. Let $1/p + 1/p' = 1$, $1/q + 1/q' = 1$ and choose $h_1, h_2 \in \mathcal{K}(G)$ with $K_1 = \operatorname{supp} h_1$, $K_2 = \operatorname{supp} h_2$. Given $\varepsilon > 0$, by Lemma 15.21 amenability of G implies the existence of $g_1, g_2 \in \mathcal{K}_+(G)$ such that $\|g_1\|_p < 1 + \varepsilon$, $\|g_2\|_{p'} < 1 + \varepsilon$ and $|g_2 * \check{g}_1(y) - 1| < \varepsilon$ whenever $y \in K_2 K_1^{-1}$. Let $f_1 = g_1 \otimes h_1$ and $f_2 = g_2 \otimes h_2$. The Marcienkiewicz interpolation theorem-type result and Hölder's inequality imply that

$$\langle T_G^G \Theta f_1, \Theta f_2 \rangle = \int_G \int_G \int_G k(y) g_1(y^{-1}x) h_1(y^{-1}xz) g_2(x) h_2(xz) \, dx \, dy \, dz$$

$$= \int_G \int_G \int_G k(y) g_1(y^{-1}x) h_1(y^{-1}z) g_2(x) h_2(z) \, dx \, dy \, dz$$

$$= \int_G \int_G \left(\int_G h_1(y^{-1}z) h_2(z) \, dz \right) k(y) g_1(y^{-1}x) g_2(x) \, dx \, dy$$

and

$$|\langle T_G^G \Theta f_1, \Theta f_2 \rangle| \leq \|T\|_{Cv^p} \|g_1\|_p \|g_2\|_{p'} \|h_1\|_q \|h_2\|_{q'}$$

$$\leq (1 + \varepsilon)^2 \|T\|_{Cv^p} \|h_1\|_q \|h_2\|_{q'}.$$

Therefore

$$|\langle T_G^G \Theta f_1, \Theta f_2 \rangle - \langle Th_1, h_2 \rangle|$$

$$= \left| \int_G \int_G k(y)(g_2 * \check{g}_1(y) - 1) h_1(y^{-1}z) h_2(z) \, dy \, dz \right|$$

$$\leq \varepsilon \|k * h_1\|_1 \|h_2\|_1$$

and

$$|\langle Th_1, h_2\rangle| \le (1+\varepsilon)^2 \|T\|_{Cv^p}\|h_1\|_q\|h_2\|_{q'} + \varepsilon\|k*h\|_1\|h_2\|_1.$$

Since $\varepsilon > 0$ may be chosen arbitrarily, $|\langle Th_1, h_2\rangle| \le \|T\|_{Cv^p}\|h_1\|_q\|h_2\|_{q'}$ and $\|Th_1\|_{Cv^q} \le \|T\|_{Cv^p}\|h_1\|_q$, that is, $T \in Cv^q$ and $\|T\|_{Cv^q} \le \|T\|_{Cv^p}$. □

Oberlin ([418] Lemma 3) established a general result which we quote here. Let G be a locally compact group and let $0 < p < q \le 2$. Consider a subspace X of $L^p(G)$ that is right translation-invariant, that is, $f_a \in X$ whenever $f \in X$, $a \in G$. Let T be a linear operator of weak (p,p)-type on X, that is, there exists $w > 0$ such that $|\{x \in G : |Tf(x)| \ge \alpha\}| \le w\|f\|_p^p/\alpha^p$ whenever $f \in X$ and $\alpha \in \mathbf{R}_+^*$. Suppose that T commutes with right translations, that is, $(Tf)_a = T(f_a)$ whenever $f \in X$, $a \in G$. There exists $c > 0$ such that the following situation holds: Let K be a compact symmetric subset of G, U a compact subset of G with $|U| > 0$, $u \in \mathcal{K}(G)$ with $u(x) = 1$ whenever $x \in K^2U$, and $h \in X \cap \mathcal{K}(G)$ with $\mathrm{supp}\, h \subset K$ such that $uh_y \in X$ whenever $y \in G$. Then we have

$$\left(\int_K |Th|^q\right)^{p/q} \le \frac{c}{|U|}\left(\int_G |u|^p\right)\left(\int_G |h|^q\right)^{p/q}.$$

We now apply this property to $X = L^p(G)$.

Proposition 18.19. *Let G be an amenable group and let $0 < p < q \le 2$. Any linear operator T on $L^p(G)$ that is of weak (p,p)-type and commutes with right translations belongs to $\mathcal{L}(L^q(G))$.*

Proof. Let $h \in \mathcal{K}(G)$ and let K be any symmetric compact subset of G containing $\mathrm{supp}\, h$. Since G is amenable, Theorem 7.9 and Proposition 7.11 imply the existence of a compact subset U of G such that $0 < |U|$ and $|K^2U|/|U| < 2$. Choose an open subset U' in G such that $K^2U \subset U'$ and $|U' \setminus K^2U| \le |U|$. By Urysohn's lemma there exists $u \in \mathcal{K}(G)$ such that $0 \le u(x) \le 1$ whenever $x \in G$ and $u(x) = 1$ for $x \in K^2U$, $u(x) = 0$ for $x \in G \setminus U'$. Then

$$\int_G u^p \le |K^2U| + |U| < 3|U|$$

hence

$$\left(\int_K |Th|^q\right)^{p/q} \le 3c\left(\int_G |h|^q\right)^{p/q}$$

and $\|Th\|_q \le (3c)^{1/p}\|h\|_q$. The statement follows by density of $\mathcal{K}(G)$ in $L^q(G)$. □

Let G be a discrete countable group admitting an aperiodic probability measure μ and let $\theta : [0, 1] \to \mathbf{R}_+$ be defined by

$$\theta(0) = 0$$

$$\theta(t) = -t \log t, \ 0 < t \leq 1.$$

One defines an *entropy*

$$h = \lim_{n \to \infty} \frac{1}{n} \sum_{x \in G} \theta(\mu^{*n}(x)) \in \overline{\mathbf{R}_+}$$

corresponding to μ ([23] II.3).

Proposition 18.20. *If G is a discrete countable group admitting an aperiodic hermitian probability measure μ with entropy $h = 0$, then G is amenable.*

Proof. Since log is a concave function, for every $n \in \mathbf{N}^*$, we have

$$\sum_{x \in G} \mu^{*n}(x) \log \mu^{*n}(x) \leq \log \sum_{x \in G} \mu^{*n}(x) \mu^{*n}(x).$$

Since μ is hermitian,

$$\sum_{x \in G} \mu^{*n}(x) \mu^{*n}(x) = \sum_{x \in G} \mu^{*n}(x) \mu^{*n}(x^{-1}) = \mu^{*2n}(e)$$

so

$$\frac{1}{n} \sum_{x \in G} \mu^{*n}(x) \log \mu^{*n}(x) \leq \log\left(\mu^{*2n}(e)^{1/n}\right),$$

and

$$\left(\exp\left(\frac{1}{n} \sum_{x \in G} \mu^{*n}(x) \log \mu^{*n}(x)\right)\right)^{1/2} \leq \mu^{*2n}(e)^{1/2n} \leq 1.$$

Thus by hypothesis $\lim_{n \to \infty} \mu^{*2n}(e)^{1/2n} = 1$ and amenability of G follows from Proposition 18.16. □

NOTES

The results stated in 18.2–6 were obtained independently by Berg and Christensen [32], and by Derriennic and Guivarc'h [134]. The properties 18.10–12 are due to Berg and Christensen [33]. Faraut exposed the subject in

[178]. Gilbert [199] had shown that a locally compact group G is amenable if and only if there exists $f = f^* \in L^1_+(G)$ such that $f(x) \neq 0$ for almost every $x \in G$ and $\|L_f\|_{Cv^p} = \int f$ for some $p \in]1, \infty[$. Feichtinger [180] established the following result: Let G be an amenable unimodular group and let $p \in]1, \infty[$. If $f \in L^p_+(G)$ is tempered, that is, $\sup\{\|\varphi * f\|_p : \varphi \in L^1(G) \cap L^p(G), \|\varphi\|_p \leq 1\} < \infty$, then $f \in L^1(G)$. Flory [186] gave supplementary information in the discrete case.

Characterizations of amenability for discrete groups in terms of convolution operators were formulated by Day in [118]. For 18.9 see Day [119]. By direct computations Kesten [311, 312] had shown that, for a discrete countable group G admitting an aperiodic hermitian probability measure μ, $\limsup_{n \to \infty} \mu^{*n}(e)^{1/n} = 1$ if and only if $l^\infty_{\mathbf{R}}(G)$ admits a left invariant mean. The proofs of 18.15–16 are due to Gerl [196]. A well-known classical fact says that, for any sequence (α_n) in $\mathbf{R}^*_+$, if $\alpha = \lim_{n \to \infty} \alpha_{n+1}/\alpha_n \in \mathbf{R}^*_+$, then also $\alpha = \lim_{n \to \infty} \alpha_n^{1/n}$, so one may formulate a sufficient condition for (K_μ) to hold. Gerl [196] considered a discrete countable group G such that, for an aperiodic hermitian probability measure μ and $n_0 \in \mathbf{N}^*$, $\mu^{*n}(e) > 0$ whenever $n > n_0$. He showed that G is amenable if and only if, for one (for every) $x \in G$, $\lim_{n \to \infty} \mu^{*(n+1)}(x)/\mu^{*n}(x) = 1$ holds. Avez [21] had proved that if an amenable discrete group G admits an aperiodic hermitian probability measure μ and there exists an even $r \in \mathbf{N}^*$ such that $\mu^{*r}(e) > 0$, then for every $x \in G$, $\lim_{n \to \infty} \mu^{*n}(x)/\mu^{*n}(e) = 1$. See Gerl [198] for further comments on the subject and a study of related questions concerning especially the existence of the limit considered by Avez.

Herz and Rivière [262] established a general version of 18.18 applying to a semidirect product of an arbitrary locally compact group by an amenable group; see also Herz [258, 261]. Cowling and Fournier [106] showed that, if $1 < p < 2$, for any infinite locally compact group there exist convolution operators of weak (p, p)-type that are not of strong (p, p)-type. Cowling [103] conjectured that, at least for amenable groups, no such operators exist in case $p > 2$. Oberlin [418] established 18.19. References for estimations of convolutor norms for general locally compact groups are Oberlin [417] and Flory [186]. Avez [23] gave 18.20.

If G is a locally compact group, H a closed subgroup, and $p \in]1, \infty[$ there exists an isometry i of $Cv^p(H)$ into $Cv^p(G)$; see Herz [259]. Derighetti [132] proved that, if H is normal or amenable, then there exists a linear mapping P of $Cv^p(G)$ into $Cv^p(H)$ such that $\|P(T)\|_{Cv^p(H)} \leq \|T\|_{Cv^p(G)}$ whenever $T \in Cv^p(G)$ and $P(i(U)) = U$ whenever $U \in Cv^p(H)$. In case G is a σ-compact group and H is an amenable subgroup, the property had been obtained by Lohoué [368]. If G is a locally compact group and $p \in]1, \infty[$, let $cv^p(G)$ be the closure in $Cv^p(G)$ of the subset of all convolutors having compact support. For any amenable group G admitting a closed normal subgroup H, Anker [12] constructed a continuous linear mapping A of $Cv^p(G)$ into $Cv^p(G/H)$. Derighetti [133] proved that $cv^p(G/H)$ is isometric to the quotient of $cv^p(G)$ by the kernel of A.

Bożejko [52] considered, for a discrete group G, the closure $l^1[G]$ of $\{L_{|f|} : f \in \mathcal{K}(G)\}$ in $PM(G)$; $l^1[G]$ is a Banach algebra. Via $(Cv_2^{\prime *})$ he showed that G is amenable if and only if $l^1(G) \simeq l^1[G]$. More generally, if E is a subset of an arbitrary discrete group G, he termed E an amenable subset in case $l^1(E) \simeq \{L_{|f|} : f \in \mathcal{K}(G), \text{supp}\, f \subset E\}$. He studied the class of these subsets which is $\mathfrak{P}(G)$ in any amenable discrete group G.

Rosenblatt [488] noticed that if G is a nonamenable σ-compact locally compact group, $\mu \in M^1(G)$, and $1 < p < \infty$, then $\lim_{n\to\infty} \|f * (1/n)\sum_{i=1}^n \mu^{*i}\|_p = 0$ uniformly for all elements f of the unit ball in $L^p(G)$. Faraut [179] studied a nonamenable locally compact group G admitting a maximal closed subgroup K that is compact and a semigroup $\{\mu_t : t \in \mathbf{R}_+\}$ in $\mathcal{M}_+(G)$ such that μ_0 is concentrated on K where it coincides with λ_K, $\lim_{t\to 0}\mu_t = \mu_0$ vaguely and $\mu_t \neq \mu_0$, $\int d\mu_t \leq 1$ whenever $t \in \mathbf{R}_+^*$; 18.3 implies that $\lim_{t\to\infty} \|L_{\mu_t}\|_{Cv^2}^{1/t} < 1$.

19. FOURIER ALGEBRA PROPERTIES

The classical Fourier analysis on **R** may be carried over to locally compact abelian groups ([HR] Chapter Eight); on the other hand Fourier analysis may be developed on compact groups ([HR] Chapter Nine). We show that the notions of Fourier algebra and Fourier–Stieltjes algebra allow transpositions of standard results to amenable groups.

A. Fourier Algebras and Fourier–Stieltjes Algebras

Let G be a locally compact group. We consider the C^*-algebra $C^*(G)$ of G defined by the seminorm

$$\|f\|_{\hat{G}} = \sup\{\|T_f\| : T \in \hat{G}\},$$

$f \in L^1(G)$ (Section 8A). The topological dual space $B(G)$ of $C^*(G)$ is spanned by $P(G)$. We have $B(G) \subset \mathscr{C}(G)$ and

$$\|u\|_B = \|u\|_{B(G)} = \sup\left\{\left|\int_G fu\right| : f \in L^1(G), \|f\|_{\hat{G}} \leq 1\right\}.$$

For the pointwise addition and multiplication, $B(G)$ is a commutative Banach algebra called *Fourier–Stieltjes algebra*. It admits the Fourier algebra $A(G)$ of G as a closed ideal. In $B(G)$, $A(G)$ is the closure of the vector space spanned by $P(G) \cap \mathcal{K}(G)$.

The classical reference is Eymard [171], who generalized many results of Fourier analysis on locally compact abelian groups to arbitrary locally compact groups. Consider a locally compact abelian group G and its character group $\hat{G}$ formed by all continuous homomorphisms of G into the torus $\mathbf{T}$ (Section 8A); $\hat{\hat{G}} = G$ ([HR] 24.8). For the commutative Banach algebra $\mathcal{M}^1(G)$ one considers the *Fourier–Stieltjes transform* associating to any $\mu \in \mathcal{M}^1(G)$ the mapping

$$\hat{\mu} : \chi \mapsto \int_G \bar{\chi}\, d\mu$$

$$\hat{G} \to \mathbf{C}.$$

In particular, if $f \in L^1(G)$, the Fourier–Stieltjes transform of $f d\lambda$ denoted by $\hat{f}$ is the *Fourier transform* of f; $B(G)$ is isometrically isomorphic to $\{\hat{\mu} : \mu \in \mathcal{M}^1(\hat{G})\}$ and $A(G)$ is isometrically isomorphic to $\{\hat{f} : f \in L^1(\hat{G})\}$ ([171] 2.5, 3.6.2).

If G is any locally compact group, a complex-valued function defined on G belongs to $B(G)$ if and only if it is continuous and it belongs to $B(G_d)$; also $\|u\|_{B(G)} = \|u\|_{B(G_d)}$ ([171] 2.24). If G is amenable, by Theorem 8.9 we have $C_L^*(G) = C^*(G)$. For $\alpha_1, \ldots, \alpha_n \in \mathbf{C}$, $a_1, \ldots, a_n \in G$, we have $\mu = \sum_{i=1}^n \alpha_i \delta_{a_i} \in L^1(G_d)$ and $\|\mu\|_{\hat{G}} = \|\mu\|_L = \|L_\mu\|_{Cv^2}$. Hence by density, for $u \in B(G)$,

$$\|u\|_B = \sup\{|\langle u, \mu\rangle| : \mu \in \mathcal{M}_f^1(G), \|L_\mu\|_{Cv^2} \leq 1\}. \tag{1}$$

Proposition 19.1. *Let G be an amenable group and let $u \in \mathscr{C}(G)$ such that $uf \in B(G)$ whenever $f \in A(G)$. Then $u \in B(G)$ and*

$$\|u\|_B = \sup\{\|uf\|_B : f \in A(G), \|f\|_A \leq 1\}.$$

Proof. (a) By the closed graph theorem there exists $c > 0$ such that $\|uf\|_B \leq c\|f\|_B = c\|f\|_A$ whenever $f \in A(G)$. Let $\varepsilon > 0$ and consider $a_1, \ldots, a_n \in G$. Since G is amenable, by Lemma 15.21 there exists $v \in A(G)$ such that $\|v\|_A < 1 + \varepsilon$ and $v(a_i) = 1$ whenever $i = 1, \ldots, n$. For all $\alpha_1, \ldots, \alpha_n \in \mathbf{C}$ and $\mu = \sum_{i=1}^n \alpha_i \delta_{a_i}$, by (1) we have

$$\left|\sum_{i=1}^n \alpha_i u(a_i)\right| = \left|\sum_{i=1}^n \alpha_i (uv)(a_i)\right| \leq \|uv\|_B \|L_\mu\|_{Cv^2} \leq c(1+\varepsilon)\|L_\mu\|_{Cv^2}.$$

Since $\varepsilon > 0$ may be chosen arbitrarily,

$$\left|\sum_{i=1}^n \alpha_i u(a_i)\right| \leq c\|L_\mu\|_{Cv^2}$$

and therefore $u \in B(G_d)$. Since also $u \in \mathscr{C}(G)$, we have $u \in B(G)$.

(b) Let $\|\|u\|\| = \sup\{\|uf\|_B : f \in A(G), \|f\|_A \leq 1\}$. For every $f \in A(G)$, $\|uf\|_B \leq \|u\|_B \|f\|_B = \|u\|_B \|f\|_A$, hence $\|\|u\|\| \leq \|u\|_B$.

Let $h \in \mathscr{K}(G)$ such that $\|h\|_{\hat{G}} \leq 1$. If $\operatorname{supp} h = \varnothing$, choose k to be the null function defined on G. Otherwise, for every $\varepsilon > 0$, by Lemma 15.21 let

$k \in A(G) \cap \mathcal{K}_+(G)$ such that $\|k\|_A < 1 + \varepsilon$ and $k(x) = 1$ whenever $x \in \mathrm{supp}\, h$. Then

$$\left|\int uh\right| = \left|\int uhk\right| \leq \|uk\|_B \leq (1+\varepsilon)\|\|u\|\|.$$

As $\varepsilon > 0$ may be chosen arbitrarily, we have $|\int uh| \leq \|\|u\|\|$, for every $h \in \mathcal{K}(G)$ with $\|h\|_{\hat{G}} \leq 1$; hence $\|u\|_B \leq \|\|u\|\|$. □

Corollary 19.2. *Let G be an amenable group. Then $B_2(G) \subset B(G)$. More precisely, if T is a multiplier of $A_2(G) = A(G)$, then there exists a unique $w \in B(G)$ such that $T(v) = vw$ whenever $v \in A(G)$ and $\|T\| = \|w\|_B$.*

Proof. With respect to Theorem 10.4 choose a net $(u_i)_{i \in I}$ of approximate units bounded by 1 in $A(G)$. For every $i \in I$, $\|T(u_i)\|_A \leq \|T\|$ and there exists a subnet $(u_{i_j})_{j \in J}$ of $(u_i)_{i \in I}$ such that $w = \lim_j T(u_{i_j}) \in A(G)$. Since T is continuous, for every $v \in A(G)$,

$$wv = \lim_j T(u_{i_j})v = \lim_j T(u_{i_j}v) = T(v) \in A(G).$$

By Proposition 19.1 we have then $w \in B(G)$ and $\|T\| = \|w\|_B$. □

We now determine a Lebesgue-type decomposition of the Fourier–Stieltjes algebra of an amenable group.

Definition 19.3. *Let G be a locally compact group. We define $B^{(1)}(G)$ to be the subspace of $B(G)$ formed by all $u \in B(G)$ having the following property: If $\varepsilon > 0$, there exists a compact subset K in G such that, for $\mu = \sum_{i=1}^n \alpha_i \delta_{a_i}$ with $\alpha_1, \ldots, \alpha_n \in \mathbf{C}$, $a_1, \ldots, a_n \in G \setminus K$ and $\|L_\mu\|_{Cv^2} \leq 1$, the relation $|\sum_{i=1}^n \alpha_i u(a_i)| < \varepsilon$ holds. We define $B^{(2)}(G)$ to be the subspace of $B(G)$ formed by all $u \in B(G)$ having the following property: If $\varepsilon > 0$ and K is a compact subset of G, there exists $\mu = \sum_{i=1}^n \alpha_i \delta_{a_i}$ with $\alpha_1, \ldots, \alpha_n \in \mathbf{C}$, $a_1, \ldots, a_n \in G \setminus K$ such that $\|L_\mu\|_{Cv^2} \leq 1$ and $|\sum_{i=1}^n \alpha_i u(a_i)| \geq \|u\|_B - \varepsilon$.*

Proposition 19.4. *If G is an amenable group, then (a) $A(G) = B^{(1)}(G)$ and (b) $B(G) = B^{(1)}(G) \oplus B^{(2)}(G)$.*

Proof. (a) Let $u \in A(G)$ and $\varepsilon \in \mathbf{R}_+^*$. There exists $u_0 \in A(G) \cap \mathcal{K}(G)$ with $K = \mathrm{supp}\, u_0$ such that $\|u - u_0\|_A < \varepsilon$. For all $\mu = \sum_{i=1}^n \alpha_i \delta_{a_i}$ with $\alpha_1, \ldots, \alpha_n \in \mathbf{C}$ and $a_1, \ldots, a_n \in G \setminus K$ such that $\|L_\mu\|_{Cv^2} \leq 1$, by (1) we have

$$\left|\sum_{i=1}^n \alpha_i u(a_i)\right| = \left|\sum_{i=1}^n \alpha_i(u - u_0)(a_i)\right| \leq \|u - u_0\|_B = \|u - u_0\|_A < \varepsilon.$$

Thus $u \in B^{(1)}(G)$.

FOURIER ALGEBRA PROPERTIES

Let now $u \in B^{(1)}(G)$ and $\varepsilon \in \mathbf{R}^*_+$, $\delta = \varepsilon/9$. There exists a compact subset K in G such that, for every $\nu \in \mathcal{M}_f^1(G)$ with $\operatorname{supp}\nu \subset G \setminus K$, $|\langle u, \nu\rangle| < \delta \|L_\nu\|_{Cv^2}$ holds. By Lemma 15.21 there exists $v \in A(G) \cap \mathcal{K}_+(G)$ with $K_1 = \operatorname{supp} v$ such that $\|v\|_A < 2$ and $v(x) = 1$ whenever $x \in K$. There also exists $w \in A(G) \cap \mathcal{K}_+(G)$ such that $\|w\|_A < 2$ and $w(x) = 1$ whenever $x \in K_1$. Let $\mu \in \mathcal{M}_f^1(G)$ such that $\|L_\mu\|_{Cv^2} \leq 1$. We have

$$\|L_{(1_G - v)\mu}\|_{Cv^2} \leq (1 + \|v\|)\|L_\mu\|_{Cv^2} \leq 1 + \|v\|_A < 3$$

and

$$\|L_{w(1_G - v)\mu}\|_{Cv^2} \leq \|w\|\, \|L_{(1_G - v)\mu}\|_{Cv^2} \leq \|w\|_A \|L_{(1_G - v)\mu}\|_{Cv^2} < 6.$$

Notice that $v = vw$ and $(\operatorname{supp}(1_G - v)\mu) \cap K = \varnothing$, $(\operatorname{supp} w(1_G - v)\mu) \cap K = \varnothing$. Therefore

$$|\langle u - wu, \mu\rangle| \leq |\langle u - wu, v\mu\rangle| + |\langle u - wu, (1_G - v)\mu\rangle|$$

$$\leq |\langle u(v - vw), \mu\rangle| + |\langle u, (1_G - v)\mu\rangle| + |\langle u, w(1_G - v)\mu\rangle|$$

$$< \delta(3 + 6) = 9\delta = \varepsilon.$$

We conclude from (1) that $\|u - wu\|_B \leq \varepsilon$. Since $\varepsilon > 0$ is arbitrary and $A(G)$ is a closed ideal in $B(G)$, we must have $u \in A(G)$.

(b) Denote by P the projection of $B(G)$ onto $B^{(1)}(G) = A(G)$ and by $\mathbf{1}$ the identity operator on $B(G)$; $B(G) = B^{(1)}(G) \oplus (\mathbf{1} - P)B(G)$.

Let $u \in B^{(2)}(G)$; $u = u_1 + u_2$ with $u_1 \in B^{(1)}(G) = A(G)$, $u_2 \in (\mathbf{1} - P)B(G)$. For every $\varepsilon \in \mathbf{R}^*_+$, there exists $u' \in A(G) \cap \mathcal{K}(G)$ such that $\|u_1 - u'\|_A < \varepsilon$. Let $K = \operatorname{supp} u'$ and $v = u - u' = (u_1 - u') + u_2 \in B(G)$. As $u \in B^{(2)}(G)$, there exists $\alpha_1, \ldots, \alpha_n \in \mathbf{C}$, $a_1, \ldots, a_n \in G \setminus K$, $\mu = \sum_{i=1}^n \alpha_i \delta_{a_i}$ such that $\|L_\mu\|_{Cv^2} \leq 1$ and

$$\left|\sum_{i=1}^n \alpha_i u(a_i)\right| \geq \|u\|_B - \varepsilon. \tag{2}$$

Then by (1) also

$$\left|\sum_{i=1}^n \alpha_i u(a_i)\right| = \left|\sum_{i=1}^n \alpha_i v(a_i)\right| \leq \left|\sum_{i=1}^n \alpha_i (u_1 - u')(a_i)\right| + \left|\sum_{i=1}^n \alpha_i u_2(a_i)\right|$$

$$\leq \|u_1 - u'\|_B + \|u_2\|_B = \|u_1 - u'\|_A + \|u_2\|_B < \varepsilon + \|u_2\|_B.$$

$$\tag{3}$$

From (2) and (3) we deduce that

$$\|u_1\|_A + \|u_2\|_B = \|u\|_B \leq \|u_2\|_B + 2\varepsilon.$$

As $\varepsilon \in \mathbf{R}_+^*$ is arbitrary, we have $\|u_1\|_A = 0$ and $u \in (\mathbf{1} - P)B(G)$.

Let now $u \in (\mathbf{1} - P)B(G)$ with $u \neq 0$. Let $\varepsilon \in]0, \|u\|_B[$ and consider a compact subset K of G. By (1) there exist $\alpha_1, \ldots, \alpha_n \in \mathbf{C}$, $a_1, \ldots, a_n \in G$, $\mu = \sum_{i=1}^n \alpha_i \delta_{a_i}$ such that $\|L_\mu\|_{Cv^2} \leq 1$ and $|\sum_{i=1}^n \alpha_i u(a_i)| \geq \|u\|_B - \varepsilon$. Let $K_1 = K \cup \{a_1, \ldots, a_n\}$. By Lemma 15.21 there exists $v \in A(G)$ such that $\|v\|_A < 1 + \varepsilon$ and $v(x) = 1$ whenever $x \in K_1$. Then $uv \in A(G)$,

$$\left|\sum_{i=1}^n \alpha_i(uv)(a_i)\right| = \left|\sum_{i=1}^n \alpha_i u(a_i)\right| \geq \|u\|_B - \varepsilon$$

and, by (1), $\|uv\|_B \geq \|u\|_B - \varepsilon$. Let $w = u - uv$. Since $uv \in A(G)$ and $u \in (\mathbf{1} - P)B(G)$, we have

$$\|w\|_B = \|uv\|_B + \|u\|_B \geq 2\|u\|_B - \varepsilon.$$

By (1) there also exists $\nu \in M_f^1(G)$ such that $\|L_\nu\|_{Cv^2} \leq 1$ and

$$|\langle w, \nu \rangle| \geq \|w\|_B - \varepsilon \geq 2(\|u\|_B - \varepsilon).$$

Let $\sigma = (2 + \varepsilon)^{-1}(\nu - v\nu) \in \mathscr{M}_f^1(G)$; $\operatorname{supp}\sigma \subset G \setminus K$. We have

$$\|L_\sigma\|_{Cv^2} \leq (2 + \varepsilon)^{-1}(1 + \|v\|)\|L_\nu\|_{Cv^2}$$
$$\leq (2 + \varepsilon)^{-1}(1 + \|v\|_A)\|L_\nu\|_{Cv^2} \leq \|L_\nu\|_{Cv^2} \leq 1$$

and

$$|\langle u, \sigma \rangle| = (2 + \varepsilon)^{-1}|\langle u - vu, \nu \rangle| = (2 + \varepsilon)^{-1}|\langle w, \nu \rangle|$$
$$\geq 2(2 + \varepsilon)^{-1}(\|u\|_B - \varepsilon).$$

We conclude that $u \in B^{(2)}(G)$. $\square$

B. The Space of p-Pseudomeasures of an Amenable Group

If G is a locally compact group, we consider $PM(G) = PM^2(G) = A_2(G)^* = A(G)^*$. As $A(G)$ is an ideal in $B(G)$, we obtain a $B(G)$-module structure on $PM(G)$ by associating to $T \in PM(G)$ and $v \in B(G)$ the element $vT \in PM(G)$ defined by

$$\langle u, vT \rangle = \langle uv, T \rangle,$$

$u \in A(G)$. Then $\|vT\|_{PM} \leq \|T\|_{PM}\|v\|_B$.

Proposition 19.5. *Let G be a locally compact group and let X be an $A(G)$-submodule of $PM(G)$. Then $\overline{A(G)X}$ is a vector space. In particular, if G is amenable and X is closed, then $A(G)X$ is a Banach subspace of $PM(G)$.*

Proof. Let $T_1, T_2 \in X$ and $u_1, u_2 \in A(G)$. For every $\varepsilon > 0$, we may determine $u_1', u_2' \in A(G) \cap \mathcal{K}(G)$ such that $\|(u_1 - u_1')T_1 + (u_2 - u_2')T_2\|_{PM} < \varepsilon$. Let $K = (\operatorname{supp} u_1') \cup (\operatorname{supp} u_2')$. By Lemma 15.21 there exists $u \in A(G)$ such that $u(x) = 1$ whenever $x \in K$. Therefore $u_1'T_1 + u_2'T_2 = u(u_1'T_1 + u_2'T_2) \in A(G)X$ and then also $u_1T_1 + u_2T_2 \in \overline{A(G)X}$. In the particular case, as $A(G)$ admits bounded approximate units by Theorem 10.4 and X is a Banach $A(G)$-module, Cohen's factorization theorem yields that $A(G)X$ is closed ([HR] 32.22). □

Recall that, if G is a locally compact group, $p \in \,]1, \infty[$ and $1/p + 1/p' = 1$, then $\mathcal{L}(L^p(G))$ may be identified with $(L^p(G) \hat{\otimes} L^{p'}(G))^*$ (Section 10B).

Lemma 19.6. *Let G be a locally compact group and $p \in \,]1, \infty[$. If $g, h, k, j \in \mathcal{K}(G)$ and $\varphi \in L^p(G)$ such that $L_\varphi \in Cv^p(G)$, then*

$$\langle \Gamma(j * \check{k})g \otimes h, L_\varphi \rangle = \langle g \otimes h, L_{(k * \check{j})\varphi} \rangle,$$

where $\Gamma f(x, y) = f(xy^{-1})$ for $f : G \times G \to \mathbf{C}$.

Proof. If $x, y \in G$, by (10.2)

$$\Gamma(j * \check{k})(x, y) = \int j_z(x) k_z(y)\, dz$$

so we have

$$\langle \Gamma(j * \check{k})g \otimes h, L_\varphi \rangle = \int \langle \varphi * (j_z g), k_z h \rangle\, dz$$

$$= \iiint \varphi(x) j(x^{-1}yz) g(x^{-1}y) k(yz) h(y)\, dx\, dy\, dz$$

$$= \iiint \varphi(yx) j(x^{-1}z) g(x^{-1}) k(yz) h(y)\, dx\, dy\, dz$$

$$= \iint \left(\int j(x^{-1}z) k(yz)\, dz \right) \varphi(yx) g(x^{-1}) h(y)\, dx\, dy$$

$$= \iint \left(\int j(x^{-1}y^{-1}z) k(z)\, dz \right) \varphi(yx) g(x^{-1}) h(y)\, dx\, dy$$

$$= \iint \left(\int \check{j}(z^{-1}yx) k(z)\, dz \right) \varphi(yx) g(x^{-1}) h(y)\, dx\, dy$$

$$= \iint k * \check{j}(yx) g(x^{-1}) \varphi(yx) h(y)\, dx\, dy$$

$$= \langle ((k * \check{j})\varphi) * g, h \rangle = \langle g \otimes h, L_{(k * \check{j})\varphi} \rangle. \quad \square$$

If G is a locally compact group and $p \in]1, \infty[$, by definition we have $PM^p(G) \subset Cv^p(G)$.

Proposition 19.7. *If G is an amenable group, for every $p \in]1, \infty[$, $Cv^p(G)$ is isometrically isomorphic to $PM^p(G)$.*

Proof. As strong convergence implies ultraweak convergence, by Lemma 18.17 it suffices to show that, for every $\varphi \in L^p(G)$ such that $L_\varphi \in Cv^p(G)$, one has $L_\varphi \in PM^p(G)$.

Let $1/p + 1/p' = 1$. We consider the directed set $I = \Re(G) \times \mathbf{R}_+^*$ where, for $i = (K, \varepsilon) \in I$, $i' = (K', \varepsilon') \in I$, $i \prec i'$ in case $K \subset K'$ and $\varepsilon' < \varepsilon$. Since G is amenable, by Lemma 15.21, given $i = (K, \varepsilon) \in I$, there exist $j \in L^p(G)$, $k \in L^{p'}(G)$, $v_i = k * \check{j} \in A_{p'}(G) \cap \mathcal{K}(G)$ such that $\|v_i\|_{A_{p'}} < 1$ and $v_i(x) = (1 + \varepsilon)^{-1}$ whenever $x \in K$; also $\check{v}_i \in A_p(G)$, $\|\check{v}_i\|_{A_p} = \|v_i\|_{A_{p'}} < 1$ and $v_i \varphi \in L^1(G)$. Lemma 19.6 implies that, for $g, h \in \mathcal{K}(G)$,

$$\langle \Gamma(\check{v}_i) g \otimes h, L_\varphi \rangle = \langle g \otimes h, L_{v_i \varphi} \rangle.$$

So by density, for every $w \in L^p(G) \hat{\otimes} L^{p'}(G)$,

$$\langle \Gamma(\check{v}_i) w, L_\varphi \rangle = \langle w, L_{v_i \varphi} \rangle.$$

Since $(L^p(G) \hat{\otimes} L^{p'}(G))^*$ may be identified with $\mathcal{L}(L^p(G))$, we have $\|L_{v_i \varphi}\|_{Cv^p} = \sup \{|\langle \Gamma(\check{v}_i) w, L_\varphi \rangle| : w \in L^p(G) \hat{\otimes} L^{p'}(G), \|w\|_{\hat{\otimes}} \leq 1\}$. By (10.4), for every $w \in L^p(G) \hat{\otimes} L^{p'}(G)$,

$$|\langle \Gamma(\check{v}_i) w, L_\varphi \rangle| \leq \|L_\varphi\|_{Cv^p} \|\Gamma(\check{v}_i) w\|_{\hat{\otimes}} \leq \|L_\varphi\|_{Cv^p} \|\check{v}_i\|_{A_p} \|w\|_{\hat{\otimes}}.$$

Hence $\|L_{v_i \varphi}\|_{Cv^p} \leq \|L_\varphi\|_{Cv^p}$.

With respect to Proposition 10.3 it suffices to show that $(L_{v_i \varphi})$ converges ultraweakly to L_φ in $Cv^p(G)$. Since $\sup \{\|L_{v_i \varphi}\|_{Cv^p} : i \in I\} \leq \|L_\varphi\|_{Cv^p}$, we have to establish weak convergence (Section 2B), that is, for all $f, g \in \mathcal{K}(G)$,

$$\lim_i \langle (v_i \varphi) * f, g \rangle = \langle \varphi * f, g \rangle.$$

Since (v_i) converges to 1_G uniformly on compacta and $\|v_i\| < 1$ whenever $i \in I$, we have $\lim_i \|v_i \varphi - \varphi\|_p = 0$. So $\lim_i \langle (v_i \varphi - \varphi) * f, g \rangle = 0$ whenever $f, g \in \mathcal{K}(G)$. □

C. The Multiplier Algebras of an Amenable Group

Proposition 19.8. *Let G be an amenable group and $p \in]1, \infty[$. Then the Banach algebra $A_p(G)$ is a (closed) ideal in the Banach algebra $B_p(G)$.*

Proof. By the very definition of $B_p(G)$, $A_p(G)$ is an ideal. Let $u \in A_p(G)$. We have

$$\|u\|_{B_p} = \sup\{\|uv\|_{A_p} : v \in A_p(G), \|v\|_{A_p} \leq 1\}$$

$$\leq \sup\{\|u\|_{A_p}\|v\|_{A_p} : v \in A_p(G), \|v\|_{A_p} \leq 1\} \leq \|u\|_{A_p}.$$

By Theorem 10.4 there exists a net $(u_i)_{i \in I}$ of approximate units bounded by 1 in $A_p(G)$. For every $i \in I$, $\|uu_i\|_{A_p} \leq \|u\|_{B_p}\|u_i\|_{A_p} \leq \|u\|_{B_p}$, hence also $\|u\|_{A_p} \leq \|u\|_{B_p}$. □

Recall that if G is a locally compact group and $p \in {]}1, \infty{[}$, then $C_p(G) \subset B_p(G)$ and $\|\cdot\|_{B_p} \leq \|\cdot\|_{C_p}$ (Section 10B).

Proposition 19.9. *If G is an amenable group and $p \in {]}1, \infty{[}$, then $B_p(G)$ is isometrically isomorphic to $C_p(G)$.*

Proof. Let $v \in B_p(G)$. If $\varphi \in L^p(G) \hat{\otimes} L^{p'}(G)$ $(1/p + 1/p' = 1)$, for every $\varepsilon > 0$, there exists $\psi \in \mathcal{K}(G \times G) \cap (L^p(G) \hat{\otimes} L^{p'}(G))$ such that $\|\varphi - \psi\|_{\hat{\otimes}} < \varepsilon$. Let K be a compact subset of G such that $xy^{-1} \in K$ whenever $(x, y) \in \mathrm{supp}\,\psi$. Since G is amenable, by Lemma 15.21 there exists $u \in A_p(G)$ such that $\|u\|_{A_p} < 1 + \varepsilon$ and $u(x) = 1$ whenever $x \in K$. We have $vu \in A_p(G)$ and, by Proposition 10.2, $vu \in C_p(G)$; for the mapping Γ considered in Section 10B, $\Gamma(vu)\psi \in L^p(G) \hat{\otimes} L^{p'}(G)$. As $\Gamma(u)(x, y) = u(xy^{-1}) = 1$ whenever $(x, y) \in \mathrm{supp}\,\psi$, we have

$$\Gamma(vu)\psi = \Gamma(v)\Gamma(u)\psi = \Gamma(v)\psi.$$

Then by Proposition 10.2 again

$$\|\Gamma(v)\psi\|_{\hat{\otimes}} = \|\Gamma(vu)\psi\|_{\hat{\otimes}} \leq \|vu\|_{C_p}\|\psi\|_{\hat{\otimes}} \leq \|vu\|_{A_p}\|\psi\|_{\hat{\otimes}}$$

$$\leq \|v\|_{B_p}\|u\|_{A_p}\|\psi\|_{\hat{\otimes}} \leq (1 + \varepsilon)\|v\|_{B_p}\|\psi\|_{\hat{\otimes}}$$

and

$$\|\Gamma(v)\varphi\|_{\hat{\otimes}} \leq \|\Gamma(v)(\varphi - \psi)\|_{\hat{\otimes}} + \|\Gamma(v)\psi\|_{\hat{\otimes}}$$

$$\leq \|v\|_{C_p}\|\varphi - \psi\|_{\hat{\otimes}} + (1 + \varepsilon)\|v\|_{B_p}\|\psi\|_{\hat{\otimes}}$$

$$\leq \|v\|_{C_p}\varepsilon + (1 + \varepsilon)\|v\|_{B_p}(\|\varphi\|_{\hat{\otimes}} + \varepsilon).$$

As $\varepsilon > 0$ may be chosen arbitrarily, we obtain

$$\|\Gamma(v)\varphi\|_{\hat{\otimes}} \leq \|v\|_{B_p}\|\varphi\|_{\hat{\otimes}},$$

that is, $v \in C_p(G)$ and $\|v\|_{C_p} \leq \|v\|_{B_p}$. □

Comments

Definition 19.10. *Let G be a locally compact group and let $p \in]1, \infty[$. We denote by $PF^p(G)$ the Banach space of p-pseudofunctions on G, that is, the closure of $\{L_f : f \in L^1(G)\}$ in $Cv^p(G)$. We also define $W^p(G) = PF^p(G)^*$.*

Via proposition 10.3 we have $PF^p(G) \subset PM^p(G) = A_p(G)^*$ and $W^p(G) \subset L^\infty(G)$. For any compact subset K of G, we consider the space $A_p(K) = \{f|_K : f \in A_p(G)\}$ with the induced norm. From general considerations on operator norms Cowling ([105] Theorem 4) deduced the following result: Let $v \in L^\infty(G)$; then $v \in W^p(G)$, $\|v\|_{W^p} \leq c$ if and only if, for every compact subset K of G, $v|_K \in A_p(K)$ and $\|v|_K\|_{A_p(K)} \leq c$. Therefore, as $A_p(G)$ is a Banach algebra, so is $W^p(G)$.

Proposition 19.11. *Let G be a locally compact group and $p \in]1, \infty[$.*
(a) *We have $W^p(G) \subset B_p(G)$ and $\|v\|_{B_p} \leq \|v\|_{W^p}$ whenever $v \in W^p(G)$.*
(b) *$W^p(G) \simeq B_p(G)$ if and only if G is amenable.*

Proof. (a) Let $v \in W^p(G)$. If $u \in A^p(G) \cap \mathcal{K}(G)$ with compact support K, $(uv)|_K = u|_K \cdot v|_K \in A_p(K)$; also $\|(uv)|_K\|_{A_p(K)} \leq \|u\|_{A_p}\|v\|_{W^p}$. By density we conclude that $uv \in A_p(G)$ whenever $u \in A_p(G)$; $v \in B_p(G)$ and $\|v\|_{B_p} \leq \|v\|_{W^p}$.

(b) Assume G to be amenable. By Lemma 15.21, given $\alpha > 1$, for every compact subset K of G, we may determine $u_K \in A_p(G)$ such that $\|u_K\|_{A_p} \leq \alpha$ and $u_K(x) = 1$ whenever $x \in K$. Let $v \in B_p(G)$; then $vu_K \in A_p(G)$, $v|_K = (vu_K)|_K \in A_p(K)$, and $\|v|_K\|_{A_p(K)} \leq \|v\|_{B_p}\alpha$. Therefore $v \in W^p(G)$ and $\|v\|_{W^p} \leq \|v\|_{B_p}\alpha$. Since $\alpha > 1$ may be chosen arbitrarily, we have $\|v\|_{W^p} \leq \|v\|_{B_p}$ and then, by (a), $\|v\|_{W^p} = \|v\|_{B_p}$.

Assume G to be nonamenable and let

$$\beta = \sup\{\|1_G|_K\|_{A_p(K)} : K \in \mathfrak{K}(G)\}.$$

If we had $\beta < \infty$, then to every $K \in \mathfrak{K}(G)$ we could associate $u_K \in A_p(G)$ such that $u_K(x) = 1$ whenever $x \in K$ and $\|u_K\|_{A_p} \leq 2\beta$; $(u_K)_{K \in \mathfrak{K}(G)}$ would constitute a net of bounded approximate units, directed by $\subset$, in $A_p(G)$. Therefore G would be amenable by Theorem 10.4. We must have $\beta = \infty$ and then $1_G \notin W^p(G)$. But obviously $1_G \in B_p(G)$. □

Corollary 19.12. *If G is an amenable group, $B_2(G) \simeq B(G)$.*

Proof. By definition $C_L^*(G) = PF^2(G)$, so Theorem 8.9 implies that $C^*(G) = PF^2(G)$. Now notice that $B(G)$ is the dual of $C^*(G)$, whereas, by Proposition 19.11, $B_2(G)$ is the dual of $PF^2(G)$. □

Definition 19.13. *If G is a locally compact group, let $Q(G)$ be the Banach space consisting of all functions $h = \sum_{i=1}^{\infty} u_i g_i$, where $u_i \in A(G)$, $g_i \in \mathscr{C}_0(G)$,*

$i \in \mathbf{N}^*$ such that $\sum_{i=1}^{\infty} \|u_i\|_A \|g_i\| < \infty$ with

$$\|h\| = \inf \left\{ \sum_{i=1}^{\infty} \|u_i\|_A \|g_i\| : u_i \in A(G), g_i \in \mathscr{C}_0(G), i \in \mathbf{N}^*; h = \sum_{i=1}^{\infty} u_i g_i \right\}.$$

We have $Q(G) \subset \mathscr{C}_0(G)$. The dual of $Q(G)$ consists of all $A(G)$-multipliers of the Banach $A(G)$-module $\mathscr{C}_0(G)$ into the Banach $A(G)$-module $PM(G)$, that is, all $\Phi \in \mathscr{L}(\mathscr{C}_0(G), PM(G))$ such that $\Phi(uf) = u\Phi(f)$ whenever $f \in \mathscr{C}_0(G)$ and $u \in A(G)$, the duality being ruled by

$$\langle h, \Phi \rangle = \sum_{i=1}^{\infty} \langle u_i, \Phi(g_i) \rangle$$

for $h \in Q(G)$, $\Phi \in Q(G)^*$ with $h = \sum_{i=1}^{\infty} u_i g_i$, $u_i \in A(G)$, $g_i \in \mathscr{C}_0(G)$, $i \in \mathbf{N}^*$; $Q(G)^*$ may be identified with the subspace of $PM(G)$ consisting of all $T \in PM(G)$ such that $\|T\|_{Q^*} = \sup\{\|uT\|_{PM} : u \in A(G), \|u\| \leq 1\} < \infty$ ([413]). As $A(G)^* = PM(G)$ and $\mathscr{C}_0(G)^* = \mathscr{M}^1(G)$, $Q(G)^*$ is also isomorphic to the space of all $A(G)$-multipliers of $A(G)$ into $\mathscr{M}^1(G)$.

Proposition 19.14. *If G is a locally compact group, the following properties are equivalent:*

(i) G is amenable.

(ii) For every $f \in \mathscr{C}_0(G)$, there exist $u \in A(G)$ and $g \in \mathscr{C}_0(G)$ such that $f = ug$.

(iii) $\mathscr{M}^1(G)$ is isomorphic to the space of all $A(G)$-multipliers of $A(G)$ into $\mathscr{M}^1(G)$.

(iv) $L^1(G)$ is isomorphic to the space of all $A(G)$-multipliers of $A(G)$ into $L^1(G)$.

Proof. (i) $\Rightarrow$ (ii)

As $A(G)$ admits bounded approximate units by Theorem 10.4 and $\mathscr{C}_0(G)$ is a Banach $A(G)$-module, Cohen's factorization theorem ([HR] 32.22) yields (ii).

(ii) $\Rightarrow$ (iii)

By hypothesis, the sets $Q(G)$ and $\mathscr{C}_0(G)$ coincide. If $f \in \mathscr{C}_0(G)$, $\|f\| \leq \|f\|_Q$. Therefore $Q(G) \simeq \mathscr{C}_0(G)$ ([HS] 14.18). Thus also $\mathscr{M}^1(G) = \mathscr{C}_0(G)^* = Q(G)^*$.

(iii) $\Rightarrow$ (iv)

For every $A(G)$-multiplier Φ of $A(G)$ into $L^1(G)$, there exists $\mu \in \mathscr{M}^1(G)$ such that $\Phi(u) = u\mu \in L^1(G)$ whenever $u \in A(G)$. Therefore we have $\mu \in L^1(G)$.

(iv) $\Rightarrow$ (i)

We establish (Cv_2'). Amenability of G then follows from Theorem 9.6.

By hypothesis there exists $c > 0$ such that $\|\varphi\|_1 \leq c\|L_\varphi\|_{Q(G)^*}$ whenever $\varphi \in L_+^1(G)$. For every $u \in A(G)$, $\|uL_\varphi\|_{Cv^2} \leq \|u\|_A \|L_\varphi\|_{Cv^2} \leq \|u\| \, \|L_\varphi\|_{Cv^2}$; hence $\|L_\varphi\|_{Q(G)^*} \leq \|L_\varphi\|_{Cv^2}$ and $\|\varphi\|_1 \leq c\|L_\varphi\|_{Cv^2}$.

If $f \in L_+^1(G)$, we have

$$\|f * f^*\|_1 = \iint f(y) f^*(y^{-1}x) \, dy \, dx = \iint f(y) f(x^{-1}y) \Delta(x^{-1}y) \, dx \, dy$$

$$= \int f(y) \left(\int f(xy) \Delta(y) \, dx \right) dy = \|f\|_1^2.$$

Therefore $\|f\|_1 = \|f * f^*\|_1^{1/2} \leq c^{1/2} \|L_{f * f^*}\|_{Cv^2}^{1/2} = c^{1/2} \|L_f\|_{Cv^2} \leq c^{1/2} \|f\|_1$, and by induction we obtain $\|f\|_1 \leq c^{1/2^n} \|L_f\|_{Cv^2}$ for every $n \in \mathbf{N}^*$. Then

$$\|f\|_1 \leq \lim_{n \to \infty} c^{1/2^n} \|L_f\|_{Cv^2} = \|L_f\|_{Cv^2}$$

and $\|L_f\|_{Cv^2} = \|f\|_1$. □

If G is a locally compact abelian group, the Fourier transform $\mathscr{F}$ on $L^1(\hat{G})$ defines an isometric isomorphism of $L^1(\hat{G})$ onto $A(G) = A_2(G)$ ([HR] Section 31, [171] 3.6), so $\mathscr{F}$ induces an isometric isomorphism $\mathscr{F}^\sim$ of $A(G)^* = PM(G)$ onto $L^1(\hat{G})^* = L^\infty(\hat{G})$ via the duality formula

$$\langle \mathscr{F}f, T \rangle = \langle f, \mathscr{F}^\sim T \rangle,$$

$f \in L^1(\hat{G})$, $T \in PM(G)$. Let $v \in A(G)$, $T \in PM(G)$ with $v = \mathscr{F}g$, and $g \in L^1(\hat{G})$. For every $f \in L^1(\hat{G})$, by (2.12) we have

$$\langle f, \mathscr{F}^\sim(vT) \rangle = \langle \mathscr{F}f, vT \rangle = \langle \mathscr{F}f, (\mathscr{F}g)T \rangle = \langle (\mathscr{F}f)(\mathscr{F}g), T \rangle$$

$$= \langle \mathscr{F}(g *_{\hat{G}} f), T \rangle = \langle g *_{\hat{G}} f, \mathscr{F}^\sim T \rangle = \langle f, \overline{g^*} *_{\hat{G}} \mathscr{F}^\sim T \rangle$$

and $\mathscr{F}^\sim(vT) = \overline{g^*} *_{\hat{G}} \mathscr{F}^\sim T$. Since $\mathscr{F}^\sim$ is an isometric isomorphism we conclude that in case G is abelian, $A(G)PM(G)$ is isometrically isomorphic to $L^1(\hat{G}) *_{\hat{G}} L^\infty(\hat{G})$, hence to $\mathscr{RUC}(\hat{G}) = \mathscr{UC}(\hat{G})$ ([HR] 32.45b). These considerations suggest the following general definition.

Definition 19.15. *If G is a locally compact group, we define $\mathscr{UC}(\hat{G})$ to be the closure of $A(G)PM(G)$ in $PM(G)$.*

In case G is amenable, by Proposition 19.5, $\mathscr{UC}(\hat{G}) = A(G)PM(G)$.

Proposition 19.16. *If G is an amenable group, $C^*(G) \subset \mathscr{UC}(\hat{G})$. There exists a linear isometry $\Theta: f \mapsto f'$ of $\mathscr{UC}(\hat{G})$ into a closed subspace of $C^*(G)^{**}$*

defined by $\langle v, f' \rangle = \langle uv, T \rangle$ for $f = uT$, $u \in A(G)$, $T \in PM(G)$ and $v \in B(G)$; the isometry extends the canonical embedding of $C^*(G)$ into its bidual space.

Proof. (a) Let $\varphi \in \mathcal{K}(G)$ with $K = \text{supp}\,\varphi$. By Lemma 15.21 choose $u \in A(G)$ such that $u(x) = 1$ whenever $x \in K$. For every $v \in A(G)$, by Proposition 10.3 we have

$$\langle v, L_\varphi \rangle = \int v(x)\varphi(x)\,dx = \int u(x)v(x)\varphi(x)\,dx = \langle v, uL_\varphi \rangle$$

hence $L_\varphi = uL_\varphi \in A(G)PM(G) = \mathcal{UC}(\hat{G})$. Thus we proved that $C_L^*(G) \subset \mathcal{UC}(\hat{G})$. As G is amenable, by Theorem 8.9 we have then $C^*(G) \subset \mathcal{UC}(\hat{G})$.

(b) We show that, if $f = uT \in \mathcal{UC}(\hat{G})$, the element f' of $B(G)^* = C^*(G)^{**}$ is well defined. As a matter of fact, let $f = u_1 T_1 = u_2 T_2$ with $u_1, u_2 \in A(G)$ and $T_1, T_2 \in PM(G)$. For every $v \in A(G)$,

$$\langle u_1 v, T_1 \rangle = \langle v, u_1 T_1 \rangle = \langle v, u_2 T_2 \rangle = \langle u_2 v, T_2 \rangle.$$

If $v \in P(G)$, let (v_i) be a net in $P(G) \cap \mathcal{K}(G)$ converging to v in the compact-open topology; then also $\lim_i \|w(v_i - v)\|_A = 0$ whenever $w \in A(G)$ ([384] Theorem 5.5, [340] Theorem 3.2, [225] Theorem A, Corollary 1). So we have

$$\langle u_1 v, T_1 \rangle = \lim_i \langle u_1 v_i, T_1 \rangle = \lim_i \langle u_2 v_i, T_1 \rangle = \langle u_2 v, T_2 \rangle.$$

Recall then that $P(G)$ spans $B(G)$. So Θ is well defined; the mapping is obviously linear.

(c) We show that Θ is an isometry.
Let $f = uT \in \mathcal{UC}(\hat{G})$ with $u \in A(G)$, $T \in PM(G)$. Then

$$\|f'\| = \|\Theta(f)\| = \sup\{|\langle v, f' \rangle| : v \in B(G), \|v\|_B \le 1\}$$

$$= \sup\{|\langle uv, T \rangle| : v \in B(G), \|v\|_B \le 1\}$$

$$\ge \sup\{|\langle v, uT \rangle| : v \in A(G), \|v\|_A \le 1\} = \|uT\|_{PM} = \|f\|_{PM}.$$

On the other hand, as by Theorem 10.4 amenability of G implies the existence of approximate units bounded by 1 in $A(G)$, given $\varepsilon > 0$, there exist $u_0 \in A(G)$, $T_0 \in PM(G)$ such that $uT = u_0 T_0$, $\|u_0\|_A \le 1$, $\|f - T_0\|_{PM} < \varepsilon$ ([HR] 32.50). Thus for all $v \in B(G)$ with $\|v\|_B \le 1$, we have

$$|\langle v, f' \rangle| = |\langle u_0 v, T_0 \rangle| \le \|u_0 v\|_A \|T_0\|_{PM} \le \|f\|_{PM} + \varepsilon.$$

As $\varepsilon > 0$ is arbitrary, $|\langle v, f' \rangle| \le \|f\|_{PM}$, hence $\|f'\| \le \|f\|_{PM}$.

(d) Let $\psi \in \mathcal{K}(G)$ with $K = \text{supp}\,\psi$ and let $f = L_\psi$. By Lemma 15.21 choose $u_0 \in A(G)$ such that $u_0(x) = 1$ whenever $x \in K$. As in (a), we have

$L_\psi = u_0 L_\psi \in A(G)PM(G) = \mathscr{U}\mathscr{C}(\hat{G})$ and, for every $v \in B(G)$,

$$\langle v, f' \rangle = \langle u_0 v, L_\psi \rangle = \langle v, L_\psi \rangle = \langle v, f \rangle.$$

By density we conclude that the restriction of Θ to $C_L^*(G) = C^*(C)$ is the canonical embedding. □

Proposition 19.17. *If G is an amenable group, there exists an isometric isomorphism of $\mathscr{U}\mathscr{C}(\hat{G})^*$ onto the subspace of $\mathscr{L}(PM(G))$ formed by all $A(G)$-multipliers of $PM(G)$.*

Proof. To $\Phi \in \mathscr{U}\mathscr{C}(\hat{G})^*$ we associate $F(\Phi) \in \mathscr{L}(PM(G))$ defined by

$$\langle u, F(\Phi)(T) \rangle = \langle uT, \Phi \rangle,$$

$u \in A(G)$, $T \in PM(G)$. For all $u \in A(G)$, $v \in A(G)$, $T \in PM(G)$, we have

$$\langle u, vF(\Phi)(T) \rangle = \langle uv, F(\Phi)(T) \rangle = \langle uvT, \Phi \rangle = \langle u, F(\Phi)(vT) \rangle,$$

that is, $vF(\Phi)(T) = F(\Phi)(vT)$. Furthermore

$$\|F(\Phi)\| = \sup\{\|F(\Phi)(T)\|_{PM} : T \in PM(G), \|T\|_{PM} \leq 1\}$$

$$= \sup\{|\langle u, F(\Phi)T \rangle| : T \in PM(G), \|T\|_{PM} \leq 1, u \in A(G), \|u\|_A \leq 1\}$$

$$= \sup\{|\langle uT, \Phi \rangle| : T \in PM(G), \|T\|_{PM} \leq 1, \|u\|_A \leq 1\}$$

$$= \sup\{|\langle S, \Phi \rangle| : S \in \mathscr{U}\mathscr{C}(\hat{G}), \|S\| \leq 1\} = \|\Phi\|.$$

The mapping F is injective as, if $\Phi \in \mathscr{K}\!\mathit{er}\, F$, then for all $T \in PM(G)$ and $u \in A(G)$,

$$\langle uT, \Phi \rangle = \langle u, F(\Phi)T \rangle = 0$$

hence $\Phi = 0$. If Ψ is an $A(G)$-multiplier in $\mathscr{L}(PM(G))$, let

$$\langle uT, \Phi \rangle = \langle u, \Psi(T) \rangle$$

for $u \in A(G)$, $T \in PM(G)$; then $\Phi \in \mathscr{U}\mathscr{C}(\hat{G})^*$, $F(\Phi) = \Psi$, that is, F is surjective. □

We now indicate some *spectral synthesis* properties related to amenability. We first recall some important general definitions ([B TS] Chap. I, Section 5; [HR] Section 39) restricting ourselves to Banach algebras of continuous bounded functions defined on a locally compact space, the algebraic operations being pointwise addition and multiplication.

Let X be a locally compact space and let A be a commutative Banach algebra formed by elements of $\mathscr{C}(X)$. We assume the following conditions to hold: (a) If F is any continuous multiplicative linear functional on A admitting a support reduced to a singleton $\{x\}$, then $F = \delta_x$; (b) for every compact subset K of X and every closed subset E of X disjoint from K, there exists $f \in A$ such that $f(K) = \{1\}$ and $f(E) = \{0\}$ (A is *regular*); (c) $\mathscr{K}(X) \cap A$ is dense in A (A is *tauberian*). Then A is said to be a *regular tauberian algebra of functions* on X. The *kernel* of a closed subset E in X is defined to be the closed ideal k^E of all $f \in A$ such that $f(x) = 0$ whenever $x \in E$. Let also k_c^E be the set of all $f \in A$ having compact support disjoint from E. The closed subset E of X is called *spectral set* or *set of spectral synthesis* in case $\overline{k_c^E} = k^E$. The algebra A is said to satisfy Ditkin's condition at $x \in X$ if, for every $f \in A$ with $f(x) = 0$, there exists a sequence $(f_n)_{n \in \mathbf{N}^*}$ in A such that $\lim_{n \to \infty} \|ff_n - f\|_A = 0$ and, for every $n \in \mathbf{N}^*$, f_n vanishes on a neighborhood of x; A is said to satisfy Ditkin's condition at infinity if, for every $f \in A$, there exists a sequence $(f_n)_{n \in \mathbf{N}^*}$ in A such that $\lim_{n \to \infty} \|ff_n - f\|_A = 0$ and, for every $n \in \mathbf{N}^*$, f_n has compact support. If Ditkin's conditions hold at each point and at infinity, we say that A satisfies *Ditkin's conditions*; then every closed subset E of X, such that the boundary of E does not contain a nonvoid perfect subset, is a spectral set ([HR] 39.26).

If G is a locally compact group and $p \in]1, \infty[$, $A_p(G)$ is a regular tauberian algebra of functions on G ([259] Proposition 3).

Lemma 19.18. *Let G be a locally compact group admitting a closed subgroup H. If K is any compact subset of H and U is any compact neighborhood of e in G, there exists $c = c_{K,U} \in \mathbf{R}_+^*$ such that to every neighborhood V of e in G there corresponds a relatively compact open neighborhood V' of e in G for which $V' \subset U$, $HV' \subset HV$, and $|KV'| \leq c|V'|$.*

Proof. There exists an open neighborhood W_1 of e in G and a compact neighborhood W_2 of e in H such that $W_2 W_1 \subset U$. As KW_2 is compact in H, there exist $n = n_{K,U} \in \mathbf{N}^*$ and $a_1, \ldots, a_n \in H$ such that $KW_2 \subset \bigcup_{i=1}^n a_i W_2$. Consider the neighborhood $V' = W_2 W_1 \cap H\mathring{V}$ of e in G; $HV' \subset HV$. As $V' \subset U$, V' is relatively compact. We have

$$KV' \subset \left(\bigcup_{i=1}^n a_i W_2 W_1 \right) \cap H\mathring{V} \subset \bigcup_{i=1}^n a_i V'.$$

Hence $|KV'| \leq n|V'|$. □

Proposition 19.19. *Let G be an amenable group and let $p \in]1, \infty[$. Every closed subgroup H of G is a spectral set for $A_p(G)$.*

Proof. Let $f \in A_p(G)$, $f \neq 0$ such that $f|_H = 0$ and let $\varepsilon \in \mathbf{R}_+^*$. As G is amenable, by Theorem 10.4 there exists $u \in A_p(G) \cap \mathscr{K}(G)$ such that $\|fu - f\|_{A_p} < \varepsilon$. We put $g = fu \in A_p(G) \cap \mathscr{K}(G)$; $g|_H = 0$ and $\|g - f\|_{A_p} < \varepsilon$.

If $g = 0$, we have $\inf \{\|f - h\|_{A_p} : h \in k_c^H\} < \varepsilon$; so $f \in \overline{k_c^H}$.

Assume now that $g \neq 0$ and let $\delta \in]0, \|g\|[$. We consider a compact neighborhood U of e in G such that $\|g_y - g\|_{A_p} < \varepsilon$ whenever $y \in U$. Consider also the compact subset $K = ((\operatorname{supp} g)U^{-1}) \cap H$ of H, and let $c_{K,U} \in \mathbf{R}_+^*$ be associated by Lemma 19.18. If V is a neighborhood of e such that $\|g_y - g\|_{A_p} < \delta$ whenever $y \in V$, let V' be associated to V by Lemma 19.18. We put

$$g'(x) = g(x) \quad \text{if } x \in HV'$$
$$= 0 \quad \text{if } x \in G \setminus HV'$$

and define $g'' = (g - g') * \check{\xi}_{V'} \in A_p(G)$. For every $a \in G$, $g''(a) = \int (g - g')(ax)\xi_{V'}(x) \, dx = 0$ whenever $aV' \subset HV'$; so $g'' \in k_c^H$. For every $x \in G$,

$$(g - g * \check{\xi}_{V'})(x) = g(x) - \int g(xy)\xi_{V'}(y) \, dy = \int (g(x) - g(xy))\xi_{V'}(y) \, dy;$$

as $V' \subset U$, $\|g - g * \check{\xi}_{V'}\|_{A_p} < \varepsilon$. If $1/p + 1/p' = 1$,

$$\|g' * \check{\xi}_{V'}\|_{A_p} \leq \|g'\|_p \|\xi_{V'}\|_{p'} = \left(\int_{HV'} |g|^p\right)^{1/p} |V'|^{-1/p}.$$

As $g - g'' = (g - g * \check{\xi}_{V'}) + g' * \check{\xi}_{V'}$ and $g'' \in k_c^H$, we obtain then

$$\inf\{\|g - h\|_{A_p} : h \in k_c^H\} < \varepsilon + \left(\int_{HV'} |g|^p\right)^{1/p} |V'|^{-1/p}.$$

As $HV' \subset HV$, given $x \in H$ and $y \in V'$, there exist $x' \in H$, $y' \in V$ such that $xy = x'y'$; so $|g(xy)| = |g(x'y') - g(x')| < \delta$ and we have $(\int_{HV'} |g|^p)^{1/p} < \delta |E|^{1/p}$, where $E = HV' \cap \operatorname{supp} g$. If $a \in E$, there exist $b \in H$, $b' \in V'$ such that $a = bb' \in \operatorname{supp} g$; so $b = ab'^{-1} \in (\operatorname{supp} g)V'^{-1} \subset (\operatorname{supp} g)U^{-1}$, $b \in K$ and $E \subset KV'$, $|E| \leq c_{K,U}|V'|$. Therefore

$$\inf\{\|g - h\|_{A_p} : h \in k_c^H\} < \varepsilon + \delta c_{K,U}^{1/p}$$

and

$$\inf\{\|f - h\|_{A_p} : h \in k_c^H\} < 2\varepsilon + \delta c_{K,U}^{1/p}.$$

As $\varepsilon \in \mathbf{R}_+^*$ and $\delta \in]0, \|g\|[$ are arbitrary, we conclude that $f \in \overline{k_c^H}$. □

Finally, if G is a locally compact group and $p \in]1, \infty[$, we consider a particular subalgebra of $A_p(G)$. Note that $uv \in L^1(G)$ whenever $u \in A_p(G) \subset \mathscr{C}_0(G)$ and $v \in L^1(G)$. So the following definition makes sense.

FOURIER ALGEBRA PROPERTIES

Definition 19.20. *Let G be a locally compact group and let $p \in \,]1, \infty[$. We denote by $A_p^1(G)$ the Banach algebra formed by $L^1(G) \cap A_p(G)$ with the norm $\|\cdot\|_{1,p} = \|\cdot\|_1 + \|\cdot\|_{A_p}$.*

Proposition 19.21. *Let G be a locally compact group and let $p \in \,]1, \infty[$. The subalgebra $A_p^1(G)$ of $A_p(G)$ is proper if and only if G is noncompact.*

Proof. (a) If G is compact, $\mathscr{C}_0(G) \subset L^1(G)$; so $A_p(G) \subset L^1(G)$.

(b) Assume G to be noncompact. Then there exist a compact neighborhood U of e in G and an infinite sequence (a_n) in G such that $(a_iU) \cap (a_jU) = \varnothing$ for all $i, j \in \mathbf{N}^*$, $i \neq j$. For every $n \in \mathbf{N}^*$,

$$\left\| \sum_{k=1}^{n} \frac{1}{k} 1_{a_kU} \right\|_p = \left(\int \left(\sum_{k=1}^{n} \frac{1}{k^p} 1_{a_kU} \right) \right)^{1/p} = \left(\sum_{k=1}^{n} \frac{1}{k^p} \right)^{1/p} |U|^{1/p}.$$

As the series $\sum_{k=1}^{\infty} 1/k^p$ converges, we conclude that $u = (\sum_{k=1}^{\infty} (1/k) 1_{a_kU}) * \check{1}_U \in A_p(G)$. On the other hand, for every $n \in \mathbf{N}^*$,

$$\left\| \left(\sum_{k=1}^{n} \frac{1}{k} 1_{a_kU} \right) * \check{1}_U \right\|_1 = \sum_{k=1}^{n} \frac{1}{k} \iint 1_{a_kU}(y) 1_U(x^{-1}y) \, dy \, dx$$

$$= \sum_{k=1}^{n} \frac{1}{k} \int \left(\int 1_{yU^{-1}}(x) \, dx \right) 1_{a_kU}(y) \, dy = \sum_{k=1}^{n} \frac{1}{k} |U^{-1}| \, |U|.$$

As the series $\sum_{k=1}^{\infty} 1/k$ diverges, $u \notin L^1(G)$. □

Proposition 19.22. *If G is a (noncompact) amenable group and $p \in \,]1, \infty[$, any net of approximate units in $A_p(G) \cap \mathscr{K}(G)$ for $A_p(G)$, bounded by 1, constitutes a net of (unbounded) units for $A_p^1(G)$.*

Proof. Let $(u_i)_{i \in I}$ be a net of approximate units in $A_p(G) \cap \mathscr{K}(G)$ for $A_p(G)$ bounded by 1.

(a) Given $u \in A_p^1(G)$ and $\varepsilon > 0$, there exists $i_0 \in I$ such that

$$\|uu_i - u\|_{A_p} < \frac{\varepsilon}{2} \tag{4}$$

whenever $i \in I$, $i \succ i_0$. For every $i \in I$, $u_i \in \mathscr{K}(G)$, so $uu_i - u \in L^1(G)$; $\lim_i \|uu_i - u\| \leq \lim_i \|uu_i - u\|_{A_p} = 0$ and there exists a compact subset K in G such that $|K| > 0$ and

$$\int_{G \setminus K} |uu_i - u| < \frac{\varepsilon}{4}$$

whenever $i \succ i_0$. By Lemma 15.21 we may determine $v \in A_p(G)$ such that

$v(x) = 1$ whenever $x \in K$. For all $x \in K$ and $i \in I$,

$$|u_i(x) - 1| = |u_i(x)v(x) - v(x)| \leq \|u_iv - v\| \leq \|u_iv - v\|_{A_p}$$

and there exists $i_1 \succ i_0$ such that, for all $x \in K$ and $i \succ i_1$,

$$|u_i(x) - 1| < \varepsilon(4|K|(\|u\| + 1))^{-1}.$$

Hence

$$\int_K |uu_i - u| = \int_K |u_i(x) - 1||u(x)| \, dx < \frac{\varepsilon}{4}$$

and then

$$\int_G |uu_i - u| < \frac{\varepsilon}{2}. \tag{5}$$

By (4) and (5) we obtain $\|uu_i - u\|_{1,p} < \varepsilon$ for $i \in I$, $i \succ i_1$. Therefore (u_i) constitutes a net of approximate units in $A_p^1(G)$.

(b) Let $\alpha \in \,]0,1[$. If C is any compact subset of G, by Lemma 15.21 we choose $w \in A_p(G)$ such that $w(x) = 1$ whenever $x \in C$. There exists $j \in I$ such that, for every $i \in I$, $i \succ j$ and every $x \in C$,

$$|u_i(x) - 1| = |u_i(x)w(x) - w(x)| \leq \|u_iw - w\| \leq \|u_iw - w\|_{A_p} < \alpha.$$

Hence $|u_i(x)| > 1 - \alpha$ and

$$\int_C |u_i(x)| \geq (1 - \alpha)|C|.$$

Thus the net (u_i) cannot admit a bound in $A_p^1(G)$ in case G is noncompact. □

If G is any locally compact group and $p \in \,]1, \infty[$, $A_p^1(G)$ is a regular tauberian algebra of functions on G ([329] Theorem 2.5).

Lemma 19.23. *Let G be an amenable group and let $p \in \,]1, \infty[$. If $u \in A_p^1(G)$ and $x_0 \in G$ with $u(x_0) = 0$, there exists a net (v_j) in $A_p^1(G)$ such that $\lim_j \|uv_j\|_{1,p} = 0$ and, for every $j \in J$, there exists a compact neighborhood V_j of x_0 such that $v_j(x) = 1$ whenever $x \in V_j$.*

Proof. For every $v \in A_p^1(G)$ and every $a \in G$, $_av \in A_p^1(G)$ and $\|_av\|_{1,p} = \|v\|_{1,p}$, hence we may assume that $x_0 = e$.

Let $(U_j)_{j \in J}$ be a system of neighborhoods of e in G such that $|U_j| \leq 1$ for every $j \in J$. As $A_p^1(G) \subset A_p(G) \subset \mathscr{C}_0(G)$, given $\varepsilon > 0$, there exists a neighbor-

FOURIER ALGEBRA PROPERTIES

hood W of e in G such that $|u(x) - u(e)| < \varepsilon/2$ whenever $x \in W$. For every $j \in J$, we consider the neighborhood $U_j' = U_j \cap W$ of e. The regularity of the Haar measure on G implies the existence of a compact symmetric neighborhood W_j of e and a compact neighborhood V_j of e such that $V_j W_j^2 \subset U_j'$ and $|V_j W_j| < 2|W_j|$, $j \in J$. If $j \in J$, let

$$v_j = 1_{V_j W_j} * \xi_{W_j} \in A_p^1(G).$$

We have

$$\|v_j\|_1 \leq |V_j W_j| \leq |U_j| \leq 1$$

and, if $1/p + 1/p' = 1$,

$$\|v_j\|_{A_p} \leq |V_j W_j|^{1/p} \frac{1}{|W_j|} |W_j|^{1/p'} = |V_j W_j|^{1/p} |W_j|^{-1/p} < 2^{1/p} < 2.$$

For every $x \in G$,

$$v_j(x) = \frac{1}{|W_j|} \int_G 1_{V_j W_j}(xy) 1_{W_j}(y)\, dy.$$

If $x \in V_j$, $v_j(x) = 1$ and if $x \notin V_j W_j^2$, $v_j(x) = 0$, thus a fortiori $v_j(x) = 0$ whenever $x \notin U_j'$.

As $\{e\}$ is a spectral set in $A_p(G)$ by Proposition 19.19 and $u(e) = 0$, given $\varepsilon > 0$, there exists $u_0 \in A_p(G) \cap \mathcal{K}(G)$ such that $e \notin \mathrm{supp}\, u_0$ and $\|u - u_0\|_{A_p} < \varepsilon/4$; hence

$$\|uv_j - u_0 v_j\|_{A_p} \leq \|u - u_0\|_{A_p} \|v_j\|_{A_p} < \frac{\varepsilon}{4} 2 = \frac{\varepsilon}{2}.$$

We may choose $j_0 \in J$ such that $U_{j_0} \cap \mathrm{supp}\, u_0 = \emptyset$, and for every $j \in J$ with $j \succ j_0$, we have $U_j \subset U_{j_0}$, $U_j' \cap \mathrm{supp}\, u_0 = \emptyset$, $u_0 v_j = 0$; so

$$\|uv_j\|_{A_p} < \frac{\varepsilon}{2}.$$

As $u(e) = 0$, also

$$\|uv_j\|_1 = \int_G |u(x) v_j(x)|\, dx = \int_G |u(x) - u(e)| v_j(x)\, dx$$

$$= \int_{U_j'} |u(x) - u(e)| v_j(x)\, dx < \frac{\varepsilon}{2} \|v_j\|_1 \leq \frac{\varepsilon}{2}.$$

We finally have $\|uv_j\|_{1,p} < \varepsilon$ and $\lim_j \|uv_j\|_{1,p} = 0$. □

Proposition 19.24. *Let G be an amenable group and $p \in \,]1, \infty[$. Then $A_p^1(G)$ satisfies Ditkin's conditions.*

Proof. (a) Let $u \in A_p^1(G)$ and $x_0 \in G$ such that $u(x_0) = 0$. We consider a net $(u_i)_{i \in I}$ of approximate units for $A_p^1(G)$ obtained in Proposition 19.22 and the net $(v_j)_{j \in J}$ determined in Lemma 19.23. If $(i, j) \in I \times J$, let $w_{i,j} = u_i - u_i v_j \in A_p^1(G)$. By Lemma 19.23, $w_{i,j}$ vanishes on a neighborhood of x_0.

Let $n \in \mathbf{N}^*$. There exists $i_n \in I$ such that $\|uu_{i_n} - u\|_{i,p} < 1/2n$. For every $j \in J$,

$$\|uw_{i_n, j} - u\|_{1,p} \leq \|uu_{i_n} - u\|_{1,p} + \|uu_{i_n} v_j\|_{1,p}$$

$$< 1/2n + \|uv_j\|_{1,p} \|u_{i_n}\|_{1,p}.$$

By Lemma 19.23 we may choose $j_n \in J$ such that

$$\|uv_{j_n}\|_{1,p} \|u_{i_n}\|_{1,p} < 1/2n; \text{ so } \|uw_{i_n, j_n} - u\|_{1,p} < 1/n.$$

Thus we have

$$\lim_{n \to \infty} \|uw_{i_n j_n} - u\|_{1,p} = 0.$$

Ditkin's condition holds at x_0.

(b) By Proposition 19.22, $A_p^1(G)$ admits approximate units with compact supports. Therefore $A_p^1(G)$ also verifies Ditkin's condition at infinity. □

NOTES

Derighetti [125] established 19.1. The proof of 19.2 was given by Renaud [457]. He showed the isometric multipliers of the Fourier algebra of the amenable group G to be the isometries T of $A(G)$ that are determined by multiplier functions $\alpha\chi$, where $\alpha \in \mathbf{T}$ and χ is a continuous homomorphism of G into $\mathbf{T}$. Flory [184, 185] established 19.4; the property had been proved in the abelian case by Doss [143, 144]. Arsac [20] performed a general version of 19.4 applying to an arbitrary locally compact group. If G is a locally compact group, let $u \in B(G)$. By a result due to McKennon [384], if $u \in A(G)$, then there exists a net (f_i) in $\{f \in P(G): f(e) = 1\}$ that is w*-converging to 1_G in $B(G)$ such that $\lim_i \|f_i u - u\|_B = 0$. The converse was established for amenable groups by Akeman and Walter [5]. The latter authors also proved that if a locally compact group G is amenable, then it satisfies the following Riemann-Lebesgue type property: If U is an open neighborhood of 1_G in the w*-topology of $B(G)$ and $\varepsilon > 0$, there exists a compact subset K of G such that to every $x \in G \setminus K$ there corresponds a positive-definite function f in U that is pure (i.e., every $g \in P(G)$ majorized by f is of the form αf, $0 \leq \alpha \leq 1$) and satisfies $\mathcal{R}e\, f(x) \leq \varepsilon$. The property is a nonsufficient condition for amenability. Let G be a locally compact group, $p \in \,]1, \infty[$ and $v \in B_p(G)$. Graniner [222] showed that if $v \in A_p(G)$, then for every net $(u_i)_{i \in I}$ in $B_p(G)$ bounded by 1 and w*-converging to 1_G, one has $\lim_i \|v(v_i - 1_G)\|_{B_p} = 0$. He proved that in the case that G is amenable, the converse holds because of the existence of approximate units in $A_p(G)$. With respect to 19.12 the statement improves the results of Akeman and Walter. It had been studied in the abelian case by Lohoué [367].

Granirer [218] established 19.5 for $A(G)PM(G)$; also see Lau [340]. The proofs of 19.6–7 are due to Eymard; 19.7 holds for arbitrary locally compact groups in case $p = 2$ [172]. Cowling [102] demonstrated 19.7 for $G = SL(2,\mathbf{R})$. The property holds for $G = F_2$; see Haagerup [245], and [102].

For the properties given in Section 19C we refer to Herz [260].

The results stated in 19.11–12 are also due to Herz [260]; our proofs are taken from Cowling [105]. Other demonstrations of 19.12 were established by McKennon [384] and Racher [441]. Leinert [347] gave a proof of $C_2(F_2) \neq B(F_2)$ in a general context. Figà-Talamanca and Picardello [183] showed that if the discrete group G contains F_2, the $B_2(G) \setminus B(G) \neq \varnothing$. Bożejko [51] proved that $B_p(F_2) \setminus C_p(F_2) \neq \varnothing$ whenever $1 < p < \infty$. Nebbia [413] established 19.14 and gave supplementary information in the discrete case. Granirer defined $\mathscr{UC}(\hat{G})$ for a general locally compact group in [218]. Lau [340] proved 19.16 and obtained a large collection of further interesting properties of $\mathscr{UC}(\hat{G})$. The proof of 19.17 is due to Cecchini [65]; independently Lau [340] demonstrated the statement in a general setting concerning closed subspaces of $PM(G)$.

Herz [257, 259] gave 19.18–19 and formulated general statements. The algebras A_p^1 were studied by Lai and I. Chen, who proved 19.21–24 [329]. Relying on the earlier work of Lai [328] concerning abelian groups, they established the following properties for an amenable group G [329]: If I is a closed ideal in $A_p(G)$, then $I \cap \overline{A_p^1(G)}^{A_p(G)} = I$; if J is a closed ideal in $A_p^1(G)$, then $J = \overline{J}^{A_p(G)} \cap A_p^1(G)$.

Gilbert [200, 201] took an alternative approach to the subject starting from tensor products of $\mathscr{C}_0(G)$ for a locally compact group G. He could use Banach space theory rather than group structure properties. Gilbert established new proofs, some of them in a broader context. He showed, for instance, the following result enounced by Herz [258]: If G is an amenable group and $1 < p \leq q \leq 2$ or $2 \leq q \leq p < \infty$, then $A_q(G)$ may be injected continuously into $A_p(G)$. For applications of Gilbert's methods to the study of amenable groups see also Dreseler and Schempp [147]. Herz [256, 259] proved that, for every closed subgroup H of a locally compact group G and $p \in]1, \infty[$, $A_p(H)$ may be identified isometrically with the algebra of restrictions of $A_p(G)$ to H; also see Eymard [172] and Spector [516]. We quote a statement due to Carling [64]: Let G be a locally compact group admitting a closed normal subgroup H and a closed connected subgroup K such that $H \cap K = \{e\}$; let (H_n) be the central descending series of H and consider $K_n = H_n/H_{n+1}$ ($n \in \mathbf{N}^*$). Suppose that $p = 2$, or $p \in]1, 2[\cup]2, \infty[$ and G is amenable. If, for any $n \in \mathbf{N}^*$, the semidirect product generated by K and K_n is not the direct product of these groups, then the restriction map of $B_p(G)$ to $B_p(H)$ is not surjective. De Michele and Soardi [385] studied symbolic calculus on Fourier algebras: Let G

be an infinite amenable discrete group containing abelian subgroups of arbitrary large order and let $F\colon [-1,1] \mapsto \mathbf{C}$. Then F operates on $A_p(G)$ ($1 < p < \infty$), that is, $F \circ f \in A_p(G)$ whenever $f\colon A_p(G) \mapsto [-1,1]$ if and only if $F(0) = 0$ and F is real-analytic in a neighborhood of 0. If G is a locally compact group, the classes of Haar integrable functions on G with values in a Banach algebra determine a generalized L^1-algebra introduced by Leptin [351, 352]. Kugler [325] showed that, if G is an amenable group, then existence of bounded approximate units in $A(G)$ implies that the corresponding generalized L^1-algebra is simple.

In [341] Lau showed, via an Arens multiplication on $PM(G)^*$, that for an amenable group G, every regular maximal left ideal in $PM(G)^*$ contains a unique $I_a = \{\varphi \in A(G)\colon \varphi(a) = 0\}$ ($a \in G$) if and only if G is compact. Following a suggestion made by Dunkl, Granirer [220] defined, for an arbitrary locally compact group G, the C^*-subalgebra $\mathscr{C}(\hat{G})$ of $PM(G)$ formed by all $C_L^*(G)$-multipliers in $PM(G)$; $\mathscr{UC}(\hat{G}) \subset \mathscr{C}(\hat{G})$. If G is abelian, $\mathscr{C}(\hat{G})$ consists of all continuous bounded functions defined on $\hat{G}$; in the case that G is amenable, $\mathscr{C}(\hat{G}) = PM(G)$ if and only if G is compact. Granirer proved in [221] that, for a general locally compact group G, $\mathscr{C}(\hat{G})$ coincides with $PM(G)$ if and only if G is compact. He also mentioned the existence of another proof due to Barnes. Additional ideas were developed in [223].

5

COMPLEMENTS ON INVARIANT MEANS PROPERTIES

This chapter should provide an outlook on the large amount of properties concerning invariant means, that is, the notions that gave rise to the concept of amenability originally.

20. EXISTENCE OF INVARIANT MEANS

We study characterizations of amenability in terms of existence properties of invariant means and in terms of the Dixmier criterion. We also investigate existence of discontinuous left invariant functionals on left invariant spaces.

A. Weak Invariant Means Properties

The following results allow the formal weakening of certain conditions stated in Theorem 4.19. We only consider the complex case and restrict ourselves to one-sided versions.

Lemma 20.1. *Let G be a locally compact group, A a closed $*$-subalgebra of $\mathcal{M}^1(G)$, and $A^1 = A \cap M^1(G)$. Let X be a subspace of $L^\infty(G)$ that admits 1_G, is closed under complex conjugation, and is also A-invariant. Assume that the following conditions hold*:
 (a) *X is A-introverted.*
 (b) *For all $M \in \mathfrak{M}(X)$, $f \in X$ such that $M(\mu * f) = M(f)$ whenever $\mu \in A^1$, one has $M \star f = M \star (\mu * f)$ whenever $\mu \in A^1$.*
 (c) *For every $f \in X$ there exists a mean M_f on X such that $M_f(\mu * f) = M_f(f)$ whenever $\mu \in A^1$.*

Then X admits an A^1-invariant mean, that is, there exists $N \in \mathfrak{M}(X)$ such that $N(\mu * f) = N(f)$ whenever $f \in X$ and $\mu \in A^1$.

Proof. If $f \in X$, we denote by $\mathfrak{M}_f(X)$ the set of all means M on X such that $M(\mu * f) = M(f)$ whenever $\mu \in A^1$. By (c) this set is nonvoid, and by Proposition 3.3 it is w*-compact in $\mathfrak{M}(X)$. We establish the lemma by showing that the family $\{\mathfrak{M}_f : f \in X\}$ has the finite intersection property. We prove that if, for $f_1, \ldots, f_{n+1} \in X$, $\cap_{i=1}^n \mathfrak{M}_{f_i} \neq \varnothing$, then also $\cap_{i=1}^{n+1} \mathfrak{M}_{f_i} \neq \varnothing$.

Let $M' \in \cap_{i=1}^n \mathfrak{M}_{f_i}$. As X is A-introverted, $g = M' * f_{n+1} \in X$. We choose $M'' \in \mathfrak{M}_g$ and define $N = M'' \odot M'$. By (b), $M' * f_i = M' * (\mu * f_i)$ whenever $\mu \in A^1$ and $i = 1, \ldots, n$. Then for every $i = 1, \ldots, n$ and every $\mu \in A^1$,

$$\langle \mu * f_i, N \rangle = \langle M' * (\mu * f_i), M'' \rangle$$
$$= \langle M' * f_i, M'' \rangle = \langle f_i, N \rangle.$$

For every $\mu \in A^1$, by (2.16),

$$M' * (\mu * f_{n+1}) = \mu * (M' * f_{n+1}) = \mu * g$$

and

$$\langle \mu * f_{n+1}, N \rangle = \langle M' * (\mu * f_{n+1}), M'' \rangle = \langle \mu * g, M'' \rangle$$
$$= \langle g, M'' \rangle = \langle M' * f_{n+1}, M'' \rangle = \langle f_{n+1}, N \rangle.$$

Therefore $N \in \cap_{i=1}^{n+1} \mathfrak{M}_{f_i}$. □

Proposition 20.2. *Let G be a locally compact group. Each of the following conditions is equivalent to amenability*:
  (i) *For every $f \in \mathcal{RUC}(G)$, there exists a mean M_f on $\mathcal{RUC}(G)$ such that $M_f({}_a f) = M_f(f)$ whenever $a \in G$.*
  (ii) *For every $f \in L^\infty(G)$ [resp. $f \in \mathcal{RUC}(G)$] there exists a mean M_f on $L^\infty(G)$ [resp. $\mathcal{RUC}(G)$] such that $M_f(\varphi * f) = M_f(f)$ whenever $\varphi \in P^1(G)$.*

Proof. If G is amenable, the properties are obviously verified by Theorem 4.19.

(a) Assume that (i) holds. Let $M \in \mathfrak{M}(\mathcal{RUC}(G))$ and $f \in \mathcal{RUC}(G)$ such that $M({}_a f) = M(f)$ whenever $a \in G$. By (2.15), for every $a \in G$ and every $x \in G$,

$$M * f(x) = \langle {}_x f, M \rangle = \langle {}_{ax} f, M \rangle = \langle {}_x ({}_a f), M \rangle = M * ({}_a f)(x),$$

that is, $M * f = M * ({}_a f)$ whenever $a \in G$. We now apply Lemma 20.1 to the

EXISTENCE OF INVARIANT MEANS

space $\mathcal{RUC}(G)$ that is $\mathcal{M}_d^1(G)$-introverted (Section 2G.4). We have LIM$(\mathcal{RUC}(G)) \neq \emptyset$, and amenability follows from Theorem 4.19.

(b) Assume that (ii) holds for $X = L^\infty(G)$ [resp. $\mathcal{RUC}(G)$]. Let $M \in \mathfrak{M}(X)$ and $f \in X$ such that $M(\varphi * f) = M(f)$ whenever $\varphi \in P^1(G)$. For every $\psi \in P^1(G)$ and every $\varphi \in P^1(G)$,

$$\langle \psi, M \star f \rangle = \langle \psi^* * f, M \rangle = \langle (\psi^* * \varphi) * f, M \rangle$$
$$= \langle \psi^* * (\varphi * f), M \rangle = \langle \psi, M \star (\varphi * f) \rangle.$$

As $L^1(G)$ is the closure of the space generated by $P^1(G)$, we have $M \star f = M \star (\varphi * f)$ whenever $\varphi \in P^1(G)$. We then apply Lemma 20.1 to X that is $L^1(G)$-introverted (Section 2G.4). We have TLIM$(X) \neq \emptyset$, and amenability follows again from Theorem 4.19. □

Corollary 20.3. *A locally compact group G is amenable if and only if, for any $f \in X = L^\infty(G)$ [resp. $\mathcal{RUC}(G)$], there exists a net (φ_i) in $P^1(G)$ such that $(f * \check{\varphi}_i)$ converges to a constant function in the weak-$*$-topology of $L^\infty(G)$. More precisely, for any $f \in X$ and any $\alpha \in \mathbf{C}$, the constant function $\alpha 1_G$ belongs to the weak-$*$-closure of $\{f * \check{\varphi} : \varphi \in P^1(G)\}$ in $L^\infty(G)$ if and only if there exists a topologically left invariant mean N on X such that $N(f) = \alpha$.*

Proof. Let $f \in X$.

(a) Assume that there exists a net (φ_i) in $P^1(G)$ such that $(f * \check{\varphi}_i)$ converges to the constant function $\alpha 1_G$ in the weak-$*$-topology of $L^\infty(G)$. By Proposition 3.3 we consider a subnet (φ_{i_j}) of (φ_i) that is w*-converging to a mean M on X. For any $g \in L^1_\mathbf{R}(G)$, by (2.12),

$$\langle g, M \star f \rangle = \langle g^* * f, M \rangle = \lim_j \langle g^* * f, \varphi_{i_j} \rangle = \lim_j \langle \varphi_{i_j}, g^* * f \rangle$$
$$= \lim_j \langle g, f * \check{\varphi}_{i_j} \rangle = \langle g, \alpha 1_G \rangle.$$

Hence $M \star f = \alpha 1_G$ and $\langle f, M \odot M \rangle = \langle M \star f, M \rangle = \alpha$.

For any $\psi \in P^1(G)$, we have $\psi * (M \star f) = \alpha 1_G$ and, by (2.16),

$$\langle \psi * f, M \odot M \rangle = \langle M \star (\psi * f), M \rangle = \langle \psi * (M \star f), M \rangle = \alpha.$$

So we obtain

$$M \odot M(\psi * f) = \alpha = M \odot M(f).$$

Property (ii) of Proposition 20.2 holds for $M \odot M \in \mathfrak{M}(X)$ and thus G is amenable.

Let $M' \in \text{TLIM}(X)$; by Proposition 4.26, $N = M' \odot M \in \text{TLIM}(X)$ and $N(f) = \langle M \star f, M' \rangle = \langle \alpha 1_G, N \rangle = \alpha$.

(b) Let $M \in \text{TLIM}(L^\infty(G))$. By Proposition 3.3 there exists a net (φ_i) in $P^1(G)$ that is w*-converging to M. For any $f \in X$ and any $\psi \in P^1(G)$, by (2.12),

$$\lim_i \langle \psi, f * \check{\varphi}_i \rangle = \lim_i \langle \varphi_i, \psi^* * f \rangle = \lim_i \langle \psi^* * f, \varphi_i \rangle$$

$$= \langle \psi^* * f, M \rangle = M(\psi^* * f) = M(f) = \langle \psi, M(f) 1_G \rangle.$$

As $P^1(G)$ spans $L^1(G)$, we conclude that $M(f) 1_G$ belongs to the w*-closure of $\{ f * \check{\varphi} : \varphi \in P^1(G) \}$. □

The averaging property of Corollary 20.3 is expressed by saying that the amenability of a locally compact group G is characterized by the *stationarity* of $L^\infty(G)$ [resp. $\mathscr{RUC}(G)$].

B. Characterizations of Amenability Following from Invariant Means Properties

Lemma 20.4. *Let G be a locally compact group, A a closed $*$-subalgebra of $\mathscr{M}^1(G)$, $A^1 = A \cap M^1(G)$, and X a subspace of $L^\infty(G)$ that admits 1_G and is A-invariant. For every $f \in X_{\mathbf{R}}$, the following conditions are equivalent:*

(i) *There exists a mean M_f on X such that $M_f(\mu * f) = M_f(f)$ whenever $\mu \in A^1$.*

(ii) $\text{ess sup}(\mu_1 - \mu_2) * f \geq 0$ *whenever* $\mu_1, \mu_2 \in A^1$.

Proof. (i) ⇒ (ii)
For all $\mu_1, \mu_2 \in A^1$,

$$\text{ess sup}(\mu_1 - \mu_2) * f \geq M_f((\mu_1 - \mu_2) * f) = 0.$$

(ii) ⇒ (i)

Let Y be the real linear span of $\{ f - \mu * f : \mu \in A^1 \}$. Suppose that there exist $\alpha_1, \ldots, \alpha_m \in \mathbf{R}_+$, $\alpha_{m+1}, \ldots, \alpha_n \in \mathbf{R}_-$, and $\mu_1, \ldots, \mu_n \in A^1$ such that $\sum_{i=1}^n |\alpha_i| > 0$ and $\text{ess sup} \sum_{i=1}^n \alpha_i (f - \mu_i * f) < 0$. Choose $\mu_0 \in A^1$; then

$$\text{ess sup} \sum_{i=1}^n \alpha_i (\mu_0 * f - \mu_0 * \mu_i * f) < 0. \tag{1}$$

We have

$$\nu' = \left(\sum_{i=1}^n |\alpha_i| \right)^{-1} \left(\sum_{i=1}^m \alpha_i \mu_0 - \sum_{i=m+1}^n \alpha_i \mu_0 * \mu_i \right) \in A^1,$$

$$\nu'' = -\left(\sum_{i=1}^n |\alpha_i| \right)^{-1} \left(\sum_{i=m+1}^n \alpha_i \mu_0 - \sum_{i=1}^m \alpha_i \mu_0 * \mu_i \right) \in A^1;$$

(1) expresses that
$$\operatorname{ess\,sup}(\nu' - \nu'') * f < 0.$$

The hypothesis would be contradicted. Hence for every $g \in Y$, we must have $\operatorname{ess\,sup} g \geq 0$. By Lemma 4.27 there exists a mean M_f on X such that $M_f(g) = 0$ whenever $g \in Y$, so $M_f(\mu * f - f) = 0$ whenever $\mu \in A^1$. □

Proposition 20.5. *In a locally compact group G, amenability is characterized by each of the following conditions*:

$$(\forall f \in \mathscr{RUC}_{\mathbf{R}}(G))\,(\forall \varphi \in P^1(G))\,(\forall \psi \in P^1(G)) \sup (\varphi * f - \psi * f) \geq 0.$$

$$(\forall f \in L^\infty_{\mathbf{R}}(G))\,(\forall \varphi \in P^1(G))\,(\forall \psi \in P^1(G)) \operatorname{ess\,sup}(\varphi * f - \psi * f) \geq 0.$$

Proof. The statement is a consequence of Lemma 20.4 and Proposition 20.2. □

Proposition 20.6. *Let G be a locally compact group. Each of the following conditions characterizes amenability of G:*

(i) *For all $f \in \mathscr{RUC}_{\mathbf{R}}(G)$, $a_1, \ldots, a_n \in G$, $b_1, \ldots, b_n \in G$, $n \in \mathbf{N}^*$,*

$$\inf \frac{1}{n} \sum_{i=1}^n ({}_{a_i}f - {}_{b_i}f) < 1.$$

(ii) *For all $f \in \mathscr{RUC}_{\mathbf{R}}(G)$, $\varphi \in P^1(G)$, $\psi \in P^1(G)$,*

$$\inf(\varphi * f - \psi * f) < 1.$$

Proof. (a) If G is amenable, by Corollary 4.14, we consider $M \in \operatorname{LIM}(\mathscr{RUC}(G)) = \operatorname{TLIM}(\mathscr{RUC}(G))$. Given $f \in \mathscr{RUC}_{\mathbf{R}}(G)$, $a_1, \ldots, a_n \in G$, $b_1, \ldots, b_n \in G$, and $\varphi, \psi \in P^1(G)$, we have

$$\inf \sum_{i=1}^n ({}_{a_i}f - {}_{b_i}f) \leq M\left(\sum_{i=1}^n ({}_{a_i}f - {}_{b_i}f)\right) = \sum_{i=1}^n M({}_{a_i}f - {}_{b_i}f) = 0$$

and

$$\inf(\varphi * f - \psi * f) \leq M(\varphi * f - \psi * f) = 0.$$

(b) Let G be nonamenable. By Proposition 20.5 there exist $f \in \mathscr{RUC}_{\mathbf{R}}(G)$ and $\varphi, \psi \in P^1(G)$ such that

$$\alpha = \sup(\varphi * f - \psi * f) < 0;$$

then

$$\inf(\varphi * (-f) - \psi * (-f)) = -\alpha > 0.$$

Let $g = 2\alpha^{-1}f \in \mathscr{RUC}_\mathbf{R}(G)$. We have

$$\inf(\varphi * g - \psi * g) = 2 \qquad (2)$$

and (ii) is violated.

Given $\varepsilon \in {]0,1[}$, via Proposition 3.6 we may determine $n \in \mathbf{N}^*$, $a_1, \ldots, a_n \in G$, and $b_1, \ldots, b_n \in G$ such that

$$\left\| \frac{1}{n} \sum_{i=1}^n {}_{a_i}g - \varphi * g \right\| < \frac{\varepsilon}{2}, \left\| \frac{1}{n} \sum_{i=1}^n {}_{b_i}g - \psi * g \right\| < \frac{\varepsilon}{2}.$$

Therefore

$$\frac{1}{n} \sum_{i=1}^n ({}_{a_i}g - {}_{b_i}g) \geq \varphi * g - \psi * g - \varepsilon 1_G.$$

By (2) we have then

$$\inf \frac{1}{n} \sum_{i=1}^n ({}_{a_i}g - {}_{b_i}g) \geq 2 - \varepsilon > 1;$$

(i) is violated. □

Proposition 20.7. *Let G be a locally compact group and let $X = \mathscr{S}_\mathbf{R}(G)$ or a left invariant subspace of $L_\mathbf{R}^\infty(G)$ containing $\mathscr{UC}_\mathbf{R}(G)$ such that $|f| \in X$ whenever $f \in X$. Then the following conditions are equivalent:*
  (i) *G is amenable.*
  (ii) *$\inf\{\|1_G - h\|_\infty : h \in \mathscr{N}(X)\} = 1$.*
  (iii) *$\overline{\mathscr{N}(X)} \neq X$.*

If these conditions hold and $f_0 \in X \setminus \overline{\mathscr{N}(X)}$, then there exists $M \in \mathrm{LIM}(X)$ such that $M(f_0) \neq 0$.

Proof. (i) ⇒ (ii)
By Theorem 4.19 let $M \in \mathrm{LIM}(X)$. For every $h \in \mathscr{N}(X)$, ess sup$(1_G - h) \geq M(1_G - h) = 1$ and a fortiori $\|1_G - h\|_\infty \geq 1$. For the null function h_0 on G, $h_0 \in \mathscr{N}(X)$ and $\|1_G - h_0\|_\infty = 1$.

(ii) ⇒ (iii)

This implication is obvious because by (ii) we have $1_G \in X \setminus \overline{\mathscr{N}(X)}$.

(iii) ⇒ (i)

We show that $\mathrm{LIM}(X) \neq \emptyset$, thus establishing amenability via Theorem 4.19.

Let $f_0 \in X \setminus \overline{\mathcal{N}(X)}$. By the Hahn–Banach theorem there exists a continuous linear functional N on X such that $N(f_0) = 1$ and $N(h) = 0$ whenever $h \in \mathcal{N}(X)$. There also exist nonnegative linear functionals N_+, N_- on X such that $N = N_+ - N_-$ and, for every $f \in X_+$,

$$N_+(f) = \sup\{N(g) : g \in X_+, g \leq f\}$$

([HR] B.37). If $f \in X_+$ and $a \in G$,

$$N_+({}_af) = \sup\{N(g) : g \in X_+, g \leq {}_af\} = \sup\{N({}_{a^{-1}}g) : g \in X_+, {}_{a^{-1}}g \leq f\}$$
$$= \sup\{N(g) : g \in X_+, g \leq f\} = N_+(f).$$

So also, for all $f \in X$ and $a \in G$, we have

$$N_+({}_af) = N_+({}_af^+) - N_+({}_af^-) = N_+(f^+) - N_+(f^-) = N_+(f)$$

where $f^+ = \sup\{f, 0\}$ and $f^- = \sup\{-f, 0\}$; moreover $N({}_af) = N(f)$ and therefore $N_-({}_af) = N_-(f)$.

As $N(f_0) = 1$, we must have $\|f_0\|_\infty > 0$ and $|N_+(f_0)| + |N_-(f_0)| > 0$, say $N_+(f_0) \neq 0$ for instance. As $|N_+(f_0)| \leq \|f_0\|_\infty N_+(1_G)$, we have $N_+(1_G) > 0$. We define

$$M(f) = \frac{1}{N_+(1_G)} N_+(f),$$

$f \in X$. If $f \in X_+$, $M(f) \geq 0$; $M(1_G) = 1$ and, for all $f \in X$, $a \in G$,

$$M({}_af) = \frac{1}{N_+(1_G)} N_+({}_af) = \frac{1}{N_+(1_G)} N_+(f) = M(f).$$

So $M \in \text{LIM}(X)$. We have $M(f_0) \neq 0$. □

Note that Proposition 20.7 applies to $L^\infty(G)$, $\mathscr{C}(G)$, $\mathscr{RUC}(G)$, and $\mathscr{UC}(G)$.

We add some characterizations for amenability on discrete groups expressing that the groups are not free.

Definition 20.8. *If G is a discrete group, let $\mathfrak{S}(G)$ be the set of all pairs (a, A), where $a \in G$ and $A \in \mathfrak{P}(G)$. If Q is a finite collection of (nonnecessarily distinct) elements (a_i, A_i) ($i = 1, \ldots, n$) in $\mathfrak{S}(G)$, for $x \in G$, let*

$$s_Q(x) = \text{card}\{i \in \{1, \ldots, n\} : x \in A_i\},$$

$$t_Q(x) = \text{card}\{i \in \{1, \ldots, n\} : a_i x \in A_i\}.$$

Proposition 20.9. *The discrete group G is amenable if and only if, for all finite collections Q, R of (nonnecessarily distinct) elements in $\mathfrak{S}(G)$,*

$$\sup\left(s_Q - s_R - (t_Q - t_R)\right) \geq 0.$$

Proof. (a) If G is nonamenable, by Proposition 4.29 there exist $f_1, \ldots, f_n \in l_\mathbf{R}^\infty(G)$ and $a_1, \ldots, a_n \in G$ such that

$$\alpha = \sup \sum_{i=1}^{n} (f_i - {}_{a_i}f) < 0.$$

Let $\beta = -\alpha^{-1}(n+1) > 0$. For $i = 1, \ldots, n$ we define

$$g_i(x) = [\beta f_i(x)] \in \mathbf{Z},$$

$x \in G$, [ ] denoting the integral part. For every $x \in G$, $g_i(x) < \beta f_i(x) + 1$. We have

$$\sup \sum_{i=1}^{n} (g_i - {}_{a_i}g_i) < \beta \sup \sum_{i=1}^{n} (f_i - {}_{a_i}f_i) + n = \beta\alpha + n = -1. \quad (3)$$

As all g_i ($i = 1, \ldots, n$) take their values in $\mathbf{Z}$, there exist $h_1, \ldots, h_p \in l^\infty(G)$ [resp. $k_1, \ldots, k_q \in l^\infty(G)$] taking their values in $\{0, 1\}$ [resp. $\{0, -1\}$] and $b_1, \ldots, b_p, c_1, \ldots, c_q \in \{a_1, \ldots, a_n\}$ such that

$$\sum_{j=1}^{p} (h_j - {}_{b_j}h_j) + \sum_{j=1}^{q} (k_j - {}_{c_j}k_j) = \sum_{i=1}^{n} (g_i - {}_{a_i}g_i). \quad (4)$$

Let $B_j = h_j^{-1}(\{1\})$ for $j = 1, \ldots, p$ and $C_j = k_j^{-1}(\{-1\})$ for $j = 1, \ldots, q$. Let also Q be the collection of all (b_j, B_j) ($j = 1, \ldots, p$) and R the collection of all (c_j, C_j) ($j = 1, \ldots, q$). Then (3) and (4) imply that, for every $x \in G$,

$$s_Q(x) - t_Q(x) - (s_R(x) - t_R(x)) < -1;$$

therefore the condition does not hold.

(b) Assume that the condition does not hold. There exist a collection Q of pairs $(a_i, A_i) \in \mathfrak{S}(G)$ ($i = 1, \ldots, m$) and a collection R of pairs $(b_j, B_j) \in \mathfrak{S}(G)$ ($j = 1, \ldots, n$) such that

$$s_Q(x) - t_Q(x) - (s_R(x) - t_R(x)) < 0$$

whenever $x \in G$. Then

$$\sup\left(\sum_{i=1}^{m} (1_{A_i} - {}_{a_i}1_{A_i}) + \sum_{j=1}^{n} ((-1_{B_j}) - {}_{b_j}(-1_{B_j}))\right) < 0$$

and, by Proposition 4.29, G is nonamenable. □

Proposition 20.10. *The discrete group G is amenable if and only if, for every finite collection Q in $\mathfrak{S}(G)$, $\sup(s_Q - t_Q) \geq 0$.*

Proof. (a) Assume G to be amenable.

We consider the singleton $Q_0 = \{(e, \{e\})\}$ in $\mathfrak{S}(G)$; then $s_{Q_0}(e) = t_{Q_0}(e) = 1$ and, for $x \in G \setminus \{e\}$, $s_{Q_0}(x) = t_{Q_0}(x) = 0$. If Q is any finite collection in $\mathfrak{S}(G)$, by Proposition 20.9 we have

$$\sup(s_Q - t_Q) = \sup\left(s_Q - s_{Q_0} - (t_Q - t_{Q_0})\right) \geq 0.$$

(b) Assume G to be nonamenable.

By Proposition 20.9 there exists a finite collection Q in $\mathfrak{S}(G)$ and a collection R in $\mathfrak{S}(G)$ consisting of the pairs (a_i, A_i) ($i = 1, \ldots, n$) such that $s_Q(x) - s_R(x) < t_Q(x) - t_R(x)$ whenever $x \in G$. Let U be the collection in $\mathfrak{S}(G)$ formed by Q and the pairs $(a_i^{-1}, a_i^{-1}A_i)$ ($i = 1, \ldots, n$). Then for every $x \in G$, $s_U(x) = s_Q(x) + t_R(x)$ and $t_U(x) = t_Q(x) + s_R(x)$, hence $s_U(x) - t_U(x) < 0$ and the condition does not hold. □

C. Existence of Discontinuous Left Invariant Functionals on Left Invariant Spaces

If G is a noncompact locally compact group we consider left invariant Banach spaces X of complex-valued functions defined on G that are closed under complex conjugation, left invariant, and do not contain 1_G. The subspace $\mathcal{N}(X)$ generated by the functions $f - {}_af$, where $f \in X$ and $a \in G$, is the intersection of the kernels of all *left invariant functionals* on X, that is, all $F \in X'$ such that $F({}_af) = F(f)$ whenever $f \in X$ and $a \in G$. Therefore, if $\mathcal{N}(X)$ is not closed, at least one of these kernels is not closed and there exists a discontinuous left invariant functional on X. Proposition 20.12 should be compared with Proposition 20.7.

Lemma 20.11. *If G is a noncompact locally compact group and K is a compact subset of G, there exists an infinite sequence $(a_n)_{n \in \mathbf{N}}$ in G such that $a_0 = e$ and $(a_iK) \cap (a_jK) = \emptyset$ whenever $i, j \in \mathbf{N}$, $i \neq j$.*

Proof. We may assume that $K \neq \emptyset$. There exists $a_1 \in G$ such that $a_1K \cap K = \emptyset$ because otherwise $G = KK^{-1}$ and G would be compact. If the lemma did not hold, there would exist $a_0 = e, a_1, \ldots, a_m \in G$ such that $(a_iK) \cap (a_jK) = \emptyset$ whenever $i, j \in \{0, 1, \ldots, m\}$, $i \neq j$ and, for every $a \in G$, $aK \cap (\cup_{i=0}^m a_iK) \neq \emptyset$, that is, $a \in (\cup_{i=0}^m a_iK)K^{-1}$. Again G would be compact.

Proposition 20.12. *Let G be a noncompact locally compact group and let X be one of the Banach spaces $X = \mathscr{C}_0(G)$, $L^p(G)$ where $1 < p < \infty$. Then $\overline{\mathcal{N}(X)} = X$.*

Proof. (a) Consider $\mu \in \mathcal{M}^1(G) = \mathscr{C}_0(G)^*$ such that $\langle {}_ag, \mu \rangle = \langle g, \mu \rangle$ whenever $g \in \mathscr{C}_0(G)$ and $a \in G$. Let $f \in \mathscr{C}_0(G)$. For each $n \in \mathbf{N}^*$, we choose a compact subset K_n in G such that $|f(x)| < (n + 1)^{-1}$ whenever $x \in G \setminus K_n$. By Lemma 20.11 we may determine $a_0 = e, a_1, \ldots, a_n \in G$ such that $a_iK_n \cap a_jK_n = \emptyset$ whenever $i, j \in \{0, 1, \ldots, n\}$, $i \neq j$. If $i \in \{0, 1, \ldots, n\}$ and $x \in G$

$\setminus a_i K_n$, we have $|_{a_i^{-1}} f(x)| < (n+1)^{-1}$. We conclude that, for every $x \in G$,

$$\sum_{i=0}^{n} |_{a_i^{-1}} f(x)| \leq \sup \left\{ \|f\| + n(n+1)^{-1}, (n+1)(n+1)^{-1} \right\} \leq \|f\| + 1.$$

Let $f_n = (n+1)^{-1} \sum_{i=0}^{n} {}_{a_i^{-1}} f$, $n \in \mathbf{N}^*$. We have

$$|\langle f, \mu \rangle| = |\langle f_n, \mu \rangle| \leq \|\mu\| \|f_n\| \leq (n+1)^{-1}(\|f\| + 1)\|\mu\|.$$

As $n \in \mathbf{N}^*$ may be chosen arbitrarily, we must have $\langle f, \mu \rangle = 0$. Hence $\mu = 0$ and $\mathscr{C}_0(G) = \mathscr{N}(\mathscr{C}_0(G))$.

(b) If $1/p + 1/p' = 1$, let $g \in L^{p'}(G)$ such that $\langle h, g \rangle = 0$ whenever $h \in \mathscr{N}(L^p(G))$. For every $f \in L^p(G)$ and every $a \in G$,

$$\langle f, g - {}_a g \rangle = \langle f - {}_{a^{-1}} f, g \rangle = 0.$$

If g were not an almost everywhere constant function, there would exist a measurable subset A of G and $a \in G$ such that $0 < |A| < \infty$ and $\langle 1_A, g - {}_a g \rangle \neq 0$. So there must exist $\alpha \in \mathbf{C}$ such that $g(x) = \alpha$ for almost every $x \in G$. As G is noncompact, necessarily $\alpha = 0$ and $\mathscr{N}(L^p(G)) = L^p(G)$. □

Lemma 20.13. *Let G be an amenable σ-compact group and $p \in \,]1, \infty]$. There exists a nondecreasing sequence $(U_n)_{n \in \mathbf{N}^*}$ of compact neighborhoods of e in G such that $\cup_{n \in \mathbf{N}^*} U_n = G$ and the following property holds: For every $h \in \mathscr{N}(L^p(G))$, there exists $\alpha_h > 0$ such that $|\int_{U_n} h| \leq \alpha_h (2^{-n} |U_n|)^{1/p'}$ whenever $n \in \mathbf{N}^*$ with $1/p + 1/p' = 1$ if $p < \infty$ and $\infty' = 1$.*

Proof. As G is σ-compact, there exists a nondecreasing sequence $(V_n)_{n \in \mathbf{N}^*}$ of compact neighborhoods of e in G such that $\cup_{n \in \mathbf{N}^*} V_n = G$. Let $U_0 = V_1$. By Proposition 16.5 we may determine a sequence $(U_n)_{n \in \mathbf{N}^*}$ of compact neighborhoods of e such that, for every $n \in \mathbf{N}^*$, $V_n U_{n-1} \subset U_n$ and $|xU_n \triangle U_n| < 2^{-n} |U_n|$ whenever $x \in V_n$. Then $\cup_{n \in \mathbf{N}} U_n = G$.

Let $h \in \mathscr{N}(L^p(G))$. There exist $f_1, \ldots, f_m \in L^p(G)$ and $a_1, \ldots, a_m \in G$ such that $h = \sum_{i=1}^{m} (f_i - {}_{a_i} f_i)$. Choose $n_0 \in \mathbf{N}^*$ such that $\{a_1, \ldots, a_m\} \subset V_{n_0}$. For every $n \geq n_0$, by Hölder's inequality we obtain

$$\left| \int_{U_n} h \right| \leq \sum_{i=1}^{m} \left| \int_{U_n} f_i(x)\, dx - \int_{U_n} f_i(a_i x)\, dx \right|$$

$$\leq \sum_{i=1}^{m} \int_{U_n \triangle a_i^{-1} U_n} |f_i| \leq \sum_{i=1}^{m} \|f_i\|_p |U_n \triangle a_i^{-1} U_n|^{1/p'}$$

$$= \sum_{i=1}^{m} \|f_i\|_p |a_i U_n \triangle U_n|^{1/p'} \leq \left(\sum_{i=1}^{m} \|f_i\|_p \right) (2^{-n} |U_n|)^{1/p'}. \quad \square$$

EXISTENCE OF INVARIANT MEANS　　　　　　　　　　　　　　　　　　　　　　239

Proposition 20.14. *Let G be a noncompact amenable σ-compact group and let $X = \mathscr{C}_0(G), \mathscr{C}(G), L^p(G)$ ($1 < p \leq \infty$). Then $\mathscr{N}(X) \neq \mathcal{N}(X)$ and there exists a discontinuous left invariant functional on X.*

Proof. (a) Let (U_n) be the sequence constructed in Lemma 20.13 for $p = \infty$ and let V be an arbitrary compact symmetric neighborhood of e contained in U_1. We put $f(x) = 1$ if $x \in VU_1$ and define inductively $f(x) = n^{-1}$ if $x \in VU_n \setminus VU_{n-1}$ for $n = 2, 3, \ldots$. Then $f \in L^\infty(G)$ and $g = \xi_V * f \in \mathscr{RUC}(G)$. If $x \in G \setminus V^2U_{n-1}$ ($n \geq 2$) and $y \in V$, then $y^{-1}x \notin VU_{n-1}$, $f(y^{-1}x) \leq n^{-1}$ and

$$g(x) = \frac{1}{|V|} \int 1_V(y) f(y^{-1}x) \, dy \leq n^{-1};$$

so $g \in \mathscr{C}_0(G)$. If $x \in U_n$ ($n \in \mathbf{N}^*$), $g(x) \geq n^{-1}$ and $|\int_{U_n} g| \geq n^{-1}|U_n|$. If we had $g \in \mathscr{N}(L^\infty(G))$, Lemma 20.13 would imply that $1 \leq \alpha_g n 2^{-n}$ whenever $n \in \mathbf{N}^*$. We would come to a contradiction; therefore necessarily $g \notin \mathscr{N}(L^\infty(G))$ and a fortiori $g \notin \mathscr{N}(\mathscr{C}_0(G))$, $g \notin \mathscr{N}(\mathscr{C}(G))$. But by Proposition 20.12 we have $g \in \mathcal{N}(\mathscr{C}_0(G))$ and a fortiori $g \in \mathcal{N}(\mathscr{C}(G))$, $g \in \mathcal{N}(L^\infty(G))$.

(b) Let $p \in]1, \infty[$ and consider again the sequence (U_n) constructed in Lemma 20.13. For $n \in \mathbf{N}^*$, let $u_n = n^{-1}|U_n|^{-1/p}$; (u_n) is decreasing. We put $f(x) = u_1$ if $x \in U_1$, $f(x) = u_n$ if $x \in U_n \setminus U_{n-1}$ for $n = 2, 3, \ldots$. As $p > 1$, $f \in L^p(G)$ with $\|f\|_p^p \leq \sum_{n \in \mathbf{N}^*} 1/n^p$ and, by Proposition 20.12, $f \in \mathcal{N}(L^p(G))$. But for every $n \in \mathbf{N}^*$,

$$\int_{U_n} f \geq u_n |U_n| = n^{-1}|U_n|^{1/p'}.$$

So if we had $f \in \mathscr{N}(L^p(G))$, Lemma 20.13 would imply that $1 \leq \alpha_f n 2^{-n/p'}$ whenever $n \in \mathbf{N}^*$, and a contradiction would arise again. Hence $f \notin \mathscr{N}(L^p(G))$.　□

Comments

We consider one more condition that is directly related to invariant means properties.

Let $\mathscr{H}$ be a Hilbert space. If $U \in \mathscr{L}(\mathscr{H})$ and $\xi \in \mathscr{H}$, $(UTU^*\xi|\xi) = (TU^*\xi|U^*\xi)$ whenever $T \in \mathscr{L}(\mathscr{H})$; hence UTU^* is a positive-definite operator whenever T is. Let A be a $*$-subalgebra of $\mathscr{L}(\mathscr{H})$. We consider the commutant $A\tilde{\,}$ of A and to every $T \in \mathscr{L}(\mathscr{H})$ we associate $A^T = \overline{\mathrm{co}}\{UTU^* : U \in A \cap \mathscr{L}(\mathscr{H})_u\}$. Any $*$-subalgebra A of $\mathscr{L}(\mathscr{H})$ is called a *von Neumann algebra* if (a) $\mathrm{id}_{\mathscr{H}} \in A$; (b) A is closed for the strong topology (or equivalently the weak topology, the ultrastrong topology, the ultraweak topology); and (c) A coincides with its bicommutant $A\tilde{\,}\tilde{\,}$. In (b) of this definition, A may be replaced by the unit ball of A. In particular, $\mathbf{C}\mathrm{id}_{\mathscr{H}}$ and

$\mathscr{L}(\mathscr{H})$ are von Neumann algebras. Any $*$-subalgebra of $\mathscr{L}(\mathscr{H})$ is a von Neumann algebra if and only if it is a C^*-algebra and a dual of a Banach space; the predual space is necessarily unique.

Standard references for von Neumann algebras are Dixmier [140] and S. Sakai [501]; also see Strătilă ([521] A.16). If G is a locally compact group, $PM(G)$ is a von Neumann algebra, and by Proposition 10.3, $A(G)$ constitutes the predual of $PM(G)$.

Definition 20.15. *Let $\mathscr{H}$ be a Hilbert space. The von Neumann algebra A in $\mathscr{L}(\mathscr{H})$ is said to have property (P) if the following condition holds*:

$$(\forall T \in \mathscr{L}(\mathscr{H})) A^T \cap A^{\sim} \neq \varnothing .$$

Lemma 20.16. *If $\mathscr{H}$ is a Hilbert space and A is a von Neumann algebra in $\mathscr{L}(\mathscr{H})$ having property (P), then there exists a linear mapping P of $\mathscr{L}(\mathscr{H})$ onto $A^{\sim}$ such that the following properties hold*:
  (i) $\|P\| \leq 1$, $P \geq 0$, $P(id_{\mathscr{H}}) = id_{\mathscr{H}}$.
  (ii) *For every* $T \in \mathscr{L}(\mathscr{H})$, $P(T) \in A^T$.
  (iii) *For every* $T \in \mathscr{L}(\mathscr{H})$ *and all* $U, V \in A^{\sim}$, $P(UTV) = UP(T)V$.

Proof. Let $\mathscr{S}$ denote the w-compact convex hull of the mappings

$$T \mapsto UTU^*$$

$$\mathscr{L}(\mathscr{H}) \to \mathscr{L}(\mathscr{H})$$

($U \in A_u$); $\mathscr{S}$ is contained in the unit ball of $\mathscr{L}(\mathscr{L}(\mathscr{H}))$ and, for every $S \in \mathscr{S}$, $S \geq 0$, $S(id_{\mathscr{H}}) = id_{\mathscr{H}}$. We introduce a partial ordering on $\mathscr{S}$ by putting $S' \prec S''$ for $S', S'' \in \mathscr{S}$ if we have $A^{S'(T)} \supset A^{S''(T)}$ for every $T \in \mathscr{L}(\mathscr{H})$. Consider a net $(S_i)_{i \in I}$ in $\mathscr{S}$; if $j \in I$, let $\mathscr{S}_j$ be the w-closure of $\{ S_i \in \mathscr{S} : S_j \prec S_i, i \in I \}$. By compactness of $\mathscr{S}$ we may choose $S \in \cap_{j \in I} \mathscr{S}_j$. For every $i \in I$ and every $T \in \mathscr{L}(\mathscr{H})$, $A^{S(T)} \subset A^{S_i(T)}$, and S is an upper bound for $(S_i)_{i \in I}$. By Zorn's lemma there exists a maximal element P in $\mathscr{S}$; $\|P\| \leq 1$, $P \geq 0$, $P(id_{\mathscr{H}}) = id_{\mathscr{H}}$.

We prove now that, for every $T \in \mathscr{L}(\mathscr{H})$, $A^{P(T)}$ is a singleton. Assume that there exists $T_0 \in \mathscr{L}(\mathscr{H})$ such that $A^{P(T_0)}$ admits at least two elements. As $A^{P(T_0)} \cap A^{\sim} \neq \varnothing$, there exists $S_0 \in \mathscr{S}$ such that $S_0(T_0) = T_1 \in A^{P(T_0)} \cap A^{\sim}$; then $A^{T_1} = \{T_1\}$ and we have $A^{T_1} \subsetneq A^{P(T_0)}$, that is, $A^{S_0(T_0)} \subsetneq A^{P(T_0)}$. We contradict the maximality of P. Therefore we conclude that, for every $T \in \mathscr{L}(\mathscr{H})$, $A^{P(T)}$ is the singleton $\{P(T)\}$ and $P(T) \in A^T \cap A^{\sim}$. For all $U \in A^{\sim}$, we have $P(U) = U$; P maps $\mathscr{L}(\mathscr{H})$ onto $A^{\sim}$. Finally for all $U, V \in A^{\sim}$ and $T \in A$, we have $P(UTV) = UP(T)V$. $\square$

Proposition 20.17. *Let G be a discrete group. Then G is amenable if and only if the von Neumann algebra $PM(G)$ in $\mathscr{L}(l^2(G))$ generated by $\{ L_a : a \in G \}$ has property (P).*

EXISTENCE OF INVARIANT MEANS   241

Proof. (a) Let G be amenable. By Theorem 4.19 there exists $M \in \text{RIM}(l^\infty(G))$. Fix $T \in \mathcal{L}(l^2(G))$. If $\varphi, \psi \in l^2(G)$, we define

$$T_{\varphi,\psi} : x \mapsto \left(L_x^* T L_x \varphi | \psi \right)$$

$$G \to \mathbf{C};$$

$$\|T_{\varphi,\psi}\| \leq \|T\| \|\varphi\|_2 \|\psi\|_2,$$

$T_{\varphi,\psi} \in l^\infty(G)$. We may consider $T' \in \mathcal{L}(l^2(G))$ defined by

$$(T'\varphi|\psi) = M(T_{\varphi,\psi}),$$

$\varphi, \psi \in l^2(G)$.

For all $\varphi, \psi \in l^2(G)$, $a \in G$ and every $x \in G$,

$$T_{L_a\varphi, L_a\psi}(x) = \left(L_x^* T L_x L_a \varphi | L_a \psi \right) = \left(L_a^* L_x^* T L_x L_a \varphi | \psi \right) = \left(L_{xa}^* T L_{xa} \varphi | \psi \right)$$

so $M(T_{L_a\varphi, L_a\psi}) = M(T_{\varphi,\psi})$, that is,

$$\left(L_a^* T' L_a \varphi | \psi \right) = (T' L_a \varphi | L_a \psi) = (T'\varphi|\psi).$$

Hence $L_a^* T' L_a = T'$, $T' L_a = L_a T'$, that is, $T' \in PM(G)^\sim$. As $M_f^1(G)$ is w*-dense in $\mathfrak{M}(l^\infty(G))$ by Corollary 3.4, the definition of T' implies that $T' \in PM(G)^T$. Therefore property (P) is satisfied.

(b) Conversely let us assume that there exists a linear mapping P of $\mathcal{L}(l^2(G))$ onto $PM(G)^\sim$ satisfying the conditions of Lemma 20.16. The von Neumann algebra $PM(G)$ and the von Neumann algebra generated by the right regular representation R are commutants of each other in $\mathcal{L}(l^2(G))$ ([140] Chap. I, Section 5, 2, théorème 1). Therefore there exists a tracial state τ on $PM(G)^\sim$, that is, a nonnegative functional on $PM(G)^\sim$ such that $\tau(id_{l^2(G)}) = 1$ and $\tau(U^*U) = \tau(UU^*)$ whenever $U \in PM(G)^\sim$ ([501] p. 182). To every $g \in l^\infty(G)$ associate

$$\Lambda_g : \varphi \mapsto g\varphi$$

$$l^2(G) \to l^2(G);$$

$\Lambda_g \in \mathcal{L}(l^2(G))$. If $f \in l_\mathbf{R}^\infty(G)$, let

$$M(f) = \tau(P(\Lambda_f));$$

then $M \in \mathfrak{M}(l_\mathbf{R}^\infty(G))$.

For every $a \in G$, $R_a \in PM(G)^\sim$. For all $f \in l_\mathbf{R}^\infty(G)$, $a \in G$, and $\varphi \in l^2(G)$, we have

$$R_a \Lambda_f R_a^* \varphi = R_a \Lambda_f \varphi_a = R_a(f\varphi_a) = f_{a^{-1}}\varphi,$$

that is, $R_a \Lambda_f R_a^* = \Lambda_{f_a-1}$. Via Lemma 20.16 (iii) we obtain

$$P(\Lambda_{f_a-1}) = P(R_a \Lambda_f R_a^*) = R_a P(\Lambda_f) R_a^*.$$

If $f \in l_+^\infty(G)$, $P(\Lambda_f)$ is a positive-definite operator in $PM(G)^\sim$ and admits a square root Q in $PM(G)^\sim$ ([HR] C.35). We have

$$M(f_{a^{-1}}) = \tau(R_a P(\Lambda_f) R_a^*) = \tau(R_a Q (R_a Q)^*) = \tau((R_a Q)^* R_a Q)$$

$$= \tau(Q^* Q) = \tau(P(\Lambda_f)) = M(f).$$

Therefore $M \in \mathrm{RIM}(l_\mathbf{R}^\infty(G))$. By Theorem 4.19, G is amenable. □

Let G be a locally compact group and let X be a (nonzero) w*-closed subalgebra of $L^\infty(G)$ that is closed under complex conjugation. Then X is left invariant if and only if there exists a (unique) closed subgroup N_X of G such that $X = \{f \in L^\infty(G) : (\forall a \in N_X) f_a = f\}$ ([342] lemma 3.2).

Proposition 20.18. *A locally compact group G is amenable if and only if, for every w*-closed left invariant subalgebra X of $L^\infty(G)$ that is closed under complex conjugation, there exists a continuous projection of $L^\infty(G)$ onto X that commutes with left translations, that is, X admits a left invariant complement.*

Proof. (a) Assume G to be amenable and let X be a nonzero w*-closed left invariant subalgebra of $L^\infty(G)$ that is closed under complex conjugation. For $f \in L^\infty(G)$, let C_f be the w*-closed convex hull of $\{f_a : a \in G\}$ and consider the continuous affine action $(x, f) \mapsto f_x$ of G on $L^\infty(G)$ equipped

$$G \times L^\infty(G) \to L^\infty(G)$$

with the w*-topology. By Proposition 13.3, N_X is amenable. Hence Theorem 5.4 implies the existence of $h \in C_f$ such that $h_x = h$ whenever $x \in N_X$, that is, $C_f \cap X \neq \emptyset$. An abstract version of Lemma 20.16 ([575] Proposition) yields the existence of a w*-continuous projection P of $L^\infty(G)$ onto X that commutes with any w*-w* continuous linear operator on $L^\infty(G)$ commuting with the right translations by the elements of N_X. In particular, P commutes with the left translations on $L^\infty(G)$.

(b) Conversely consider the trivial subalgebra $X = \mathbf{C} 1_G$. Let I be a continuous projection of $L^\infty(G)$ onto X that commutes with left translations. Let $\Phi(f) = \alpha$ for $f \in L^\infty(G)$ with $I(f) = \alpha 1_G$, $\alpha \in \mathbf{C}$. Consider $\Psi = \frac{1}{2}(\Phi + \Phi')$, where $\Phi'(f) = \overline{\Phi(\bar{f})}$, $f \in L^\infty(G)$, and $\Psi = \Psi_+ - \Psi_-$, where Ψ_+, Ψ_- are nonnegative linear functionals. Suppose that, for instance, $\Psi_+ \neq 0$, then $\Psi_+(1_G) \neq 0$ and $M = \Psi_+(1_G)^{-1} \Psi \in \mathrm{LIM}(L_\mathbf{R}^\infty(G))$. By Theorem 4.19, the group G is amenable. □

EXISTENCE OF INVARIANT MEANS

In Corollary 14.3 we showed that $SO(3, \mathbf{R})_d$ is nonamenable. Hausdorff's [253] original proof established the nonamenability of the discrete rotation group of the unit sphere S_2 in $\mathbf{R}^3$ by producing ingenious "paradoxical decompositions" implying the nonexistence of a nontrivial finite nonnegative finitely additive measure on all subsets of S_2 that is invariant under all rotations (Section 1). We describe some general situations where this phenomenon occurs. The properties emphasize the characterization of amenability in terms of the Dixmier criterion stated in Section 4D.

Definition 20.19. *Let G be a locally compact group. If $n \in \mathbf{N}^*$ and $E_1, \ldots, E_n$ are nonnecessarily distinct Borel subsets of G, then $S = (E_1, \ldots, E_n)$ is called multiset. To S we associate the multiset function $\chi_S = \sum_{i=1}^{n} 1_{E_i}$.*

If G is a locally compact group, we denote by $\mathscr{D}(G)$ the set of all classes of multisets for G where we identify multisets S and T admitting the same multiset function. If $S = (E_1, \ldots, E_n)$, $S' = (E'_1, \ldots, E'_{n'}) \in \mathscr{D}(G)$, let $S \vee S' = (E_1, \ldots, E_n; E'_1, \ldots, E'_{n'})$. If E is a Borel subset of G, the n-tuple $(E, \ldots, E)$ is denoted by $\vee_n E$; E is identified with $\vee_1 E$. We consider the partial order $\ll$ defined for $\mathscr{D}(G)$ in the following way: If $S, S' \in \mathscr{D}(G)$, $S \ll S'$ signifies that $\chi_S \leq \chi_{S'}$. We also define an equivalence relation. For $S, T \in \mathscr{D}(G)$, $S \sim T$ signifies that there exist $a_1, \ldots, a_n \in G$, $S_0 = (E_1, \ldots, E_n) \in \mathscr{D}(G)$, $T_0 = (a_1 E_1, \ldots, a_n E_n) \in \mathscr{D}(G)$ such that $\chi_S = \chi_{S_0}$ and $\chi_T = \chi_{T_0}$. We denote by $\mathscr{D}(G)^{\cdot}$ the set of the corresponding equivalence classes.

Proposition 20.20. *A locally compact group G is amenable if and only if there exists $F \in \mathscr{D}(G)^{\cdot} \to \mathbf{R}_+$ such that $F(\dot{G}) > 0$ and $F((S \vee T)^{\cdot}) = F(\dot{S}) + F(\dot{T})$ whenever $\dot{S}, \dot{T} \in \mathscr{D}(G)^{\cdot}$.*

Proof. If G is amenable and $M \in \mathrm{LIM}(L^\infty_{\mathbf{R}}(G))$, it suffices to put

$$F(\dot{S}) = M(\chi_S)$$

for $\dot{S} \in \mathscr{D}(G)^{\cdot}$.

Conversely if the condition holds, let $M(\chi_S) = F(\dot{G})^{-1} F(\dot{S})$ for $S \in \mathscr{D}(G)$. Then $M(1_G) = M(\chi_G) = 1$. If $S, T \in \mathscr{D}(G)$ and $S \sim T$, we have $M(\chi_S) = M(\chi_T)$. We extend M to a left invariant mean on the space of all bounded real-valued Borel measurable functions on G. Therefore G is amenable by Theorem 4.19. □

Proposition 20.21. *Let G be a locally compact group. Then G is nonamenable if and only if there exist $n \in \mathbf{N}^*$ and $S, T \in \mathscr{D}(G)$ such that $S \sim T$ and $S \vee (\vee_n G) \ll T$.*

Proof. (a) If G is nonamenable, as $\mathrm{LIM}(\mathscr{C}_{\mathbf{R}}(G)) = \varnothing$, a fortiori there does not exist a left invariant mean on the real vector space spanned by all real-valued simple Borel measurable functions defined on G. By Lemma 4.28 there exist $S, T \in \mathscr{D}(G)$ and $n \in \mathbf{N}^*$ such that $S \sim T$ and $\sup(\chi_S - \chi_T) \leq$

$-n$. We have then $\chi_S + n1_G \leq \chi_T$, that is, $S \vee (\vee_n G) \ll T$.

(b) Assume that the condition holds. There exist $E_1, \ldots, E_m \in \mathcal{B}(G)$ and $a_1, \ldots, a_m \in G$ such that

$$\sum_{i=1}^m 1_{E_i} + n1_G \leq \sum_{i=1}^m 1_{a_i E_i},$$

that is,

$$\sum_{i=1}^m \left(1_{E_i} - _{a_i^{-1}}1_{E_i}\right) \leq -n1_G.$$

We conclude from Proposition 4.29 that G is nonamenable. □

Proposition 20.22. *A locally compact group G is nonamenable if and only if there exists $n_0 \in \mathbf{N}^*$ such that, for every $n \in \mathbf{N}^*$, one may determine $V_n \in \mathcal{D}(G)$ satisfying $V_n \sim (\vee_{n_0} G)$ and $\vee_n G \ll V_n$.*

Proof. (a) Assume that the condition holds. If G were amenable, with the notations of Proposition 20.20 we would have

$$F(\dot{V}_n) = n_0 F(\dot{G})$$

and

$$nF(\dot{G}) \leq F(\dot{V}_n)$$

giving rise to a contradiction.

(b) Assume G to be nonamenable. By Proposition 20.21 there exist S, $T_1 \in \mathcal{D}(G)$ such that $S \sim T_1$ and $S \vee (\vee_1 G) \ll T_1$. Therefore there exists $S' \in \mathcal{D}(G)$ such that $S \vee S' = T_1$ and $\vee_1 G \ll S'$. Then also $S \sim T_1 \sim T_1 \vee S' \sim S \vee (\vee_2 S') = T_2$ and $S \vee (\vee_2 G) \ll T_2$. Inductively, for every $n \in \mathbf{N}^*$, we may determine $T_n \in \mathcal{D}(G)$ such that $T_n \sim S$ and $S \vee (\vee_n G) \ll T_n$.

There exist $D \in \mathcal{D}(G)$ and $n_0 \in \mathbf{N}^*$ such that $S \vee D = \vee_{n_0} G$. Then for every $n \in \mathbf{N}^*$, $V_n = T_n \vee D \sim S \vee D = \vee_{n_0} G$ and $\vee_n G \ll S \vee (\vee_n G) \ll T_n \ll V_n$. □

Proposition 20.23. *A locally compact group G is nonamenable if and only if there exist $n_0 \in \mathbf{N}^*$ and a sequence (W_n) in $\mathcal{D}(G)$ such that, for every $n \in \mathbf{N}^*$, $W_n \sim G$ and $W_1 \vee \cdots \vee W_n \ll \vee_{n_0} G$.*

Proof. The condition is stronger than the one considered in Proposition 20.22. Therefore it implies nonamenability.

Now assume G to be nonamenable. By Proposition 20.21 there exist S, $T \in \mathcal{D}(G)$ such that $S \sim T$ and $S \vee G \ll T$. One may therefore determine $D \in \mathcal{D}(G)$ such that $T \sim S \vee D$ and $G \ll D$. Thus $S \sim S \vee D$, and there exist $S_1, G_1 \in \mathscr{D}(G)$ satisfying $S_1 \sim S$, $G_1 \sim D$, $G \ll G_1$, $S_1 \vee G_1 = S$. We have $S_1 \sim (S_1 \vee G_1)$ and there exist $S_2, G_2 \in \mathscr{D}(G)$ such that $S_2 \sim S$, $G_2 \sim G_1$, and $S_2 \vee G_2 = S_1$. Continuing this procedure inductively we determine sequences $(S_n), (G_n)$ in $\mathscr{D}(G)$ such that, for every $n \in \mathbf{N}^*$, $S_n \sim S$, $G_n \sim D$, and $S_n \vee G_n = S_{n-1}$. So $S = S_n \vee G_1 \vee \cdots \vee G_n$.

As for every $n \in \mathbf{N}^*$, $G_n \sim D$, and $G \ll D$, there also exists a sequence (W_n) of multisets such that, for every $n \in \mathbf{N}^*$, $W_n \sim G$, and $W_n \ll G_n$; $W_1 \vee \cdots \vee W_n \ll G_1 \vee \cdots \vee G_n \ll S$ and one may determine $n_0 \in \mathbf{N}^*$ such that $S \ll \vee_{n_0} G$. □

NOTES

Stationarity properties had been considered originally for amenable discrete semigroups by Mitchell [391] and Day [120]. Versions of the properties 20.2-3 were first given by Granirer and Lau [224], and Wong [560]. Emerson [167] also proved 20.2 for $L^\infty(G)$. Yeadon [575] and Day [121] obtained important results concerning these conditions on discrete semigroups and groups. For topological semigroups such properties were established by Granirer and Lau [224], Berglund, Junghenn, and Milnes [34] and, in a very broad setting, by Junghenn [301].

The proofs of 20.4-6 are essentially due to Emerson [167] who examined in detail this type of property stemming initially from the Dixmier criterion. In the discrete case 20.7 (i) ⇔ (iii) had been obtained by Day [114] for bounded functions. Raimi [442] established (iii) ⇒ (i) for a general lattice. Hewitt and Ross [HR I] proved 20.7 for bounded functions on semigroups, assumed to be left cancellative in the demonstration of (iii) ⇒ (i). Melven and Myren Krom [323] gave 20.9-10. The proof of the Dixmier criterion developed in 4.27-28 makes use of the Hahn-Banach theorem in view of extending an invariant functional on a subspace to an invariant functional over the whole space. Various general extension properties linked to amenability were studied by Dixmier [138], Silverman [512], Day [120], and Wong [561].

G. Woodward [572] proved 20.11 and the existence properties of discontinuous left invariant functionals stated in 20.12-14.

Property (P) was introduced by Schwartz [509]. He showed that (P) implies amenability for countable discrete groups. Our demonstration of 20.16 is taken from S. Sakai [501] who also established the proof of 20.17 and examined amenability of the discrete group G in connection with the structure of $PM(G)$. Also see Milnes [387]. Lau [342] established 20.18 and added further comments.

As we stated already in Section 1, Hausdorff [253] produced paradoxical decompositions for $\mathbf{R}^3$. Tarski [534] showed nonexistence of paradoxical decompositions for arbitrary bounded subsets of $\mathbf{R}^2$. In their joint work Banach and Tarski [29] proved via the axiom of choice that in $\mathbf{R}^3$ any sphere can be decomposed into a finite number of pieces and reassembled into two spheres with the same radius. They showed two arbitrary bounded subsets in $\mathbf{R}^n$ ($n \geq 3$) having nonvoid interiors to be equivalent by finite decomposition. The latter general result is in fact due to Tarski; see G. Moore [400] for this historical point. The important later contributions on paradoxical decompositions are Tarski [535] and von Neumann [414]. Emerson [168] gave 20.20–23 and carefully studied the subject after the discrete case had been investigated by Sherman [511]. For further studies and complementary historical comments see Wagon [546, 547].

K. Sakai [498] generalized the considerations on (P) to the following situation. Let G be a topological group, X a closed invariant subspace of $l^\infty(G)$ that contains 1_G and is closed under complex conjugation. If T' [resp. T''] is a unitary representation of G on a Hilbert space $\mathscr{H}_1$ [resp. $\mathscr{H}_2$], let $I(T', T'') = \{A \in \mathscr{L}(\mathscr{H}_1, \mathscr{H}_2) : (\forall x \in G)\ T_x''A = AT_x'\}$. Suppose that, for each $A \in \mathscr{L}(\mathscr{H}_1, \mathscr{H}_2)$, the mapping $U: x \mapsto T_x''AT_{x^{-1}}'$ is X-admissible, that is, the mappings $x \mapsto \langle U_x \xi, \eta \rangle$ belong to X whenever $\xi \in \mathscr{H}_1$ and $\eta \in \mathscr{H}_2$. If X admits a left invariant mean, then there exists an endomorphism P of $\mathscr{L}(\mathscr{H}_1, \mathscr{H}_2)$ such that $\|P\| = 1$, for all $A \in \mathscr{L}(\mathscr{H}_1, \mathscr{H}_2)$, $P(A)$ belongs to $\overline{\mathrm{co}}\,\{T_x''AT_{x^{-1}}' : x \in G\}$, $P(A) \in I(T', T'')$, $P(P(A)) = P(A)$, and $P(B''AB') = B''P(A)B'$ whenever $B' \in \mathscr{L}(\mathscr{H}_1)$ and $B'' \in \mathscr{L}(\mathscr{H}_2)$ commute with T', T''. If T is a unitary representation of G on a Hilbert space $\mathscr{H}$ such that the mapping $x \mapsto T_x A T_x^{-1}$ of G into $\mathscr{L}(\mathscr{H})$ is X-admissible for every $A \in \mathscr{L}(\mathscr{H})$ and X admits a left invariant mean, then the von Neumann algebra generated by $\{T_x : x \in G\}$ has property (P). In particular, if G is amenable, that is, $\mathrm{LIM}(\mathscr{C}(G)) \neq \varnothing$, for every continuous unitary representation T of G on a Hilbert space, the von Neumann algebra generated by $\{T_x : x \in G\}$ has property (P).

21. KERNELS AND RANGES OF INVARIANT MEANS

We give indications on the kernels of invariant means and equivalently determine under certain conditions the subsets on which invariant means are concentrated. The interplay between the problems of existence of invariant means for a group G and existence of G-invariant finitely additive measures, that is, the question which led to the notion of amenability originally, facilitates the comparison of the kernels of different invariant means. Upper and lower bounds for the sets of invariant means are investigated. We also determine subspaces of functions on which all invariant means agree. Certain amenable groups admit nested families of subsets such that all invariant means on the corresponding characteristic functions agree and admit ranges $[0, 1]$.

A. Kernels of Invariant Means

We first state a technical lemma.

Lemma 21.1. *Let G be an amenable group. For every $A \in \mathfrak{H}(G)$ and every $M \in \mathrm{TLIM}(L^\infty(G))$, $M(1_A) \leq |A^{-1}|$.*

Proof. Let K be a compact subset of G such that $|K| \geq 1$; $\xi_K \leq 1_G$. By Proposition 3.3 there exists a net $(\varphi_i)_{i \in I}$ in $P^1(G)$ such that

$$\lim_i \langle \xi_K * 1_A, \varphi_i \rangle = M(\xi_K * 1_A) = M(1_A).$$

On the other hand, for every $i \in I$,

$$\langle \xi_K * 1_A, \varphi_i \rangle = \iint \xi_K(xy) 1_A(y^{-1}) \varphi_i(x) \, dx \, dy$$

$$\leq \int \varphi_i(x) \, dx \int 1_A(y^{-1}) \, dy = |A^{-1}|. \qquad \Box$$

The next fundamental proposition is quite immediate.

Proposition 21.2. *In a noncompact amenable group G the following properties hold:*
   (a) *For every compact subset K of G and every left invariant mean on $L^\infty(G)$, $M(1_K) = 0$.*
   (b) *For every measurable subset A of G such that $|A^{-1}| < \infty$ and every topologically left invariant mean M on $L^\infty(G)$, $M(1_A) = 0$.*
   (c) *If $f \in \mathcal{K}(G)$ and M is a left invariant mean on $\mathcal{C}(G)$, then $M(f) = 0$.*
   (d) *If $f \in \mathcal{C}_0(G)$ and M is a left invariant mean on $L^\infty(G)$, then $M(f) = 0$.*

Proof. (a) To K we associate the infinite sequence (a_n) determined in Lemma 20.11. For any $n \in \mathbf{N}^*$, let $B_n = \bigcup_{i=1}^n a_i K$; then $nM(1_K) = M(1_{B_n}) \leq M(1_G) = 1$. So we must have $M(1_K) = 0$.
   (b) For every $\varepsilon \in \mathbf{R}_+^*$, regularity of the Haar measure implies the existence of a compact subset K in G such that $|A^{-1} \setminus K^{-1}| < \varepsilon$. By Lemma 21.1 we have

$$0 \leq M(1_A) = M(1_{A \setminus K}) + M(1_K) \leq |A^{-1} \setminus K^{-1}| + M(1_K) < \varepsilon + M(1_K).$$

By Corollary 4.12, $M \in \mathrm{LIM}(L^\infty(G))$; so we deduce from (a) that $0 \leq M(1_A) < \varepsilon$ and, as $\varepsilon \in \mathbf{R}_+^*$ is arbitrary, we have $M(1_A) = 0$.
   (c) Let $K = \operatorname{supp} f$. To K we associate again the infinite sequence (a_n) determined in Lemma 20.11 and let $g_n = \sum_{i=1}^n {}_{a_i} f$, $n \in \mathbf{N}^*$. For any $n \in \mathbf{N}^*$,

$$n|M(f)| = |nM(f)| = |M(g_n)| \leq \|g_n\| = \|f\|$$

hence necessarily $M(f) = 0$.
   (d) For every $\varepsilon \in \mathbf{R}_+^*$, there exists a compact subset K in G such that $|f(x)| < \varepsilon$ whenever $x \in G \setminus K$. Then $|f| \leq \|f\| 1_K + \varepsilon 1_{G \setminus K} \leq \|f\| 1_K + \varepsilon 1_G$. By (a), $|M(f)| \leq \varepsilon$; as $\varepsilon \in \mathbf{R}_+^*$ is arbitrary, $M(f) = 0$. $\qquad \Box$

Definition 21.3. *Let G be a locally compact group and let A be a measurable subset of G. The mean M on $L^\infty(G)$ is said to be concentrated on A if $M(1_A) = 1$.*

Proposition 21.4. *Let G be an amenable group and let A be a measurable subset of G. Then there exists a topologically left invariant [resp. topologically right invariant] mean on $L^\infty(G)$, that is concentrated on A, if and only if for one (every) compact subset K of G with $|K| > 0$, $\sup\{|K \cap Ax| : x \in G\} = |K|$ [resp. $\sup\{|K \cap xA| : x \in G\} = |K|$] holds.*

Proof. If K is any compact subset of G such that $|K| > 0$, consider $\xi_K^* * 1_A, \xi_K * 1_{A^{-1}} \in \mathcal{RUC}(G)$. For every $x \in G$,

$$\xi_K^* * 1_A(x) = \frac{1}{|K|} \int 1_K(y^{-1}) \Delta(y^{-1}) 1_A(y^{-1}x) \, dy$$

$$= \frac{1}{|K|} \int 1_K(y) 1_A(yx) \, dy = \frac{|K \cap Ax^{-1}|}{|K|}$$

and

$$\xi_K * 1_{A^{-1}}(x) = \frac{1}{|K|} \int 1_K(y) 1_{A^{-1}}(y^{-1}x) \, dy = \frac{1}{|K|} \int 1_K(y) 1_A(x^{-1}y) \, dy$$

$$= \frac{1}{|K|} \int 1_K(y) 1_{xA}(y) \, dy = \frac{|K \cap xA|}{|K|}.$$

(a) If there exists a compact subset K of G with $|K| > 0$ that does not satisfy the condition, then $\|\xi_K^* * 1_A\| < 1$ [resp. $\|\xi_K * 1_{A^{-1}}\| < 1$]. For every $M \in \text{TLIM}(L^\infty(G))$, $0 \leq M(1_A) = M(\xi_K^* * 1_A) < 1$ [resp. for every $M \in \text{TRIM}(L^\infty(G))$, with respect to Proposition 4.7, $0 \leq M(1_A) = \check{M}(\check{1}_A) = \check{M}(1_{A^{-1}}) = \check{M}(\xi_K * 1_{A^{-1}}) < 1$].

(b) Assume that there exists a compact subset K of G with $|K| > 0$ satisfying the condition. Let $u = \xi_K^* * 1_A$ [resp. $u = \xi_K * 1_{A^{-1}}$]. For every $n \in \mathbb{N}^*$, there exist $x_n \in G$ and a compact symmetric neighborhood U_n of e in G such that $1 - u(x_n x) < 1/2n$ whenever $x \in U_n$. Therefore

$$1 - u * \xi_{U_n}(x_n) = \int \xi_{U_n}(x) \, dx - \int u(x_n x) \xi_{U_n}(x) \, dx$$

$$= \int (1 - u(x_n x)) \xi_{U_n}(x) \, dx < \frac{1}{2n}.$$

As $u * \xi_{U_n}$ is also a continuous function, there exists a compact neighborhood V_n of x_n such that $1 - u * \xi_{U_n}(x) < 1/n$ whenever $x \in V_n$; thus

$$\langle \xi_{V_n}, 1_G - u * \check{\xi}_{U_n} \rangle = \langle \xi_{V_n}, 1_G - u * \xi_{U_n} \rangle < 1/n.$$

KERNELS AND RANGES OF INVARIANT MEANS 249

By Corollary 20.3 there exists $M \in \text{TLIM}(L^\infty(G))$ such that $M(u) = 1$. Hence $M(1_A) = M(\xi_K^* * 1_A) = 1$ [resp. by Proposition 4.7, for $\check{M} \in \text{TRIM}(L^\infty(G))$, $\check{M}(1_A) = M(\check{1}_A) = M(1_{A^{-1}}) = M(\xi_K * 1_{A^{-1}}) = 1$]. □

Corollary 21.5. *Let G be an amenable group and let A be a measurable subset of G such that $M(1_A) > 0$ whenever $M \in \text{TLIM}(L^\infty(G))$. Then there exists a compact subset K in G such that $A \cap Kx \neq \emptyset$ whenever $x \in G$.*

Proof. If the property did not hold, then we could associate to every compact subset K of G an element a of G such that $Ka \subset G \setminus A$. Proposition 21.4 would imply the existence of $M \in \text{TLIM}(L^\infty(G))$ such that $M(1_{G \setminus A}) = 1$. We would come to a contradiction. □

Proposition 21.6. *If G is an amenable group that is generated by a measurable subsemigroup S and there exists $M \in \text{LIM}(L^\infty(G))$ such that $M(1_S) > 0$, then there exists a left invariant mean N on $L^\infty(G)$ that is concentrated on S. It is defined by $M'(f) = M(1_S)^{-1} M(1_S f)$, $f \in L^\infty(G)$.*

Proof. Obviously $M' \in \mathfrak{M}(L^\infty(G))$. Let $N(f) = M(1_S f)$, $f \in L^\infty(G)$. If $a \in S$, for every $n \in \mathbf{N}$, consider $A_n = a^n S \setminus a^{n+1} S$. The sets $A_n (n \in \mathbf{N})$ form a partition of S and $aA_n = A_{n+1}$ for every $n \in \mathbf{N}$. Therefore $M(1_{A_0}) = 0$. Thus for every $f \in L^\infty(G)$,

$$N(_{a^{-1}}f) = M(1_S {_{a^{-1}}f}) = M(1_{A_0 a^{-1}} f) + M(1_{aS a^{-1}} f) = M(1_{aS a^{-1}} f)$$

$$= M(_{a^{-1}}(1_S f)) = M(1_S f) = N(f).$$

Now let $f \in L^\infty(G)$ and $a, b \in S$. We have

$$N(_a f) = N(_{a^{-1}}(_a f)) = N(f)$$

and

$$N(_{ab} f) = N(_b (_a f)) = N(_a f) = N(f).$$

So for all $x \in G$, $N(_x f) = N(f)$. Therefore $M' \in \text{LIM}(L^\infty(G))$, and $M'(1_S) = 1$. □

Proposition 21.7. *Let G be a locally compact group and let $\mathfrak{I}(G)$ be a set of measurable subsets of G with positive measure such that (a) $A \cap B \in \mathfrak{I}(G)$ whenever $A, B \in \mathfrak{I}(G)$, (b) $aA \in \mathfrak{I}(G)$ whenever $A \in \mathfrak{I}(G)$ and $a \in G$. If G_d is amenable [resp. G is amenable], there exists $M \in \text{LIM}(L^\infty(G))$ [resp. $M \in \text{LIM}(\mathcal{RUC}(G))$] such that $M(f) = 0$ whenever $f \in L^\infty(G)$ [resp. $f \in \mathcal{RUC}(G)$] and $\mathcal{K}\!\mathit{er}\, f$ contains an element of $\mathfrak{I}(G)$.*

Proof. Let X be $L^\infty(G)$ [resp. the $\mathcal{M}_d^1(G)$-introverted subspace $\mathcal{RUC}(G)$ of $L^\infty(G)$ (Section 2G.4)]. We consider the set I of all $f \in X$ such that there exists $A \in \mathfrak{F}(G)$ with $A \subset \mathscr{K}\!er\, f$. If $f_1, f_2 \in X$, $\mathscr{K}\!er\,(f_1 + f_2) \supset \mathscr{K}\!er\, f_1 \cap \mathscr{K}\!er\, f_2$, $\mathscr{K}\!er\,(f_1 f_2) \supset \mathscr{K}\!er\, f_1$; I is an ideal in X by (a). As $1_G \notin I$, I is contained in a proper maximal ideal J of the commutative Banach algebra X, and J is the kernel of a (unique) multiplicative linear functional F on X; $\|F\| = F(1_G) = 1$ ([HR] C. 17, 21). We define $F' \in \mathscr{L}(L^\infty(G), l^\infty(G))$ [resp. $F' \in \mathscr{L}(\mathscr{RUC}(G), \mathscr{RUC}(G))$] by putting

$$F'(f)(x) = F(_x f)$$

for $f \in X$, $x \in G$. Let $N \in \mathrm{LIM}(l^\infty(G_d))$ [resp. $N \in \mathrm{LIM}(\mathscr{RUC}(G))$] and

$$M(f) = \langle F'(f), N \rangle,$$

$f \in X$. Then $M(1_G) = 1$, $\|M\| = 1$ and, by Proposition 3.2, $M \in \mathfrak{M}(X)$. If $f \in X$ and $a, x \in G$, we have

$$F'(_a f)(x) = F(_x(_a f)) = F(_{ax} f),$$

that is, $F'(_a f) = {_a}F'(f)$. As N is left invariant, we conclude that $M \in \mathrm{LIM}(X)$. If $f \in I$, for all $x \in G$, $_x f \in I$ by (b); $F(_x f) = 0$. Hence $F'(f) = 0$ and $M(f) = 0$. □

Corollary 21.8. *Let G be a compact group such that G_d is amenable. If $A \in \mathfrak{F}(G)$ such that $|A| < 1$ and $|\bigcap_{j=1}^m a_j A| > 0$ whenever $a_1, \cdots, a_m \in G$, then $1_A \notin \mathscr{C}_\mathbf{R}(G) + \mathscr{N}(L^\infty_\mathbf{R}(G))$.*

Proof. We may apply Proposition 21.7 to the particular collection $\mathfrak{F}(G)$ that consists of all subsets $\bigcap_{x \in F} xA (F \in \mathfrak{F}(G))$; we determine $M \in \mathrm{LIM}(L^\infty(G))$ such that $M(1_{G \setminus A}) = 0$, that is, $M(1_A) = 1$. Let $f \in \mathscr{C}_\mathbf{R}(G)$ and $h \in \mathscr{N}(L^\infty_\mathbf{R}(G))$. By Proposition 12.1 we have $M(f) = \int f$; moreover $M(h) = 0$ and $\int h = 0$. Therefore

$$\|1_A - (f + h)\|_\infty \geq \tfrac{1}{2}\left[|M(1_A - (f+h))| + \left|\int(1_A - (f+h))\right|\right]$$

$$\geq \tfrac{1}{2}\left|M(1_A - f - h) - \int(1_A - f - h)\right|$$

$$= \tfrac{1}{2}[M(1_A) - |A|] = \tfrac{1}{2}[1 - |A|] > 0. \quad \square$$

We establish another consequence of Proposition 21.7 that should be compared with Proposition 9.3. We first prove a technical lemma.

Lemma 21.9. *Let G be a nondiscrete σ-compact locally compact group. For every $\varepsilon > 0$, there exists an open subset U in G such that $\overline{U} = G$ and $|U| < \varepsilon$, that is, there exists a Borel subset D in G such that $\overline{D} = G$ and $|D| = 0$.*

Proof. (a) Assume G to be compactly generated. As G is nondiscrete, $|\{e\}| = 0$. For every $n \in \mathbf{N}^*$, there exists a compact neighborhood U_n of e in G such that $|U_n| < 1/n$. Then there also exists a compact normal subgroup K of the compactly generated group G such that $K \subset \cap_{n \in \mathbf{N}^*} U_n$; $|K| = 0$ and the metrizable set G/K admits a countable basis for its open subsets. If π is the canonical homomorphism of G onto G/K, we may find $A = \{x_n : n \in \mathbf{N}^*\}$ in G such that $\overline{\pi(A)} = G/K$ ([HR] 8.3,7). Then AK is dense in G and, as $|K| = 0$, we have $|AK| = 0$.

(b) We now come to the general case.

There exists a sequence (K_n) of compact subsets in G such that $G = \cup_{n \in \mathbf{N}^*} K_n$. For every $n \in \mathbf{N}^*$, we consider a compact symmetric neighborhood U_n of e such that $K_n \subset U_n$. Then $G_n = \cup_{p \in \mathbf{N}^*} U_n^p$ is a compactly generated open subgroup of G and the left Haar measure λ_{G_n} on G_n may be considered to be induced by λ_G.

For every $n \in \mathbf{N}^*$, we deduce from (a) the existence of a dense Borel subset D_n in G_n such that $\lambda(D_n) = \lambda_{G_n}(D_n) = 0$. Let $D = \cup_{n \in \mathbf{N}^*} D_n$. As $G = \cup_{n \in \mathbf{N}^*} G_n$ and $\overline{D} \supset \cup_{n \in \mathbf{N}^*} \overline{D_n}$, we have $|D| = 0$ and $\overline{D} = G$. □

Proposition 21.10. *Let G be a nondiscrete locally compact group such that G_d is amenable. Then there exists $T \in \mathscr{L}(L^\infty(G))$ having the following properties:*

(a) *For all $f \in L^\infty(G)$ and $a \in G$, $T(f_a) = (Tf)_a$.*

(b) *There exist $g \in \mathscr{K}(G)$, $h \in L^\infty(G)$ with $(Th) * g \neq T(h * g)$.*

Proof. Let $g \in \mathscr{K}_+(G)$ such that $\int g = 2$ and put $\alpha = \|g\|^{-1}$. For every $B \in \mathfrak{H}(G)$ such that $|B| < \alpha$, we have $\int_B g < 1$. Let V be a relatively compact open subset of G containing $\operatorname{supp} g$. On the open subgroup H of G generated by V we consider the Haar measure induced by λ_G. Lemma 21.9 implies the existence of a dense open subset U in H such that $|U| < \alpha$. If $\{x_i : i \in I\}$ is a system of representatives of G/H, $A = (\cup_{i \in I} x_i U)^{-1}$ is open and dense in G. Then for all $a_1, \cdots, a_m \in G$, we have $\cap_{j=1}^m a_j A \neq \varnothing$. We apply Proposition 21.7 to $\mathfrak{F}(G)$ formed by the subsets $\cap_{x \in F} xA (F \in \mathfrak{F}(G))$. There exists $M \in \operatorname{LIM}(L^\infty(G))$ such that $M(1_{G \setminus A}) = 0$.

We define $T \in \mathscr{L}(L^\infty(G))$ by putting

$$Tf(x) = M(\check{f}),$$

$f \in L^\infty(G)$, $x \in G$. As M is left invariant, for all $f \in L^\infty(G)$, $a \in G$, by (2.1) we have $M((f_a)^\vee) = M(_{a^{-1}}\check{f}) = M(\check{f})$, hence $T(f_a) = Tf$. As Tf is a constant function, we also have $(Tf)_a = Tf$, and therefore $T(f_a) = (Tf)_a$.

From the fact that $M(1_{G \setminus A}) = 0$ we deduce that

$$T(1_{G \setminus A^{-1}}) * \check{g} = 0. \tag{1}$$

For every $x \in G$,

$$1_{G\setminus A^{-1}} * \check{g}(x) = \int 1_{G\setminus A^{-1}}(xy) g(y) \, dy = \int_G g(y) \, dy - \int_{x^{-1}A^{-1}} g(y) \, dy$$

$$= 2 - \int_{x^{-1}A^{-1}} g(y) \, dy.$$

Moreover

$$|H \cap x^{-1} A^{-1}| = |xH \cap A^{-1}| = \left| xH \cap \left(\bigcup_{i \in I} x_i U \right) \right| \leq |U| < \alpha$$

and then $\int_{x^{-1}A^{-1}} g(y) \, dy < 1$. So for every $x \in G$, we have $1_{G\setminus A^{-1}} * \check{g}(x) > 1$, hence $g * 1_{G\setminus A} \geq 1_G$. As M is mean, we obtain $M(g * 1_{G\setminus A}) \geq 1$ and

$$T(1_{G\setminus A^{-1}} * \check{g}) \geq 1_G. \tag{2}$$

The relations (1) and (2) establish (b). □

We now consider more specifically invariant means on continuous bounded functions.

Definition 21.11. *Let G be a locally compact group. A sequence $(A_n)_{n \in \mathbb{N}^*}$ of pairwise disjoint subsets of G is called scattered if, to every compact subset K of G, there corresponds $n_K \in \mathbb{N}^*$ such that, for every $x \in G$, $Kx \cap A_n \neq \emptyset$ for at most one $n \geq n_K$.*

Proposition 21.12. *Let G be a noncompact amenable group and let $f \in \mathscr{C}(G)$ such that $|f| \leq \sum_{n \in \mathbb{N}^*} 1_{U_n}$, where $(U_n)_{n \in \mathbb{N}^*}$ is a scattered sequence of open relatively compact subsets of G. If $\alpha = \sup\{|U_n^{-1}| : n \in \mathbb{N}^*\} < \infty$, then $M(|f|) = 0$ whenever $M \in \mathrm{TLIM}(\mathscr{C}_\mathbf{R}(G))$.*

Proof. Let $\varepsilon > 0$. As G is noncompact, there exists a compact symmetric neighborhood K of e such that $|K| > \alpha \varepsilon^{-1}$. There exists $n_K \in \mathbb{N}^*$ such that, for every $x \in G$, $Kx \cap U_n \neq \emptyset$ for at most one $n \geq n_K$. Let $V = \bigcup_{n=n_K}^\infty U_n$ and $g = |f| 1_V$; g is continuous on $\bigcup_{n=1}^\infty U_n$. If $x_0 \in G \setminus \bigcup_{n=1}^\infty U_n$, $f(x_0) = 0$ and $g(x_0) = 0$. As f is continuous, for every $\eta > 0$, there exists a neighborhood W of x_0 such that $|f(x)| < \eta$ whenever $x \in W$, so also $0 \leq g(x) < \eta$ whenever $x \in W$. Therefore $g \in \mathscr{C}(G)$ and, by the definition of f, we have $h = |f| - g \in \mathscr{K}(G)$. By Proposition 21.2, for any $M \in \mathrm{LIM}(\mathscr{C}_\mathbf{R}(G))$, $M(|f|) = M(g)$. For the case $M \in \mathrm{TLIM}(\mathscr{C}_\mathbf{R}(G))$, we have

$$M(|f|) = M(\xi_K * g). \tag{3}$$

As K is symmetric, for every $x \in K$, there exists at most one $p_x \geq n_K$ such that $x^{-1}K \cap U_{p_x}^{-1} \neq \varnothing$. Then $1_{x^{-1}K}\check{g} \leq 1_{U_{p_x}^{-1}}$ and

$$\xi_K * g(x) = |K|^{-1}\int 1_K(xy)g(y^{-1})\,dy = |K|^{-1}\int 1_{x^{-1}K}(y)\check{g}(y)\,dy$$

$$\leq |K|^{-1}\int 1_{U_{p_x}^{-1}}(y)\,dy = |K|^{-1}|U_{p_x}^{-1}| \leq |K|^{-1}\alpha < \varepsilon.$$

As $\varepsilon > 0$ is arbitrary, we deduce from (3) that $M(|f|) = 0$. □

B. Singularity and Mutual Singularity of Invariant Means

Let G be a locally compact group and let X be a (commutative) Banach subalgebra of $L_\mathbf{R}^\infty(G)$ such that $1_G \in X$. By Gelfand's theory the structure space $\hat{X}$ of X may be identified with the set of all (closed) proper maximal ideals of X, each such ideal being the kernel of a unique nontrivial continuous multiplicative linear functional on X ([B TS] Chap. I, Section 3, No. 1, théorème 2; [HR] C.17). The isomorphism of X with $\mathscr{C}(\hat{X})$ gives rise to an isomorphism of $\mathfrak{M}(X)$ with $M^1(\hat{X})$ (Section 3A) and allows to term *support* of $M \in \mathfrak{M}(X)$ the support of $\hat{M} \in M^1(\hat{X})$; we denote it by $\operatorname{supp} M$. If I is a closed ideal in X, we consider the closed set

$$h(I) = \{F \in \hat{X} : I \subset \operatorname{\mathscr{K}\!er} F\}$$

called the *hull* of I. We have $I = X$ if and only if $h(I) = \varnothing$ and $I = \{0\}$ if and only if $h(I) = \hat{X}$. Let I_1 and I_2 be closed ideals in X; $I_1 \subset I_2$ if and only if $h(I_1) \supset h(I_2)$ and $I_1 + I_2 = X$ if and only if $h(I_1) \cap h(I_2) = \varnothing$. If A is a measurable subset of G, $1_A \in L_\mathbf{R}^\infty(G)$ and $\widehat{1_A}$ has range $\{0,1\}$. As $\widehat{1_A} \in \mathscr{C}(\widehat{L_\mathbf{R}^\infty(G)})$, there exists an open and closed subset $\hat{A}$ of $\widehat{L_\mathbf{R}^\infty(G)}$ such that $1_{\hat{A}} = \widehat{1_A}$. Properties concerning the kernels of invariant means may be cast into the language of singular measures.

Definition 21.13. *Let G be a locally compact group. A mean M on $L^\infty(G)$ is called singular if, for every finitely additive measure μ on $\mathfrak{H}(G)$ such that $0 \leq \mu(A) \leq M(1_A)$ whenever $A \in \mathfrak{H}(G)$, one has $\mu = 0$. Two means M, N on $L_\mathbf{R}^\infty(G)$ are called mutually singular if the measures $\hat{M}, \hat{N}$ in $M^1(\widehat{L^\infty(G)})$ are concentrated on disjoint Borel subsets of $\widehat{L^\infty(G)}$.*

In particular, if G is a compact group, for $M \in \operatorname{LIM}(L_\mathbf{R}^\infty(G))$, Lebesgue's decomposition of $\hat{M}$ with respect to $\hat{\lambda}$ insures that M is singular if and only if M and λ are mutually singular.

Proposition 21.14. *Let G be a compact group. If there exist a measurable subset A of G and $M \in \operatorname{LIM}(L_\mathbf{R}^\infty(G))$ such that $|A| > 0$ and $M(1_A) = 0$, then M is singular.*

Proof. Consider the Lebesgue decomposition $\hat{M} = \mu^a + \mu'$ of $\hat{M}$ with respect to $\hat{\lambda}$; μ^a is absolutely continuous with respect to $\hat{\lambda}$ and by uniqueness of Haar measure there exist $\alpha \geq 0$, $N \in \text{LIM}(L_{\mathbf{R}}^{\infty}(G))$ such that $\hat{N} = \mu^a$, $N = \alpha\lambda$. If we had $\alpha > 0$, we should also have

$$\langle 1_A, M \rangle = \langle \widehat{1_A}, \hat{M} \rangle \geq \langle \widehat{1_A}, \mu^a \rangle = \langle 1_A, N \rangle = \alpha|A| > 0$$

and would come to a contradiction. Therefore $\alpha = 0$, $\hat{M} = \mu'$, and M is singular. □

Proposition 21.15. *Let G be an amenable group and let A be a measurable subset of G. If $M \in \text{LIM}(L_{\mathbf{R}}^{\infty}(G))$ with $|G \setminus A^{-1}| < 1$, $1_A \in \mathcal{K}\!er\, M$ and $N \in \text{TLIM}(L_{\mathbf{R}}^{\infty}(G))$, then M and N are mutually singular.*

Proof.

CASE 1: G IS COMPACT. By hypothesis $|A| > 0$. As also $M(1_A) = 0$ we deduce from Proposition 21.14 that M and λ are mutually singular. Therefore the statement follows from the fact that $\text{TLIM}(L_{\mathbf{R}}^{\infty}(G)) = \{\lambda\}$ by Proposition 12.1.

CASE 2: G IS NONCOMPACT. For any $N \in \text{TLIM}(L_{\mathbf{R}}^{\infty}(G))$, by Proposition 21.2 we have $N(1_{G \setminus A}) = 0$. Let $\hat{M} = \mu^a + \mu'$ be the Lebesgue decomposition of $\hat{M}$ with respect to $\hat{N}$. Since μ^a is absolutely continuous with respect to $\hat{N}$, we must have $\langle \widehat{1_{G \setminus A}}, \mu^a \rangle = \langle \widehat{1_{G \setminus A}}, \mu^a \rangle = 0$. But on the other hand,

$$\langle \widehat{1_{G \setminus A}}, \hat{M} \rangle = \langle \widehat{1_{G \setminus A}}, \hat{M} \rangle = \langle 1_{G \setminus A}, M \rangle = 1 - \langle 1_A, M \rangle = 1.$$

Therefore $\hat{M} = \mu'$; M and N are mutually singular. □

Proposition 21.16. *Let G be a locally compact group such that G_d is amenable. If I is a proper closed left invariant ideal in $L_{\mathbf{R}}^{\infty}(G)$, then there exists a G-invariant probability measure on $h(I)$ and there exists $M \in \text{LIM}(L_{\mathbf{R}}^{\infty}(G))$ such that $\text{supp}\, M \subset h(I)$ and $I \subset \mathcal{K}\!er\, M$.*

Proof. As the set $h(I)$ is right invariant and nonvoid, G acts affinely on $M^1(h(I))$. As G_d is amenable, Theorem 5.4 implies the existence of a fixed point μ_0 in $M^1(h(I))$. We extend μ_0 to $\nu \in M^1(\overline{L_{\mathbf{R}}^{\infty}(G)})$ by putting

$$\langle 1_B, \nu \rangle = \langle 1_{B \cap h(I)}, \mu_0 \rangle$$

for every Borel subset B of $\overline{L^{\infty}(G)}$; ν is a G-invariant probability measure on $\overline{L_{\mathbf{R}}^{\infty}(G)}$, that is, there exists $M \in \text{LIM}(L_{\mathbf{R}}^{\infty}(G))$ such that $\nu = \hat{M}$. For every $f \in L_{\mathbf{R}}^{\infty}(G)$,

$$\langle f, M \rangle = \langle \hat{f}, \hat{M} \rangle = \langle \hat{f}, \nu \rangle = \int_{h(I)} \hat{f} d\nu;$$

$\text{supp}\, M \subset h(I)$. If $f \in I$, $M(f) = 0$. □

C. Upper and Lower Bounds of Invariant Means

Definition 21.17. *Let G be a locally compact group, A a closed $*$-subalgebra of $\mathcal{M}^1(G)$ and $A^1 = A \cap M^1(G)$. Let X be a subspace of $L^\infty_{\mathbf{R}}(G)$ that contains 1_G and is A-invariant. If $P \subset A^1$, we denote by $\mathcal{N}(X, P)$ the real vector space spanned by $\{f - \mu * f : f \in X, \mu \in P\}$. We consider the set of A^1-invariant means on X:*

$$\mathfrak{M}(X, A^1) = \{M \in \mathfrak{M}(X) : (\forall f \in X)(\forall \mu \in A^1) M(\mu * f - f) = 0\}.$$

If $\mathfrak{M}(X, A^1) \neq \emptyset$, we define the mappings

$$S_{X, A^1} : f \mapsto \sup \{M(f) : M \in \mathfrak{M}(X, A^1)\}$$

$$I_{X, A^1} : f \mapsto \inf \{M(f) : M \in \mathfrak{M}(X, A^1)\}$$

$$X \to \mathbf{R}.$$

For every $f \in X$, we have $\operatorname{ess\,inf} f \leq I_{X, A^1}(f) \leq S_{X, A^1}(f) \leq \operatorname{ess\,sup} f$. In accordance with previously used notations, we write $\mathcal{N}(X)$ for $\mathcal{N}(X, M^1_d(G))$. Also in case $X = L^\infty_{\mathbf{R}}(G)$ and $A^1 = M^1_d(G)$, we write simply S [resp. I] for S_{X, A^1} [resp. I_{X, A^1}].

The next proposition makes sense for the function spaces on amenable groups considered in Theorem 4.19.

Proposition 21.18. *Let G be a locally compact group, A a closed $*$-subalgebra of $\mathcal{M}^1(G)$, $A^1 = A \cap M^1(G)$, X a subspace of $L^\infty_{\mathbf{R}}(G)$ that contains $\mathcal{UC}_{\mathbf{R}}(G)$ and is A-invariant. If $\mathfrak{M}(X, A^1) \neq \emptyset$, then for every $f \in X$,*

$$S_{X, A^1}(f) = \inf \{\operatorname{ess\,sup}(f + h) : h \in \mathcal{N}(X, A^1)\}$$

and also

$$I_{X, A^1}(f) = \sup \{\operatorname{ess\,inf}(f + h) : h \in \mathcal{N}(X, A^1)\}.$$

Proof. (a) If $f \in X$, let $N(f) = \inf \{\operatorname{ess\,sup}(f + h) : h \in \mathcal{N}(X, A^1)\}$. Obviously N is positively homogenous. If $f_1, f_2 \in X$, for every $\varepsilon > 0$, there exist $h_1, h_2 \in \mathcal{N}(X, A^1)$ such that $N(f_i) > \operatorname{ess\,sup}(f_i + h_i) - \varepsilon$; $i = 1, 2$. So $N(f_1) + N(f_2) \geq N(f_1 + f_2) - 2\varepsilon$ and, as $\varepsilon > 0$ is arbitrary, we have $N(f_1) + N(f_2) \geq N(f_1 + f_2)$, that is, N is subadditive. If $h \in \mathcal{N}(X, A^1)$, $N(h) \leq \operatorname{ess\,sup}(h - h) = 0$.

Let $f_0 \in X$. By the Hahn–Banach theorem there exists $M_0 \in X^*$ such that $M_0(f) \leq N(f)$ whenever $f \in X$ and $M_0(f_0) = N(f_0)$. For every $f \in X$, $M_0(f) \leq \operatorname{ess\,sup}(f + 0) = \operatorname{ess\,sup} f$, that is, M_0 is a mean on X. For every $h \in \mathcal{N}(X, A^1)$, $M_0(h) \leq N(h) \leq 0$, and $-M_0(h) = M_0(-h) \leq N(-h) \leq 0$, and hence $M_0 \in \mathfrak{M}(X, A^1)$. Therefore

$$N(f_0) = M_0(f_0) \leq S_{X, A^1}(f_0). \tag{4}$$

On the other hand, for every $M \in \mathfrak{M}(X, A^1)$ and every $h \in \mathcal{N}(X, A^1)$, $M(f_0) = M(f_0 + h) \le \operatorname{ess\,sup}(f_0 + h)$, hence $S_{X, A^1}(f_0) \le \operatorname{ess\,sup}(f_0 + h)$ and

$$S_{X, A^1}(f_0) \le N(f_0). \tag{5}$$

The relations (4) and (5) imply that $N = S_{X, A^1}$.

(b) For every $f \in X$, by (a) we have

$$I_{X, A^1}(f) = \inf \{ M(f) : M \in \mathfrak{M}(X, A^1) \}$$

$$= -\sup \{ M(-f) : M \in \mathfrak{M}(X, A^1) \} = -S_{X, A^1}(-f)$$

$$= -\inf \{ \operatorname{ess\,sup}(-f + h) : h \in \mathcal{N}(X, A^1) \}$$

$$= \sup \{ -\operatorname{ess\,sup}(-f + h) : h \in \mathcal{N}(X, A^1) \}$$

$$= \sup \{ \operatorname{ess\,inf}(f + h) : h \in \mathcal{N}(X, A^1) \}. \quad \square$$

Proposition 21.19. *Let G be a locally compact group such that G_d is amenable. Then for every $f \in L_{\mathbb{R}}^{\infty}(G)$,*

$$S(f) = \inf \left\{ \operatorname{ess\,sup} \left((\operatorname{card} F)^{-1} \sum_{a \in F} {}_a f \right) : F \in \mathcal{F}(G) \right\}.$$

Proof. For every nonvoid finite subset F of G,

$$f + (\operatorname{card} F)^{-1} \sum_{a \in F} ((-f) - {}_a(-f)) = (\operatorname{card} F)^{-1} \sum_{a \in F} {}_a f;$$

hence by Proposition 21.18 we have

$$S(f) \le \inf \left\{ \operatorname{ess\,sup} \left((\operatorname{card} F)^{-1} \sum_{a \in F} {}_a f \right) : F \in \mathcal{F}(G) \right\}. \tag{6}$$

Let $\varepsilon > 0$. By Proposition 21.18 there exists $h_0 = \sum_{i=1}^{n} (f_i - {}_{a_i} f_i)$ with $f_i \in L^{\infty}(G)$, $a_i \in G$, $i \in \{1, \ldots, n\}$ such that

$$S(f) + \frac{\varepsilon}{2} \ge \operatorname{ess\,sup}(f + h_0)$$

so locally for almost every $x \in G$,

$$S(f) + \frac{\varepsilon}{2} \ge (f + h_0)(x)$$

and then also

$$S(f) + \frac{\varepsilon}{2} \geq \operatorname{ess\,sup}\left((\operatorname{card} F)^{-1} \sum_{a \in F} (f +_a h_0)\right) \quad (7)$$

whenever F is a nonvoid finite subset of G. Let

$$\delta = \varepsilon(2n(1 + \sup\{\|f_i\|_\infty : i = 1, \ldots, n\}))^{-1}.$$

By (F) there exists a nonvoid finite subset E in G such that, for every $i = 1, \ldots, n$, $\operatorname{card}(a_i E \triangle E)/\operatorname{card} E < \delta$. Then for every $i = 1, \ldots, n$ and locally for almost every $x \in G$,

$$\frac{1}{\operatorname{card} E}\left|\sum_{a \in E} {}_a(f_i - {}_{a_i}f_i)(x)\right| = \frac{1}{\operatorname{card} E}\left|\sum_{a \in E} ({}_a f_i - {}_{a_i a}f_i)(x)\right|$$

$$\leq \frac{\operatorname{card}(a_i E \triangle E)}{\operatorname{card} E}\|f_i\|_\infty < \frac{\varepsilon}{2n}$$

and

$$\frac{1}{\operatorname{card} E}\left|\sum_{a \in E} h_0(ax)\right| < \frac{\varepsilon}{2}. \quad (8)$$

We deduce from (7) and (8) that

$$S(f) + \varepsilon > \operatorname{ess\,sup}\left((\operatorname{card} E)^{-1} \sum_{a \in E} {}_a f\right). \quad (9)$$

As $\varepsilon > 0$ may be chosen arbitrarily, (6) and (9) yield the statement. □

Proposition 21.20. *Let G be an amenable group and let $a_1, \ldots, a_n$ be distinct elements in G. Then there exists a Borel subset E in G such that $I(1_E) > 0$ and $a_i E \cap a_j E = \emptyset$ whenever $i, j = 1, \ldots, n$, $i \neq j$.*

Proof. The family of all Borel subsets B in G such that $a_i B \cap a_j B = \emptyset$ whenever $i, j = 1, \ldots, n$ with $i \neq j$ is nonvoid as it contains $\{e\}$. Let E be a maximal element of that family; necessarily $E \neq G$. Consider $a \in G \setminus E$ and $E' = E \cup \{a\}$. There exist $i, j = 1, \ldots, n$ such that $i \neq j$ and $a_i E' \cap a_j E' \neq \emptyset$. Hence there exists $a' \in E$ such that $a = a_i^{-1} a_j a' \in a_i^{-1} a_j E$ or $a = a_j^{-1} a_i a' \in a_j^{-1} a_i E$. We have $G = \bigcup_{i,j=1}^n a_i^{-1} a_j E$ and, for every $M \in \operatorname{LIM}(L^\infty(G))$, $1 = M(1_G) \leq n^2 M(1_E)$; $M(1_E) \geq n^{-2}$ and $I(1_E) \geq n^{-2}$. □

Proposition 21.21. *Let G be a locally compact group such that G_d is amenable and let E be a nonvoid measurable subset in G. Then $S(1_E) = 1$ if and only if $|\bigcap_{i=1}^n a_i E| > 0$ whenever $a_1, \ldots, a_n \in G$, $n \in \mathbf{N}^*$.*

Proof. By Proposition 21.19, $S(1_E) = 1$ if and only if, for every nonvoid finite subset F in G, ess sup $(\text{card } F)^{-1}\sum_{a \in F} {_a}1_E = 1$. The condition stated in the proposition fails if and only if there exists a finite subset F in G such that $|\cap_{a \in F} a^{-1}E| = 0$, that is,

$$\text{ess sup} \sum_{a \in F} {_a}1_E = \text{ess sup} \sum_{a \in F} 1_{a^{-1}E} < \text{card } F. \qquad \square$$

We now consider the function spaces where the upper and lower bound functions we just defined coincide, that is, all invariant means agree.

Definition 21.22. *Let G be an amenable group, A a closed $*$-subalgebra of $\mathscr{M}^1(G)$, $A^1 = A \cap M^1(G)$ and X an A-invariant subspace of $L^\infty_\mathbf{R}(G)$ with $1_G \in X$. We define $\mathscr{V}(X, A^1)$ to be the set of all $f \in X$ such that, for every $M \in \mathfrak{M}(X, A^1)$, $M(f)$ admits a constant value $V_{X, A^1}(f) = V(f)$.*

Proposition 21.23. *Let G be an amenable group and $X = L^\infty_\mathbf{R}(G)$ or $\mathscr{RUC}_\mathbf{R}(G)$. If $f \in X$ and $c \in [0, 1]$, the following conditions are equivalent:*
  (i) $f \in \mathscr{V}(X, P^1(G))$ *and* $V(f) = c$.
  (ii) $f \in \{\text{cl}_G + h : h \in \mathscr{N}(X, P^1(G))\}^-$.

We have $\mathscr{V}(X, P^1(G)) = \mathbf{R}1_G \oplus \overline{\mathscr{N}(X, P^1(G))}$.

Proof. (i) $\Rightarrow$ (ii)

We have $V(f - \text{cl}_G) = 0$. If $f - \text{cl}_G \notin \overline{\mathscr{N}(X, P^1(G))}$, by the Hahn–Banach theorem there exists $F \in X^*$ such that $F(f - \text{cl}_G) \neq 0$ and $F(g) = 0$ whenever $g \in \overline{\mathscr{N}(X, P^1(G))}$. There also exist $M_1, M_2 \in \mathfrak{M}(X)$ and $c_1, c_2 \in \mathbf{R}_+$ such that $F = c_1 M_1 - c_2 M_2$ (Section 3A). Let $M_0 \in \text{TLIM}(X)$. By Proposition 4.26, $F = M_0 \odot F$ and $M_0 \odot M_1, M_0 \odot M_2 \in \text{TLIM}(X)$; therefore

$$F(f - \text{cl}_G) = c_1 M_0 \odot M_1(f - \text{cl}_G) - c_2 M_0 \odot M_2(f - \text{cl}_G) = 0.$$

We come to a contradiction. So we must have $f - \text{cl}_G \in \overline{\mathscr{N}(X, P^1(G))}$ and $f \in \{\text{cl}_G + h : h \in \mathscr{N}(X, P^1(G))\}^-$.

The implication (ii) $\Rightarrow$ (i) is trivial. $\qquad \square$

D. Nested Families

If G is a locally compact group and $M \in \mathfrak{M}(L^\infty_\mathbf{R}(G))$, we have $M(1_\varnothing) = 0$, $M(1_G) = 1$ and, for measurable subsets A and B such that $A \subset B$, $M(1_A) \leq M(1_B)$. We examine a condition expressing regular increase of M in case G is amenable.

Definition 21.24. *Let G be an amenable group. A family $(A_t)_{0 \leq t \leq 1}$ of Borel subsets in G is called nested family if $A_{t'} \subset A_{t''}$ whenever $0 \leq t' \leq t'' \leq 1$ and $M(1_{A_t}) = t$ whenever $M \in \text{LIM}(L^\infty(G))$, $0 \leq t \leq 1$.*

Lemma 21.25. *An amenable group G admits a nested family if there exists a countable dense subset D in $[0,1]$ and a family $(A_t)_{t \in D}$ of Borel subsets in G such that $A_{t'} \subset A_{t''}$ whenever $t', t'' \in D$ with $t' \leq t''$ and $M(1_{A_t}) = t$ whenever $M \in \mathrm{LIM}(L^\infty(G))$, $t \in D$.*

Proof. Assume that the condition holds. Let $B_0 = \varnothing$ and $B_1 = G$. If $t \in\,]0,1[$, we put $B_t = \bigcup_{\substack{s \leq t \\ s \in D}} A_s$; B_t is a Borel subset of G. For any $\varepsilon > 0$, there exist $s', s'' \in D$ such that $s' \leq t \leq s''$ with $s'' - s' < \varepsilon$. Also for every $M \in \mathrm{LIM}(L^\infty(G))$,

$$s' = M(1_{A_{s'}}) \leq M(1_{B_t}) \leq M(1_{A_{s''}}) = s''.$$

As $\varepsilon > 0$ may be chosen arbitrarily, $M(1_{B_t}) = t$. Therefore $(B_t)_{0 \leq t \leq 1}$ is a nested family. □

Proposition 21.26. *Let G be an amenable group admitting a decreasing infinite sequence (H_n) of closed normal subgroups such that G/H_n is finite for every $n \in \mathbf{N}^*$. Then G admits a nested family.*

Proof. For $n \in \mathbf{N}^*$, we put $p_n = \operatorname{card} G/H_n$, and then $p_n | p_{n+1}$. We also define $Q_n = \{k/p_n : k = 1, \ldots, p_n - 1\}$ and $D = \bigcup_{n \in \mathbf{N}^*} Q_n$. As $\lim_{n \to \infty} p_n = \infty$, $\overline{D} = [0,1]$.

(a) We establish the following claim: Given $n \in \mathbf{N}^*$, if $\{a_1, \ldots, a_{p_n}\}$ is a system of representatives for G/H_n, then there exists a system of representatives $\{b_1, \ldots, b_{p_{n+1}}\}$ for G/H_{n+1} such that, for all $k'/p_n, k''/p_n \in Q_n$ and $k/p_{n+1} \in Q_{n+1}$ with $k'/p_n \leq k/p_{n+1} \leq k''/p_n$, we have

$$\bigcup_{i'=1}^{k'} a_{i'} H_n \subset \bigcup_{i=1}^{k} b_i H_{n+1} \subset \bigcup_{i''=1}^{k''} a_{i''} H_n.$$

Let $\{c_1, \ldots, c_m\}$ be a system of representatives for H_n/H_{n+1}. As $p_{n+1} = mp_n$, every $i \in \{1, \ldots, p_{n+1}\}$ may be written in a unique manner

$$i = \alpha_i m + \beta_i$$

where $\alpha_i \in \{0, \ldots, p_n - 1\}$ and $\beta_i \in \{1, \ldots, m\}$; we put $b_i = a_{\alpha_i + 1} c_{\beta_i}$. Then $\{b_i : i = 1, \ldots, p_{n+1}\}$ is a system of representatives for G/H_{n+1}. By assumption $mk' \leq k \leq mk''$, and we have

$$\bigcup_{i'=1}^{k'} a_{i'} H_n = \bigcup_{i'=1}^{k'} a_{i'} \left(\bigcup_{j=1}^{m} c_j H_{n+1} \right) = \bigcup_{i'=1}^{k'} \bigcup_{j=1}^{m} a_{i'} c_j H_{n+1}$$

$$\subset \bigcup_{i=1}^{k} b_i H_{n+1} \subset \bigcup_{i''=1}^{k''} \bigcup_{j=1}^{m} a_{i''} c_j H_{n+1}$$

$$= \bigcup_{i''=1}^{k''} a_{i''} \left(\bigcup_{j=1}^{m} c_j H_{n+1} \right) = \bigcup_{i''=1}^{k''} a_{i''} H_n.$$

(b) If $n \in \mathbf{N}^*$ and $k/p_n \in Q_n$, let $A_{k/p_n} = \cup_{i=1}^k a_i H_n$. For every $M \in$ LIM$(L^\infty(G))$, $M(1_{H_n}) = 1/p_n$, so we have $M(1_{A_{k/p_n}}) = k/p_n$. Thus we inductively construct a family $(A_t)_{t \in D}$ of Borel subsets in G such that $A_{t'} \subset A_{t''}$ whenever $t', t'' \in D$ with $0 \le t' \le t'' \le 1$, and $M(1_{A_t}) = t$ whenever $M \in$ LIM$(L^\infty_{\mathbf{R}}(G))$, $t \in D$. We conclude via Lemma 21.25. □

Corollary 21.27. *A discrete abelian group G that is generated by an infinite sequence (a_n) of elements of finite order admits a nested family.*

Proof. For $n \in \mathbf{N}^*$, let H_n be the subgroup of G that is generated by the elements a_i $(i \ge n)$. We may apply Proposition 21.26. □

Proposition 21.28. *A discrete abelian group G having an element a of infinite order admits a nested family.*

Proof. (a) Let H be the subgroup of G generated by the element a. If $n \in \mathbf{N}^*$, for every $k \in \{0, 1, \ldots, 2^n - 1\}$, we consider $B_{k,2^n} = \{a^{k+m2^n} : m \in \mathbf{Z}\}$. As the sets $B_{k,2^n}$ $(k = 0, 1, \ldots, 2^n - 1)$ form a partition of H, for every $M \in$ LIM$(l^\infty_{\mathbf{R}}(H))$ and every $k = 0, 1, \ldots, 2^n - 1$,

$$M(1_{B_{k,2^n}}) = M(1_{B_{0,2^n}}) = 2^{-n}.$$

Let $k = 1, \ldots, 2^n - 1$ and $C_{k/2^n} = \cup_{i=0}^{k-1} B_{i,2^n}$. For every $M \in$ LIM$(l^\infty_{\mathbf{R}}(H))$, $M(1_{C_{k/2^n}}) = k/2^n$. As $\{k/2^n : k = 1, \ldots, 2^n - 1; n \in \mathbf{N}^*\}$ is dense in $[0,1]$, Lemma 21.25 yields the existence of a nested family $(A_t)_{0 \le t \le 1}$ in H.

(b) Let $\{x_j : j \in J\}$ be a class of representatives of G/H and consider $A'_t = \{x_j : j \in J\} A_t \subset G$ for $t \in [0, 1]$. Obviously $A'_t \subset A'_u$ whenever $t, u \in [0, 1]$ with $t \le u$. If $f \in l^\infty(H)$, let

$$f'(x_j y) = f(y)$$

for $j \in J$ and $y \in H$, so $f' \in l^\infty(G)$ and, in particular, $1_{A'_t} = 1'_{A_t}$ whenever $t \in [0, 1]$. If $N \in$ IM$(l^\infty(G))$, let

$$N'(f) = N(f'),$$

$f \in l^\infty(H)$; then $N' \in$ IM$(l^\infty(H))$. For every $t \in [0, 1]$, $N(1_{A'_t}) = N'(1_{A_t}) = t$. □

Proposition 21.29. *The group $\mathbf{T}$ admits a nested family.*

Proof. If $n \in \mathbf{N}^*$ and $k \in \{1, \ldots, n\}$, consider the Borel subset $A_{k/n} = \{e^{2\pi i x} : 0 \le x < k/n\}$ of $\mathbf{T}$. For every $M \in$ IM$(L^\infty(\mathbf{T}))$, $M(1_{A_{1/n}}) = 1/n$ and $M(1_{A_{k/n}}) = k/n$. We obtain a nested family via Lemma 21.25. □

Proposition 21.30. *Let G be an amenable group and let H be a closed normal subgroup. If G/H admits a nested family, so does G.*

Proof. Let π be the canonical homomorphism of G onto G/H. If $(A_t)_{0 \le t \le 1}$ is a nested family for G/H, define $B_t = \pi^{-1}(A_t)$, $t \in [0,1]$. If $M \in \text{LIM}(L^\infty(G))$, we consider $M' \in \text{LIM}(L^\infty(G/H))$ defined by

$$M'(f) = M(f \circ \pi),$$

$f \in L^\infty(G/H)$. For every $t \in [0,1]$, we have $M(1_{B_t}) = M(1_{A_t} \circ \pi) = M'(1_{A_t}) = t$; $(B_t)_{0 \le t \le 1}$ is a nested family for G. □

Proposition 21.31. *Any compactly generated locally compact infinite abelian group G admits a nested family.*

Proof. (a) As $\mathbf{T}$ admits a nested family by Proposition 21.29 and $\mathbf{T} \simeq \mathbf{R}/\mathbf{Z}$, we deduce from Proposition 21.30 that $\mathbf{R}$ admits a nested family.

(b) There exists a compact subgroup H of G such that $G/H \simeq \mathbf{R}^m \times \mathbf{Z}^n \times \mathbf{T}^p \times F$, where $m, n, p \in \mathbf{N}$ and F is a finite group ([HR] 9.6). If we may have $m \ne 0$ [resp. $n \ne 0$; $p \ne 0$], the statement follows from (a) [resp. Proposition 21.28; Proposition 21.29] and Proposition 21.30. Otherwise we may determine a sequence (H_n) of open normal subgroups in G satisfying the hypothesis of Proposition 21.26. Let $H_1 = G$ and assume that H_k have been obtained for $k \le n$. A compact normal subgroup H_{n+1} exists in the open subgroup H_n such that H_n/H_{n+1} is a finite group ([HR] 9.6). As H_{n+1} is closed and there exists only a finite number of cosets, H_{n+1} is open. As G is infinite, the sequence (H_n) is infinite. The statement now follows from Proposition 21.26. □

Corollary 21.32. *Any nonunimodular amenable group G admits a nested family.*

Proof. Consider the closed normal subgroup $H = \{x \in G : \Delta(x) = 1\}$ of G; G/H is isomorphic to the multiplicative group $\mathbf{R}^*_+$ and it admits a nested family by Proposition 21.31 as G then does by Proposition 21.30. □

Comments

We show that certain amenable groups G admit functions in $\mathscr{C}_\mathbf{R}(G)$ that are close to 1_G and are nevertheless annihilated by all topologically left invariant means.

Lemma 21.33. *Let G be a noncompact σ-compact locally compact group and let (K_m) be a sequence of compact subsets in G. Then there exists a sequence $(a_m)_{m \in \mathbf{N}^*}$ in G such that $(K_m a_m)$ is scattered.*

Proof. Let $L_n = \cup_{m=1}^n K_m$, $n \in \mathbf{N}^*$. As G is σ-compact and noncompact, there exists an increasing sequence (U_n) of nonvoid open relatively compact subsets in G such that $\cup_{n \in \mathbf{N}^*} U_n = G$. Choose a_1 arbitrarily and assume that $a_1, \ldots, a_n$ have been determined. As G is noncompact, there exists

$$a_{n+1} \in G \setminus \bigcup_{j=1}^n L_{n+1}^{-1} U_n U_n^{-1} L_{n+1} a_j.$$

For every $x \in G$ and every $m \in \mathbf{N}^*$, at most one of the sets $K_n a_n \cap U_n x$ ($n \geq m$) is nonvoid. Otherwise there would exist n', $n'' \in \mathbf{N}^*$ such that $n' \geq n'' \geq m$ and $K_{n'+1} a_{n'+1} \cap U_m x \neq \varnothing$, $K_{n''} a_{n''} \cap U_m x \neq \varnothing$; then a fortiori $L_{n'+1} a_{n'+1} \cap U_m x \neq \varnothing$ and $L_{n''} a_{n''} \cap U_m x \neq \varnothing$. So there should exist $l' \in L_{n'+1}$, $l'' \in L_{n''} u'$, $u'' \in U_m$ such that $l' a_{n'+1} = u'x$, $l'' a_{n''} = u''x$ and

$$a_{n'+1} = l'^{-1} u' u''^{-1} l'' a_{n''} \in L_{n'+1}^{-1} U_m U_m^{-1} L_{n'+1} a_{n''} \subset L_{n'+1}^{-1} U_n U_{n'}^{-1} L_{n'+1} a_{n''};$$

we would contradict the choice of $a_{n'+1}$.

For any compact subset C in G, there exists $m_0 \in \mathbf{N}^*$ such that $C \subset U_{m_0}$. We conclude that, for every $x \in G$, $K_m a_m \cap Cx \neq \varnothing$ for at most one $m \geq m_0$. The sets $K_m a_m$ ($m \in \mathbf{N}^*$) are also pairwise disjoint. Otherwise there would exist m', $m'' \in \mathbf{N}^*$ with $m' \geq m''$ such that $K_{m'+1} a_{m'+1} \cap K_{m''} a_{m''} \neq \varnothing$. Then, as $e \in U_{m'} U_{m'}^{-1}$,

$$a_{m'+1} \in K_{m'+1}^{-1} K_{m''} a_{m''} \subset L_{m'+1}^{-1} U_{m'} U_{m''} L_{m''} a_{m''}.$$

We would contradict the choice of $a_{m'+1}$. Hence $(K_m a_m)$ is scattered. □

Proposition 21.34. *Let G be a noncompact nondiscrete amenable σ-compact group. There exists $f \in \mathscr{C}_+(G)$ such that (a) $f \leq 1_G$, (b) $M(f) = 0$ whenever $M \in \mathrm{TLIM}(\mathscr{C}_\mathbf{R}(G))$, and (c) $\inf \{\sup \{1 - f(a_i x) : i = 1, \ldots, p\} : x \in G\} = 0$ whenever $a_1, \ldots, a_p \in G$.*

Proof. As G is noncompact and σ-compact, there exists an increasing sequence (U_m) of nonvoid open relatively compact subsets in G such that $G = \cup_{m \in \mathbf{N}^*} U_m$. As G is nondiscrete and σ-compact, by Lemma 21.9 there exists an open dense subset V in G such that $|V^{-1}| < 1$. If $m \in \mathbf{N}^*$, let $V_m = U_m \cap V$; $|V_m^{-1}| \leq |V^{-1}| < 1$. By regularity of the Haar measure, for every $m \in \mathbf{N}^*$, there exists a sequence $(V'_{m,n})$ of compact subsets in V_m such that $|V'_{m,n}| > 0$ and $|V_m \setminus V'_{m,n}| < 1/n$ whenever $n \in \mathbf{N}^*$. Urysohn's lemma implies that, for all $m, n \in \mathbf{N}^*$, there exists $f_{m,n} \in \mathscr{C}_+(G)$ such that $f_{m,n} \leq 1_G$, $\mathrm{supp}\, f_{m,n} \subset V_m$, and $f_{m,n}(V'_{m,n}) = \{1\}$. Thus for every $m \in \mathbf{N}^*$, $\lim_{n \to \infty} f_{m,n} = 1_{V_m}$. Applying a diagonalization process to Lemma 21.33 we may choose $(a_{m,n})_{m,n \in \mathbf{N}^*}$ in G such that $(\overline{U}_m a_{m,n})_{m,n \in \mathbf{N}^*}$ is a double-indexed scattered sequence, that is, the sets $\overline{U}_m a_{m,n}$ ($m, n \in \mathbf{N}^*$) are pairwise disjoint and to every compact subset K in G there correspond $m_K, n_K \in \mathbf{N}^*$ such that, for every $x \in G$, $Kx \cap \overline{U}_m a_{m,n} \neq \varnothing$ for at most one $(m,n) \in \mathbf{N}^* \times \mathbf{N}^*$ with $m \geq m_K$, $n \geq n_K$. We put $f = \Sigma_{m,n \in \mathbf{N}^*} (f_{m,n})_{a_{m,n}^{-1}}$. By the choice of $(a_{m,n})$ and the fact that $V_m \subset \overline{U}_m$ whenever $m \in \mathbf{N}^*$, we have $f \in \mathscr{C}_+(G)$ and $f \leq 1_G$.

Moreover $f \leq \Sigma_{m,n \in \mathbf{N}^*} 1_{V_m a_{m,n}}$ and

$$\sup \left\{ \left| (V_m a_{m,n})^{-1} \right| : m, n \in \mathbf{N}^* \right\} = \sup \left\{ |V_m^{-1}| : m \in \mathbf{N}^* \right\} < 1.$$

Therefore Proposition 21.12 implies that $M(f) = 0$ whenever $M \in \mathrm{TLIM}(\mathscr{C}_\mathbf{R}(G))$.

Let $a_1, \ldots, a_p \in G$. There exists $m_1 \in \mathbf{N}^*$ such that $\{a_1, \ldots, a_p\} \subset U_{m_1}$, so $e \in \cap_{i=1}^{p} a_i^{-1} U_{m_1}$. As V is dense in G, $\cap_{i=1}^{p} a_i^{-1} V_{m_1} \neq \emptyset$. Choose $a \in \cap_{i=1}^{p} a_i^{-1} V_{m_1}$. If $n_1 \in \mathbf{N}^*$, let $b_{n_1} = aa_{m_1, n_1}$. For every $i = 1, \ldots, p$,

$$_{a_i}f(b_{n_1}) = \sum_{m,n \in \mathbf{N}^*} f_{m,n}(a_i b_{n_1} a_{m,n}^{-1}) = \sum_{m,n \in \mathbf{N}^*} f_{m,n}(a_i aa_{m_1, n_1} a_{m,n}^{-1})$$

and $a_i a \in V_{m_1}$. For $(m, n) \in \mathbf{N}^* \times \mathbf{N}^*$, $\operatorname{supp} f_{m,n} \subset V_m$. So if $f_{m,n}(a_i a a_{m_1, n_1} a_{m,n}^{-1}) \neq 0$, we must have

$$\left(V_{m_1} a_{m_1, n_1} a_{m,n}^{-1}\right) \cap V_m \neq \emptyset,$$

that is,

$$V_{m_1} a_{m_1, n_1} \cap V_m a_{m,n} \neq \emptyset.$$

Therefore necessarily $m = m_1$ and $n = n_1$. Hence for all $i = 1, \ldots, p$ and all $n \in \mathbf{N}^*$, $_{a_i}f(b_n) = f_{m_1, n}(a_i a)$. As $\lim_{n \to \infty} f_{m_1, n} = 1_{V_{m_1}}$, given $\varepsilon \in]0, 1[$, there exists $n_0 \in \mathbf{N}^*$ such that $1 - \varepsilon < {}_{a_i}f(b_{n_0}) \leq 1$ whenever $i = 1, \ldots, p$. □

Proposition 21.35. *Let G be an amenable group. If $\varepsilon > 0$ and $M \in \operatorname{LIM}(L_\mathbf{R}^\infty(G))$ is singular, there exists $A \in \mathfrak{H}(G)$ such that $|A| > 1 - \varepsilon$ and $\mathcal{K}\!\mathit{er}\, M$ contains the closed left invariant ideal of $L_\mathbf{R}^\infty(G)$ generated by 1_A.*

Proof. If $A \in \mathfrak{H}(G)$, we denote by $\mathcal{I}_A$ the closed left invariant ideal in $L_\mathbf{R}^\infty(G)$ generated by 1_A.

CASE 1: G IS COMPACT. Consider the nonnegative finitely additive measure μ on $\mathfrak{H}(G)$ defined by

$$\mu(A) = \inf\left\{\sum_{i=1}^{\infty} M(1_{A_i}) : A_i \in \mathfrak{H}(G), i \in \mathbf{N}^*, \bigcup_{i=1}^{\infty} A_i \supset A\right\},$$

$A \in \mathfrak{H}(G)$. As M is singular, $\mu = 0$ and there exists a decreasing sequence (B_k) in $\mathfrak{H}(G)$ such that $|B_k| > 1 - \varepsilon$ and $\langle 1_{B_k}, M \rangle \leq 1/2^k$ whenever $k \in \mathbf{N}^*$. For $B = \cap_{k=1}^{\infty} B_k$, we have $|B| \geq 1 - \varepsilon$ and $M(1_B) = 0$. Also $\mathcal{I}_B \subset \mathcal{K}\!\mathit{er}\, M$.

CASE 2: G IS NOT COMPACT. For any compact subset K in G, $M(1_K) = 0$ by Proposition 21.2. Then also $\mathcal{I}_K \subset \mathcal{K}\!\mathit{er}\, M$. □

NOTES

The properties 21.1 and 21.2 are folklore. For 21.4 see Day [120, 122]. Mitchell [391] had considered properties of this type on discrete semigroups. Chou [80] stated 21.5 in the discrete case. A discrete version of 21.6 is due to Douglas [145]. Rudin [491] gave 21.7 for $L^\infty(G)$ and proved 21.8–10. Rosenblatt [484] introduced scattered sequences and established 21.12.

Rosenblatt was the first to study singularity properties systematically. He proved 21.14–16 [480].

Granirer established 21.18 [214] and 21.19 [216]. Chou [84] proved that if, for an amenable σ-compact group G, the condition 21.19 holds, then G_d is amenable. Chou [80] stated 21.20 in the discrete case. Wong [560] essentially proved 21.23; see Day [120] for the discrete version and Lust-Piquard [374] for the case of a locally compact abelian group.

Nested families had first been considered by Granirer [212] in the context of discrete semigroups. He showed their existence in a large class of groups including all discrete abelian infinite groups and conjectured that, on other infinite semigroups S, for every left invariant mean M on $l^\infty(S)$, $M(l^\infty(S)) = [0,1]$. Chou [77] established Granirer's conjecture for amenable discrete right cancellative infinite semigroups. The results stated in 21.25–32 go back to Snell [515]. In fact, Proposition 21.31 remains valid even if G is not compactly generated. Let U then be a compact symmetric neighborhood of e. It generates a subgroup H of G and the abelian quotient group G/H is discrete. If G/H admits an element of infinite order, it has a nested family by Proposition 21.28. Otherwise by Granirer's result [212], it still admits a nested family. Snell [514] produced, for every amenable infinite group G and every $M \in \text{LIM}(L^\infty(G))$, a family $(A_t)_{0 \le t \le 1}$ of Borel subsets of G depending on M such that $A_{t'} \subset A_{t''}$ for $t', t'' \in [0,1]$, $t' \le t''$ and $M(1_{A_t}) = t$ for $t \in [0,1]$. He gave a stronger version applying to semigroups. Adler [3] established a general version of Granirer's conjecture concerning an arbitrary measure space (X, μ), where μ is a finitely additive probability measure that is invariant under a family of mappings from X into X.

Rosenblatt gave 21.33–34 [484] and 21.35 [480].

Let G be a noncompact amenable σ-compact group and let $\mathscr{V}_0(G)$ be the set of all $f \in \mathscr{UC}_R(G)$ such that $M(|f|) = 0$ whenever $M \in \text{LIM}(\mathscr{UC}_R(G))$. Chou [81, 82] showed that any element $g \in \mathscr{UC}_R(G)$ belongs to $\mathbf{R}1_G \oplus \mathscr{V}_0$ if and only if it is a multiplier function of $\mathscr{V}(\mathscr{UC}_R(G), P^1(G))$; Chou [80] proved that, for certain amenable discrete groups, invariant means of multiplier functions are approximately evaluated on the subsets where the functions are nearly constant. Let G be an amenable discrete group and put $\mathscr{V} = \mathscr{V}(l^\infty_R(G), M^1_d(G))$. Assume that the following condition holds: For every subset A of G such that $I(1_A) > 0$, there exists a subset B of A such that $1_B \in \mathscr{V}$ and $V(1_B) > 0$. The property holds for all elementary groups [86]. If f is a real-valued multiplier function of $\mathscr{V}$, there exists $c \in \mathbf{R}_+$ such that, for every $\varepsilon > 0$, one may determine a subset C of G satisfying $V(1_C) = 1$ and $\sup\{|f(x) - c| : x \in C\} < \varepsilon$.

22. THE SETS OF INVARIANT MEANS

We compare the sets of means admitting the different types of invariance properties and consider relations existing between these sets for groups, subgroups, or quotient groups. A formal description of the set of topologically left invariant means for general amenable groups is given. It takes a more explicit form for amenable σ-compact groups. We study equivalence and nonequivalence of left and right invariance properties. We also give examples of interesting particular invariant means. Finally we examine the cardinalities of certain sets of invariant means for amenable groups showing that, unless the groups are compact, these sets are very big. Since refined statements may be obtained only by long technical developments, we merely state some basic results.

A. Different Types of Invariance Properties

Proposition 22.1. *Let G be an amenable discrete group. If X is a left invariant real subspace of $l_\mathbf{R}^\infty(G)$ containing 1_G, then every left invariant mean M on X may be extended to a left invariant mean on $l_\mathbf{R}^\infty(G)$.*

Proof. By the Hahn–Banach theorem there exists $M_1 \in l_\mathbf{R}^\infty(G)^*$ extending M such that $\|M_1\| = 1$, $M_1(1_G) = 1$; hence $M_1 \in \mathfrak{M}(l_\mathbf{R}^\infty(G))$. Let $N \in \mathrm{LIM}(l_\mathbf{R}^\infty(G))$. By Proposition 4.26, $N \odot M_1 \in \mathrm{LIM}(l_\mathbf{R}^\infty(G))$. If $f \in X$, $x \in G$, then $M_1({}_x f) = M({}_x f) = M(f)$ and by (2.15), $M_1 * f(x) = M(f)1_G$, so

$$\langle f, N \odot M_1 \rangle = \langle M_1 * f, N \rangle = M(f)$$

and $N \odot M_1$ extends M. □

Proposition 22.2. *Let G be a locally compact group and let X be a topologically left invariant subspace of $L_\mathbf{R}^\infty(G)$ with $1_G \in X$. If $M \in \mathrm{LIM}(X)$ and there exists $\varphi_0 \in P^1(G)$ such that $M(\varphi_0 * f) = M(f)$ whenever $f \in X$, then also $M \in \mathrm{TLIM}(X)$.*

Proof. For every $f \in X$ consider

$$I_f : \psi \mapsto M(\psi * f)$$

$$L_\mathbf{R}^1(G) \to \mathbf{R}.$$

For all $\psi \in L^1(G)$ and $a \in G$,

$$I_f({}_a\psi) = M(({}_a\psi) * f) = M({}_a(\psi * f)) = M(\psi * f) = I_f(\psi).$$

By uniqueness of the left Haar measure up to a multiplicative positive factor ([HR] 15.5) there exists $c_f > 0$ such that $I_f(\psi) = c_f \int \psi$ whenever $\psi \in L_\mathbf{R}^1(G)$. We conclude that, for all $f \in X$ and $\varphi \in P^1(G)$,

$$M(\varphi * f) = c_f \int \varphi = c_f \int \varphi_0 = M(\varphi_0 * f) = M(f). \quad \square$$

Proposition 22.3. *Let G be a σ-compact locally compact group such that G_d is amenable. Then $\mathrm{LIM}(L^\infty(G)) = \mathrm{TLIM}(L^\infty(G))$ if and only if G is discrete. If*

G is nondiscrete, noncompact, there exist a measurable subset A of G and $M_0 \in \text{LIM}(L^\infty(G))$ such that $M_0(1_A) = 0$, but $M(1_A) = 1$ whenever $M \in \text{TLIM}(L^\infty(G))$.

Proof. If G is discrete, the equality holds trivially.

Assume G to be nondiscrete. By Lemma 21.9 we may determine an open dense subset U in G such that $|U| < \frac{1}{2}$. Let $A = G \setminus U^{-1}$. If G is compact, as $|A| \geq \frac{1}{2}$, we deduce from Proposition 12.1 that, for the unique element M of $\text{TLIM}(L^\infty(G))$, $M(1_A) = |A| \geq \frac{1}{2}$. If G is noncompact, by Proposition 21.2, for every $M \in \text{TLIM}(L^\infty(G))$, $M(1_{U^{-1}}) = 0$, hence $M(1_A) = 1$. We claim that there exists $M_0 \in \text{LIM}(L^\infty(G))$ such that $M_0(1_A) = 0$. Otherwise, by Proposition 21.19,

$$\sup \left\{ \operatorname*{ess\,inf} \left((\operatorname{card} F)^{-1} \sum_{a \in F} {}_{a^{-1}}1_A \right) : F \in \mathfrak{F}(G) \right\} > 0;$$

in other words there would exist $\alpha > 0$ and $a_1, \ldots, a_n \in G$ such that $n^{-1} \sum_{i=1}^n 1_{a_i A}(x) \geq \alpha$ locally for almost every $x \in G$. Let $B = \cap_{i=1}^n a_i U^{-1} = G \setminus (\cup_{i=1}^n a_i A)$. By density of the open subset U in G, $B \neq \emptyset$. As B is also open, $|B| > 0$. But for every $x \in B$, $\sum_{i=1}^n 1_{a_i A}(x) = 0$. We would come to a contradiction. □

Corollary 22.4. *If G is a nondiscrete, compact, locally compact group such that G_d is amenable, there exists a left invariant mean on $L^\infty(G)$ that is not the normalized Haar measure on G.*

Proof. The corollary is a consequence of Propositions 12.1 and 22.3.

Lemma 22.5. *Let G be a locally compact group that is neither compact nor discrete. Given $\varphi \in P^1(G) \cap \mathcal{K}(G)$, there exists $g \in \mathscr{C}(G)$ having the following properties*:

(i) $\|\varphi * g\| \leq \frac{1}{2}$.
(ii) *If $a_1, \ldots, a_p \in G$, there exists a nonvoid open subset V in G such that, for every $x \in V$ and every $i = 1, \ldots, p$, $g(a_i x) = 1$.*

Proof. Let m be the least cardinal such that there exists a covering $\mathscr{R}$ of G consisting of compact subsets with card $\mathscr{R} = m$. As G is noncompact, m is infinite. Let α be the least ordinal with cardinal m.

As G is nondiscrete, the set $\mathscr{E}$ of all nonvoid relatively compact open subsets S in G with $|\overline{S^{-1}}| < (2\|\varphi\|)^{-1}$ is nonvoid. For $n \in \mathbf{N}^*$, we consider $\mathscr{R}_n = \{K_1 \times \cdots \times K_n : K_1, \ldots, K_n \in \mathscr{R}\}$. Then $\mathscr{R}_n$ is a compact covering of the cartesian product H_n of n copies of G with card $\mathscr{R}_n = m$; H_n admits an open covering formed by a subset of n copies of $\mathscr{E}$ such that every element of $\mathscr{R}_n$ is included in a finite number of members of that open covering.

So there exists a subset $\mathscr{E}_n$ of $\mathscr{E}$ such that card $\mathscr{E}_n = m$ and

$\cup_{S^{(1)},\ldots,S^{(n)} \in \mathscr{E}_n} S^{(1)} \times \cdots \times S^{(n)} = H_n$. We consider in $\mathscr{E}$ a well-ordered family $(S_\beta)_{\beta \leq \alpha}$ such that $\cup_{n \in \mathbf{N}^*} \mathscr{E}_n = \{S_\beta : \beta \leq \alpha\}$.

If $\beta \leq \alpha$, via Urysohn's lemma we may choose $g_\beta \in \mathscr{K}_+(G)$ such that $\|g_\beta\| \leq 1$, $g_\beta(x) = 1$ whenever $x \in S_\beta$ and $\int_G g_\beta(x^{-1}) dx \leq (2\|\varphi\|)^{-1}$. Then we have

$$\|\varphi * g_\beta\| = \sup\left\{\int \varphi(yx) g_\beta(x^{-1}) dx : y \in G\right\} \leq \tfrac{1}{2}. \tag{1}$$

As $\varphi \in \mathscr{K}(G)$, there exists an open relatively compact neighborhood U of e in G such that $\varphi(x) = 0$ whenever $x \notin U$. Let $T_\beta = \operatorname{supp} g_\beta$, $\beta \leq \alpha$. If $\alpha_1 \leq \alpha$, choose $a_\beta \in G$ for $\beta < \alpha_1$. Then $\{(T_{\alpha_1}^{-1} U^{-1} U U^{-1} U T_\beta a_\beta)^- : \beta < \alpha_1\}$ is a family of compact subsets of G that does not constitute a covering of G. Hence there exists $a_{\alpha_1} \in G$ such that

$$a_{\alpha_1} \notin T_{\alpha_1}^{-1} U^{-1} U U^{-1} U T_\beta a_\beta,$$

that is,

$$\left(U^{-1} U T_\beta a_\beta\right) \cap \left(U^{-1} U T_{\alpha_1} a_{\alpha_1}\right) = \varnothing$$

whenever $\beta \leq \alpha_1$. So by transfinite induction we may determine $(a_\beta)_{\beta \leq \alpha}$ in G such that $(U^{-1} U T_\beta a_\beta) \cap (U^{-1} U T_{\beta'} a_{\beta'}) = \varnothing$ whenever $\beta \leq \alpha$, $\beta' \leq \alpha$, $\beta \neq \beta'$.

We now put

$$g(x) = g_\beta(x a_\beta^{-1}) \quad \text{if } x \in U^{-1} U T_\beta a_\beta, \beta \leq \alpha,$$

$$g(x) = 0 \quad \text{if } x \notin \bigcup_{\beta \leq \alpha} T_\beta a_\beta.$$

For any $z \in G$, $U^{-1}z$ intersects, at most, one of the sets $T_\beta a_\beta$ ($\beta \leq \alpha$) because otherwise there would exist $\beta, \beta' \leq \alpha$, $\beta \neq \beta'$ such that $(U^{-1}z) \cap (T_\beta a_\beta) \neq \varnothing$, $(U^{-1}z) \cap (T_{\beta'} a_{\beta'}) \neq \varnothing$, and then $z \in (U T_\beta a_\beta) \cap (U T_{\beta'} a_{\beta'})$. We would contradict the fact that $(U^{-1} U T_\beta a_\beta) \cap (U^{-1} U T_{\beta'} a_{\beta'}) = \varnothing$. Consider $z_0 \in G \setminus \cup_{\beta \leq \alpha}(T_\beta a_\beta)$. If $U^{-1} z_0$ intersects only $T_{\beta_1} a_{\beta_1}$ with $\beta_1 \leq \alpha$, then

$$\left(U^{-1} z_0\right) \cap \left(G \setminus \bigcup_{\beta \leq \alpha}(T_\beta a_\beta)\right) = \left(U^{-1} z_0\right) \cap \left(G \setminus T_{\beta_1} a_{\beta_1}\right)$$

is an open neighborhood of z_0. If $U^{-1} z_0$ does not intersect any $T_\beta a_\beta$ with $\beta \leq \alpha$, then

$$U^{-1} z_0 \subset G \setminus \bigcup_{\beta \leq \alpha} T_\beta a_\beta.$$

Therefore $G \setminus \cup_{\beta \leq \alpha}(T_\beta a_\beta)$ is open. The very definition of g insures then that g must be continuous. As for every $\beta \leq \alpha$, $\|g_\beta\| \leq 1$, we also have $\|g\| \leq 1$.

Let $a \in G$. If there exists $\beta \leq \alpha$ such that $a \in UT_\beta a_\beta$, then $U^{-1}a \subset U^{-1}UT_\beta a_\beta$ and, for every $x \in G$ such that $\varphi(x) \neq 0$, we have $x \in U$, $x^{-1}a \in U^{-1}UT_\beta a_\beta$ and $g(x^{-1}a) = g_\beta(x^{-1}aa_\beta^{-1})$. We deduce now from (1) that

$$\varphi * g(a) = \varphi * g_\beta(aa_\beta^{-1}) \leq \|\varphi * g_\beta\| \leq \tfrac{1}{2}.$$

If $a \in G \setminus \cup_{\beta \leq \alpha} UT_\beta a_\beta$, for every $\beta \leq \alpha$, $(U^{-1}a) \cap (T_\beta a_\beta) = \varnothing$. For every $x \in G$ such that $\varphi(x) \neq 0$, $x \in U$, $x^{-1}a \notin \cup_{\beta \leq \alpha} T_\beta a_\beta$, and $g(x^{-1}a) = 0$, hence $\varphi * g(a) = 0$. Thus (i) is satisfied.

There exist S_β containing $\{a_1, \ldots, a_p\}$ and an open neighborhood U' of e such that $a_i U' \subset S_\beta$ whenever $i = 1, \ldots, p$. We put $V = U'a_\beta$. For any $x \in V$ and any $i \in \{1, \ldots, p\}$, $a_i x \in a_i U' a_\beta \subset S_\beta a_\beta \subset T_\beta a_\beta \subset U^{-1}UT_\beta a_\beta$. Hence for every $i = 1, \ldots, p$,

$$g(a_i x) = g_\beta(a_i x a_\beta^{-1}) = 1$$

and (ii) is satisfied. $\square$

Proposition 22.6. *Let G be a locally compact group that is neither compact nor discrete and assume G_d to be amenable. Then* $\mathrm{LIM}(\mathscr{C}_\mathbf{R}(G)) \neq \mathrm{TLIM}(\mathscr{C}_\mathbf{R}(G))$.

Proof. Let $\varphi \in P^1(G) \cap \mathscr{K}(G)$, and let $g \in \mathscr{C}(G)$ be the associated function determined in Lemma 22.5. Consider $A = \{x \in G : g(x) = 1\}^\circ \neq \varnothing$. We denote by E the closed real vector subspace of $L^\infty_\mathbf{R}(G)$ generated by all the functions $f \in L^\infty_\mathbf{R}(G)$ such that there exist $a_1, \ldots, a_p \in G$ with $\cap_{i=1}^p a_i A \subset \mathscr{K}\!er f$. We deduce from (ii) of Lemma 22.5 that there exists a nonvoid open subset V of G such that $g(a_i x) = 1$ whenever $x \in V$ and $i = 1, \ldots, p$. Hence $\cup_{i=1}^p a_i V \subset A$ and $V \subset \cap_{i=1}^p a_i^{-1} A$. Applying Proposition 21.7 to the family $\{\cap_{a \in F} aA : F \in \mathscr{F}(G)\}$ we determine $M \in \mathrm{LIM}(L^\infty_\mathbf{R}(G))$ such that $M(f) = 0$ whenever $f \in E$. In particular, $1_G - g \in E$ and $M(g) = 1$. But by (i) of Lemma 22.5 we have $M(\varphi * g) \leq \|\varphi * g\| \leq \tfrac{1}{2}$, and hence $M \notin \mathrm{TLIM}(\mathscr{C}_\mathbf{R}(G))$. $\square$

B. Affine Mappings Between the Sets of Invariant Means

Proposition 22.7. *Let G be an amenable group admitting an open (and closed) subgroup H. Let $X = \mathscr{C}_\mathbf{R}(G)$ and $Y = \mathscr{C}_\mathbf{R}(H)$. Then there exists an affine injection Φ of $\mathrm{LIM}(Y)$ into $\mathrm{LIM}(X)$. If $\{x_i : i \in I\}$ is a family of representatives of the homogeneous space G/H, for every $g \in Y$, let $g'(x_i y) = g(y)$, $i \in I$, $y \in H$. Then, for every $N \in \mathrm{LIM}(Y)$, $\Phi(N)(g') = N(g)$.*

Proof. Choose a fixed element M_0 in $\mathrm{LIM}(X)$. For $N \in \mathrm{LIM}(Y)$, $f \in X$ and $x \in G$, we put

$$N_f(x) = N((_x f)|_H).$$

THE SETS OF INVARIANT MEANS

The function $N_f: G \to \mathbf{R}$ is constant on the left cosets of H. We have $N_f \in X$. Let

$$M_N(f) = \langle N_f, M_0 \rangle,$$

$f \in X$; then $M_N \in \mathfrak{M}(X)$. The mapping $\Phi: N \mapsto M_N$ of $\mathrm{LIM}(Y)$ into $\mathfrak{M}(X)$ is affine. For all $f \in X$, $a \in G$, and $x \in G$, we have $N_{af}(x) = N((_{ax}f)|_H)$, therefore $\langle N_{af}, M_0 \rangle = \langle N_f, M_0 \rangle$ and Φ maps $\mathrm{LIM}(Y)$ into $\mathrm{LIM}(X)$. In particular, for $N \in \mathrm{LIM}(Y)$, $x \in G$, we have $N_{g'}(x) = N(g)$ and $M_N(g') = N(g)$.

If $N^{(1)}$ and $N^{(2)}$ are distinct elements in $\mathrm{LIM}(Y)$, there exists $g_0 \in Y$ such that $N^{(1)}(g_0) \neq N^{(2)}(g_0)$. Then $N^{(1)}_{g_0'}$ and $N^{(2)}_{g_0'}$ are distinct constant functions on G. Therefore $M_{N^{(1)}}(g_0') \neq M_{N^{(2)}}(g_0')$ and Φ is injective. □

Proposition 22.8. *Let G be an amenable group admitting a closed normal subgroup H. There exists an affine mapping of $\mathrm{LIM}(\mathcal{RUC}_\mathbf{R}(G))$ [resp. $\mathrm{LIM}(\mathcal{UC}_\mathbf{R}(G))$] onto $\mathrm{LIM}(\mathcal{RUC}_\mathbf{R}(G/H))$ [resp. $\mathrm{LIM}(\mathcal{UC}_\mathbf{R}(G/H))$]. There also exists an affine mapping of $\mathrm{IM}(\mathcal{UC}_\mathbf{R}(G))$ onto $\mathrm{IM}(\mathcal{UC}_\mathbf{R}(G/H))$.*

Proof. (a) Let $X = \mathcal{RUC}_\mathbf{R}(G)$ [resp. $X = \mathcal{UC}_\mathbf{R}(G)$] and $Y = \mathcal{RUC}_\mathbf{R}(G/H)$ [resp. $Y = \mathcal{UC}_\mathbf{R}(G/H)$]. If $g \in Y$, we put

$$F(g)(x) = g(\dot{x}),$$

$x \in G$; then $F(g) \in X$. We may define an affine mapping $\Phi: \mathrm{LIM}(X) \to \mathrm{LIM}(Y)$ by putting

$$\langle g, \Phi(M) \rangle = \langle F(g), M \rangle$$

for $M \in \mathrm{LIM}(X)$, $g \in Y$.

By Corollary 13.2 there exists $N_0 \in \mathrm{LIM}(Y)$. We define a left invariant mean N_0' on the left invariant [resp. invariant] subspace $F(Y)$ of X which contains 1_G by putting

$$N_0'(F(g)) = \langle g, N_0 \rangle,$$

$g \in Y$. We consider

$$\mathfrak{M}_0 = \{ M \in \mathfrak{M}(X) : M|_{F(Y)} = N_0' \};$$

$\mathfrak{M}_0$ is a nonvoid convex w*-closed, hence w*-compact, subset of $\mathfrak{M}(X)$. The group G acts affinely and continuously on $\mathfrak{M}_0$ via the mapping defined by

$$\langle f, xM \rangle = \langle _x f, M \rangle,$$

for $M \in \mathfrak{M}_0$, $x \in G$, and $f \in X$. From Theorem 5.4 we conclude that there exists $M_0 \in \mathfrak{M}_0$ such that $xM_0 = M_0$ whenever $x \in G$, that is, $M_0 \in \text{LIM}(X)$; $M_0|_{F(Y)} = N_0'$ and $\Phi(M_0) = N_0$.

(b) If $X = \mathcal{UC}_{\mathbf{R}}(G)$ and $Y = \mathcal{UC}_{\mathbf{R}}(G/H)$, by Theorem 4.19 we may choose $N_0 \in \text{IM}(Y)$; G acts also affinely and continuously on $\mathfrak{M}_0$ via the mapping defined by

$$\langle f, x \cdot M \rangle = \langle f_{x^{-1}}, M \rangle$$

for $M \in \mathfrak{M}_0$, $x \in G$, and $f \in X$. Then (FP) yields the existence of $M_1 \in \mathfrak{M}_0$ such that $x \cdot M_1 = M_1$ whenever $x \in G$, that is, $M_1 \in \text{RIM}(X)$; $M_1|_{F(Y)} = N_0'$ and $\Phi(M_1) = N_0$.

If $f \in X$, for every $x \in G$, $_xf \in X$. Let $f'(x) = M_1(_xf)$. For every $a \in G$, $(_af)'(x) = M_1(_{ax}f)$; $(f_a)'(x) = M_1(_xf_a) = M_1(_xf) = f'(x)$. Let $M(f) = M_0(f')$. So for all $f \in X$ and $a \in G$, $M(_af) = M_0(f') = M(f_a)$, that is, $M \in \text{IM}(X)$. Moreover for all $g \in Y$ and $a \in G$,

$$M_1(_aF(g)) = M_1(F(_{\dot a}g)) = N_0(_{\dot a}g) = N_0(g)$$

and $M(F(g)) = N_0(g)$, therefore $\Phi(M) = N_0$. □

C. The Set of Topologically Left Invariant Means

If G is a locally compact group, X is a topologically left invariant subspace of $L_{\mathbf{R}}^{\infty}(G)$ with $1_G \in X$, and U is a compact subset of G such that $|U| > 0$, we consider

$$\Gamma_{X,U} : f \mapsto \xi_U^* * f$$

$$X \to \mathcal{RUC}_{\mathbf{R}}(G).$$

For $f \in X$, $a \in G$, let

$$M_{U,a}(f) = \xi_U^* * f(a) = \langle \Gamma_{X,U}(f), \delta_a \rangle = \langle f, \tilde\Gamma_{X,U}(\delta_a) \rangle;$$

$$M_{U,a}(f) = \int_G \xi_U(x) f(xa) \, dx = \frac{1}{|U|} \int_U f(xa) \, dx. \tag{2}$$

We have $M_{U,a} = \tilde\Gamma_{X,U}(\delta_a) \in \mathfrak{M}(X)$.

Definition 22.9. *Let G be a locally compact group and let X be a topologically left invariant subspace of $L_{\mathbf{R}}^{\infty}(G)$ containing $\mathcal{UC}_{\mathbf{R}}(G)$. We denote by $\mathcal{TN}(X)$ the subset of $\mathfrak{M}(X)$ that consists of all w*-cluster points of nets $(\Gamma_{X,U_i}^{\sim}(\delta_{a_i}))$, where (U_i) is a net of $\mathcal{AN}(G)$ and (a_i) is a net in G. We denote by $\mathcal{T}(X)$ the subset of $\mathfrak{M}(X)$ that consists of all w*-cluster points of sequences $(\Gamma_{X,U_n}(\delta_{a_n}))$, where $(U_n) \in \mathcal{A}_c(G)$ is formed by compact subsets of G and (a_n) is a sequence in*

G. We consider also the subset $\mathcal{T}_s\mathcal{N}(X)$ [resp. $\mathcal{T}_s(X)$] of $\mathcal{T}\mathcal{N}(X)$ [resp. $\mathcal{T}(X)$] determined by symmetric members of $\mathcal{A}\mathcal{N}(G)$ [resp. $\mathcal{A}_c(G)$].

If G is an amenable group [resp. an amenable σ-compact group], then $\mathcal{T}_s\mathcal{N}(X) \neq \emptyset$ [resp. $\mathcal{T}_s(X) \neq \emptyset$] by Propositions 16.31 and 16.10.

The next result goes along the lines of Corollary 20.3.

Proposition 22.10. *If G is an amenable group and X is a topologically left invariant subspace of $L^\infty_\mathbf{R}(G)$ containing $\mathcal{UC}_\mathbf{R}(G)$, then $\overline{\mathrm{co}}\,\mathcal{T}_s\mathcal{N}(X) = \overline{\mathrm{co}}\,\mathcal{T}\mathcal{N}(X) = \mathrm{TLIM}(X)$. In particular, if G is an amenable σ-compact group, then $\overline{\mathrm{co}}\,\mathcal{T}_s(X) = \overline{\mathrm{co}}\,\mathcal{T}(X) = \mathrm{TLIM}(X)$.*

Proof. (a) Let $(U_i)_{i \in I} \in \mathcal{A}\mathcal{N}(G)$ and let $(a_i)_{i \in I}$ be a net in G. We consider a w*-cluster point M of $(\Gamma_{X,U_i}^\sim(\delta_{a_i}))$. For every $i \in I$ and all $f \in X$, $\varphi \in P^1(G)$, we have

$$\left|M_{U_i,a_i}(\varphi * f - f)\right| = \left|\xi^*_{U_i} * (\varphi * f - f)(a_i)\right| \leq \|\xi^*_{U_i} * \varphi - \xi^*_{U_i}\|_1 \|f\|_\infty$$

$$= \|\varphi^* * \xi_{U_i} - \xi_{U_i}\|_1 \|f\|_\infty.$$

Therefore, by Proposition 16.32, $M \in \mathrm{TLIM}(X)$ and then also $\overline{\mathrm{co}}\,\mathcal{T}\mathcal{N}(X) \subset \mathrm{TLIM}(X)$.

(b) Assume that there exists $M_0 \in \mathrm{TLIM}(X) \setminus \overline{\mathrm{co}}\,\mathcal{T}_s\mathcal{N}(X)$. The Hahn–Banach theorem implies the existence of $g \in X$ such that

$$\alpha = M_0(g) - \sup\left\{M(g) : M \in \overline{\mathrm{co}}\,\mathcal{T}_s\mathcal{N}(X)\right\} > 0. \tag{3}$$

Via Propositions 16.10 and 16.31 consider $(U_i)_{i \in I} = (U_i^{-1})_{i \in I} \in \mathcal{A}\mathcal{N}(X)$. For every $i \in I$, $M_0(\xi^*_{U_i} * g) = M_0(g)$. As $\xi^*_{U_i} * g \in \mathcal{C}(G)$ and M_0 is a mean on X, there exists $a_i \in G$ such that $\xi^*_{U_i} * g(a_i) \geq M_0(g) - \alpha/2$. So for every w*-cluster point M of $(\Gamma_{X,U_i}^\sim(\delta_{a_i}))$ we should have $M(g) \geq M_0(g) - \alpha/2$, and we would contradict (3). □

We add an application of the preceding proposition.

Proposition 22.11. *Let G be an amenable σ-compact group and let $(U_n) \in \mathcal{A}_c(G)$ consist of compact subsets in G. For every $f \in L^\infty_\mathbf{R}(G)$, $S_{L^\infty_\mathbf{R}(G), P^1(G)}(f) = \lim_{n \to \infty}(\sup \xi^*_{U_n} * f)$.*

Proof. We write $S'(f)$ for $S_{L^\infty_\mathbf{R}(G), P^1(G)}(f)$.

For all $M \in \mathrm{TLIM}(L^\infty_\mathbf{R}(G))$ and $n \in \mathbf{N}^*$,

$$M(f) = M(\xi^*_{U_n} * f) \leq \sup \xi^*_{U_n} * f.$$

Hence

$$S'(f) \leq \liminf_{n \to \infty}\left(\sup \xi^*_{U_n} * f\right).$$

On the other hand, for every $\alpha < \limsup_{n \to \infty} (\sup \xi_{U_n}^* * f)$, by Proposition 22.10 we have $S'(f) \geq \alpha$. We conclude that $S'(f) = \lim_{n \to \infty} (\sup \xi_{U_n}^* * f)$. □

Proposition 22.12. *Let G be an amenable σ-compact group and let X be a topologically left invariant subalgebra of $L^\infty(G)$ that is closed under complex conjugation and contains $\mathcal{UC}(G)$. If $M \in \mathrm{TLIM}(X)$, the following conditions are equivalent:*
  (i) *M is an extreme point in $\mathrm{TLIM}(X)$.*
  (ii) *If $g_1, \ldots, g_p \in X$, there exists (φ_n) in $P^1(G)$ such that, for all $f \in X$, $\lim_{n \to \infty} M((\varphi_n * f)g_k) = M(f)M(g_k)$ whenever $k = 1, \ldots, p$.*
  (iii) *For any $f \in X$, there exists a sequence (ψ_p) in $P^1(G)$ such that*

$$\lim_{p \to \infty} M(\psi_p * f \psi_p * g) = M(f)M(g)$$

  whenever $g \in X$.

Proof. (i) ⇒ (ii)
By Propositions 16.31 and 16.32 there exists a sequence (χ_m) in $P^1(G)$ such that $\lim_{m \to \infty} \|\varphi * \chi_m - \chi_m\|_1 = 0$ whenever $\varphi \in P^1(G)$.

If $g_k = 0$, the relation (ii) is trivially verified. Let now $g \in X_+$, $g \neq 0$. For every $m \in \mathbf{N}^*$, we put

$$M_m(f) = \|g\|_\infty^{-1} \big(M((\chi_m^* * f)g) - M(f)M(g) \big),$$

$f \in X$. If $f \in X_+$,

$$(M + M_m)(f) = M(f)\big(1 - \|g\|_\infty^{-1}M(g)\big) + \|g\|_\infty^{-1}M((\chi_m^* * f)g) \geq 0$$

and, as

$$M((\chi_m^* * f)g) \leq \|g\|_\infty M(\chi_m^* * f) = \|g\|_\infty M(f),$$

$$(M - M_m)(f) = \|g\|_\infty^{-1}M(f)M(g) + \big(M(f) - \|g\|_\infty^{-1}M((\chi_m^* * f)g)\big) \geq 0.$$

Moreover $M_m(1_G) = 0$, hence $M \pm M_m \in \mathfrak{M}(X)$. Via Proposition 3.3 we determine a w*-cluster point N of $(M + M_m)$ in $\mathfrak{M}(X)$. Then $N' = N - M$ is a w*-cluster point of (M_m), that is, there exists a subsequence (M_{m_n}) of (M_m) such that $\lim_{n \to \infty} M_{m_n}(f) = N'(f)$ whenever $f \in X$. We have $M \pm N' \in \mathfrak{M}(X)$.

Let $f \in X$ and $\varphi \in P^1(G)$. For every $n \in \mathbf{N}^*$, topological left invariance of M implies that

$$\big|M_{m_n}(\varphi * f - f)\big| = \|g\|_\infty^{-1}\big|M\big((\chi_{m_n}^* * (\varphi * f - f)) \cdot g\big)\big|$$

$$\leq \|\chi_{m_n}^* * \varphi - \chi_{m_n}^*\|_1 \|f\|_\infty = \|\varphi^* * \chi_{m_n} - \chi_{m_n}\|_1 \|f\|_\infty.$$

Therefore we obtain

$$N'(\varphi * f) = \lim_{n \to \infty} M_{m_n}(\varphi * f) = \lim_{n \to \infty} M_{m_n}(f) = N'(f).$$

We thus showed that $M \pm N' \in \text{TLIM}(X)$. As M is extremal, necessarily $N' = 0$. Hence for all $f \in X$,

$$\lim_{n \to \infty} M\big((\chi_{m_n}^* * f) g\big) = M(f) M(g).$$

Now let $g \in X_{\mathbf{R}}$. There exists $\alpha > 0$ such that $\operatorname{ess\,sup}(g + \alpha 1_G) \geq 0$. Applying the foregoing result to $g + \alpha 1_G$ and using topological left invariance of M, we obtain

$$M(f)M(g) = M(f)M(g + \alpha 1_G) - \alpha M(f)$$

$$= \lim_{n \to \infty} M\big((\chi_{m_n}^* * f)(g + \alpha 1_G)\big) - \alpha M(f)$$

$$= \lim_{n \to \infty} M(\chi_{m_n}^* * f \cdot g)$$

whenever $f \in X$. An iterating process then yields the property.

$$(\text{ii}) \Rightarrow (\text{iii})$$

If $f, g \in X$, by the Cauchy–Schwarz inequality we have

$$M(|f + g|^2) \leq M(|f|^2) + M(|g|^2) + 2M(|f|^2)^{1/2} M(|g|^2)^{1/2}.$$

Hence $I = \{f \in X : M(|f|^2) = 0\}$ is an ideal in X, and we may define a Hilbert space $\mathcal{H}$ that is the completion of X/I by putting

$$(\dot{f} | \dot{g}) = M(f\bar{g})$$

for $\dot{f}, \dot{g} \in X/I$.

Let $f \in X_1 z = M(f) \dot{1}_G \in \mathfrak{H}$. If $\varepsilon \in \mathbf{R}^*$ and $g_1, \ldots, g_p \in X_1$ there exists $n_0 \in \mathbf{N}^*$ such that, for $n \geq n_0$ and $k = 1, \ldots, p$,

$$|M(f)M(g_k) - M((\varphi_n * f) \cdot g_k)| < \varepsilon;$$

so

$$|(z - (\varphi_n * f)\dot{}\,|\dot{g}_k)| = |(z|\dot{g}_k) - M(\varphi_n * f \cdot g_k)| < \varepsilon.$$

Hence z belongs to the convex subset $\{(\varphi * f)\dot{} : \varphi \in P^1(G)\}^-$ of $\mathcal{H}$ and there exists a sequence (θ_m) in $P^1(G)$ such that

$$\lim_{m \to \infty} \|(\theta_m * f)^{\cdot} - z\|_{\mathcal{H}} = 0.$$

Thus given $c > 0$, for every $g \in X$ with $\|g\|_\infty \leq c$, the sequence $M(\theta_m * f \cdot g)$ converges uniformly to $M(f)M(g)$. As for every $j \in \mathbf{N}^*$, $\|\theta_j * g\| \leq \|g\|_\infty \leq c$, we have also

$$\lim_{m \to \infty} M(\theta_m * f \cdot \theta_m * g) = M(f)M(g).$$

We conclude that given $p \in \mathbf{N}^*$, there exists $\psi_p \in P^1(G)$ such that, for every $g \in X$ with $\|g\|_\infty \leq p$,

$$\left| M(\psi_p * f \cdot \psi_p * g) - M(f)M(g) \right| < \frac{1}{p}.$$

Then

$$\lim_{p \to \infty} M(\psi_p * f \cdot \psi_p * g) = M(f)M(g)$$

whenever $g \in X$.

$$\text{(iii)} \Rightarrow \text{(i)}$$

If M were not an extreme point in $\mathrm{TLIM}(X)$, there would exist $\alpha \in\ ]0, 1[$ and extreme points M', $M'' \in \mathrm{TLIM}(X)$ such that $M = \alpha M' + (1 - \alpha)M''$.

Let $f_0 \in X_\mathbf{R}$ with $M(f_0) = 0$. By hypothesis, for every $\varepsilon > 0$, there exists $p \in \mathbf{N}^*$ such that

$$0 \leq M(|\psi_p * f_0|^2) < \varepsilon.$$

Then by the Cauchy–Schwarz inequality,

$$0 \leq |M'(f_0)|^2 = |M'(\psi_p * f_0)|^2 \leq M'(1_G) M'(|\psi_p * f_0|^2)$$

$$\leq \alpha^{-1} M(|\psi_p * f_0|^2) < \varepsilon \alpha^{-1}.$$

As $\varepsilon > 0$ may be chosen arbitrarily, we conclude that $M'(f) = 0$ whenever $M(f) = 0$, $f \in X$. Thus M' belongs to the one-dimensional vector subspace of X' spanned by M; there must exist $k \in \mathbf{N}^*$ such that $M' = kM$. But $k = kM(1_G) = M'(1_G) = 1$; hence $M = M'$. □

D. Left and Right Invariance

The problem that exists in the comparison of left and right invariance conditions may be tackled in the important class [IN] formed by locally compact groups G admitting a compact neighborhood U of e that is invariant by inner automorphisms, that is, $xUx^{-1} = U$ whenever $x \in G$. General references for the study of this class are Grosser and Moskowitz [234], Palmer [425], and Moran and Williamson ([402]. A locally compact group G belongs to [IN] if and only if, for one (for every) compact neighborhood U of e, $\inf\{|U \cap x^{-1}Ux| : x \in G\} > 0$ ([402] Theorem 1.9). All groups belonging to [IN] are unimodular ([425] p. 718). The class $[FC]^-$ is contained in [IN]. For a survey of the properties of $[FC]^-$ we refer to Kaniuth and Schlichting [307]. Recall that by Corollary 14.26 every group in $[FC]^-$ is amenable.

Proposition 22.13. *If $G \in [FC]^-$, then* $\mathrm{TLIM}(L_{\mathbf{R}}^\infty(G)) \subset \mathrm{RIM}(L_{\mathbf{R}}^\infty(G))$.

Proof. Let $(K_j)_{j \in J}$ be the net of all σ-compact open subgroups of G, ordered by inclusion. Let $M \in \mathcal{T}_s\mathcal{N}(L^\infty(G))$. We may consider M to be defined by the nets $(U_{j,n})_{j \in J, n \in \mathbf{N}^*}$ and $(a_{j,n})_{j \in J, n \in \mathbf{N}^*}$, where (a) for every $j \in J$, $(U_{j,n}) = (U_{j,n}^{-1}) \in \mathcal{A}(K_j)$ and, for $(j, n), (j', n') \in J \times \mathbf{N}^*$ with $(j, n) < (j', n')$ we have $j \prec j'$, $n \leq n'$, $U_{j,n} \subset U_{j',n'}$, (b) for every $j \in J$, $(a_{j,n})_{n \in \mathbf{N}^*}$ is a sequence in K_j. As K_j belongs to $[FC]^-$ whenever $j \in J$, by Corollary 16.20 we have $(a_{j,n}^{-1}U_{j,n}^{-1}) \in \mathcal{AN}(G)$. So for all $a \in G$, unimodularity of G implies that

$$\lim_{(j,n)} \frac{|U_{j,n}a_{j,n}a \triangle U_{j,n}a_{j,n}|}{|U_{j,n}|} = \lim_{(j,n)} \frac{|U_{j,n}a_{j,n}a \triangle U_{j,n}a_{j,n}|}{|U_{j,n}a_{j,n}|}$$

$$= \lim_{(j,n)} \frac{|a^{-1}a_{j,n}^{-1}U_{j,n}^{-1} \triangle a_{j,n}^{-1}U_{j,n}^{-1}|}{|a_{j,n}^{-1}U_{j,n}^{-1}|} = 0.$$

Let $f \in L_{\mathbf{R}}^\infty(G)$ and $a \in G$. For all $(j, n) \in J \times \mathbf{N}^*$, by (2)

$$\left| M_{U_{j,n}, a_{j,n}}(f_a - f) \right| \leq \frac{1}{|U_{j,n}|} \int_{U_{j,n}} |f(xa_{j,n}a) - f(xa_{j,n})| \, dx$$

$$\leq \|f\|_\infty \frac{|U_{j,n}a_{j,n}a \triangle U_{j,n}a_{j,n}|}{|U_{j,n}|}.$$

Hence $M(f_a) = M(f)$. Then by Proposition 22.10 we have $\mathrm{TLIM}(L_{\mathbf{R}}^\infty(G)) \subset \mathrm{RIM}(L_{\mathbf{R}}^\infty(G))$. □

Lemma 22.14. *Let G be an amenable unimodular σ-compact group such that $G \notin [FC]^-$. Then $\mathcal{T}_s(L_{\mathbf{R}}^\infty(G)) \setminus \mathrm{RIM}(L_{\mathbf{R}}^\infty(G)) \neq \emptyset$.*

Proof. Via Proposition 16.10 we determine $(U_n) \in \mathcal{A}_c(G)$ formed by compact symmetric subsets of G. By hypothesis there exists $a \in G$ such that $\overline{K_a}$ is noncompact. We construct inductively a sequence (x_n) in G such that, for $n \geq 2$, $x_n a x_n^{-1} \notin U_n^2$, $x_n \notin U_n(\cup_{m=1}^{n-1}(U_n x_m a))$, and $x_n \notin U_n(\cup_{m=1}^{n-1}(U_m x_m a^{-1}))$.

Choose $x_1 = e$ and assume that $x_1, \ldots, x_{n-1}$ have been determined. Let

$$C_n = \left(U_n\left(\bigcup_{m=1}^{n-1}(U_m x_m a)\right)\right) \cup \left(U_n\left(\bigcup_{m=1}^{n-1}(U_m x_m a^{-1})\right)\right).$$

As $\overline{K_a}$ is noncompact, $(C_n a C_n^{-1}) \cup U_n^2$ is compact and $\overline{K_a} \supset C_n a C_n^{-1}$, we must have $(\overline{K_a} \setminus C_n a C_n^{-1}) \setminus U_n^2 \neq \varnothing$, and there exists $x_n \in G \setminus C_n$ such that

$$x_n a x_n^{-1} \notin U_n^2. \tag{4}$$

Then also

$$x_n \notin U_n\left(\bigcup_{m=1}^{n-1}(U_m x_m a)\right), \tag{5}$$

$$x_n \notin U_n\left(\bigcup_{m=1}^{n-1}(U_m x_m a^{-1})\right). \tag{6}$$

Let now $U = \cup_{n=1}^{\infty} U_n x_n$. If we had $Ua \cap U \neq \varnothing$, there would exist n', $n'' \in \mathbf{N}^*$, $u' \in U_{n'}$, and $u'' \in U_{n''}$ such that $u' x_{n'} a = u'' x_{n''}$ and $u' x_{n'} = u'' x_{n''} a^{-1}$. By (5) [resp. (6)] we should have $n'' \leq n'$ [resp. $n' \leq n''$], so $n' = n''$. But then we should have $x_{n'} a x_{n'}^{-1} = u'^{-1} u'' \in U_{n'}^2$, and we would contradict (4). Therefore $Ua \cap U = \varnothing$ and, for every $n \in \mathbf{N}^*$, by unimodularity of G,

$$M_{U_n, x_n}((1_U)_a - 1_U) = \int \xi_{U_n}(x)(1_U(xx_n a) - 1_U(xx_n))\,dx$$

$$= \int (\xi_{U_n x_n a}(x) - \xi_{U_n x_n}(x)) 1_U(x)\,dx$$

$$= -\int \xi_{U_n x_n}(x)\,dx = -1.$$

Thus no w^*-cluster point of (M_{U_n, x_n}) belongs to $\mathrm{RIM}(L_{\mathbf{R}}^{\infty}(G))$. □

Proposition 22.15. *Let G be an amenable unimodular σ-compact group. If $\mathrm{LIM}(L_{\mathbf{R}}^{\infty}(G)) = \mathrm{RIM}(L_{\mathbf{R}}^{\infty}(G))$, then G belongs to $[FC]^-$.*

Proof. Assume that $G \notin [FC]^-$. There then exists $a \in G$ such that the conjugacy class K_a of a is not relatively compact. We claim that G admits a

THE SETS OF INVARIANT MEANS 277

σ-compact open subgroup L containing a such that $C_a = \{xax^{-1} : x \in L\}$ is not relatively compact. As a matter of fact, if H is any σ-compact open subgroup of G and $\{x_i : i \in I\}$ is a family of representatives of the homogeneous space G/H, then either $\overline{K_a}$ meets at least one $x_i H$ ($i \in I$) for which the intersection is noncompact or $\overline{K_a}$ intersects infinitely many cosets $x_i H$ ($i \in I$). So, we may determine a sequence $\{y_n : n \in \mathbf{N}^*\}$ in G such that $\{y_n a y_n^{-1} : n \in \mathbf{N}^*\}^-$ is not compact and the set $\{a\} \cup \{y_n : n \in \mathbf{N}^*\}$ is contained in a σ-compact open subgroup L of G such that $C_a = \{xax^{-1} : x \in L\}^-$ is not compact.

By Proposition 13.3, L is amenable. Lemma 22.14 and Proposition 22.10 imply that there exists $M \in \text{TLIM}(L_\mathbf{R}^\infty(L)) \setminus \text{RIM}(L_\mathbf{R}^\infty(L))$. Then by Proposition 22.7 there exists a left invariant mean on $L_\mathbf{R}^\infty(G)$ that is not right invariant. □

Lemma 22.16. *Let G be an amenable σ-compact group. If $G \notin [IN]$, then* $\text{TLIM}(L_\mathbf{R}^\infty(G)) \neq \text{TRIM}(L_\mathbf{R}^\infty(G))$.

Proof. (a) As $G \notin [IN]$, for every compact neighborhood U of e, $\inf\{|U \cap x^{-1}Ux| : x \in G\} = 0$. If K and L are compact subsets of G, there exists a compact neighborhood W of e such that $K \cup L \subset W$. For every $x \in G$, $|K \cap x^{-1}Lx| \leq |W \cap x^{-1}Wx|$, hence also

$$\inf\{|K \cap x^{-1}Lx| : x \in G\} = 0.$$

(b) Let K be any compact subset of G with $|K| > 0$ and let (K_n) be a sequence of compact subsets of G such that $G = \bigcup_{n \in \mathbf{N}^*} K_n$. For every $n \in \mathbf{N}^*$, by (a) there exists $x_n \in G$ such that

$$\left| K^{-1}K \cap x_n^{-1}\left(\bigcup_{m=1}^n K_m\right)^{-1}\left(\bigcup_{m=1}^n K_m\right)x_n \right| < 2^{-(n+1)}|K|.$$

We consider $A = \bigcup_{n=1}^\infty ((\bigcup_{m=1}^n K_m)x_n)$.

Let $x \in G$ and, for every $n \in \mathbf{N}^*$, let $C_n = (xK) \cap ((\bigcup_{m=1}^n K_m)x_n)$. If $C_n \neq \emptyset$, choose $c_n \in C_n$. Thus

$$|C_n| = |c_n^{-1}C_n| \leq \left| K^{-1}K \cap x_n^{-1}\left(\bigcup_{m=1}^n K_m\right)^{-1}\left(\bigcup_{m=1}^n K_m\right)x_n \right| < 2^{-(n+1)}|K|.$$

The inequality holds trivially in case $C_n = \emptyset$. Hence

$$|K \cap x^{-1}A| = |xK \cap A| \leq \sum_{n=1}^\infty 2^{-(n+1)}|K| = \frac{|K|}{2}.$$

By Proposition 21.4, for every $M \in \text{TRIM}(L_\mathbf{R}^\infty(G))$, we have $M(1_A) < 1$.

(c) For every $n \in \mathbf{N}^*$,

$$|K \cap Ax_n^{-1}| \geq \left|K \cap \left(\bigcup_{m=1}^n K_m\right)\right|.$$

Hence

$$\sup\{|K \cap Ax| : x \in G\} \geq \limsup_{n \to \infty} \left|K \cap \left(\bigcup_{m=1}^n K_m\right)\right| = |K|.$$

Then by Proposition 21.4 there exists $N \in \mathrm{TLIM}(L_\mathbf{R}^\infty(G))$ such that $N(1_A) = 1$. □

Proposition 22.17. *Let G be an amenable σ-compact group. Then $\mathrm{TLIM}(L_\mathbf{R}^\infty(G)) = \mathrm{TRIM}(L_\mathbf{R}^\infty(G))$ if and only if $G \in [FC]^-$.*

Proof. (a) If $\mathrm{TLIM}(L_\mathbf{R}^\infty(G)) = \mathrm{TRIM}(L_\mathbf{R}^\infty(G))$, by Lemma 22.16, $G \in [IN]$ and therefore G is unimodular. If we had $G \notin [FC]^-$, by Lemma 22.14 and Proposition 22.10 we would have $\mathrm{TLIM}(L_\mathbf{R}^\infty(G)) \setminus \mathrm{RIM}(L_\mathbf{R}^\infty(G)) \neq \emptyset$; then by Corollary 4.12 we should have $\mathrm{TLIM}(L_\mathbf{R}^\infty(G)) \setminus \mathrm{TRIM}(L_\mathbf{R}^\infty(G)) \neq \emptyset$.

(b) Let $G \in [FC]^-$. Consider $M \in \mathcal{T}_s(L_\mathbf{R}^\infty(G))$ determined by $(U_n) = (U_n^{-1}) \in \mathcal{A}_c(G)$ and a sequence (a_n) in G. By Corollary 16.20 we have $(a_n^{-1}U_n) \in \mathcal{A}_c(G)$.

Let $f \in L_\mathbf{R}^\infty(G)$, $f \neq 0$, and $\varphi \in \mathcal{K}(G) \cap P^1(G)$ with $K = \mathrm{supp}\,\varphi$. If $\varepsilon > 0$, by unimodularity of G, there exists $n_0 \in \mathbf{N}^*$ such that

$$\frac{|U_n a_n x \triangle U_n a_n|}{|U_n a_n|} = \frac{|x^{-1} a_n^{-1} U_n^{-1} \triangle a_n^{-1} U_n^{-1}|}{|a_n^{-1} U_n^{-1}|} < \frac{\varepsilon}{\|f\|_\infty}$$

whenever $n \geq n_0$ and $x \in K$. Then also by (2) and unimodularity we obtain

$$M_{U_n, a_n}(f * \check{\varphi} - f)$$

$$= \frac{1}{|U_n|} \int_{U_n} (f * \check{\varphi}(ya_n) - f(ya_n))\, dy$$

$$= \frac{1}{|U_n|} \int_{U_n} \left(\int_G f(x)\check{\varphi}(x^{-1}ya_n)\, dx - \int_G f(ya_n)\varphi(x)\, dx\right) dy$$

$$= \frac{1}{|U_n|} \int_{U_n} \left(\int_G f(x)\varphi(a_n^{-1}y^{-1}x)\, dx - \int_G f(ya_n)\varphi(x)\, dx\right) dy$$

$$= \frac{1}{|U_n|} \int_{U_n} \left(\int_G (f(ya_n x) - f(ya_n))\varphi(x)\, dx\right) dy$$

$$= \frac{1}{|U_n|} \int_G \int_G 1_{U_n}(ya_n^{-1})(f(yx) - f(y))\varphi(x)\, dx\, dy$$

$$= \frac{1}{|U_n|} \int_G \int_G \left(1_{U_n}(yx^{-1}a_n^{-1}) - 1_{U_n}(ya_n^{-1})\right) f(y)\varphi(x)\, dx\, dy.$$

THE SETS OF INVARIANT MEANS

Hence

$$\left|M_{U_n,a_n}(f*\check{\varphi}-f)\right| \le \|f\|_\infty \int_G \frac{|U_n a_n x \triangle U_n a_n|}{|U_n|} \varphi(x)\, dx < \varepsilon.$$

Thus we have $M(f*\check{\varphi}) = M(f)$. So by Proposition 22.10 we have $\mathrm{TLIM}(L^\infty_\mathbf{R}(G)) \subset \mathrm{TRIM}(L^\infty_\mathbf{R}(G))$. By Proposition 4.7 these subsets must be equal. □

E. Examples of Invariant Means and Invariant Measures

As we noted several times, the problem of the existence of invariant means amounts to the determination of invariant probability measures. Via Corollary 4.20 amenability of a discrete group is characterized by the existence of a left invariant, finitely additive probability measure on the algebra of all subsets of the group. In the next proposition we give a direct proof based on the axiom of choice showing that left Haar measure for a locally compact group may be extended to a left invariant, finitely additive measure on all subsets in case the underlying discrete group is amenable.

Proposition 22.18. *Let G be a locally compact group such that G_d is amenable. Then there exists a left invariant finitely additive real-valued measure defined on the algebra $\mathfrak{P}(G)$ of all subsets of G and extending the left Haar measure; more precisely there exists $\mu: \mathfrak{P}(G) \to \overline{\mathbf{R}}_+$ having the following properties*:
  (i) *If $A, B \in \mathfrak{P}(G)$ and $A \cap B = \varnothing$, then $\mu(A \cup B) = \mu(A) + \mu(B)$*.
  (ii) *If $A \in \mathfrak{P}(G)$ and $a \in G$, $\mu(aA) = \mu(A)$*.
  (iii) *If A is a Haar measurable subset of G, $\mu(A) = \lambda(A)$*.

Proof. We may assume that not all subsets of G belong to $\mathfrak{H}(G)$.
  (a) Let Γ be the smallest ordinal with corresponding cardinal $2^{\mathrm{card}\, G} = \mathrm{card}\, \mathfrak{P}(G)$. We assume $\mathfrak{P}(G) = \{A_\sigma; \le\}$ to be well ordered in such a way that the λ-measurable subsets constitute the first elements: for an ordinal Γ_0, A_σ is λ-measurable whenever $\sigma < \Gamma_0$. We now define $\Phi: \mathfrak{P}(G) \to \overline{\mathbf{R}}_+$ such that the following property holds: If $\alpha_1,\ldots,\alpha_m$, $\alpha'_1,\ldots,\alpha'_{m'} \in \mathbf{R}^*_+$ and $\tau_1,\ldots,\tau_m$, $\tau'_1,\ldots,\tau'_{m'}$ are ordinals with $\tau_i < \Gamma$ and $\tau'_{i'} < \Gamma$ whenever $i = 1,\ldots,m$ and $i' = 1,\ldots,m'$, then the inequality

$$\sum_{i=1}^m \alpha_i 1_{A_{\tau_i}} \le \sum_{i'=1}^{m'} \alpha'_{i'} 1_{A_{\tau'_{i'}}} \tag{7}$$

implies that

$$\sum_{i=1}^m \alpha_i \Phi(A_{\tau_i}) \le \sum_{i'=1}^{m'} \alpha'_{i'} \Phi(A_{\tau'_{i'}}).$$

We proceed to the definition of Φ.

If $\tau < \Gamma_0$, we put $\Phi(A_\tau) = |A_\tau|$. Then if $\tau_i < \Gamma_0$, $\tau'_{i'} < \Gamma_0$ for $i = 1, \ldots, m$ and $i' = 1, \ldots, m'$, the assumption (7) implies that

$$\sum_{i=1}^{m} \Phi(A_{\tau_i}) = \sum_{i=1}^{m} \alpha_i |A_{\tau_i}| \leq \sum_{i'=1}^{m'} \alpha'_{i'} |A_{\tau'_{i'}}| = \sum_{i'=1}^{m'} \alpha'_{i'} \Phi(A_{\tau'_{i'}}).$$

We now extend the definition of Φ by transfinite induction.

Assume that $\Phi(A_\sigma)$ has been defined for $\sigma < \tau$ where $\Gamma_0 \leq \tau < \Gamma$ and that the property holds up to that level. If

$$\inf \left\{ \sum_{i'=1}^{m'} \alpha'_{i'} \Phi(A_{\tau'_{i'}}) : 1_{A_\tau} \leq \sum_{i'=1}^{m'} \alpha'_{i'} 1_{A_{\tau'_{i'}}}; \alpha'_{i'} \in \mathbf{R}^*_+; \right.$$

$$\left. \tau'_{i'} < \tau; i' = 1, \ldots, m'; m' \in \mathbf{N}^* \right\} = \infty,$$

we put $\Phi(A_\tau) = \infty$. The property then obviously holds up to τ. Otherwise there exist β_j, $\beta'_{j'} \in \mathbf{R}^*_+$ and ordinals σ_j, $\sigma'_{j'} < \tau$ ($j = 1, \ldots, n$, $j' = 1, \ldots, n'$) such that

$$\sum_{j=1}^{n} \beta_j 1_{A_{\sigma_j}} + 1_{A_\tau} \leq \sum_{j'=1}^{n'} \beta'_{j'} 1_{A_{\sigma'_{j'}}}$$

and

$$\sum_{j'=1}^{n'} \beta'_{j'} \Phi(A_{\sigma'_{j'}}) < \infty.$$

We define $\Phi(A_\tau)$ to be the greatest lower bound of the set of the corresponding numbers $\sum_{j'=1}^{n'} \beta'_{j'} \Phi(A_{\sigma'_{j'}}) - \sum_{j=1}^{n} \beta_j \Phi(A_{\sigma_j})$. In order to establish the property up to τ it suffices then to consider the two following cases for γ_k, $\gamma'_{k'} \in \mathbf{R}^*_+$ and ordinals ρ_k, $\rho'_{k'} < \tau$ ($k = 1, \ldots, p$; $k' = 1, \ldots, p'$):

$$(\alpha) \quad \sum_{k=1}^{p} \gamma_k 1_{A_{\rho_k}} + 1_{A_\tau} \leq \sum_{k'=1}^{p'} \gamma'_{k'} 1_{A_{\rho'_{k'}}};$$

$$(\beta) \quad \sum_{k=1}^{p} \gamma_k 1_{A_{\rho_k}} \leq \sum_{k'=1}^{p'} \gamma'_{k'} 1_{A_{\rho'_{k'}}} + 1_{A_\tau}.$$

In case (α), by the definition of $\Phi(A_\tau)$ we obviously have

$$\Phi(A_\tau) \leq \sum_{k'=1}^{p'} \gamma'_{k'} \Phi(A_{\rho'_{k'}}) - \sum_{k=1}^{p} \gamma_k \Phi(A_{\rho_k}).$$

In case (β),

$$\sum_{j=1}^{n}\beta_j 1_{A_{\sigma_j}} + \sum_{k=1}^{p}\gamma_k 1_{A_{\rho_k}} \le \sum_{j'=1}^{n'}\beta'_{j'} 1_{A_{\sigma'_{j'}}} + \sum_{k'=1}^{p'}\gamma'_{k'} 1_{A_{\rho'_{k'}}}$$

so the induction hypothesis yields that

$$\sum_{k=1}^{p}\gamma_k \Phi(A_{\rho_k}) \le \Phi(A_\tau) + \sum_{k'=1}^{p'}\gamma'_{k'}\Phi(A_{\rho'_{k'}}).$$

Hence the property holds in both cases.

We conclude that if A and B are disjoint subsets of G, $1_{A\cup B} = 1_A + 1_B$ implies that

$$\Phi(A\cup B) = \Phi(A) + \Phi(B).$$

(b) For $A \in \mathfrak{P}(G)$, we consider

$$F_A: \mapsto \Phi(xA)$$

$$G \to \overline{\mathbf{R}}_+.$$

If A and B are disjoint subsets of G, $F_{A\cup B} = F_A + F_B$. By Theorem 4.19 choose $M \in \mathrm{RIM}(l_\mathbf{R}^\infty(G))$. We put

$$\mu(A) = \lim_{n\to\infty} M(\min\{F_A, n1_G\})$$

for $A \in \mathfrak{P}(G)$ and claim that if A and B are disjoint subsets of G, then for every $n \in \mathbf{N}^*$,

$$\min\{F_{A\cup B}, n1_G\} \le \min\{F_A, n1_G\} + \min\{F_B, n1_G\}$$

$$\le \min\{F_{A\cup B}, 2n1_G\}. \tag{8}$$

Let $x \in G$. If $F_A(x) \le n$ and $F_B(x) \le n$,

$$\min\{F_{A\cup B}, n1_G\}(x) \le F_{A\cup B}(x) = F_A(x) + F_B(x)$$

$$= \min\{F_A, n1_G\}(x) + \min\{F_B, n1_G\}(x) \le 2n.$$

If $F_A(x) \le n < F_B(x)$, we have

$$n \le F_A(x) + n < F_A(x) + F_B(x) = F_{A\cup B}(x)$$

and $F_A(x) + n \leq 2n$. If $n < F_A(x)$ and $n < F_B(x)$, we have

$$n < n + n = 2n < F_A(x) + F_B(x) = F_{A \cup B}(x).$$

The relation (8) implies that $\mu(A \cup B) = \mu(A) + \mu(B)$ and (i) holds.
If $A \subset G$ and $a \in G$, $F_{aA} = (F_A)_a$. As M is right invariant,

$$\mu(aA) = \lim_{n \to \infty} M(\min\{F_{aA}, n1_G\}) = \lim_{n \to \infty} M((\min\{F_A, n1_G\})_a)$$

$$= \lim_{n \to \infty} M(\min\{F_A, n1_G\}) = \mu(A)$$

and (ii) holds.

Finally if $A \in \mathfrak{H}(G)$ and $x \in G$, $\Phi(xA) = |xA|$ and $F_A(x) = |xA| = |A|$. Therefore $\mu(A) = |A|$ and (iii) holds. □

We list some particular invariant means.

If $f \in \mathscr{C}_\mathbf{R}(\mathbf{R})$ such that $\lim_{n \to \pm\infty} f(x) = l \in \mathbf{R}$, then $f - l1_G \in \mathscr{C}_0(\mathbf{R})$ and by Proposition 21.2, for every $M \in \mathrm{IM}(L^\infty_\mathbf{R}(\mathbf{R}))$, we have $M(f) = l$. We indicate a one-sided version of this result; the proof is modeled after 12.2(a).

Proposition 22.19. *There exists $M \in \mathrm{IM}(\mathscr{C}_\mathbf{R}(\mathbf{R}))$ such that $M(f) = l$ whenever $f \in \mathscr{C}_\mathbf{R}(\mathbf{R})$ with $\lim_{n \to \infty} f(x) = l$.*

Proof. Consider $f_0 \in \mathscr{C}_\mathbf{R}(\mathbf{R})$ with $\lim_{n \to \infty} f_0(x) = 0$ and let $\alpha \in \mathbf{R}$ such that $|f_0(x)| \leq \frac{1}{2}$ whenever $x \geq \alpha$. Consider also $f_1, \ldots, f_n \in \mathscr{C}_\mathbf{R}(\mathbf{R})$, $a_1, \ldots, a_n \in \mathbf{R}$, $n \in \mathbf{N}^*$ and let

$$\beta = \sup\left\{\sum_{i=1}^n (f_i - {_{a_i}}f_i)(x) : x \geq \alpha\right\}.$$

For every $i = 1, \ldots, n$, $f_i - {_{a_i}}f_i = (-{_{a_i}}f_i) - {_{-a_i}}(-{_{a_i}}f)$, so we may suppose that $a_i \geq 0$ whenever $i = 1, \ldots, n$. Let $p \in \mathbf{N}^*$ and denote by S the set of the p^n mappings from $\{1, \ldots, n\}$ into $\{1, \ldots, p\}$. We consider the mapping

$$\theta : S \to \mathbf{R}$$

$$s \mapsto \alpha + \sum_{j=1}^n s(j)a_j.$$

Let $i \in \{1, \ldots, n\}$ be fixed. If $k = 1, \ldots, p$, we denote by S_k the set of the p^{n-1} mappings $s \in S$ such that $s(i) = k$. Observe that, for $k = 2, \ldots, p$,

$$\left\{\alpha + \sum_{j=1}^n s(j)a_j : s \in S_k\right\} = \left\{\alpha + a_i + \sum_{j=1}^n s(j)a_j : s \in S_{k-1}\right\}.$$

Therefore in the sum $\sum_{s \in S}(f_i(\theta(s)) - f_i(a_i + \theta(s)))$ at most $2p^{n-1}$ terms do not cancel. Hence

$$\beta p^n \geq \sup \sum_{s \in S} \sum_{i=1}^{n} (f_i(\theta(s)) - f_i(a_i + \theta(s)))$$

$$\geq -2p^{n-1} n \sup \{\|f_i\| : i = 1, \ldots, n\}$$

and

$$\beta \geq -2np^{-1} \sup \{\|f_i\| : i = 1, \ldots, n\}.$$

As $p \in \mathbf{N}^*$ may be chosen arbitrarily, we have $\beta \geq 0$. Hence for every $h = f_0 + g$ where $g \in \mathcal{N}(\mathcal{C}_\mathbf{R}(\mathbf{R}))$, we have

$$\sup \{h(x) : x \geq \alpha\} \geq -\tfrac{1}{2} \text{ and } \|1_G + h\| \geq \tfrac{1}{2}.$$

Let Z be the subspace of $\mathcal{C}_\mathbf{R}(\mathbf{R})$ formed by the functions $h = f + g$, where $f \in \mathcal{C}_\mathbf{R}(\mathbf{R})$ with $\lim_{n \to \infty} f(x) = 0$ and $g \in \mathcal{N}(\mathcal{C}_\mathbf{R}(\mathbf{R}))$. As $1_G \notin Z$, by the Hahn–Banach theorem there exists $M \in \mathcal{C}_\mathbf{R}(\mathbf{R})^*$ such that $\|M\| = M(1_G) = 1$ and $M(h) = 0$ whenever $h \in Z$; M is a mean on $\mathcal{C}_\mathbf{R}(\mathbf{R})$ that is left invariant. Finally if $f \in \mathcal{C}_\mathbf{R}(\mathbf{R})$ and $l = \lim_{x \to \infty} f(x)$, then also $\lim_{x \to \infty} (f - l1_G)(x) = 0$, $f - l1_G \in Z$, and $M(f) = M(f - l1_G) + M(l1_G) = l$. □

Proposition 22.20. *Let* $M \in \text{IM}(L_\mathbf{R}^\infty(\mathbf{R}))$. *For every* $f \in \mathcal{C}_\mathbf{R}(\mathbf{R})$ *such that* $l = \lim_{x \to \infty} f(x)$ *and* $m = \lim_{x \to -\infty} f(x) \in \mathbf{R}$, *we have*

$$\inf \{l, m\} \leq M(f) \leq \sup \{l, m\}.$$

Proof. Let $g \in L_\mathbf{R}^\infty(\mathbf{R})$ such that

$$g(x) = \sup \{l, m\} \quad \text{if } x \geq 0,$$

$$g(x) = \inf \{l, m\} \quad \text{if } x < 0.$$

Then

$$\inf \{l, m\} \leq M(g) \leq \sup \{l, m\}. \tag{9}$$

For every $\varepsilon > 0$, there exists a compact subset K of G such that

$$|f - g| \leq \|f - g\|_\infty 1_K + \varepsilon 1_{\mathbf{R} \setminus K}.$$

Hence by Proposition 21.2 we have

$$|M(f - g)| \leq M(|f - g|) \leq \varepsilon.$$

As $\varepsilon > 0$ may be chosen arbitrarily, we obtain

$$M(f) = M(g). \tag{10}$$

The relations (9) and (10) establish the statement. □

Almost periodic functions constitute a space of functions admitting a unique invariant mean. Recall that if G is a group and $f \in l^\infty(G)$, the following properties are equivalent: $\{f_a : a \in G\}^-$ is compact in $l^\infty(G)$, $\{_a f : a \in G\}^-$ is compact in $l^\infty(G)$, and $\{_a f_b : a, b \in G\}^-$ is compact in $l^\infty(G)$. As $l^\infty(G)$ is a complete metric space, we may replace relative compactness by total boundedness in the conditions ([HR] 18.1; 3.7). A function satisfying these equivalent conditions is called *almost periodic*. The theory of almost periodic functions was inaugurated by Bohr [39]. Important surveys are due to Weil ([552] Section 34), Maak [376], Loomis ([370] Section 41), Corduneanu [101]; see also ([DS] IV. 7), ([HR] Section 18), ([D] Section 16), Alfsen and Holm [6], Burckel [61], and, for the particular case of abelian groups, Dhombres [135].

The set of almost periodic functions on G forms a vector space, contains 1_G, and is closed under mutliplication. It is invariant and admits a unique left invariant mean M; M is invariant, and $M(\check{f}) = M(f)$ whenever f is almost periodic. Also $M(f) > 0$ for every nonzero almost periodic function f in $l^\infty_+(G)$. If G is a topological group, the set $AP(G)$ consisting of all almost periodic continuous functions on G constitutes a commutative C^*-algebra with unit; moreover $AP(G) \subset \mathcal{UC}(G)$. To every topological group G is associated its Bohr compactification, that is, a compact group $B_G(\simeq \overline{AP(G)})$ such that G is dense in B_G and $f \in \mathcal{C}(G)$ belongs to $AP(G)$ if and only if f may be extended to an element of $\mathcal{C}(B_G) = \mathcal{UC}(B_G)$. Via Gelfand's theory there exists a one-to-one correspondence between $AP(G)$ and $\mathcal{C}(B_G)$ given by a continuous homomorphism θ of G onto a dense subset of B_G; $f \in AP(G)$ if and only if there exits $g \in \mathcal{C}(B_G)$ such that $f = g \circ \theta$ ([370] Section 41). The unique invariant mean on $AP(G)$ is induced by the unique element of $\mathrm{IM}(\mathcal{C}(B_G))$, that is, the normalized Haar measure on B_G. In an amenable group G, the restrictions to $AP(G)$ of all invariant means on $\mathcal{UC}(G)$ coincide.

Proposition 22.21. *Let G be an amenable σ-compact group and let $(U_n) \in \mathcal{A}(G)$. If M is the invariant mean on $AP(G)$ and $f \in AP(G)$, then*

$$M(f) = \lim_{n \to \infty} \frac{1}{|U_n|} \int_{U_n} f(x) \, dx.$$

Proof. If $f \in AP_{\mathbf{R}}(G)$, $a \in G$ and $n \in \mathbf{N}^*$,

$$\left| \frac{1}{|U_n|} \int_{U_n} (_a f - f) \right| \leq \frac{|a^{-1} U_n \triangle U_n|}{|U_n|} \|f\|$$

hence
$$\lim_{n\to\infty} \frac{1}{|U_n|} \int_{U_n} (_a f - f) = 0. \tag{11}$$

Now let
$$p(f) = \limsup_{n\to\infty} \frac{1}{|U_n|} \int_{U_n} f(x)\, dx,$$

$f \in AP_{\mathbf{R}}(G)$; p is a sublinear functional on $AP_{\mathbf{R}}(G)$. By the Hahn–Banach theorem ([HR] B.13) there exists a linear functional M_0 on $AP_{\mathbf{R}}(G)$ such that
$$-p(-f) \le M_0(f) \le p(f)$$
whenever $f \in AP_{\mathbf{R}}(G)$. Obviously $M_0(1_G) = 1$ and $M_0 \ge 0$; $M_0 \in \mathfrak{M}(AP_{\mathbf{R}}(G))$. Moreover, by (11), for every $f \in AP_{\mathbf{R}}(G)$ and every $a \in G$, $M_0(_a f - f) = 0$. Hence M_0 is the unique (left) invariant mean on $AP_{\mathbf{R}}(G)$.

Assume that there exists $g \in AP_{\mathbf{R}}(G)$ satisfying
$$-p(-g) = \liminf_{n\to\infty} \frac{1}{|U_n|} \int_{U_n} g(x)\, dx$$
$$< \alpha < \limsup_{n\to\infty} \frac{1}{|U_n|} \int_{U_n} g(x)\, dx = p(g)$$

with $\alpha \ne M_0(g)$. Let S be the subspace of $AP_{\mathbf{R}}(G)$ generated by $\{g, 1_G\}$. By one form of the Hahn–Banach theorem ([HR] B.12) there exists a linear functional N_1 on S such that $N_1(1_G) = 1$, $N_1(g) = \alpha$, and $-p(-f) \le N_1(f) \le p(f)$ whenever $f \in S$; $N_1 \ge 0$. Via the Hahn–Banach theorem again we may extend N_1 to a linear functional M_1 on $AP_{\mathbf{R}}(G)$ such that $-p(-f) \le M_1(f) \le p(f)$ whenever $f \in AP_{\mathbf{R}}(G)$. By (11), M_1 is a left invariant mean on $AP_{\mathbf{R}}(G)$; $M_1(g) = \alpha \ne M_0(g)$, and we obtain a contradiction as there exists a unique left invariant mean on $AP_{\mathbf{R}}(G)$. We conclude that, for every $f \in AP(G)$,
$$M(f) = \lim_{n\to\infty} \frac{1}{|U_n|} \int_{U_n} f(x)\, dx. \qquad \square$$

F. Cardinalities of the Sets of Invariant Means

We recall some properties of the Stone–Čech compactification $\beta \mathbf{N}$ of $\mathbf{N}$; a reference is Gillman and Jerison [202]. Every $f \in l^{\infty}_{\mathbf{R}}(\mathbf{N})$ may be extended to $f_{\beta} \in \mathscr{C}_{\mathbf{R}}(\beta \mathbf{N})$; $l^{\infty}_{\mathbf{R}}(\mathbf{N})$ and $\mathscr{C}_{\mathbf{R}}(\beta \mathbf{N})$ are isomorphic (Section 2A). If $f \in l^{\infty}_{\mathbf{R}}(\mathbf{N})$ and $w \in \beta \mathbf{N}$, we put
$$\langle f, w_{\beta} \rangle = f_{\beta}(w) = \langle f_{\beta}, \delta_w \rangle. \tag{12}$$

For $A \subset \beta\mathbf{N}$, let $A_\beta = \{w_\beta \in l_\mathbf{R}^\infty(\mathbf{N})' : w \in A\}$. The compact space $\beta\mathbf{N}$ is isomorphic to the Gelfand structure space $\widehat{\mathscr{C}_\mathbf{R}(\beta\mathbf{N})} \simeq \widehat{l_\mathbf{R}^\infty(\mathbf{N})}$ (Section 2B.3). Moreover $\mathfrak{M}(l_\mathbf{R}^\infty(\mathbf{N})$ is affinely $w^* - w^*$ homeomorphic to $M^1(\widehat{\mathscr{C}_\mathbf{R}(\beta\mathbf{N})})$ and $M^1(\beta\mathbf{N})$ (Section 3A). So $\mathfrak{M}(l_\mathbf{R}^\infty(\mathbf{N}))$ is $w^* - w^*$ homeomorphic to $\overline{\mathrm{co}}\{\delta_w : w \in \beta\mathbf{N}\}$ ([HR] B. 25, 30). Via (12) we obtain $\mathfrak{M}(l_\mathbf{R}^\infty(\mathbf{N})) = \overline{\mathrm{co}}(\beta\mathbf{N})_\beta$.

There exists a copy of $\beta\mathbf{N}$ in $\beta\mathbf{N} \setminus \mathbf{N}$ and we have $\mathrm{card}(\beta\mathbf{N} \setminus \mathbf{N}) = 2^{\mathfrak{c}}$ ([202] 6.10 (a)); then also $\mathrm{card}\,\mathfrak{M}(l_\mathbf{R}^\infty(\mathbf{N})) \geq 2^{\mathfrak{c}}$. As $\mathrm{card}\,l_\mathbf{R}^\infty(\mathbf{N}) = 2^{\aleph_0} = \mathfrak{c}$, we have exactly $\mathrm{card}\,\mathfrak{M}(l_\mathbf{R}^\infty(\mathbf{N})) = 2^{\mathfrak{c}}$.

Lemma 22.22. *In $l_\mathbf{R}^\infty(\mathbf{N})$ we consider the subspace*

$$Z = \{f \in l_\mathbf{R}^\infty(\mathbf{N}) : \lim_{n \to \infty} f(n) \in \mathbf{R}\}.$$

Let M_0 be the mean on Z that is defined by $M_0(f) = \lim_{n\to\infty} f(n)$. Then $\mathfrak{M}(l_\mathbf{R}^\infty(\mathbf{N}))$ is a subset of $\mathfrak{M}_0 = \{M \in l_\mathbf{R}^\infty(\mathbf{N})^* : \|M\| = 1, M|_Z = M_0\}$.

Proof. (a) Let $w \in \beta\mathbf{N} \setminus \mathbf{N}$. There exists a net (n_i) in $\mathbf{N}$ such that $\lim_i n_i = \infty$ and also $\lim_i n_i = w$ in $\beta\mathbf{N}$. So for every $f \in Z$,

$$\langle f, w_\beta \rangle = f_\beta(w) = \lim_i f_\beta(n_i) = \lim_{n \to \infty} f(n),$$

that is, $w_\beta \in \mathfrak{M}_0$ and then also $\overline{\mathrm{co}}((\beta\mathbf{N}\setminus\mathbf{N})_\beta) \subset \mathfrak{M}_0$.

(b) As $\mathfrak{M}(l_\mathbf{R}^\infty(\mathbf{N})) = \overline{\mathrm{co}}(\beta\mathbf{N})_\beta$ and $\beta\mathbf{N}\setminus\mathbf{N}$ contains a copy of $\beta\mathbf{N}$, we have also $\mathfrak{M}(l_\mathbf{R}^\infty(\mathbf{N})) \subset \overline{\mathrm{co}}(\beta\mathbf{N}\setminus\mathbf{N})_\beta$. □

Lemma 22.23. *Let G be a noncompact amenable σ-compact group. The set $\mathfrak{M}(l_\mathbf{R}^\infty(\mathbf{N}))$ is affinely $w^* - w^*$ homeomorphic to a subset E of $\mathrm{TIM}(L_\mathbf{R}^\infty(G))$ such that $M(f) = M(\check{f})$ whenever $M \in E$ and $f \in L_\mathbf{R}^\infty(G)$.*

Proof. With respect to Lemma 22.22 it suffices to show that $\mathfrak{M}_0$ is affinely homeomorphic to such a subset.

By Proposition 16.10 and noncompactness of G there exists an increasing sequence $(U_n) \in \mathscr{A}(G)$ consisting of symmetric subsets such that $\lim_{n\to\infty} |U_n| = \infty$ and, for every $n \in \mathbf{N}^*$, $|U_{n+1}| > (n+1)|U_n|$. Then also

$$|U_{n+1} \setminus U_n| > n|U_n| \tag{13}$$

and

$$\frac{|U_{n+1}|}{|U_{n+1} \setminus U_n|} = 1 + \frac{|U_n|}{|U_{n+1} \setminus U_n|} < 1 + \frac{1}{n}. \tag{14}$$

Define $\Theta : L_\mathbf{R}^\infty(G) \to l_\mathbf{R}^\infty(\mathbf{N})$ by putting

$$\Theta(f)(n) = \frac{1}{|U_{n+2} \setminus U_{n+1}|} \int_{U_{n+2}\setminus U_{n+1}} f(x)\, dx,$$

$f \in L^\infty_{\mathbb{R}}(G)$, $n \in \mathbb{N}^*$; Θ is linear, $\|\Theta\| = 1$. For every $n \in \mathbb{N}^*$, $\Theta(1_{U_{n-2} \setminus U_{n-1}})(n) = 1$, hence Θ is surjective. If $M \in \mathfrak{M}_0$, consider

$$\langle f, \Theta^\sim(M) \rangle = \langle \Theta(f), M \rangle,$$

$f \in L^\infty_{\mathbb{R}}(G)$; $\Theta^\sim$ is an affine $w^* - w^*$ homeomorphism of $\mathfrak{M}_0$ onto a subset of $\mathfrak{M}(L^\infty_{\mathbb{R}}(G))$.

For $f \in L^\infty_{\mathbb{R}}(G)$, $a \in G$, and $n \in \mathbb{N}^*$, by (14) and (13) we obtain

$$|\Theta(_af - f)(n - 1)| \leq \frac{1}{|U_{n+1} \setminus U_n|} \left| \int_{U_{n+1} \setminus U_n} (f(ax) - f(x)) \, dx \right|$$

$$\leq \frac{|a^{-1}(U_{n+1} \setminus U_n) \triangle (U_{n+1} \setminus U_n)|}{|U_{n+1} \setminus U_n|} \|f\|_\infty$$

$$\leq \frac{|a^{-1}U_{n+1} \triangle U_{n+1}| + 2|U_n|}{|U_{n+1} \setminus U_n|} \|f\|_\infty$$

$$\leq \left(\frac{|a^{-1}U_{n+1} \triangle U_{n+1}|}{|U_{n+1}|} \left(1 + \frac{1}{n}\right) + \frac{2}{n} \right) \|f\|_\infty.$$

Hence $\lim_{n \to \infty} \Theta(_af - f)(n) = 0$. If $f \in L^\infty_{\mathbb{R}}(G)$, U is a compact subset of G with $\overset{\circ}{U} \neq \varnothing$ and $n \in \mathbb{N}^*$, we also therefore obtain

$$|\Theta(\xi_U * f - f)(n - 1)| = \frac{1}{|U_{n+1} \setminus U_n|} \left| \int_{U_{n+1} \setminus U_n} (\xi_U * f(x) - f(x)) \, dx \right|$$

$$= \frac{1}{|U_{n+1} \setminus U_n|} \left| \int_{U_{n+1} \setminus U_n} \left(\frac{1}{|U|} \int_U (f(y^{-1}x) - f(x)) \, dy \right) dx \right|$$

$$\leq \frac{1}{|U_{n+1} \setminus U_n|} \sup \left\{ |y^{-1}(U_{n+1} \setminus U_n) \triangle (U_{n+1} \setminus U_n)| : y \in U \right\} \|f\|_\infty$$

$$\leq \sup \left\{ \frac{|y^{-1}U_{n+1} \triangle U_{n+1}|}{|U_{n+1}|} \left(1 + \frac{1}{n}\right) + \frac{2}{n} : y \in U \right\} \|f\|_\infty;$$

by Proposition 16.16, $\lim_{n \to \infty} \Theta(\xi_U * f - f)(n) = 0$. Thus if $M \in \mathfrak{M}_0$, then for every $a \in G$ and every compact subset U of G with $\overset{\circ}{U} \neq \varnothing$,

$$\langle _af - f, \Theta^\sim(M) \rangle = \langle \Theta(_af - f), M \rangle = 0$$

and

$$\langle \xi_U * f - f, \Theta^{\sim}(M) \rangle = \langle \Theta(\xi_U * f - f), M \rangle = 0.$$

We conclude from Proposition 22.2 that $\Theta^{\sim}(M) \in \text{TLIM}(L_{\mathbf{R}}^{\infty}(G))$.

As the set U_n ($n \in \mathbf{N}^*$) are symmetric, for every $f \in L_{\mathbf{R}}^{\infty}(G)$, $\Theta(f) = \Theta(\check{f})$, and then for every $M \in \mathfrak{M}_0$,

$$\langle \check{f}, \Theta^{\sim}(M) \rangle = \langle \Theta(\check{f}), M \rangle = \langle \Theta(f), M \rangle = \langle f, \Theta^{\sim}(M) \rangle.$$

Therefore also $\Theta^{\sim}(M) \in \text{TIM}(L_{\mathbf{R}}^{\infty}(G))$ by Proposition 4.7. □

We next state a general property.

Proposition 22.24. *Every noncompact locally compact group G contains a noncompact σ-compact open subgroup H.*

Proof. Let U_1 be a compact symmetric neighborhood of e in G and let K_1 be the subgroup of G generated by U_1. If K_1 is noncompact, take $H = K_1$. Otherwise there exists a compact symmetric neighborhood U_2 of e such that $K_1 \subsetneq U_2$. Let K_2 be the subgroup generated by U_2. If K_2 is noncompact, take $H = K_2$. Otherwise the same procedure may be carried on. If it does not end after a finite number of steps, take $H = \cup_{n \in \mathbf{N}^*} K_n$; then H is a σ-compact open subgroup and H is noncompact as (K_n) is an increasing sequence of open subsets. □

Proposition 22.25. *If G is a noncompact amenable group, $\mathfrak{M}(l_{\mathbf{R}}^{\infty}(\mathbf{N}))$ may be embedded affinely and $w^* - w^*$ topologically into $\text{LIM}(\mathscr{C}_{\mathbf{R}}(G))$.*

Proof. Proposition 22.24 implies the existence of a noncompact σ-compact open subgroup H in G. This subgroup is amenable by Proposition 13.3. Via Lemma 22.23 we may embed $\mathfrak{M}(l_{\mathbf{R}}^{\infty}(\mathbf{N}))$ affinely and $w^* - w^*$ topologically into $\text{TIM}(L_{\mathbf{R}}^{\infty}(H))$, and then also into $\text{LIM}(\mathscr{C}_{\mathbf{R}}(H))$ by Corollary 4.12. The conclusion now follows from Proposition 22.7. □

In opposition to the uniqueness of invariant means for compact groups stated in Proposition 12.1 the following result says that, for noncompact amenable groups, invariant means are very numerous.

Proposition 22.26. *Let G be a noncompact amenable group. Then card $\text{LIM}(\mathscr{C}_{\mathbf{R}}(G)) \geq 2^c$. If G is either σ-compact or nonunimodular, card $\text{TIM}(L_{\mathbf{R}}^{\infty}(G)) \geq 2^c$.*

Proof. As card $\mathfrak{M}(l_{\mathbf{R}}^{\infty}(\mathbf{N})) = 2^c$, the first statement follows from Proposition 22.25; in case G is σ-compact, the second statement follows from Lemma 22.23.

If G is nonunimodular, $H = \{x \in G : \Delta(x) = 1\}$ is a closed normal subgroup of G and G/H is isomorphic to the noncompact σ-compact abelian multiplicative group $\mathbf{R}_+^*$; we have card $\mathrm{TIM}(L_{\mathbf{R}}^\infty(G/H)) \geq 2^c$. Then by Propositions 4.7, 4.17, and 22.8 we have also $\mathrm{TIM}(L_{\mathbf{R}}^\infty(G)) \geq 2^c$. □

G. Specific Invariant Means

We study invariant means that verify some particular properties.

Definition 22.27. *If G is a locally compact group, let X be a subspace of $L_{\mathbf{R}}^\infty(G)$ [resp. $L^\infty(G)$] that contains 1_G and is closed under the mapping $\iota: f \to \check{f}$ [and also under complex conjugation]. A mean M on X is said to be inversion invariant if $M = \check{M}$.*

Note that if $M \in \mathfrak{M}(X)$, $N = \frac{1}{2}(M + \check{M})$ is an inversion invariant mean. If X is left [resp. right] invariant, then by (2.1) it is also right [resp. left] invariant. If X is topologically left [resp. topologically right] invariant, then by (2.3) it is also topologically right [resp. topologically left] invariant.

Proposition 22.28. *Let G be a locally compact group and let X be a subspace of $L_{\mathbf{R}}^\infty(G)$ [resp. $L^\infty(G)$] that contains 1_G and is closed under the mapping ι [and also under complex conjugation].*

(a) *If X is left and right invariant, any left invariant [resp. right invariant] mean on X that is inversion invariant is also right invariant [resp. left invariant].*

(b) *If X is topologically left and right invariant, any topologically left invariant [resp. topologically right invariant] mean on X that is inversion invariant is also topologically right invariant [resp. topologically left invariant].*

Proof. The statement follows readily from Proposition 4.7. □

Definition 22.29. *Let G be a locally compact group and let X be a subspace of $L_{\mathbf{R}}^\infty(G)$ [resp. $L^\infty(G)$] that contains 1_G and is closed under the mapping ι [and also under complex conjugation]. We denote by $\mathrm{IIM}(X)$ the set of all means on X that are invariant and inversion invariant.*

Proposition 22.30. *The sets $\mathrm{IM}(\mathscr{C}_{\mathbf{R}}(\mathbf{R}))$ and $\mathrm{IIM}(\mathscr{C}_{\mathbf{R}}(\mathbf{R}))$ do not coincide.*

Proof. We consider the invariant mean M defined on $\mathscr{C}_{\mathbf{R}}(\mathbf{R})$ by Proposition 22.19. Let $f \in \mathscr{C}_{\mathbf{R}}(\mathbf{R})$ be defined by

$$f(x) = 1 \quad \text{if } x \geq 1,$$
$$f(x) = x \quad \text{if } 0 \leq x \leq 1,$$
$$f(x) = 0 \quad \text{if } x \leq 0.$$

Then $M(f) = 1$ and $M(\check{f}) = 0$. □

Proposition 22.31. *Let G be an infinite discrete abelian group that is not a torsion group, with $\operatorname{card} G \leq \mathfrak{c}$. If $M \in \operatorname{IIM}(l_{\mathbf{R}}^{\infty}(G))$, there exists $N \in \operatorname{IM}(l_{\mathbf{R}}^{\infty}(G))$ such that $M = \frac{1}{2}(N + \check{N})$ and $N, \check{N}$ are concentrated on disjoint subsets; also $N \in \operatorname{IM}(l_{\mathbf{R}}^{\infty}(G)) \setminus \operatorname{IIM}(l_{\mathbf{R}}^{\infty}(G))$.*

Proof. Let H be the subgroup of G generated by the elements of finite order in G. As G is not a torsion group, G/H is nontrivial and admits no nontrivial element of finite order; G/H may be embedded into a divisible group that is the direct sum of at most $\mathfrak{c}$ copies of $\mathbf{Q}$ ([HR] A.15, A.14). Therefore one may consider an archimedean order on G/H. Let S_1 be the subsemigroup of G/H formed by the nonnegative elements; $S_1 \cup (-S_1) = G/H$ and $S_1 \cap (-S_1) = \{e\}$. If π denotes the canonical homomorphism of G onto G/H, let $S = \pi^{-1}(S_1)$ and put

$$L(f) = M(1_S f), \quad f \in l_{\mathbf{R}}^{\infty}(G).$$

As G/H is infinite and M is left invariant, we must have $M(1_H) = 0$. As M is inversion invariant, we obtain $M(1_{S \setminus H}) = M(1_{(-S) \setminus H})$; hence $L(1_G) = M(1_S) = M(1_{S \setminus H}) = \frac{1}{2}$. We have $N = 2L \in \operatorname{IM}(l_{\mathbf{R}}^{\infty}(G))$ by Proposition 21.6. Also $M = \frac{1}{2}(N + \check{N})$; N is concentrated on $S \setminus H$ whereas $\check{N}$ is concentrated on $(-S) \setminus H$ and $(S \setminus H) \cap ((-S) \setminus H) = \varnothing$. The invariant mean N is not inversion invariant. □

If G is a group and $A \subset G$, let $A^{(2)} = \{x^2 : x \in A\}$.

Lemma 22.32. *Let G be a discrete abelian countable torsion group such that $G^{(2)}$ is infinite. Then there exists a subset E in G having the following properties*:

(i) $E \cap (-E) = \varnothing$.

(ii) *For every finite subset A in G, there exists $a \in G$ satisfying $a + A \subset E$.*

Proof. Choose a well ordering $e = y_0, y_1, \ldots$ on G. Let $H_0 = \{e\} = \{y_0\}$. If $n_1 \in \mathbf{N}^*$ is the smallest element in $\mathbf{N}^*$ such that $y_{n_1} + y_{n_1} \neq e$, let H_1 be the subgroup of G generated by $y_0, y_1, \ldots, y_{n_1}$. As G is a torsion group, H_1 is finite. We then consider the smallest $n_2 \in \mathbf{N}^*$ such that $n_2 > n_1$ and $y_{n_2} + y_{n_2} \notin H_1$. Let H_2 be the subgroup generated by $y_0, y_1, \ldots, y_{n_2}$.

Continuing this procedure inductively we obtain an infinite increasing sequence (H_k) of finite subgroups of G such that $G = \cup_{k \in \mathbf{N}} H_k$ and an infinite sequence (y_{n_k}) in G such that

$$y_{n_k} \in H_k, \quad y_{n_k} + y_{n_k} \notin H_{k-1} \tag{15}$$

for $k \in \mathbf{N}^*$. Let $E_k = y_{n_k} + H_{k-1}$, $k \in \mathbf{N}^*$, and put $E = \cup_{k \in \mathbf{N}^*} E_k$. As for every $k \in \mathbf{N}^*$, H_{k-1} is a subgroup and $y_{n_k} + y_{n_k} \notin H_{k-1}$, we have necessarily

$$y_{n_k} \notin H_{k-1}. \tag{16}$$

If $h, k \in \mathbf{N}^*$, then $E_h \cap (-E_k) = \emptyset$. As a matter of fact, if for $h, k \in \mathbf{N}^*$ with $h < k$ we had $E_h \cap (-E_k) \neq \emptyset$, there would exist $x'_h \in H_{h-1}$ and $x'_k \in H_{k-1}$ such that $y_{n_h} + x'_h = -y_{n_k} - x'_k$. Then

$$y_{n_k} = -y_{n_h} - x'_h - x'_k \in H_{k-1}$$

and (16) would be contradicted. If for one $k \in \mathbf{N}^*$ we had $E_k \cap (-E_k) \neq \emptyset$, there would exist $x''_k, x'''_k \in H_{k-1}$ such that $y_{n_k} + x''_k = -y_{n_k} - x'''_k$. Then

$$y_{n_k} + y_{n_k} = -x''_k - x'''_k \in H_{k-1}$$

and (15) would be contradicted. Therefore $E \cap (-E) = \emptyset$.

If A is a finite subset of G, there exists $p \in \mathbf{N}^*$ such that $A \subset H_{p-1}$. Then we have

$$y_{n_p} + A \subset y_{n_p} + H_{p-1} = E_p \subset E. \qquad \square$$

Proposition 22.33. *Let G be an abelian torsion group. If $G^{(2)}$ is infinite, then $\mathrm{IIM}(l^\infty_\mathbf{R}(G)) \neq \mathrm{IM}(l^\infty_\mathbf{R}(G))$.*

Proof. (a) Case: G is a countable group.

Let E be the subset of G determined in Lemma 22.32. By (ii) of that lemma and Proposition 21.4 there exists $M \in \mathrm{IM}(l^\infty_\mathbf{R}(G))$ such that $M(1_E) = 1$. By (i) of the lemma, M is not inversion invariant.

(b) Case: G is an uncountable group.

Let H be any countable subgroup of G; H may be embedded into a countable divisible group D ([HR] A.15). This embedding may be extended to a homomorphism Φ of G into D ([HR] A.7). The group $\Phi(G)$ is a countable torsion group such that $\Phi(G)^{(2)}$ is infinite. By (a) there exists an invariant mean M on $l^\infty_\mathbf{R}(\Phi(G))$ that is not inversion invariant. Let

$$M_1(f) = M(f \circ \Phi),$$

$f \in l^\infty_\mathbf{R}(G)$. Then $M_1 \in \mathrm{IM}(l^\infty_\mathbf{R}(G)) \setminus \mathrm{IIM}(l^\infty_\mathbf{R}(G))$. $\qquad \square$

The latter proposition applies for instance to the discrete abelian group that is freely generated by an infinity of elements each of order two.

Proposition 22.34. *Let G be a discrete abelian group such that $G^{(2)}$ is finite. Then $\mathrm{IIM}(l^\infty_\mathbf{R}(G)) = \mathrm{IM}(l^\infty_\mathbf{R}(G))$.*

Proof. Let $G^{(2)} = \{a_1, \ldots, a_n\}$. For each $x \in G$, $G^{(2)} = x + x + G^{(2)}$; hence $x + G^{(2)} = -x + G^{(2)}$. For all $f \in l^\infty_\mathbf{R}(G)$ and $x \in G$,

$$\sum_{i=1}^n {}_{a_i}f(x) = \sum_{i=1}^n f(a_i + x) = \sum_{i=1}^n f(a_i - x) = \sum_{i=1}^n {}_{-a_i}\check{f}(x).$$

If $M \in \text{IM}(l_\mathbf{R}^\infty(G))$,

$$M(f) = \frac{1}{n} M\left(\sum_{i=1}^n {}_{a_i}f\right) = \frac{1}{n} M\left(\sum_{i=1}^n {}_{-a_i}\check{f}\right) = M(\check{f}). \qquad \square$$

Corollary 22.35. *Let G be a discrete abelian group such that* $\text{card}\, G \leq \mathfrak{c}$. *Then* $\text{IM}(l_\mathbf{R}^\infty(G)) = \text{IIM}(l_\mathbf{R}^\infty(G))$ *if and only if $G^{(2)}$ is finite.*

Proof. The statement follows from Propositions 22.31, 22.33, and 22.34. $\square$

We introduce a condition that is weaker than amenability. For reasons of simplicity we consider only the discrete case. If G is any group, the *trivial mean* $M_0 = \delta_e$ on $l^\infty(G)$ satisfies $M_0({}_af_{a^{-1}}) = M(f)$ whenever $f \in l^\infty(G)$, $a \in G$. Let G be a nontrivial group. Then for any mean M on $l^\infty(G)$ we have $M \neq \delta_e$ if and only if $M(1_{\{e\}}) < 1$. We say that M is a *nontrivial mean*.

Definition 22.36. *A discrete group G is called inner amenable if $l^\infty(G)$ admits a nontrivial mean M that is inner, that is, $M({}_af_{a^{-1}}) = M(f)$ whenever $f \in l^\infty(G)$ and $a \in G$.*

If M is a nontrivial mean on $l^\infty(G)$, let

$$M_1 = \left(1 - M(1_{\{e\}})\right)^{-1}\left(M - M(1_{\{e\}})\delta_e\right).$$

We have $M_1(1_G) = 1$. If $f \in l_+^\infty(G)$, $f - f(e)1_{\{e\}} \geq 0$, $M(f) \geq f(e)M(1_{\{e\}})$ and $M_1(f) \geq 0$. Hence M_1 is a mean on $l^\infty(G)$; $M_1(1_{\{e\}}) = 0$. In particular, if M is inner, then so is M_1.

From Propositions 12.1 and 21.2 we deduce immediately that every nontrivial amenable group is also inner amenable.

Proposition 22.37. *There exists an inner amenable group that is not amenable.*

Proof. Let H be the discrete group of all permutations of $\mathbf{N}$ leaving all but a finite number of elements unchanged; H is amenable by Corollary 13.10. Let $G = H \times F_2$. As F_2 is nonamenable, G is nonamenable by Proposition 13.3.

As H is an infinite amenable group, by Theorem 4.19 and Proposition 21.2 there exists a nontrivial invariant mean M on $l^\infty(H)$. Let $e = (e_1, e_2)$ be the identity element in G. For $f \in l^\infty(G)$ and $x = (x_1, e_2) \in G$, let

$$f'(x_1) = f(x_1, e_2);$$

$f' \in l^\infty(H)$. We put $M'(f) = M(f')$ for $f \in l^\infty(G)$; M' is a mean on $l^\infty(G)$. For all $f \in l^\infty(G)$, $x = (x_1, x_2) \in G$ and $a = (a_1, a_2) \in G$, we have

$${}_af_{a^{-1}}(x) = f(a_1 x_1 a_1^{-1}, e_2)$$

and

$$M'(_af_{a^{-1}}) = M(_{a_1}f'_{a_1^{-1}}) = M(f') = M'(f).$$

Hence M' is an inner mean. We have $M'(1_{\{e\}}) = M(1_{\{e_1\}}) < 1$; M' is nontrivial and G is inner amenable. □

Proposition 22.38. *The nonamenable group F_2 is not inner amenable.*

Proof. Let a, b be the generators of F_2 and let A be the subset of F_2 formed by all reduced words of the form $\ldots a^n$, $n \in \mathbf{Z}^*$. Then $F_2 = A \cup aAa^{-1} \cup \{e\}$ and the subsets A, bAb^{-1}, $b^{-1}Ab$ are pairwise disjoint. If M were a nontrivial inner mean on $l^\infty(G)$ such that $M(1_{\{e\}}) = 0$, we should have

$$2M(1_A) = 2M(1_A) + M(1_{\{e\}}) = M(1_A) + M(1_{aAa^{-1}}) + M(1_{\{e\}})$$

$$\geq M(1_{F_2}) = 1$$

and

$$1 = M(1_{F_2}) \geq M(1_A) + M(1_{bAb^{-1}}) + M(1_{b^{-1}Ab}) = 3M(1_A).$$

We would obtain a contradiction. □

Comments

For an arbitrary topological group G, consider the set $P(G)$ of continuous (bounded) positive-definite functions on G; $P(G) \subset \mathcal{LUC}(G)$ ([HR] 32.4). Godement showed that there exists a unique left invariant mean M on the space $B(G)$ generated by $P(G)$ that is invariant and given by

$$M(\varphi) = \inf \left\{ \sum_{i,j=1}^n \alpha_i \alpha_j \varphi(a_i^{-1} a_j) : \alpha_1, \ldots, \alpha_n \in \mathbf{R}_+^*; a_1, \ldots, a_n \in G; \right.$$

$$\left. \sum_{i=1}^n \alpha_i = 1; n \in \mathbf{N}^* \right\},$$

$\varphi \in P(G)$. If $f, g \in P(G)$, let $h^{(x)}(y) = f(y)g(y^{-1}x)$ for $x, y \in G$; define

$$f \circledast g(x) = M(h^{(x)}).$$

For this *convolution product*, $f \circledast g \in AP(G)$ ([208] 23, 24).

Every topological group $(G, \mathcal{T})$ admits a finest uniform topological structure $\mathcal{T}_\mathcal{U}$, corresponding to the topology of B_G, that is coarser than $\mathcal{T}$ and for which $(G, \mathcal{T}_\mathcal{U})$ is a totally bounded topological group ([6]). In the next proposition it suffices to assume, for the topological group G, the existence of a right invariant mean on $\mathscr{LUC}_R(G)$.

Proposition 22.39. *Let $(G, \mathcal{T})$ be an amenable group. The subset V of G is a $\mathcal{T}_\mathcal{U}$-neighborhood of e if and only if there exist a subset A of G and a $\mathcal{T}$-neighborhood U of e such that a finite number of right translates of A cover G and $(U^{-1}A^{-1}AU)^2 \subset V$.*

Proof. (a) The condition is necessary. As $\mathcal{T}_\mathcal{U}$ is coarser than $\mathcal{T}$ and $(G, \mathcal{T}_\mathcal{U})$ is totally bounded, one may determine a $\mathcal{T}$-neighborhood U of e such that $(U^{-1}U^{-1}UU)^2 \subset V$ and a finite number of right translates of U cover G.

(b) Let $M \in \mathrm{RIM}(\mathscr{LUC}_R(G))$. Assume that the condition holds and $\cup_{j=1}^n Ac_j = G$ for $c_1, \ldots, c_n \in G$.

By Urysohn's lemma, there exists $f \in \mathscr{LUC}_R(G)$ such that $f(G) \subset [0,1]$, $f(e) = 1$, and $f(x) = 0$ whenever $x \in G \setminus U$. Let

$$g(x) = \sup\{f(y^{-1}x) : y \in A\},$$

$x \in G$; $g \in \mathscr{LUC}_R(G)$. For every $x \in A$, $g(x) = f(e) = 1$; thus $1_A \le g$. For all $\gamma_1, \ldots, \gamma_n \in \mathbf{R}_+^*$ with $\sum_{j=1}^n \gamma_j = 1$, we have

$$M(g) = M\left(\sum_{j=1}^n \gamma_j g_{c_j^{-1}}\right) \ge \inf \sum_{j=1}^n \gamma_j g_{c_j^{-1}} \ge \inf \sum_{j=1}^n \gamma_j 1_{Ac_j} > 0.$$

If $x \in G \setminus AU$, $g(x) = 0$.

For every $x \in G$, let $h(x) = \langle g_x g, M \rangle$; $h \in \mathscr{LUC}(G)$. If $\alpha_1, \ldots, \alpha_m \in \mathbf{C}$ and $a_1, \ldots, a_m \in G$,

$$\sum_{i,j=1}^m \alpha_i \overline{\alpha_j} h(a_i^{-1} a_j) = \left\langle \left|\sum_{i=1}^m \alpha_i g_{a_i}\right|^2, M\right\rangle \ge 0$$

hence $h \in P(G)$. In particular, if $\alpha_1, \ldots, \alpha_m \in \mathbf{R}_+^*$ with $\sum_{i=1}^m \alpha_i = 1$, the Cauchy–Schwarz inequality implies that

$$\sum_{i,j=1}^m \alpha_i \alpha_j h(a_i^{-1} a_j) \ge \left(M\left(\sum_{i=1}^m \alpha_i g_{a_i}\right)\right)^2 = M(g)^2 > 0$$

and then $M(h) > 0$.

If $z \in G \setminus U^{-1}A^{-1}AU$ and $y \in G$, $g(yz)g(y) = 0$ because otherwise we should have simultaneously $yz \in AU$ and $y \in AU$. So if $z \in G \setminus U^{-1}A^{-1}AU$, $h(z) = 0$. Consider $h' = h \circledast h \in AP(G)$. If $x \notin (U^{-1}A^{-1}AU)^2$ and $y \in U^{-1}A^{-1}AU$, we have $y^{-1}x \notin U^{-1}A^{-1}AU$, hence $h(y^{-1}x) = 0$ and also $h'(x) = 0$. As $h \in P_R(G)$, $h = \check{h}$, and the Cauchy–Schwarz inequality implies that

$$h'(e) = M(h^2) \geq M(h)^2 > 0.$$

As $h' \in AP(G)$, $V_0 = \{x \in G : |h'(x) - h'(e)| < h'(e)\}$ is a $\mathcal{T}_\mathcal{U}$-neighborhood of e. We have $G \setminus (U^{-1}A^{-1}AU)^2 \subset G \setminus V_0$; hence $V_0 \subset (U^{-1}A^{-1}AU)^2 \subset V$, and V is a $\mathcal{T}_\mathcal{U}$-neighborhood of e. □

Corollary 22.40. *Let $(G, \mathcal{T})$ be an amenable group. Then a subset V in G is a $\mathcal{T}_\mathcal{U}$-neighborhood of e if and only if there exists a symmetric $\mathcal{T}$-neighborhood W of e such that a finite number of right translates of W cover G and $W^7 \subset V$.*

Proof. The necessity of the condition is trivial as $\mathcal{T}_\mathcal{U}$ is coarser than $\mathcal{T}$ and $(G, \mathcal{T}_\mathcal{U})$ is totally bounded. Conversely if the condition holds, in Proposition 22.39 let $A = W$ and let U be a $\mathcal{T}$-neighborhood of e such that $UU^{-1} \subset W$. Then

$$(U^{-1}A^{-1}AU)^2 \subset U^{-1}A^{-1}A(UU^{-1})A^{-1}AU \subset W^7 \subset V$$

and by Proposition 22.39, V is a $\mathcal{T}_\mathcal{U}$-neighborhood of e. □

Large information on the set of invariant means may be obtained by identifying these invariant means with invariant probability measures. As the study employs very elaborate and quite technical developments, we indicate just one statement that should provide a sample of the available results on the subject. It constitutes a quantitative formulation of Proposition 22.3.

Proposition 22.41. *Let G be a nondiscrete σ-compact locally compact group such that G_d is amenable. Then $\mathrm{LIM}(L^\infty_\mathbf{R}(G)) \setminus \mathrm{TLIM}(L^\infty_\mathbf{R}(G))$ is not norm-separable in $L^\infty_\mathbf{R}(G)^*$.*

Proof. By Lemma 21.9 there exists an open dense subset V in G such that $|V^{-1}| < 1$. Let I be the closed, left invariant ideal in $L^\infty_\mathbf{R}(G)$ generated by $1_{G \setminus V}$. In the nondiscrete σ-compact locally compact group G there exists a continuum $\{N_\alpha : 0 \leq \alpha \leq 1\}$ of proper closed, left invariant ideals of $L^\infty_\mathbf{R}(G)$ such that $N_\alpha \supset I$ whenever $\alpha \in [0,1]$, and $N_\beta + N_\gamma = L^\infty_\mathbf{R}(G)$ whenever β, $\gamma \in [0,1]$, $\beta \neq \gamma$. ([480] proposition 3.5). By Proposition 21.16, for every $\alpha \in [0,1]$, there exists $M_\alpha \in \mathrm{LIM}(L^\infty_\mathbf{R}(G))$ such that $\mathrm{supp}\, M_\alpha \subset h(N_\alpha)$ and $I \subset N_\alpha \subset \mathcal{K}\!er\, M_\alpha$. Proposition 21.15 implies that $M_\alpha \notin \mathrm{TLIM}(L^\infty_\mathbf{R}(G))$. If $\beta, \gamma \in [0,1]$ with $\beta \neq \gamma$, we may determine $g \in \mathscr{C}_\mathbf{R}(\overline{L^\infty_\mathbf{R}(G)})$ with range in $[-1,1]$ such that $g(h(N_\beta)) = \{-1\}$ and $g(h(N_\gamma)) = \{1\}$. Let $f \in L^\infty_\mathbf{R}(G)$ such that $\hat{f} = g$. Then

$$|\langle f, M_\beta - M_\gamma \rangle| = |\langle g, \widehat{M_\beta} - \widehat{M_\gamma} \rangle| = 2$$

and $\|M_\beta - M_\gamma\| = 2$. Therefore $\mathrm{LIM}(L^\infty_\mathbf{R}(G)) \setminus \mathrm{TLIM}(L^\infty_\mathbf{R}(G))$ cannot be normseparable. □

Proposition 22.42. *Let G be an amenable group. If $\mathscr{RUC}_\mathbf{R}(G)$ admits a left invariant mean M that is multiplicative [that is, $M(fg) = M(f)M(g)$ whenever $f, g \in \mathscr{RUC}_\mathbf{R}(G)$], then G is necessarily trivial.*

Proof. If $G \neq \{e\}$, let V be a relatively compact symmetric open neighborhood of e with $a \in V$, $a \neq e$ and let $F = \{a\} \cup (G \setminus V)$. By Urysohn's lemma there exists $f \in \mathscr{RUC}_\mathbf{R}(G)$ such that $f(G) \subset [0,1]$, $f(e) = 1$, and $f(F) = \{0\}$. By Zorn's lemma we obtain a maximal subset $\{x_i : i \in I\}$ in G such that $Vx_{i'} \cap Vx_{i''} = \emptyset$ whenever $i', i'' \in I$, $i' \neq i''$. Let

$$g : x \mapsto \sup \{f(xx_i^{-1}) : i \in I\}$$

$$G \to \mathbf{R};$$

$g \in \mathscr{RUC}_\mathbf{R}(G)$. For all $i_0 \in I$, $ax_{i_0}x_{i_0}^{-1} = a \in F$. If $i \in I$, $i \neq i_0$, then $ax_{i_0}x_i^{-1} \notin V$ as $Vx_{i_0} \cap Vx_i = \emptyset$. We have

$$g(x_{i_0}) = f(e) = 1$$

$$g(ax_{i_0}) = 0. \tag{17}$$

As $g \in \mathscr{RUC}_\mathbf{R}(G)$, co$\{g_a : a \in G\}$ constitutes an equicontinuous family of functions. The closure of co$\{g_a : a \in G\}$ in the topology of pointwise convergence is its closure in the topology of uniform convergence on compacta ([B TG] X. 16, Section 2, No. 4, théorème 1). If $M \in \text{LIM}(\mathscr{RUC}_\mathbf{R}(G))$ and (μ_k) is a net of $M_f^1(G)$ that is w*-converging to M, then for every $f \in \mathscr{RUC}_\mathbf{R}(G)$ and every $x \in G$,

$$\lim_k \langle {}_xf, \mu_k \rangle = M({}_xf) = M(f).$$

If M is multiplicative, there must exist a net (a_k) in G such that, for every $x \in G$,

$$\lim_k g(xa_k) = \lim \langle {}_xg, \delta_{a_k} \rangle = M(g).$$

We deduce then the existence of a net $(y_j)_{j \in J}$ in G and $\alpha \in \mathbf{R}$ such that $\lim_j g_{y_j} = \alpha 1_G$ uniformly on compacta. Hence there exists $j_0 \in J$ such that, for every $x \in V^3$,

$$|g(xy_{j_0}) - \alpha| < \tfrac{1}{2}. \tag{18}$$

From (17) and (18) we conclude that $y_{j_0} \notin \{x_i : i \in I\}$ because otherwise, as $e, a \in V^3$, we should simultaneously have $|1 - \alpha| < \tfrac{1}{2}$ and $|\alpha| < \tfrac{1}{2}$. The family $\{x_i : i \in I\}$ having been chosen maximal, there exists $i_0 \in I$ such that $Vx_{i_0} \cap Vy_{j_0} \neq \emptyset$, so $x_{i_0} \in V^2 y_{j_0}$ and $Vx_{i_0} \subset V^3 y_{j_0}$. Let $s \in V^2$, $t \in V^3$ such that $x_{i_0} = sy_{j_0}$ and $ax_{i_0} = ty_{j_0}$. By (17) we have $g(sy_{j_0}) = 1$ and $g(ty_{j_0}) = 0$; then again by (18) we should simultaneously have $|1 - \alpha| < \frac{1}{2}$ and $|\alpha| < \frac{1}{2}$. Hence there cannot exist a multiplicative left invariant mean on $\mathcal{RUC}_R(G)$. □

If G is an amenable discrete group, we have obviously card LIM$(l_R^\infty(G)) \le 2^{2^{\operatorname{card} G}}$. In particular, if G is a countable infinite group, by Proposition 22.26, card LIM$(l_R^\infty(G)) = 2^c$. More precise statements are at hand.

We enounce a general property concerning an arbitrary uncountable set E. There exists a family $\{E_i : i \in I\}$ of distinct subsets in E such that card $I = 2^{\operatorname{card} E}$ and $\cap_{i \in J} \omega_{I_0}(E_i) \ne \emptyset$ whenever J is a finite subset of I, I_0 is a subset of I and $\omega_{I_0} : \{E_i : i \in I\} \to \mathfrak{P}(E)$ satisfies $\omega_{I_0}(E_i) = E_i$ for $i \in I_0$, $\omega_{I_0}(E_i) = E \setminus E_i$ for $i \in I \setminus I_0$. A proof written out for the case card $E = \mathfrak{c}$ is given in ([HR] 16.8); also see Chou [83].

Definition 22.43. *If G is a discrete group, let $m(G)$ ($\le$ card G) be the largest cardinal satisfying the following condition: For every subset F of G with* card $F < m(G)$, *there exists $x \in G$ such that $(FxF) \cap (FxF)^{-1} = \emptyset$.*

Lemma 22.44. *Let G be a discrete group with center Z. Then the following properties hold*:
  (a) $m(G) \le \operatorname{card} G^{(2)}$.
  (b) *If Z is infinite*, card $Z^{(2)} \le m(G)$.
  (c) *If G is abelian and infinite*, $m(G) = \operatorname{card} G^{(2)}$.

Proof. (a) If we had card $G^{(2)} < m(G)$, there would exist $a \in G$ such that $(G^{(2)}aG^{(2)}) \cap (G^{(2)}aG^{(2)})^{-1} = \emptyset$. As $e \in G^{(2)}$, in particular, $a = eae \notin (G^{(2)}ae)^{-1} = a^{-1}G^{(2)}$, hence $a^2 \notin G^{(2)}$, and a contradiction would arise.

(b) Let $F \subset G$ with card $F <$ card $Z^{(2)}$. As Z is infinite, card$(F^{-1})^4 <$ card $Z^{(2)}$. There exists $x \in Z$ such that $x^2 \notin (F^{-1})^4$. Hence $FxF \cap (FxF)^{-1} = \emptyset$, because otherwise we could find $a, b, c, d \in F$ with $axb = d^{-1}x^{-1}c^{-1}$ and then $x^2 = a^{-1}d^{-1}c^{-1}b^{-1} \in (F^{-1})^4$.

(c) If G is abelian, $Z = G$ and the statement follows from (a) and (b). □

Proposition 22.45. *Let G be an uncountable amenable discrete group. There exist $2^{2^{\operatorname{card} G}}$ mutually singular elements in* IIM$(l_R^\infty(G))$. *[If in addition $m(G) = $ card G, then there exists a subset $\mathcal{I}$ of* IM$(l_R^\infty(G)) \setminus $ IIM$(l_R^\infty(G))$, *for which the following properties hold*:
  (a) card $\mathcal{I} = 2^{2^{\operatorname{card} G}}$.
  (b) *The elements of $\mathcal{I}$ are mutually singular.*
  (c) *For every $M \in \mathcal{I}$, there exists $E_M \subset G$ such that $M(1_{E_M}) = 1$ and $M(\check{1}_{E_M}) = 0$*].

Proof. Let Γ be the least ordinal such that card $\Gamma = $ card G and let $\{x_\gamma : \gamma \in \Gamma\}$ be a well ordering of G admitting $x_0 = e$ as its first element. For each $\gamma \in \Gamma$, let $F_\gamma = \{x_\alpha : \alpha \le \gamma\}$ with $F_0 = \{e\}$. By transfinite induction we now determine $\{y_\gamma : \gamma \in \Gamma\} \subset G$ such that, for all $\alpha, \beta \in \Gamma$ with $\beta \neq \alpha$ we have

$$((F_\alpha y_\alpha F_\alpha) \cup (F_\alpha y_\alpha F_\alpha)^{-1}) \cap ((F_\beta y_\beta F_\beta) \cup (F_\beta y_\beta F_\beta)^{-1}) = \emptyset \quad (19)$$

$$[\text{and } (F_\alpha y_\alpha F_\alpha) \cap (F_\alpha y_\alpha F_\alpha)^{-1} = \emptyset]. \quad (20)$$

[Obviously we may determine $y_0 \in G$ satisfying (20)]. Suppose that $\gamma \in \Gamma$ and that $y_\beta (\beta \in \gamma)$ have been determined satisfying (19) [and (20)]. Let

$$F'_\gamma = \bigcup_{\substack{\beta \in \Gamma \\ \beta < \gamma}} ((F_\beta y_\beta F_\beta) \cup (F_\beta y_\beta F_\beta)^{-1}), \quad F_\gamma = \bigcup_{\substack{\beta \in \Gamma \\ \beta < \gamma}} (F_\beta y_\beta F_\beta)^{-1}$$

[and let S be the set of all $y \in G$ such that $(F_\gamma y F_\gamma) \cap (F_\gamma y F_\gamma)^{-1} = \emptyset$; put $T = F_\gamma \cup S \cup S^{-1}$. If we had card $S < m(G)$, as also card $F_\gamma < m(G)$, we would have card $T < m(G)$ and there would exist $z \in G$ such that $(TzT) \cap (TzT)^{-1} = \emptyset$. As $F_\gamma \subset T$, necessarily $z \in S$ and then

$$z \in \{z^{-1}zz\} \cap \{z^{-1}zz^{-1}\}^{-1} \subset (S^{-1}zS) \cap (S^{-1}zS^{-1})^{-1}$$

$$\subset (TzT) \cap (TzT)^{-1}.$$

In order to avoid that contradiction, we must have card $S = m(G) = $ card G]. By definition card $F_\gamma <$ card G and card $F'_\gamma <$ card G, hence also card $(F_\gamma^{-1} F'_\gamma F_\gamma^{-1}) <$ card G. Therefore we may choose

$$y_\gamma \in G \setminus F_\gamma^{-1} F'_\gamma F_\gamma^{-1} \left[y_\gamma \in S \setminus F_\gamma^{-1} F'_\gamma F_\gamma^{-1} \right].$$

As $y_\gamma \notin F_\gamma^{-1} F'_\gamma F_\gamma^{-1}$, we have

$$((F_\gamma y_\gamma F_\gamma) \cup (F_\gamma y_\gamma F_\gamma)^{-1}) \cap ((F_\beta y_\beta F_\beta) \cup (F_\beta y_\beta F_\beta)^{-1}) = \emptyset$$

whenever $\beta < \gamma$ [and, as $y_\gamma \in S$, also $(F_\gamma y_\gamma F_\gamma) \cap (F_\gamma y_\gamma F_\gamma)^{-1} = \emptyset$]. Thus we verified (19) [and (20)].

We now consider a partition $\{\pi_\gamma : \gamma \in \Gamma\}$ of Γ with card $\pi_\gamma = $ card G whenever $\gamma \in \Gamma$. Let $X_\gamma = \bigcup_{\alpha \in \pi_\gamma} F_\alpha y_\alpha F_\alpha$. From (19) [and (20)] we deduce that $(X_\beta \cup X_\beta^{-1}) \cap (X_\alpha \cup X_\alpha^{-1}) = \emptyset$ [and $X_\alpha \cap X_\alpha^{-1} = \emptyset$] whenever $\alpha, \beta \in \Gamma$ with $\alpha \neq \beta$. There exists a family $\{P_\theta : \theta \in \Theta\}$ of subsets in Γ such that card $\Theta = 2^{\text{card } G}$ and $\bigcap_{k=1}^n \omega_I(P_{\theta_k}) \neq \emptyset$ whenever $\theta_1, \ldots, \theta_n \in \Theta$ and $I \subset \Theta$. If $\theta \in \Theta$ and $I \subset \Theta$, let $Y_{\theta, I} = \bigcup_{\gamma \in \omega_I(P_\theta)} X_\gamma$ and

$$C_I = \bigcap_{\substack{x, y \in G \\ \theta \in \Theta}} (x Y_{\theta, I} y)^-,$$

the closure being taken in the Stone–Čech compactification of G.

THE SETS OF INVARIANT MEANS 299

We establish that $C_I \neq \varnothing$ by showing that the family $\{(xY_{\theta,I}y)^- : x \in G, y \in G, \theta \in \Theta\}$ has the finite intersection property. Let $z_1, \ldots, z_n, z_1', \ldots, z_n' \in G$ and $\theta_1, \ldots, \theta_n \in \Theta$. Choose $\alpha \in \cap_{k=1}^n \omega_I(P_{\theta_k})$. Let $\beta_0 \in \Gamma$ such that $\beta_0 \in \pi_\alpha$ and $z_k, z_k' \in F_{\beta_0}^{-1}$ whenever $k = 1, \ldots, n$. Then $\cap_{k=1}^n (z_k Y_{\theta_k,I} z_k')$ contains the infinite subset $\{y_\beta : \beta \in \pi_\alpha, \beta \geq \beta_0\}$ and $C_I \neq \varnothing$.

Let $I \subset \Theta$. By (FP) there exist a left invariant probability measure and a right invariant probability measure on βG having their support contained in C_I; so there exist $M_I' \in \mathrm{LIM}(l_{\mathbf{R}}^\infty(G))$ and $M_I'' \in \mathrm{RIM}(l_{\mathbf{R}}^\infty(G))$ such that $M_I'(1_{Y_{\theta,I}}) = M_I''(1_{Y_{\theta,I}}) = 1$ whenever $\theta \in \Theta$. Proposition 4.26 implies that $N_I = M_I' \odot M_I'' \in \mathrm{LIM}(l_{\mathbf{R}}^\infty(G))$. If $f \in l_{\mathbf{R}}^\infty(G)$ and $a \in G$, $x \in G$, by (2.15) we have

$$M_I'' \star f_a(x) = M_I''({}_x f_a) = M_I''({}_x f) = M_I'' \star f(x)$$

and

$$N_I(f_a) = \langle M_I'' \star f_a, M_I' \rangle = \langle M_I'' \star f, M_I' \rangle = \langle f, N_I \rangle.$$

Therefore $N_I \in \mathrm{IM}(l_{\mathbf{R}}^\infty(G))$. For every $\theta \in \Theta$ and every $x \in G$,

$$M_I'' \star 1_{Y_{\theta,I}}(x) = M_I''(1_{Y_{\theta,I}}) = 1$$

so $N_I(1_{Y_{\theta,I}}) = 1$. Then $M_I = \tfrac{1}{2}(N_I + (N_I)^{\vee}) \in \mathrm{IIM}(l_{\mathbf{R}}^\infty(G))$ and, for every $\theta \in \Theta$, $M_I(1_{Y_{\theta,I} \cup Y_{\theta,I}^{-1}}) = 1$. [As $Y_{\theta,I}^{-1} \subset G \setminus Y_{\theta,I}$, we have $(N_I)^{\vee}(1_{Y_{\theta,I}}) = 0$].

If I and J are distinct subsets of Θ, $(C_I \cup C_I^{-1}) \cap (C_J \cup C_J^{-1}) = \varnothing$; M_I and M_J as well as N_I and N_J are mutually singular. Finally notice that card $\mathfrak{P}(\Theta) = 2^{2^{\mathrm{card}\, G}}$. □

Corollary 22.46. *Let $\{G_i : i \in I\}$ be an infinite uncountable family of amenable discrete groups having weak direct product G, and let Z_i be the center of G_i $(i \in I)$. If for each $i \in I$, $\mathrm{card}\, G_i \leq \mathrm{card}\, I$ and $Z_i^{(2)}$ is nontrivial, then $\mathrm{card}\,(\mathrm{IM}(l_{\mathbf{R}}^\infty(G)) \setminus \mathrm{IIM}(l_{\mathbf{R}}^\infty(G))) = 2^{2^{\mathrm{card}\, G}}$.*

Proof. The group G is amenable by Proposition 13.8. For each $i \in I$, let $x_i \in Z_i^{(2)}$ with $x_i \neq e_i$. We may consider x_i to be an element of G. Thus card $\{x_i : i \in I\} \leq \mathrm{card}\, Z^{(2)}$, where Z is the center of G. By Lemma 22.44 (b) we then have card $I \leq m(G)$. By hypothesis card $G = \mathrm{card}\, I$. We obtain $m(G) = \mathrm{card}\, G$ and Proposition 22.45 applies. □

Remark 22.47. There exists an infinite amenable discrete group G such that $m(G) < \mathrm{card}\, G^{(2)}$ and $\mathrm{IM}(l_{\mathbf{R}}^\infty(G)) = \mathrm{IIM}(l_{\mathbf{R}}^\infty(G))$.

Let H be an abelian group and consider the semidirect product G of H by $\mathbf{Z}_2 = \{\pm 1\}$ where the action τ of $\mathbf{Z}_2$ on H is given by

$$\tau_1(y) = y, \tau_{-1}(y) = y^{-1} \quad (y \in H).$$

If $y, y' \in H$ and $u, u' \in \mathbf{Z}_2$,

$$(y,u)(y',u') = (y\tau_u(y'), uu')$$

([HR] 2.6). As $G/H \simeq \mathbf{Z}_2$ and H is amenable, G is amenable. For $y \in H$, we have the following relations

$$z_1 = (e_H, 1)(y, 1)(e_H, 1) = (y, 1),$$

$$z_2 = (e_H, 1)(y, 1)(e_H, -1) = (y, -1),$$

$$z_3 = (e_H, -1)(y, 1)(e_H, 1) = (y^{-1}, -1),$$

$$z_4 = (e_H, -1)(y, 1)(e_H, -1) = (y^{-1}, 1),$$

$z_1^{-1} = z_4$, $z_2^{-1} = z_2$, $z_3^{-1} = z_3$; also

$$(e_H, 1)(y, -1)(e_H, 1) = z_2,$$

$$(e_H, 1)(y, -1)(e_H, -1) = z_1,$$

$$(e_H, -1)(y, -1)(e_H, 1) = z_4,$$

$$(e_H, -1)(y, -1)(e_H, -1) = z_3.$$

Let $F = \{(e_H, 1), (e_H, -1)\}$. For every $x = (y, 1) \in G$, we have then $(FxF)^{-1} = FxF$, and therefore $m(G) \leq 2$. Let $M \in \mathrm{IM}(l_{\mathbf{R}}^{\infty}(G))$. For $f = f \cdot 1_{H \times \{1\}} + f \cdot 1_{H \times \{-1\}} \in l_{\mathbf{R}}^{\infty}(H)$, we have

$$M(f) = M(\check{f} \cdot 1_{H \times \{1\}}) + M(\check{f} \cdot 1_{H \times \{-1\}}) = M(\check{f}).$$

Thus $\mathrm{IM}(l_{\mathbf{R}}^{\infty}(G)) = \mathrm{IIM}(l_{\mathbf{R}}^{\infty}(G))$.

Obviously H may be chosen sufficiently large in order to guarantee that card $H^{(2)} > 2$ and then also card $G^{(2)} > 2$.

Definitions of left invariant means make sense on $\mathscr{C}(G)$, $\mathscr{RUC}(G)$, $\mathscr{UC}(G)$ for any topological group G.

Proposition 22.48. *Let G be a topological group admitting a topological subgroup H. Then every left invariant mean on $\mathscr{RUC}(H)$ may be extended to a left invariant mean on $\mathscr{RUC}(\overline{H})$.*

Proof. Every $f \in \mathscr{RUC}(H)$ may be extended uniquely to an element $f' \in \mathscr{RUC}(\overline{H})$ ([B TG] II, Section 3, No. 6, théorème 2). Conversely if $f' \in \mathscr{RUC}(\overline{H})$, $f = f'|_H \in \mathscr{RUC}(H)$. Hence if M is a left invariant mean on $\mathscr{RUC}(H)$, we may define

$$M_1(f') = M(f),$$

$f' \in \mathscr{RUC}(\overline{H})$; $M_1 \in \mathfrak{M}(\mathscr{RUC}(\overline{H}))$.

Consider $f' \in \mathcal{UC}(\overline{H})$ and $a \in \overline{H}$. For every $\varepsilon > 0$, there exists a neighborhood U of e in G such that $|f'(x) - f'(y)| < \varepsilon$ whenever $x, y \in \overline{H}$ and $xy^{-1} \in U \cap \overline{H}$. If $b \in Ua \cap H$, we have $|_bf'(x) - {}_af'(x)| < \varepsilon$ whenever $x \in \overline{H}$. Therefore

$$|M_1({}_af') - M_1(f')| \leq |M_1({}_af' - {}_bf')| + |M_1({}_bf' - f')|$$

$$< \varepsilon + |M({}_bf - f)| = \varepsilon.$$

Hence M_1 is left invariant on $\mathcal{UC}(\overline{H})$. □

Proposition 22.49. *Let A be a von Neumann algebra over a separable Hilbert space. If an increasing sequence (A_n) of finite-dimensional $*$-subalgebras of A with dense union in A exists, then the unitary group A_u of A admits an invariant mean on $\mathcal{UC}_{\mathbf{R}}(A_u)$.*

Proof. For any $n \in \mathbf{N}^*$, the unitary group $(A_n)_u$ may be considered to be a compact subgroup of A_u and is hence amenable. By Kaplansky's density theorem ([140] Chapitre I, Section 3, théorème 3) we have $\cup_{n \in \mathbf{N}^*}(A_n)_u = A_u$. Then the statement follows from Proposition 13.6. □

The next proposition shows that Theorem 4.19 cannot be carried over in its full generality to arbitrary topological groups.

Proposition 22.50. *Let $\mathcal{H}$ be a separable Hilbert space and let G be the topological group consisting of all operators in $\mathcal{L}(\mathcal{H})_u$ with the strong topology. Then $\mathcal{UC}_{\mathbf{R}}(G)$ admits an invariant mean, but there does not exist an invariant mean on $\mathcal{C}_{\mathbf{R}}(G)$.*

Proof. (a) The existence of an invariant mean on $\mathcal{UC}_{\mathbf{R}}(G)$ follows from Proposition 22.49.

(b) Assume that an invariant mean M on $\mathcal{C}_{\mathbf{R}}(G)$ exists. Consider a fixed element ξ in H such that $\|\xi\| = 1$. If $U \in \mathcal{L}(\mathcal{H})$ consider the mapping

$$f_{U,\xi}: \begin{array}{c} V \mapsto \mathcal{R}e(VUV^*\xi|\xi) \\ G\text{-}\mathbf{R} \end{array}$$

and let

$$F_\xi: U \mapsto M(f_{U,\xi})$$

$$\mathcal{L}(\mathcal{H}) \to \mathbf{R}.$$

If $V_1, V_2 \in G$,

$$f_{V_1U,\xi}(V_2) = \mathcal{R}e(V_2V_1UV_2^*\xi|\xi)$$

and

$$f_{UV_1,\xi}(V_2) = \mathscr{R}e\left(V_2UV_1V_2^*\xi|\xi\right).$$

As M is invariant, we have $F_\xi(V_1U) = F_\xi(UV_1)$, and by linearity we obtain $F_\xi(U'U) = F_\xi(UU')$ whenever $U, U' \in \mathscr{L}(\mathscr{H})$ ([140] Chap. I, Section 1, proposition 3). As $id_\mathscr{H}$ is the limit of commutators in $\mathscr{L}(\mathscr{H})$ ([54] Corollary 5.2), we should have $F_\xi(id_\mathscr{H}) = 0$. But $f_{id_\mathscr{H},\xi} = 1_G$ and, since M is a mean, $F_\xi(id_\mathscr{H}) = 1$, so a contradiction would arise. □

NOTES

For 22.2 see Greenleaf [226]. Preliminary attempts to compare the set of left invariant means and the set of topological left invariant means had been made by Renaud [458]. The property 22.3 is due to Granirer [216]. He also proved 22.4. Liu and van Rooij [362] established 22.5-6.

Chou [78] gave 22.7-8. Supplementary references for studies on affine mappings between sets of invariant or topologically invariant means are Chou [84] and Talagrand [529]. The latter author proved that if G is an amenable group admitting a compact normal subgroup H, then $\mathrm{TLIM}(L_\mathbf{R}^\infty(G))$ and $\mathrm{TLIM}(L_\mathbf{R}^\infty(G/H))$ are affinely w*-w* homeomorphic.

Chou [78] gave 22.10 in the case of a σ-compact group. The generalization was studied by Milnes [388]. A general version of 22.12 was given in Pier [435]. The proof relies on Converse, Namioka, and Phelps [98]. Talagrand [528, 529] extensively studied the set $\mathscr{E}(G)$ of extreme points in $\mathrm{TLIM}(L_\mathbf{R}^\infty(G))$ for an amenable σ-compact group G. He established a technical necessary and sufficient condition to have $M \in \mathscr{E}(G)$ for $M \in \mathrm{TLIM}(L_\mathbf{R}^\infty(G))$ in case G is noncompact. No nonvoid w*-G_δ-subset of $\mathrm{TLIM}(L_\mathbf{R}^\infty(G))$ is contained in $\mathscr{E}(G)$. He also showed that if G is a nondiscrete σ-compact locally compact group such that G_d is amenable, then $\mathrm{TLIM}(L_\mathbf{R}^\infty(G))$ does not contain a nonvoid G_δ-subset of $\mathrm{LIM}(L_\mathbf{R}^\infty(G))$. Talagrand investigated extreme capacities for locally compact spaces that are invariant under the action of a locally compact group, especially an amenable group [530, 531]. We quote one of his results: Let G be an amenable group acting continuously on a locally compact space Z, then the G-invariant $\mu_0 \in M^1(Z)$ is an extreme point in the convex set of all G-invariant $\mu \in \mathscr{M}_+^1(Z)$ such that $\|\mu\| \leq 1$ if and only if, for all $M \in \mathrm{IM}(\mathscr{C}(G))$ and $\varphi, \psi \in L^2(Z,\mu)$, $M(\Phi_{\varphi,\psi}) = (\int_Z \varphi \, d\mu)(\int_Z \psi \, d\mu)$ where $\Phi_{\varphi,\psi}: G \to \mathbf{C}, x \mapsto \int_Z \varphi(\zeta)\psi(x\zeta) \, d\mu(\zeta)$. See Moulin Ollagnier and Pinchon [404] for further studies on capacities in this context. Fairchield [177] examined extreme invariant means for particular situations including infinite discrete solvable groups and countable infinite locally finite groups.

Hewitt and Ross [HR I] established that $\text{LIM}(l_{\mathbf{R}}^\infty(G)) \neq \text{RIM}(l_{\mathbf{R}}^\infty(G))$ for the discrete free group G on two generators of order 2. An alternative proof was given by Glasner [204]. Paterson [430] demonstrated 22.16 for compactly generated groups. He proved 22.17 under that hypothesis and established 22.15 for discrete groups. Milnes [388] stated the general version of 22.16. He established 22.13–15 and 22.17. Milnes also showed that for an amenable group G having equivalent left and right uniform structures, $G \in [FC]^-$ if and only if $\text{TLIM}(L_{\mathbf{R}}^\infty(G)) \subset \text{RIM}(L_{\mathbf{R}}^\infty(G))$. For refined properties concerning the discrete case see Rosenblatt and Talagrand [489].

As we stated already in Section 1, von Neumann explained the Hausdorff–Banach–Tarski paradox by the nature of the underlying transformation group. He reduced Hausdorff's problem to the determination, for a locally compact group G, of a left invariant functional on $l_{\mathbf{R}}^\infty(G)$ that extends the Haar integral. The proof of 22.18 is an adaptation made by Hewitt and Ross [HR I] of von Neumann's original demonstration [414]. For the proofs of 22.19–20 we also refer to Hewitt and Ross [HR I]. Greenleaf [229] gave an explicit form of 22.21 for $G = F_2$.

Day [115] noted that if an amenable discrete group admits a subgroup or a quotient group having more than one left invariant mean, then the group itself admits more than one left invariant mean. He showed that a discrete abelian group G admits exactly one invariant mean on $l^\infty(G)$ if and only if it is finite. Luthar [375] proved the nonuniqueness of invariant means on $\mathscr{C}(G)$ for any Hausdorff topological abelian group G satisfying a special property that holds, for instance, if the group is noncompact. Kaur and Manocha [309] sharpened these results. By 22.4 one has, in particular, $\text{LIM}(L^\infty(\mathbf{T})) \neq \{\lambda\}$. In F_2 there exists a decreasing sequence (H_n) of normal subgroups such that $\cap_{n=1}^\infty H_n = \{e\}$ and, for every $n \in \mathbf{N}^*$, $G_n = F_2/H_n$ is finite. The direct product $G = \prod_{n=1}^\infty G_n$ is compact whereas G_d is nonamenable. Chou [84] showed that $\text{LIM}(L^\infty(G)) \neq \{\lambda\}$. The existence of a compact infinite group G such that $\text{LIM}(L^\infty(G)) = \{\lambda\}$ seems to be unknown. Chou [78] established 22.22–26; see [75] for his previous work. If G is a noncompact amenable σ-compact group, by 16.10 there exists $(U_n) \in \mathscr{A}(G)$ such that $U_n = U_n^{-1}$ and $|U_{n+1} \setminus U_n| > n|U_n|$ whenever $n \in \mathbf{N}^*$. Let $M_n(f) = (1/|U_{n+1} \setminus U_n|)\int_{U_{n+1} \setminus U_n} f(x)\,dx$, $f \in L^\infty(G)$. As in the proof of 22.23 (M_n) is w*-converging to $M \in \text{IM}(L^\infty(G))$. Also $M(\sum_{n=1}^\infty 1_{U_{n+1} \setminus U_n}) = 1 \neq 0 = \sum_{n=1}^\infty M(1_{U_{n+1} \setminus U_n})$, verifying that M is not σ-additive.

The example 22.30 is quoted by Granirer [215]. For 22.31 we refer to Douglas [145]. The proofs of 22.32–35 are taken from Rajagopalan and Witz [448]. Kaufman [308] had also given 22.34–35.

Effros [152] established the fundamental properties of inner amenability especially 22.37–38. Paschke [426] proved that an infinite discrete group G is inner amenable if and only if the projection of $l^2(G)$ onto $\mathbf{C}\delta_e$ does not belong to the C^*-algebra $C_I^*(G)$, where I is the inner representation of G on $l^2(G)$ defined by $I_x f = {}_x f_{x^{-1}}$, $f \in l^2(G)$, $x \in G$. M. Choda [70] characterized inner amenability for countable discrete groups admitting only infinite nontrivial conjugacy classes; M. Choda and Watatani [72] established sufficient conditions for inner amenability of these groups. H. Choda and M. Choda [69] investigated the relationship between inner amenability of a discrete group G, simplicity of $C_L^*(G)$, and uniqueness of tracial states on $C_L^*(G)$; also see Watatani [551]. We would like to mention that Akeman [4] studied a class of discrete groups that are not inner amenable, do not satisfy property (T), and admit a simple C_L^*-algebra with unique trace. Further results on inner amenability may be found in Pier [437]. More generally M. Choda [71] considered, for given discrete countable groups K and G, an action α of K on G that is a homomorphism of K into the group of automorphisms of G. She showed the equivalence of the following properties: (i) A sequence (φ_n) exists in the unit sphere of $l^2(G)$ satisfying $\lim_{n\to\infty}\Sigma_{x\in G}|\varphi_n(x) - \varphi_n \circ \alpha_a(x)|^2 = 0$, whenever $a \in K$, and for which $(\varphi_n(e))$ does not converge to 1; (ii) there exists a nontrivial mean M on $l^\infty(G)$ that is α-invariant, that is, $M(f) = M(f \circ \alpha_a)$ whenever $f \in l^\infty(G)$ and $a \in K$. Also if these conditions fail and K is not inner amenable, then the semidirect product of K by G is not inner amenable.

Tomter [540] gave a version of 22.39 in the abelian case; for the general statement 22.39 and 22.40 see Landstad [331]. Making use of the existence of a unique invariant mean M on $B(G)$ for a locally compact group G, Bożejko [49] showed that if the dual space of $C_L^*(G)$ is not contained in $\{u \in B(G): M(|u|^2) = 0\}$, then G is amenable. A more general statement had been obtained by Derighetti [127].

Rosenblatt [480] gave 22.41.

Granirer and Lau [224] established 22.42. They proved that if G is a locally compact group and $\mathscr{RUC}_R(G)$ admits a left invariant mean $M = \Sigma_{i=1}^n \alpha_i M_i$, where $\alpha_1,\ldots,\alpha_n \in \mathbf{R}_+^*$, $\Sigma_{i=1}^n \alpha_i = 1$ and $M_1,\ldots,M_n$ are distinct multiplicative means on $\mathscr{RUC}_R(G)$, then G is a finite group of order n. For the case in which G is a nontrivial amenable σ-compact group, whereas by 22.42 there do not exist multiplicative (topologically) left invariant means on $\mathscr{RUC}_R(G)$, 22.12 shows that the extreme points of TLIM$(\mathscr{RUC}_R(G))$ are approximately multiplicative. This characterization may be compared with the general result due to Phelps [433] stipulating that the extreme points of the convex set of positive normalized operators on function algebras are the multiplicative elements. Granirer exhaustively studied discrete semigroups admitting multiplicative left invariant means. He called them extremely amenable; basic references are [211, 213]. Other important studies on extremely amenable semigroups are Mitchell [392], Lau [332, 333], Grossman [235], and Wong [568].

A great amount of technical studies succeeded in providing much information on the cardinalities of the various sets of invariant means. Considerations on singularity properties of the associated measures led to the most satisfactory results. Glasner [203] gave a proof showing that card LIM$(l^\infty(G)) = 2^c$ for any infinite countable amenable discrete group G. Chou [83] evaluated precisely that if G is an infinite amenable discrete group, card LIM$(l_R^\infty(G)) = $ card IM$(l_R^\infty(G)) = $ card IIM$(l_R^\infty(G)) = 2^{2^{\text{card }G}}$. Rosenblatt [482] stated the first part of 22.45. Paterson [431] established 22.44–47. He obtained further important results and showed, for instance, that if $m(G) < \infty$ for an amenable discrete group G, then $\text{IM}(l_{\mathbf{R}}^\infty(G)) = \text{IIM}(l_{\mathbf{R}}^\infty(G))$. Refinements and generalizations applying to actions of discrete amenable groups on arbitrary sets were obtained by Rosenblatt and Talagrand [489]. Making heavy use of topological structure properties of metric groups, Rosenblatt [481] showed that if G is a nondiscrete σ-compact locally compact group such that G_d is amenable, then there exists at least 2^c mutually singular elements in $\text{LIM}(L_{\mathbf{R}}^\infty(G))$ each of which is singular with respect to any element of $\text{TLIM}(L_{\mathbf{R}}^\infty(G))$. Talagrand [532] stated that if G is an infinite locally compact group such that G_d is amenable and F is a nonvoid closed G-invariant subset of $\overline{L_{\mathbf{R}}^\infty(G)}$, then the set of all left invariant means on $L_{\mathbf{R}}^\infty(G)$ that are supported by F is nonnormseparable. If G is an amenable group, two elements M, N of $\text{LIM}(\mathscr{C}_{\mathbf{R}}(G))$ are called mutually singular if there exists $f \in \mathscr{C}_{\mathbf{R}}(G)$ such that $f(G) \subset [0,1]$ and $M(f) = 0$, $N(f) = 1$. Complementing Proposition 22.6 Rosenblatt [484] proved that if G is a noncompact, nondiscrete, σ-compact, locally compact group such that G_d is amenable, then there exist at least 2^c mutually singular elements in $\text{LIM}(\mathscr{C}_{\mathbf{R}}(G))$, all of which are mutually singular to all elements of $\text{TLIM}(\mathscr{C}_{\mathbf{R}}(G))$. For every locally compact group G, $\mathscr{RUC}(G)^*$ is a Banach algebra (Section 2.G). Granirer [210] proved that if G is amenable and $M_0 \in \text{LIM}(\mathscr{RUC}(G))$, then the set $\{M - M_0 : M \in \text{LIM}(\mathscr{RUC}(G))\}$ is contained in the radical of $\mathscr{RUC}(G)^*$. So one gets information about the power of the radical; see Chou [78] and Granirer [217]. Granirer [215] studied the size of $\text{LIM}(\mathscr{RUC}(G))$ for special groups G. Pachl [424] showed that if G is a separable metrizable topological group and $\text{LIM}(\mathscr{RUC}_{\mathbf{R}}(G))$ contains a G_δ-point, then $\text{LIM}(\mathscr{RUC}_{\mathbf{R}}(G))$ reduces to a singleton.

De la Harpe established 22.49–50 [249]. He studied the subject extensively in [250].

Banach's proof [28] showed the existence of a finitely additive measure over $\mathfrak{P}(\mathbf{R}^2)$ that is invariant under the group of isometries. Amenability is merely a property of the transformation group. But we would like to quote here some facts about the problems raised originally by Lebesgue [344] concerning measures that are invariant under transformation groups.

If $\mathscr{R}$ is a subring of an algebra of sets $\mathscr{B}$, then every finitely additive measure on $\mathscr{R}$ with values in $\overline{\mathbf{R}}_+$ may be extended to a finitely additive measure on $\mathscr{B}$. Mycielski [409] gave this result and stated a consequence that is an extension property applying to amenable discrete transformation groups: Let G be an amenable discrete group of automorphisms on an algebra of sets $\mathscr{B}$ and let $\mathscr{R}$ be a subring of $\mathscr{B}$ that is G-invariant, that is, $aA \in \mathscr{R}$ whenever $A \in \mathscr{R}$ and $a \in G$. Assume that there exists a finitely additive measure $\mu : \mathscr{R} \to \overline{\mathbf{R}}_+$ and a homomorphism θ of G into the multiplicative group $\mathbf{R}_+^*$ satisfying $\mu(aA) = \theta(a)\mu(A)$ whenever $A \in \mathscr{R}$ and $a \in G$. Then μ may be extended to a finitely additive real measure μ on $\mathscr{B}$ such that $\mu(aA) = \theta(a)\mu(A)$ whenever $A \in \mathscr{B}$ and $a \in G$. Rosenblatt [485] proved that a discrete group G of automorphisms of an algebra of sets $\mathscr{B}$ is amenable if and only if, for every proper G-invariant ideal $\mathscr{I}$ in $\mathscr{B}$, there exists a finitely additive G-invariant probability measure on $\mathscr{B}$ that vanishes on all elements of $\mathscr{I}$.

Lebesgue's classical result says that on **T** there exists exactly one normalized σ-additive nonnegative measure that is invariant under isometries [344]. In 1923 Ruziewicz had asked Banach whether Lebesgue's integral is the only nonnegative functional on $L^\infty(\mathbf{R}^n)$ that is normalized on a given measurable subset and is invariant under isometries [28]. Banach showed failure of unicity for $n = 1$ and $n = 2$; Banach has to be given credit for the first proof of the existence of an invariant mean on $L^\infty(\mathbf{T})$ that does not coincide with the normalized Lebesgue integral. Let $(X, \mathscr{X}, \mu)$ be a probability space that is nonatomic and let τ be a measurable measure-preserving transformation of X. One considers "almost invariant" Følner-type elements of $\mathscr{X}$; for $\varepsilon > 0$, $A \in \mathscr{X}$ is said to be ε-invariant under τ if $\mu(A \triangle \tau^{-1}(A)) < \varepsilon\mu(A)$. As we mentioned already in Section 16, del Junco and Rosenblatt [297] showed that if a finite set $\{\tau_1, \ldots, \tau_n\}$ of measurable measure-preserving transformations on X generate a discrete semigroup S that admits a right invariant mean on $l^\infty(S)$, then for every $\varepsilon > 0$, there exists $A \in \mathscr{X}$ such that $0 < \mu(A) < \varepsilon$ and $\mu(A \triangle \tau_i^{-1}(A)) < \varepsilon\mu(A)$ whenever $i = 1, \ldots, n$. Let S be a countable semigroup of measurable measure-preserving transformations on X and suppose S to be ergodic, that is, $\tau(A) = A$ whenever $\tau \in S$, if and only if $A = X$ or $\varnothing$; then there exist arbitrary small almost invariant subsets if and only if there exist S-invariant means on $L^\infty(X, \mu)$ other than the mean determined by $I(f) = \int f d\mu$, $f \in L^\infty(X)$. Rosenblatt [487] proved that, for the compact abelian groups $\mathbf{T}^n$ ($n \geq 2$), the Haar integral is the only mean on $L^\infty(\mathbf{T}^n)$ that is invariant under all topological automorphisms. He also formulated general statements coming close to a solution of Ruziewicz's problem for higher dimensions. Losert and Rindler [372] considered the following set-up: Let H be a nonamenable discrete group of continuous automorphisms on a compact abelian group G; then H acts naturally on $\hat{G}$. If $\chi \in \hat{G}$, let H_χ be the subgroup of all elements in H that leave χ unchanged. Losert and Rindler proved that if, for every nontrivial character χ in $\hat{G}$, H_χ is amenable, then G admits no ε-almost H-invariant subsets for $\varepsilon < 2$. The Haar integral is the unique H-invariant mean on $L^\infty(G)$. The statement applies to $G = \mathbf{T}^2$, $H = SL(2, \mathbf{Z})$, and a fortiori $H = SL(2, \mathbf{R})$. Margulis [380] established the following general result: If (X, μ) is a probability space and G is a group of measurable measure-preserving transformations that acts ergodically on X and has property (T), then X does not admit arbitrarily small almost G-invariant subsets. Reproducing observations of Del Junco and Rosenblatt he showed, in particular, that the Lebesgue integral is the only $SL(n, \mathbf{R})$-invariant mean on $L^\infty(\mathbf{T}^n)$ for $n \geq 3$. Via algebraic group theory he could prove that, if $n \geq 4$, the Lebesgue integral is the only $SO(n + 1, \mathbf{R})$-invariant mean on $L^\infty(S_n)$. Relying on Rosenblatt's work [487], Sullivan [524] produced a solution of Ruziewicz's problem for S_n ($n \geq 4$).

Von Neumann reduced the Hausdorff–Banach–Tarski paradox to the study of the underlying transformation group. If a discrete group G acts on a set X, via (FP) amenability of G implies the existence of a finitely additive G-invariant probability measure on X. What can be said about the converse? If X admits a subset of fixed points, any probability measure on this subset is G-invariant whether G is amenable or not. Promislow [439] established a statement generalizing a result obtained by Rosenblatt [487]: Let G be a discrete group acting on a set X; suppose that, for some $r \in \mathbf{N}^*$, G contains a free group F with rank $F \geq r + 1$ and, for every $\xi \in X$ and the corresponding stabilizer group $G_\xi = \{x \in G : x\xi = \xi\}$, one has rank $(G_\xi \cap F) \leq r$. Then there does not exist a G-invariant finitely additive probability measure on X. Several related problems are not completely solved. What is the nature of a group G of isometries on S_n if there exists a G-invariant finitely additive probability measure defined on $\mathfrak{P}(S_n)$? Wagon [546] raised the following question: Can a Lebesgue measure of $\mathbf{R}^n$ be extended to a finitely additive measure over all subsets of $\mathbf{R}^n$ that is invariant under translations but not under reflections? More generally, if μ is a finitely additive measure extending Lebesgue measure on $\mathbf{R}^n$, what can be said about the group $\operatorname{Inv}\mu$ of all isometries that leave μ invariant? Wagon proved that if G is an amenable discrete group of isometries of $\mathbf{R}^n$ ($n \in \mathbf{N}^*$), then there exists a finitely additive measure μ on $\mathfrak{P}(\mathbf{R}^n)$ that extends Lebesgue measure such that $\operatorname{Inv}\mu = G$. Any subgroup G of the group O_n of all isometries on $\mathbf{R}^n$ ($n = 1, 2$) that leave the origin fixed is amenable if and only if there exists a finitely additive measure μ on $\mathfrak{P}(\mathbf{R}^n)$ that extends Lebesgue measure such that G coincides with the subgroup of O_n consisting of the elements that leave μ invariant.

Let G be a compact group and let $f \in L_\mathbf{R}^\infty(G)$. Talagrand [533] says that f admits a unique invariant mean if $M(f) = \int f$ whenever $M \in \operatorname{IM}(L_\mathbf{R}^\infty(G))$. He calls f Riemann measurable if it is a class of functions for which the sets of discontinuity points are null, that is, for every $\varepsilon > 0$, there exist $g_1, g_2 \in \mathscr{C}_\mathbf{R}(G)$ such that $g_1 \leq f \leq g_2$ and $\int(g_2 - g_1) < \varepsilon$. Proposition 12.1 implies that every Riemann measurable function admits a unique invariant mean. The converse does not hold Talagrand producing for certain compact groups a non Riemann measurable function f such that, for every Riemann measurable function g, fg admits a unique invariant mean.

For a locally compact group G, one considers the space $\operatorname{WAP}(G)$ of all continuous weakly almost periodic functions on G, that is, all $f \in \mathscr{C}(G)$ for which $\{_x f : x \in G\}$ [or equivalently $\{f_x : x \in G\}$] is relatively compact in the weak topology of $\mathscr{C}(G)$. The reference is Burckel [61]. One has $\operatorname{WAP}(G) \subset \mathscr{UC}(G)$ and there exists a unique invariant mean M on $\operatorname{WAP}(G)$. Chou [82] proved that if G is a noncompact amenable group that satisfies a certain restrictive hypothesis which holds, for instance, in case the center is not compact, then $\mathscr{V}_0(G)/\operatorname{WAP}_0(G)$ is nonnormseparable, with $\mathscr{V}_0(G) = \{f \in$

$\mathcal{UC}_R(G): (\forall M \in \text{LIM}\,(\mathcal{UC}_R(G))M(|f|)\} = 0$ and $\text{WAP}_0(G) = \{f \in \text{WAP}(G): M(|f|) = 0\}$. Chou [85] also established the following result: Let G be a noncompact connected amenable group such that $\mathcal{C}_0(G) = \text{WAP}_0(G)$ and let K be the largest compact normal subgroup of G; then G/K is topologically isomorphic to the semidirect product of a compact connected subgroup of $SO(n, \mathbf{R})$ with $\mathbf{R}^n$ ($n \in \mathbf{N}^*$). He gave complementary information in [88].

From considerations on weakly almost periodic functions Lau [339] deduced that if G is a locally compact group and X is a closed topologically left invariant subspace of $L^\infty(G)$, closed under complex conjugation and containing 1_G, then a necessary and sufficient condition for $\text{TLIM}(X) \neq \emptyset$ is the existence of a weakly compact nonnegative operator from X into X, of norm 1, commuting with convolution from the left.

Let G be a locally compact group and let B be a convex set of means on $L^\infty(G)$; $M_0 \in B$ is called exposed point of B if there exists $f_0 \in L^\infty(G)$ such that $\mathcal{R}e\langle f_0, M_0\rangle < \mathcal{R}e\langle f_0, M\rangle$ whenever $M \in B$, $M \neq M_0$. Chou [79] had studied exposed points in the space of invariant means for discrete groups. As a consequence of very general considerations Granirer [215] obtained the following result: Let G be an amenable σ-compact group, B a convex subset of $P^1(G)$ identified with a subset of $\mathfrak{M}(L^\infty(G))$, (f_n) a sequence in $L^\infty(G)$, $\mathfrak{M}_0 = \text{TLIM}(L^\infty(G))$, $\text{TIM}(L^\infty(G))$ and

$$D = \overline{B}^{w^*} \cap \{M \in \mathfrak{M}_0 : (\forall n \in \mathbf{N}^*)\langle f_n, M\rangle = 0\}.$$

If D is normseparable or admits a w^*-exposed point M_0, then G is compact. The proposition remains true for the set of topologically left invariant means that are inversion invariant. The property applied to $B = P^1(G)$ yields that, if G is a noncompact amenable group, then $\mathcal{UC}(G)/\text{WAP}(G)$ is not normseparable.

6

THE PHENOMENON OF AMENABILITY

We give some supplementary description of the class of amenable groups and briefly indicate various directions into which amenability properties have been generalized.

23. COMPLEMENTS ON THE CLASS OF AMENABLE GROUPS

We consider stability properties of the class of amenable groups and give further examples of amenable groups. Examining properties related to amenability we obtain information on the boundaries of the class. Amenable semigroups are studied in connection with amenable groups.

A. Relativization

We call relativization problem the study of amenable subgroups in a general locally compact group. Our purpose is not a systematic investigation of that situation as made by Reiter [455]. We merely indicate some prototypes of relativization properties and develop proofs showing the technical complications to which the generalization of the standard demonstrations give rise.

We state the relativized forms of Reiter's properties: Let G be a locally compact group admitting the closed subgroup H. Then H is amenable if and only if for one (for every) $p \in [1, \infty[$, any one of the following conditions holds:

$$(\forall K \in \mathfrak{K}(H))(\forall \varepsilon > 0)(\exists \varphi \in P^p(G))(\forall a \in K)\|_a\varphi - \varphi\|_p < \varepsilon;$$

$$(\forall F \in \mathfrak{F}(H))(\forall \varepsilon > 0)(\exists \varphi \in P^p(G))(\forall a \in F)\|_a\varphi - \varphi\|_p < \varepsilon;$$

$$(\forall K \in \mathfrak{K}(H))(\forall \varepsilon > 0)(\exists \varphi \in P^p(G) \cap \mathcal{K}(G))(\forall a \in K)\|_a\varphi - \varphi\|_p < \varepsilon;$$

$$(\forall F \in \mathfrak{F}(H))(\forall \varepsilon > 0)(\exists \varphi \in P^p(G) \cap \mathcal{K}(G))(\forall a \in F)\|_a\varphi - \varphi\|_p < \varepsilon.$$

If $f \in L^1(G)$, let

$$L_{G,H}(f) = \overline{\mathrm{co}}\{L_y f : y \in H\}, \quad R_{G,H}(f) = \overline{\mathrm{co}}\{R_y f : y \in H\}$$

in $L^1(G)$. Let also $L_{G,H}$ [resp. $R_{G,H}$] denote the convex hull of the set of isometric operators L_x [resp. R_x] ($x \in H$) on $L^1(G)$. Define $J_l(G, H)$ [resp. $J_r(G, H)$] to be the Banach subspace of $L^1(G)$ generated by the functions $f - L_x f$ [resp. $f - R_x f$], where $f \in L^1(G)$, $x \in H$; then $J_r(G, H) = \mathrm{Ker}\, J_{H,\rho}$ and, in case H is normal, $J_r(G, H) = J_l(G, H)$.

If $f \in L^1(G)$, $g \in L^1(H)$, we define

$$f \square g(x) = \int_H R_y f(x) g(y)\, dy = \int_H f(xy^{-1}) \Delta_G(y^{-1}) g(y)\, dy$$

$$= \int_H f(xy) \frac{\Delta_G(y)}{\Delta_H(y)} g(y^{-1})\, dy,$$

$x \in G$; $f \square g \in L^1(G)$. With the notations of Section 14A, $J_{H,\rho}(f \square g) = (\int_H g) J_{H,\rho}(f)$, hence $\mathrm{Ker}\, J_{H,\rho}$ constitutes a right Banach $L^1(H)$-module. If $g \in L^0(H)$ and $\varepsilon \in \mathbf{R}_+^*$, there exist $u \in \mathscr{K}_+(G)$ and a compact neighborhood V of e in G such that, for $f = u \square g \in \mathrm{Ker}\, J_{H,\rho}$, $\|_x((1/\rho)f) - g\|_{H,1} < \varepsilon$ whenever $x \in V$ ([129, 131]). Proposition 3.6 admits a partial generalization: For every $f \in L^1(G)$, $\{f \square \varphi : \varphi \in P^1(H)\}^- \subset R_{G,H}(f)$.

Definition 23.1. *Let G be a locally compact group admitting a closed subgroup H. For $f \in L^1(G)$, we define*

$$d_{l,G,H}(f) = \inf\{\|g\|_1 : g \in L_{G,H}(f)\}$$

and

$$d_{r,G,H}(f) = \inf\{\|g\|_1 : g \in R_{G,H}(f)\}.$$

By (2.1) we have

$$d_{l,G,H}(f) = d_{r,G,H}(f^*).$$

By (14.6), for every $g \in R_{G,H}(f)$, $J_{H,\rho}(f) = J_{H,\rho}(g)$, hence $\|J_{H,\rho}(f)\|_1 \leq \|g\|_1$, that is,

$$\|J_{H,\rho}(f)\|_1 \leq d_{r,G,H}(f). \tag{1}$$

We now proceed to the generalization of the Glicksberg–Reiter property. The proof relies on theorem 6.21.

Proposition 23.2. *Let G be a locally compact group admitting a closed subgroup H. Then H is amenable if and only if, for any $f \in L^1(G)$,*

$$d_{r,G,H}(f) = \|J_{H,\rho}(f)\|_1.$$

Proof. (a) Assume H to be amenable.

Let $g \in \mathscr{K}(G)$ with supp $g = K$. For any $a \in G$, let $g^{(a)}$ be the restriction of the function $_a((1/\rho)g)$ to H; $g^{(a)} \in \mathscr{K}(H)$. Let π be the canonical mapping of G onto G/H, and let μ be the associated quasi-invariant measure on G/H. By Proposition 3.6, Theorem 6.21 and (6.10), given $\varepsilon > 0$, we may determine $\psi \in P^1(H)$ such that

$$\|g^{(a)} *_H \psi\|_1 < \left|\int_H g^{(a)}\right| + (\mu(\pi(K)) + 1)^{-1}\frac{\varepsilon}{2}.$$

By density of $\mathscr{K}(G)$ in $L^1(G)$, there exists $\varphi \in P^1(G) \cap K(G)$ such that

$$\|g^{(a)} *_H \varphi\|_1 \leq \|g^{(a)} *_H \psi\|_1 + (\mu(\pi(K)) + 1)^{-1}\frac{\varepsilon}{2}$$

hence

$$\|g^{(a)} *_H \varphi\|_1 < \left|\int_H g^{(a)}\right| + (\mu(\pi(K)) + 1)^{-1}\varepsilon.$$

We have

$$\|g^{(a)} *_H \varphi\|_1 = \int_H \left|\int_H g^{(a)}(xy)\varphi(y^{-1})\,dy\right| dx$$

$$= \int_H \left|\int_H \frac{g(axy)}{\rho(axy)}\varphi(y^{-1})\,dy\right| dx.$$

Note also that if, for $x, y \in H$, one has $axy \in K$, then $a \in KH$. Hence if $a \in G \setminus KH$, $g(axy) = 0$ for all $x, y \in H$. Thus for all $a \in G$,

$$\int_H \left|\int_H \frac{g(axy)}{\rho(axy)}\varphi(y^{-1})\,dy\right| dx \leq \left|\int_H g^{(a)}\right| + (\mu(\pi(K)) + 1)^{-1}\varepsilon 1_{KH}(a).$$

Then

$$\|g \square \varphi\|_1 = \int_G \left|\int_H g(zy)\varphi(y^{-1})\Delta_G(y)\Delta_H(y^{-1})\,dy\right| dz$$

$$= \int_G \left|\int_H \frac{g(zy)}{\rho(zy)}\varphi(y^{-1})\,dy\right|\rho(z)\,dz$$

$$= \int_{G/H}\left(\int_H \left|\int_H \frac{g(zxy)}{\rho(zxy)}\varphi(y^{-1})\,dy\right| dx\right) d\mu(\dot{z})$$

$$\leq \int_{G/H}\left|\int_H \frac{g(zx)}{\rho(zx)}\,dx\right| d\mu(\dot{z}) + \varepsilon = \|J_{H,\rho}g\|_1 + \varepsilon.$$

We have $d_{r,G,H}(g) \leq \|J_{H,\rho}(g)\|_1 + \varepsilon$. As $\varepsilon > 0$ may be chosen arbitrarily, we deduce from (1) that $d_{r,G,H}(g) = \|J_{H,\rho}(g)\|_1$.

If $f \in L^1(G)$ and $\delta \in \mathbf{R}_+^*$ are given, consider $h \in \mathcal{K}(G)$ such that $\|f - h\|_1 < \delta/3$. Choose $A \in R_{G,H}$ such that $\|Ah\|_1 < \|J_{H,\rho}(h)\|_1 + \delta/3$. Then

$$\|Af\|_1 \leq \|A(f-h)\|_1 + \|Ah\|_1 < \|f-h\|_1 + \|J_{H,\rho}(h)\|_1 + \frac{\delta}{3}$$

$$< \frac{\delta}{3} + \|J_{H,\rho}(h-f)\|_1 + \|J_{H,\rho}(f)\|_1 + \frac{\delta}{3} < \|J_{H,\rho}(f)\|_1 + \delta.$$

Hence $d_{r,G,H}(f) < \|J_{H,\rho}(f)\|_1 + \delta$. As $\delta \in \mathbf{R}_+^*$ is arbitrary, the statement follows from (1).

(b) Assume now that the condition holds. Let $g \in L^0(H) \cap \mathcal{K}(H)$. We show that $d_r(g) = 0$; then (6.10) and Corollary 6.23 yield amenability of H.

Let $\delta > 0$; choose $u \in \mathcal{K}_+(G)$ and a compact neighborhood V of e such that $f = u \square g \in \mathcal{K}\!er\, J_{H,\rho}$ and $\|_x((1/\rho)f) - g\|_{H,1} < \delta/2$ whenever $x \in V$. We have $\mu(\pi(V)) > 0$, μ being the quasi-invariant measure on G/H. By hypothesis there exist $\alpha_1, \ldots, \alpha_n \in \mathbf{R}_+^*$ with $\sum_{i=1}^n \alpha_i = 1$ and $y_1, \ldots, y_n \in H$ such that, for

$$A = \sum_{i=1}^n \alpha_i R_{y_i^{-1}} \in R_{G,H}, \|Af\|_1 < \frac{\delta}{2}\mu(\pi(V)).$$

Let $E = \{\dot{x} \in G/H: J_{H,\rho}(|Af|)(\dot{x}) \geq \delta/2\}$. We must have $\mu(E) < \mu(\pi(V))$ and there exists $x_0 \in V$ such that $\dot{x}_0 \notin E$, that is,

$$\int_H \left| \sum_{i=1}^n \alpha_i \frac{f(x_0 z y_i)}{\rho(x_0 z)} \Delta_G(y_i) \right| dz < \frac{\delta}{2};$$

also

$$\int_H \left| \sum_{i=1}^n \alpha_i \frac{f(x_0 z y_i)}{\rho(x_0 z y_i)} \Delta_H(y_i) \right| dz < \frac{\delta}{2}.$$

Finally we obtain

$$\int_H \left| \sum_{i=1}^n \alpha_i g(zy_i) \Delta_H(y_i) \right| dz \leq \int_H \left| \sum_{i=1}^n \alpha_i \left(g(zy_i) - \frac{f(x_0 z y_i)}{\rho(x_0 z y_i)} \right) \Delta_H(y_i) \right| dz$$

$$+ \int_H \left| \sum_{i=1}^n \alpha_i \frac{f(x_0 z y_i)}{\rho(x_0 z y_i)} \Delta_H(y_i) \right| dz$$

$$\leq \sum_{i=1}^n \alpha_i \int_H \left| g(zy_i) - \frac{f(x_0 z y_i)}{\rho(x_0 z y_i)} \right| \Delta_H(y_i) dz + \int_H \left| \sum_{i=1}^n \alpha_i \frac{f(x_0 z y_i)}{\rho(x_0 z y_i)} \Delta_H(y_i) \right| dz$$

$$< \frac{\delta}{2} + \frac{\delta}{2} = \delta. \qquad \square$$

Proposition 23.3. *Let G be a locally compact group admitting a closed normal subgroup H. Then H is amenable if and only if, for every $f \in L^1(G)$, $d_{l,G,H}(f) = d_{r,G,H}(f)$. This common value equals $\|J_H f\|_1$.*

Proof. Assume H to be amenable. By (2.1), Proposition 23.2, and (14.7) we have

$$d_{l,G,H}(f) = d_{r,G,H}(f^*) = \|J_H(f^*)\|_1 = \|(J_H f)^*\|_1 = \|J_H f\|_1 = d_{r,G,H}(f).$$

Now assume H to be nonamenable.

(a) There exist $f \in L^1(G)$ and $a \in H$ such that $d_{r,G,H}(g) > 0$ for $g = L_a f - f \in L^1(G)$.

Otherwise given $\varphi \in P^1(G)$, $a \in H$, and $\varepsilon > 0$, we could determine $A \in R_{G,H}$ such that $\|A(_a\varphi - \varphi)\|_1 < \varepsilon$. Then, given $F = \{a_1, \ldots, a_n\} \subset H$, by a generalization of Lemma 6.20 we could obtain $B \in R_{G,H}$ such that $\|B(_a\varphi - \varphi)\|_1 < \varepsilon$ whenever $i = 1, \ldots, n$. Let $\psi = B\varphi \in P^1(G)$. We would have $\|_{a_i}\psi - \psi\|_1 < \varepsilon$ for every $i = 1, \ldots, n$ and amenability of H would follow.

(b) Let $\varepsilon > 0$ and consider $m \in \mathbf{N}^*$ such that $m > 2\varepsilon^{-1}\|f\|_1$; let $T = m^{-1}\sum_{i=0}^{m-1} L_{a^i}$. Then

$$Tg = m^{-1}\sum_{i=0}^{m-1} L_{a^i}(L_a f - f) = m^{-1}(L_{a^m}f - f)$$

and $\|Tg\|_1 \leq 2m^{-1}\|f\|_1 < \varepsilon$. As $\varepsilon > 0$ is arbitrary, we must have $d_{l,G,H}(g) = 0$. □

We consider a Leptin-type condition.

Proposition 23.4. *Let G be a σ-compact locally compact group. Consider a closed subset C of G such that $\mathring{C} \neq \varnothing$ and let H be the (closed) subgroup of G generated by $C^{-1}C$. Then H is amenable if and only if, for every compact subset K of G contained in C,*

$$\inf\left\{\frac{|KU|}{|U|} : U \in \mathfrak{K}_0(G)\right\} = 1.$$

Proof. As H is open in G, we may consider λ_H to be induced on H by λ_G.

(a) Assume that the condition does not hold. There exists a nonvoid compact subset K in C such that $\inf\{|KU|/|U| : U \in \mathfrak{K}_0(G)\} > 1$ and a fortiori, for the compact subset $K' = K^{-1}K$ of H, $\inf\{|K'U|/|U| : U \in \mathfrak{K}_0(G)\} > 1$. Then Theorem 16.2 implies nonamenability of H.

(b) Assume that the condition holds. By hypothesis there exists a nondecreasing sequence $(K_n)_{n \in \mathbf{N}^*}$ of compact subsets in C such that $\mathring{K}_n \neq \varnothing$ whenever $n \in \mathbf{N}^*$ and $\cup_{n=1}^\infty K_n = C$. Choose $a \in \mathring{K}_1$. For every $n \in \mathbf{N}^*$, let $K'_n = a^{-1}K_n$, hence $K'^{-1}_n K'_n = K_n^{-1}K_n$.

Let $n \in \mathbf{N}^*$ and $\varepsilon \in \mathbf{R}^*_+$. There exists $U_n \in \mathfrak{R}_0(G)$ such that $|K'_n U_n| = |K_n U_n| < (1 + \varepsilon)|U_n|$. Let $h_n = 1_{K'_n} * 1_{U_n} \in L^1(G) \cap L^2(G)$; $h_n = h_n 1_{K'_n U_n}$, and also

$$\|h_n\|_1 = |K'_n| \, |U_n| = |K_n| \, |U_n|.$$

Via the Cauchy–Schwarz inequality we obtain

$$\|h_n\|_1 = \|h_n 1_{K'_n U_n}\|_1 \leq \|h_n\|_2 |K'_n U_n|^{1/2} < (1 + \varepsilon)^{1/2} |U_n|^{1/2} \|h_n\|_2,$$

hence

$$\|1_{K'_n} * 1_{U_n}\|_2 = \|h_n\|_2 > (1 + \varepsilon)^{-1/2} |U_n|^{-1/2} \|h_n\|_1$$
$$= (1 + \varepsilon)^{-1/2} |U_n|^{1/2} |K'_n| = (1 + \varepsilon)^{-1/2} |K'_n| \, \|1_{U_n}\|_2.$$

As $\varepsilon > 0$ may be chosen arbitrarily, we proved that $\|1_{K'_n}\|_{Cv^2} \geq |K'_n|$, so $\|L_\varphi\|_{Cv^2} = 1$ for $\varphi = \xi_{K'_n} \in P^1(G)$. As K'_n is a neighborhood of e, we deduce from Corollary 18.6 that the closed subgroup H_n generated by K'_n is amenable. For every $n \in \mathbf{N}^*$, $H_n \subset H_{n+1}$ and the group $H' = \bigcup_{n=1}^\infty H_n$ is amenable by Proposition 13.6. For every element x of $C^{-1}C$, there exists $m \in \mathbf{N}^*$ such that $x \in K_m^{-1} K_m = K'^{-1}_m K'_m$. Therefore H is a closed subgroup of H', and thus H is amenable by Proposition 13.3. $\square$

If $g \in L^1(H)$ and $a \in H$, we put

$$R_a^H g(x) = g(xa^{-1}) \Delta_H(a^{-1}),$$

$x \in G$. For $f \in L^1(G)$, $g \in L^1(H)$, $a \in H$, we have

$$R_a(f \square g) = f \square R_a^H(g). \quad (2)$$

As a matter of fact, for almost every $x \in G$,

$$R_a(f \square g)(x) = \int_H f(xa^{-1}y) \Delta_G(a^{-1}) \frac{\Delta_G(y)}{\Delta_H(y)} g(y^{-1}) \, dy$$

$$= \int_H f(xy) \Delta_G(a^{-1}) \frac{\Delta_G(a) \Delta_G(y)}{\Delta_H(a) \Delta_H(y)} g(y^{-1} a^{-1}) \, dy$$

$$= \int_H f(xy) \frac{\Delta_G(y)}{\Delta_H(y)} R_a^H g(y^{-1}) \, dy.$$

Proposition 23.5. *Let G be a locally compact group admitting a closed subgroup H. Then H is amenable if and only if the right $L^0(H)$-module $\mathcal{K}er\, J_{H,\rho}$ admits* "bounded right approximate units" in $L^0(H)$, that is, there exists $c > 0$ such that, given $f \in \text{Ker } J_{H,\rho}$ and $\varepsilon > 0$, one may determine $v \in L^0(H)$ satisfying $\|v\|_1 \leq c$ and $\|f - f \square v\|_1 < \varepsilon$.

Proof. (a) Assume H to be amenable. Let $f \in \text{Ker } J_{H,\rho}$ and $\varepsilon > 0$. By Proposition 23.2 there exist $\alpha_1, \ldots, \alpha_n \in \mathbf{R}_+^*$ and $a_1, \ldots, a_n \in H$ such that $\sum_{i=1}^n \alpha_i = 1$ and

$$\left\| \sum_{i=1}^n \alpha_i R_{a_i^{-1}} f \right\|_1 < \frac{\varepsilon}{3}. \tag{3}$$

There also exists a compact neighborhood U of e in G such that

$$\|R_{a^{-1}} f - f\|_1 = \int_G |f(xa)\Delta_G(a) - f(x)| \, dx < \frac{\varepsilon}{3}$$

whenever $a \in U$. Let $u = \xi_{U \cap H} \in P^1(H)$. We have

$$\|f \square u - f\|_1 = \int_G \left| \int_H f(xy) \frac{\Delta_G(y)}{\Delta_H(y)} u(y^{-1}) \, dy - \int_H f(x) \frac{u(y^{-1})}{\Delta_H(y)} \, dy \right| dx$$

$$\leq \int_H \left(\int_G |f(xy)\Delta_G(y) - f(x)| \, dx \right) \frac{u(y^{-1})}{\Delta_H(y)} \, dy < \frac{\varepsilon}{3} \|u\|_1 = \frac{\varepsilon}{3}. \tag{4}$$

The relations (4) and (3) imply that

$$\left\| \sum_{i=1}^n \alpha_i R_{a_i^{-1}}(f \square u) \right\|_1 \leq \sum_{i=1}^n \alpha_i \|f \square u - f\|_1 + \left\| \sum_{i=1}^n \alpha_i R_{a_i^{-1}} f \right\|_1 < \frac{2\varepsilon}{3}. \tag{5}$$

Let

$$v = u - \sum_{i=1}^n \alpha_i R_{a_i^{-1}}^H u \in L^0(H);$$

$\|v\|_1 \leq 2\|u\|_1 = 2$. From (2), and (4), (5) we deduce that

$$\|f \square v - f\|_1 \leq \|f \square u - f\|_1 + \left\| \sum_{i=1}^n \alpha_i f \square R_{a_i^{-1}}^H u \right\|_1$$

$$= \|f \square u - f\|_1 + \left\| \sum_{i=1}^n \alpha_i R_{a_i^{-1}}(f \square u) \right\|_1 < \frac{\varepsilon}{3} + \frac{2\varepsilon}{3} = \varepsilon.$$

(b) Assume that the condition holds. We show that $L^0(H)$ admits bounded right approximate units, thus amenability of H follows from Theorem 10.1.

Let $g \in L^0(H)$ and $\varepsilon \in \mathbf{R}_+^*$. We associate $u \in \mathscr{K}_+(G)$ and a compact neighborhood V of e in G such that $f = u \square g \in \mathscr{K}\!\mathit{er}\, J_{H,\rho}$ and $\|_x((1/\rho)f) - g\|_1 < \varepsilon/3(1 + c)$ whenever $x \in V$. By hypothesis there exists $v \in L^0(H)$ such that $\|v\|_1 \le c$ and $\|f - f \square v\|_1 < 1/3(1 + c)\,\mu(\pi(V))\varepsilon$, μ being the quasi-invariant measure on G/H and π the canonical mapping of G onto G/H. Let A be the μ-measurable set formed by all $\dot{x} \in G/H$ such that

$$\int_H \frac{|f(xy) - f\square v(xy)|}{\rho(xy)}\, dy \ge \frac{\varepsilon}{3(1+c)}.$$

Hence $\mu(A) < \mu(\pi(V))$ and there must exist $x_0 \in V$ with $\dot{x}_0 \notin A$, that is,

$$\int_H \frac{|f(x_0 y) - f\square v(x_0 y)|}{\rho(x_0 y)}\, dy < \frac{\varepsilon}{3(1+c)}.$$

We have

$$\int_H \left| \frac{f\square v(x_0 y)}{\rho(x_0 y)} - g * v(y) \right| dy$$

$$\le \int_H \int_H \left| \frac{f(x_0 yz)}{\rho(x_0 y)} \frac{\Delta_G(z)}{\Delta_H(z)} v(z^{-1}) - g(yz)v(z^{-1}) \right| dy\, dz$$

$$= \int_H \int_H \left| \frac{f(x_0 yz)}{\rho(x_0 yz)} v(z^{-1}) - g(yz)v(z^{-1}) \right| dy\, dz$$

$$\le \int_H \int_H \left| \frac{f(x_0 y)}{\rho(x_0 y)} - g(y) \right| |v(z^{-1})| \Delta_H(z^{-1})\, dy\, dz$$

$$= \left(\int_H \left| \frac{f(x_0 y)}{\rho(x_0 y)} - g(y) \right| dy \right) \|v\|_1 < \frac{\varepsilon c}{3(1+c)} < \frac{\varepsilon}{3}.$$

Then also

$$\|g - g * v\|_1 \le \int_H \left| \frac{f(x_0 y)}{\rho(x_0 y)} - g(y) \right| dy + \int_H \frac{|f(x_0 y) - f\square v(x_0 y)|}{\rho(x_0 y)}\, dy$$

$$+ \int_H \left| \frac{f\square v(x_0 y)}{\rho(x_0 y)} - g * v(y) \right| dy$$

$$< \frac{\varepsilon}{3(1+c)} + \frac{\varepsilon}{3(1+c)} + \frac{\varepsilon}{3} < \varepsilon. \qquad \square$$

We finally study some properties concerning closed subspaces of $L^1(G)$ and $L^1(G/H)$ for an amenable (closed) subgroup H of a locally compact group G.

Proposition 23.6. *Let G be a locally compact group admitting an amenable subgroup H. If X is a closed subspace of $L^1(G)$ such that $f_a \in X$ whenever $f \in X$ and $a \in H$, then $J_{H,\rho}(X)$ is a closed subspace of $L^1(G/H)$.*

Proof. Let $Y = J_{H,\rho}(X)$. If $f' \in \overline{Y}$, there exists a sequence $(f'_n)_{n \in \mathbf{N}^*}$ in Y such that $\sum_{n=1}^\infty f'_n = f'$ and $\sum_{n=1}^\infty \|f'_n\|_1 < \infty$. For every $n \in \mathbf{N}^*$, choose $f_n \in J_{H,\rho}^{-1}(f'_n)$. As H is amenable, by Proposition 23.2 there exists $A_n \in R_{G,H}$ such that

$$\|A_n f_n\|_1 \le \|f'_n\|_1 + \frac{1}{2^n}.$$

As

$$\sum_{n=1}^\infty \|A_n f_n\|_1 \le \sum_{n=1}^\infty \|f'_n\|_1 + 1 < \infty,$$

the series $\sum_{n=1}^\infty A_n f_n$ converges to an element f_0 in the closed subspace X of $L^1(G)$. Then

$$\|J_{H,\rho} f_0 - f'\|_1 = \left\| J_{H,\rho}\left(\sum_{n=1}^\infty A_n f_n\right) - f' \right\|_1 = \left\| \sum_{n=1}^\infty f'_n - f' \right\|_1.$$

We have $f' = J_{H,\rho} f_0 \in Y$. □

We add some immediate consequences of Proposition 23.6.

Corollary 23.7. *Let G be a locally compact group admitting the amenable closed normal subgroup H. If I is a closed right [resp. left] ideal of $L^1(G)$, then $J_H(I)$ is a closed right [resp. left] ideal of $L^1(G/H)$.*

Proof. As J_H is a surjective homomorphism, $J_H(I)$ is a right [resp. left] ideal of $L^1(G/H)$. The fact that $J_H(I)$ is closed follows from Proposition 23.6 for the right-hand version; the left-hand version holds too, by (14.7). □

Corollary 23.8. *Let G be a locally compact group admitting the closed subgroups H_1 and H_2. If one at least of these subgroups is normal and one at least is amenable, then $J_r(G, \overline{H_1 H_2})[= J_r(G, \overline{H_2 H_1})] = J_r(G, H_1) + J_r(G, H_2)$ and this left ideal is closed in $L^1(G)$.*

Proof. Assume H_1 to be normal.
(a) The subspace $J_r(G, H_2) = \mathcal{K}\!er\, J_{H_2, \rho}$ is closed. If H_1 is amenable, Proposition 23.6 implies that $J_{H_1}(J_r(G, H_2))$ is closed in $L^1(G/H_1)$.

As H_1 is normal, we have $J_r(G, H_1) = J_l(G, H_1)$. Then Proposition 23.6 implies that, if H_2 is amenable, $J_{H_2, \rho}(J_r(G, H_1))$ is closed in $L^1(G/H_2, \mu_\rho)$.

As J_{H_1} and $J_{H_2, \rho}$ are continuous, we conclude now that in both cases $J_r(G, H_1) + J_r(G, H_2)$ is closed in $L^1(G)$.

(b) For each $j = 1, 2$, $J_r(G, H_j) \subset J_r(G, \overline{H_1 H_2})$ and then also $J_r(G, H_1) + J_r(G, H_2) \subset J_r(G, \overline{H_1 H_2})$.

Consider $f \in L^1(G)$, $a_1 \in H_1$, $a_2 \in H_2$. We have

$$R_{a_1 a_2} f - f = \left(R_{a_1}(R_{a_2} f) - R_{a_2} f \right) + \left(R_{a_2} f - f \right) \in J_r(G, H_1) + J_r(G, H_2).$$

As the latter set is closed by (a) and the mapping $x \mapsto R_x$ of G into $L^1(G)$ is continuous, we obtain

$$J_r\left(G, \overline{H_1 H_2}\right) \subset J_r(G, H_1) + J_r(G, H_2). \qquad \square$$

B. Amenable Discrete Groups

We indicate complements on the stability properties of the class of amenable discrete groups.

Definition 23.9. *We denote by NF the class of all discrete groups having no free subgroup on two generators.*

Recall that all elementary discrete groups are amenable (Section 13 A). All known amenable discrete groups are elementary groups. By Propositions 14.1 and 13.3 we have $EG \subset NF$.

Let EG_0 be the class of all discrete groups that are either finite or abelian. Let Λ be an ordinal with $0 < \Lambda$ and assume that EG_Γ has been defined for each ordinal $\Gamma < \Lambda$. If $\Lambda - 1$ exists, we denote by EG_Λ the class of all groups obtained from $EG_{\Lambda - 1}$ by applying once either (III) or (IV) of Definition 13.7; otherwise we put $EG_\Lambda = \cup_{\Gamma < \Lambda} EG_\Gamma$.

Lemma 23.10. *For each ordinal Λ, EG_Λ is closed under the processes* (I) *and* (II) *of Definition 13.7.*

Proof. The demonstration is performed by transfinite induction.

All subgroups and quotient groups of abelian groups are abelian. All subgroups and quotient groups of finite groups are finite. Therefore EG_0 is closed under (I) and (II). Assume now that, for any $\Gamma < \Lambda$, EG_Γ is closed under (I) and (II). We consider $H \in EG_\Lambda$. Let K be a subgroup of H and let L be a homomorphic image of H.

If $\Lambda - 1$ exists, then either $M \simeq H/N$ for $M, N \in EG_{\Lambda - 1}$ or H is a direct union of groups (H_i) belonging to $EG_{\Lambda - 1}$. In the first case, $K/K \cap N$ is isomorphic to the subgroup KN/N of H/N; L is the extension of a homomorphic image M_1 of M by a homomorphic image N_1 of N. We have KN/N,

$K \cap N$, M_1, $N_1 \in EG_{\Lambda-1}$ and therefore $K, L \in EG_\Lambda$. In the second case, K is the direct union of the subgroups $H_i \cap K$ belonging to $EG_{\Lambda-1}$; L is the direct union of the homomorphic images of H_i belonging to $EG_{\Lambda-1}$. Therefore also $K, L \in EG_\Lambda$.

If $\Lambda - 1$ does not exist, we determine $\Gamma < \Lambda$ such that $H \in EG_\Gamma$. Then also $K, L \in EG_\Gamma$ and $K, L \in EG_\Lambda$. □

Proposition 23.11. *EG coincides with $E = \cup_{\Lambda \text{ ordinal}} EG_\Lambda$.*

Proof. We have $EG_0 \subset E \subset EG$ and show that E is closed under (I), (II), (III), and (IV) of Definition 13.7. By Lemma 23.10, E is closed under (I) and (II); obviously it is also closed under (III). In order to show that E is closed under (IV), let G be a direct union of $(G_i)_{i \in I}$ where $G_i \in E$, $i \in I$. For every $i \in I$, there exists an ordinal Λ_i such that $G_i \in EG_{\Lambda_i}$. Let Λ' be the least upper bound of $\{\Lambda_i : i \in I\}$; then $G \in EG_{\Lambda'+1}$ and $G \in E$. □

Corollary 23.12. *Every torsion group belonging to EG is locally finite.*

Proof. Every torsion group in EG_0 is of course locally finite. Assume now that for an ordinal Λ and every ordinal Γ such that $\Gamma < \Lambda$, every torsion group belonging to EG_Γ is locally finite. Consider a torsion group G in EG_Λ.

If $\Lambda - 1$ exists, either G is an extension of a torsion group by a torsion group belonging both to $EG_{\Lambda-1}$, hence an extension of a locally finite group by a locally finite group and therefore locally finite itself, or a direct union of torsion groups belonging to $EG_{\Lambda-1}$ and therefore locally finite by assumption. If $\Lambda - 1$ does not exist, we have $G \in EG_\Gamma$ for some $\Gamma < \Lambda$ and G is locally finite by assumption.

The statement now follows from Proposition 23.11 by transfinite induction. □

There exist torsion groups that are not locally finite ([416]). By Corollary 23.12 such groups cannot belong to EG whereas they obviously lie in NF.

Recall that if G is a locally compact group and G_d is amenable, then G is amenable by Proposition 4.21. Therefore the following result is of interest.

Proposition 23.13. *There exists a compact group G such that G_d is nonamenable, and G admits a dense subgroup H such that H_d is amenable.*

Proof. For every $n \in \mathbf{N}^*$, we consider the corresponding finite symmetric group $\mathfrak{S}_n$. Let G be the cartesian product $\prod_{n=1}^\infty \mathfrak{S}_n$; G is compact in the product topology. By Proposition 14.2, G_d is nonamenable. Let H be the weak direct product of the groups $\mathfrak{S}_n$ ($n \in \mathbf{N}^*$). We have $\overline{H} = G$ ([HR] 6.2). For every $n \in \mathbf{N}^*$, let $G_n = \{e\}$ and let H_n be the weak direct product of the subgroups $\mathfrak{S}_j$ ($j = 1, \ldots, n$) with the subgroups G_k ($k = n+1, \ldots$). The finite group $(H_n)_d$ is amenable, and thus H_d is amenable by Proposition 13.8. □

C. Further Examples of Amenable Groups

Definition 23.14. *A locally compact group G is said to be contractible if, for every neighborhood U of e in G and every finite subset F in G, there exists a topological automorphism $h_{U,F}$ of G such that $h_{U,F}(F) \subset U$.*

Proposition 23.15. *Any contractible group G is amenable.*

Proof. We establish (P_1^*). Amenability follows from Proposition 6.12. Let F be a finite subset of G and $\varepsilon > 0$. If ψ is any fixed element in $P^1(G)$, there exists a neighborhood U of e in G such that $\|_z\psi - \psi\|_1 < \varepsilon$ whenever $z \in U$. Let $h = h_{U,F}$. There exists $\alpha > 0$ such that

$$\alpha \int f \circ h(x)\, dx = \int f(x)\, dx$$

for every $f \in L^1(G)$ ([B INT] Chap. VII, Section 1, No. 4). Let $\psi = \alpha(\varphi \circ h) \in P^1(G)$. For every $a \in F$, we have $h(a) \in U$ and

$$\|_a\varphi - \varphi\|_1 = \alpha \int |\psi \circ h(ax) - \psi \circ h(x)|\, dx$$

$$= \alpha \int |_{h(a)}\psi - \psi| \circ h(x)\, dx = \|_{h(a)}\psi - \psi\|_1 < \varepsilon. \quad \square$$

Definition 23.16. *A locally compact group G is called uniformly distributed if there exists a sequence (a_n) in G such that, for every $f \in L^1(G)$,*

$$\lim_{n \to \infty} \left\| \frac{1}{n} \sum_{i=1}^{n} {}_{a_i}f \right\|_1 = \left| \int f \right|.$$

The standard reference for uniform distribution is Kuipers and Niederreiter [326].

Proposition 23.17. *Any uniformly distributed locally compact group G is amenable.*

Proof. Let (a_n) be the sequence given by Definition 23.16. For every $f \in L^1(G)$,

$$d_l(f) \leq \lim_{n \to \infty} \left\| \frac{1}{n} \sum_{i=1}^{n} {}_{a_i}f \right\|_1.$$

So the statement is an immediate consequence of Theorem 6.21. $\quad \square$

Proposition 23.18. *Every completely separable amenable group G is uniformly distributed.*

Proof. (a) As G is completely separable, the set $L^0(G)$ admits a dense sequence (f_k). By Lemma 6.20, for any $m \in \mathbf{N}^*$, there exists $T^{(m)} \in L_G$ such that $\|T^{(m)} f_k\|_1 < 1/m$ whenever $k = 1, \ldots, m$. By density of $\mathbf{Q}$ in $\mathbf{R}$ we may suppose that $T^{(m)}$ is of the type $T^{(m)} = \sum_{i=1}^{p_m} L_{z_i^{(m)}}/p_m$, where $z_i^{(m)} \in G$, for $i = 1, \ldots, p_m$, and (p_m) is increasing.

We put $q_1 = 1$ and choose inductively a sequence (q_m) in $\mathbf{N}^*$. If q_m has been determined, let $q_{m+1} \in \mathbf{N}^*$ such that $q_{m+1} > (m+1)p_{m+1}$ and $(q_{m+1} - q_m)/p_m \in \mathbf{N}^*$. Hence the sequence (q_m) is increasing and $\lim_{m \to \infty} p_m/q_m = 0$. We put $r_m = (q_{m+1} - q_m)/p_m$, $m \in \mathbf{N}^*$.

Each $N \geq q_1 + p_1 + 1$ may be written in the form $N = q_l + jp_l + i$, with $l \in \mathbf{N}^*$, $j \in \{0, 1, \ldots, r_l - 1\}$, $i \in \{1, \ldots, p_l\}$. We put $a_N = z_i^{(l)}$. Let $\varepsilon \in \,]0, 1[$ and $k \in \mathbf{N}^*$ such that $1/k \leq \varepsilon$. Let also $s \in \mathbf{N}^*$ such that $s \geq k$ and $p_m/q_m < \varepsilon$ for $m \geq s$. We consider $q_l > q_s/\varepsilon$. As $\varepsilon \in \,]0, 1[$, we have $q_l > q_s$ and $l > s$. Then

$$\sum_{n=1}^{N} L_{a_n} = L_{a_1} + \cdots + L_{a_{q_k}} + L_{a_{q_k+1}} + \cdots + L_{a_{q_k+p_k+1}} + \cdots + L_{a_{q_k+2p_k}}$$

$$+ \cdots + L_{a_{q_k+(r_k-1)p_k+1}} + \cdots + L_{a_{q_k+r_k p_k}} + L_{a_{q_{k+1}+1}}$$

$$+ \cdots + L_{a_{q_{k+1}+r_{k+1}p_{k+1}}} + \cdots + L_{a_{q_{l-1}+1}} + \cdots + L_{a_{q_{l-1}+r_{l-1}p_{l-1}}} + L_{a_{q_l+1}}$$

$$+ \cdots + L_{a_{q_l+jp_l}} + L_{a_{q_l+jp_l+1}} + \cdots + L_{a_{q_l+jp_l+i}}$$

$$= L_{a_1} + \cdots + L_{a_{q_k}} + r_k p_k T^{(k)} + r_{k+1} p_{k+1} T^{(k+1)}$$

$$+ \cdots + r_{l-1} p_{l-1} T^{(l-1)} + jp_l T^{(l)} + L_{a_{q_l+jp_l+1}} + \cdots + L_{a_{q_l+jp_l+i}}.$$

Hence

$$\left\| \sum_{n=1}^{N} L_{a_n} f_k \right\|_1 < q_k \|f_k\|_1 + \sum_{t=k}^{l-1} \left(r_t p_t \frac{1}{t}\right) + jp_l \frac{1}{l} + p_l \|f_k\|_1$$

$$\leq q_s \|f_k\|_1 + \left(\sum_{t=k}^{l-1} r_t p_t\right) \frac{1}{k} + jp_l \frac{1}{l} + \varepsilon q_l \|f_k\|_1$$

$$< \varepsilon q_l \|f_k\|_1 + \left(\sum_{t=k}^{l-1} r_t p_t + jp_l\right) \varepsilon + \varepsilon q_l \|f_k\|_1$$

$$= \left(q_l \|f_k\|_1 + (q_l - q_k + jp_l) + q_l \|f_k\|_1 \right) \varepsilon$$

$$< N(2\|f_k\| + 1)\varepsilon$$

and

$$\left\| \frac{1}{N} \sum_{n=1}^{N} L_{a_n} f_k \right\|_1 < (2\|f_k\| + 1)\varepsilon.$$

As $\varepsilon \in\,]0, 1[$ may be chosen arbitrarily,

$$\lim_{N \to \infty} \left\| \frac{1}{N} \sum_{n=1}^{N} L_{a_n} f_k \right\|_1 = 0.$$

Thus for any $f \in L^0(G)$ we have

$$\lim_{N \to \infty} \left\| \frac{1}{N} \sum_{n=1}^{N} L_{a_n} f \right\|_1 = 0.$$

(b) We now consider $g \in \mathcal{K}(G)$. Let K be a compact subset of G such that $\operatorname{supp} g \subset K$ and $|K| > 0$. Let $\alpha = \int g$ and $g' = \alpha \xi_K \in L^1(G)$; then $f = g - g' \in L^0(G)$. By (a) there exists a sequence (a_n) in G such that

$$\lim_{N \to \infty} \left\| \frac{1}{N} \sum_{n=1}^{N} L_{a_n} f \right\|_1 = 0.$$

Then also

$$\limsup_{N \to \infty} \left\| \frac{1}{N} \sum_{n=1}^{N} L_{a_n} g \right\|_1 \leq \lim_{N \to \infty} \left\| \frac{1}{N} \sum_{n=1}^{N} L_{a_n} f \right\|_1 + \limsup_{N \to \infty} \left\| \frac{1}{N} \sum_{n=1}^{N} L_{a_n} g' \right\|_1$$

$$\leq |\alpha| = \left| \int g \right|.$$

(c) If $h \in L^1(G)$ and $\varepsilon > 0$, let $g \in \mathcal{K}(G)$ such that $\|h - g\|_1 < \varepsilon$. Then by (b) there exists a sequence (a_n) in G such that

$$\limsup_{N \to \infty} \left\| \frac{1}{N} \sum_{n=1}^{N} L_{a_n} g \right\|_1 \leq \left| \int g \right|.$$

Therefore also

$$\limsup_{N \to \infty} \left\| \frac{1}{N} \sum_{n=1}^{N} L_{a_n} h \right\|_1 \leq \limsup_{N \to \infty} \left\| \frac{1}{N} \sum_{n=1}^{N} L_{a_n} g \right\|_1 + \varepsilon$$

$$\leq \left| \int g \right| + \varepsilon \leq \left| \int h \right| + \|h - g\|_1 + \varepsilon < \left| \int h \right| + 2\varepsilon.$$

As $\varepsilon > 0$ is arbitrary, we have

$$\limsup_{N\to\infty} \left\| \frac{1}{N} \sum_{n=1}^{N} L_{a_n} h \right\|_1 \leq \left| \int h \right|.$$

Obviously also

$$\left| \int h \right| \leq \liminf_{N\to\infty} \left\| \frac{1}{N} \sum_{n=1}^{N} L_{a_n} h \right\|_1. \qquad \square$$

We fix some notations from probability theory.

Let G be a locally compact that is completely separable and let $\mu \in M^1(G)$. If f is a bounded, Borel measurable, real-valued function defined on G and $a \in G$, we put $Pf(a) = \int_G f(ax)\,d\mu(x)$. The function f is called μ-*superharmonic* [resp. μ-*harmonic*] if $Pf \leq f$ [resp. $Pf = f$]. For an aperiodic $\mu \in M^1(G)$, we consider a sequence (Y_n) of independent random variables and the right random walk defined on G by $X_0 = e$, $X_n = Y_1 \cdots Y_n$, $n \in \mathbf{N}^*$. To $x \in G$ we associate the probability $P_x = \delta_x * \mu$. The element x is called *recurrent* if, for every neighborhood V of x in G, $P_e(\limsup_{n\to\infty} X_n \in V) = 1$ or, equivalently, $P_e(\sum_{n=0}^{\infty} 1_V(X_n) = \infty) = 1$. The random walk is called recurrent if all the elements of the group are recurrent. This situation occurs exactly when $\sum_{n=0}^{\infty} \mu^{*n}(V) = \infty$ for every open neighborhood V of e ([460] Ch. 3, Theorem 4.6). A locally compact, completely separable group admitting a recurrent random walk associated to an aperiodic probability measure is called a *recurrent group*. The recurrence of the group G is characterized by the fact that, for the aperiodic $\mu \in M^1(G)$, every continuous nonnegative μ-superharmonic function defined on G is a constant function (Choquet–Deny property) ([243] Chapitre I, proposition 45). Examples of recurrent groups are the compact groups and the groups $\mathbf{R}^{d_1} \oplus \mathbf{Z}^{d_2}$ with $d_1, d_2 \in \mathbf{N}$ and $d_1 + d_2 \in \{1, 2\}$. For details we refer to ([460] Ch. 3, proposition 5.5).

Proposition 23.19. *Any recurrent group G is amenable.*

Proof. Let (G, Z) be a flow and let μ be an aperiodic probability measure on G, for which G is recurrent. We define

$$\mu_n = \frac{1}{n} \sum_{k=1}^{n} \mu^{*k}$$

for $n \in \mathbf{N}^*$; then $\|\mu * \mu_n - \mu_n\| \leq 2/n$ and

$$\lim_{n\to\infty} \|\mu * \mu_n - \mu_n\| = 0. \qquad (6)$$

Let $\nu \in M^1(Z)$. If $n \in \mathbf{N}^*$, define $\nu_n \in M^1(Z)$ by putting

$$\nu_n(A) = \int_G \int_Z 1_A(x\zeta)\,d\mu_n(x)\,d\nu(\zeta),$$

A being a Borel subset of Z. The set $\{\nu_n : n \in \mathbf{N}^*\}$ is relatively w*-compact, so there exist a subsequence (μ_{n_p}) of (μ_n) and $\sigma \in M^1(Z)$ such that

$$\langle f, \sigma \rangle = \lim_{p \to \infty} \int_G \int_Z f(x\zeta)\, d\mu_{n_p}(x)\, d\nu(\zeta)$$

whenever $f \in \mathscr{C}(Z)$. Lebesgue's theorem and (6) imply that

$$\int_G \int_Z f(x\zeta)\, d\mu(x)\, d\sigma(\zeta) = \lim_{p \to \infty} \int_G \int_G \int_Z f(xy\zeta)\, d\mu(x)\, d\mu_{n_p}(y)\, d\nu(\zeta)$$

$$= \lim_{p \to \infty} \int_G \int_Z f(z\zeta)\, d(\mu * \mu_{n_p})(z)\, d\nu(\zeta)$$

$$= \lim_{p \to \infty} \int_G \int_Z f(z\zeta)\, d\mu_{n_p}(z)\, d\nu(\zeta) = \langle f, \sigma \rangle. \quad (7)$$

If $f \in \mathscr{C}_+(Z)$, let

$$\varphi(x) = \int_Z f(x\zeta)\, d\sigma(\zeta),$$

$x \in G$. As Z is compact, $\varphi \in \mathscr{C}_+(G)$. For every $a \in G$, by (7) we have

$$P\varphi(a) = \int_G \varphi(ax)\, d\mu(x) = \int_G \int_Z f(ax\zeta)\, d\sigma(\zeta)\, d\mu(x)$$

$$= \int_Z f(a\eta)\, d\sigma(\eta) = \varphi(a).$$

As the μ-harmonic function $\varphi \in \mathscr{C}_+(G)$ must be a constant function, we conclude that, if $f \in \mathscr{C}_+(Z)$ and $a \in G$,

$$\int_Z f(a\zeta)\, d\nu(\zeta) = \int_Z f(\zeta)\, d\nu(\zeta).$$

The equality holds for every $f \in \mathscr{C}(Z)$ and therefore σ is G-invariant. Amenability follows from Theorem 5.4. □

D. Properties Related to Amenability

In Section 12B we showed the intimate relationship of amenability to growth conditions. We now mention a couple of other classes of locally compact groups that are of interest for the study of amenability.

Kunze and Stein ([327] Theorem 9) established an important result concerning the nonamenable group $SL(2, \mathbf{R})$: Let $G = SL(2, \mathbf{R})$, $p \in [1, 2[$, and $f \in L^p(G)$; then $L_f \in Cv^2(G)$. This property leads to the following definition.

Definition 23.20. *Let G be a locally compact group. Let $p \in [1, 2[$; G is said to have property (KS_p) if $L_f \in Cv^2(G)$ whenever $f \in L^p(G)$. We say that G has property (KS) if (KS_p) holds for all $p \in [1, 2[$.*

Property (KS_1) holds for every locally compact group.

If G is any locally compact group, $L^1(G)$ is a Banach algebra. If G is a compact group and $p \in]1, \infty[$, $L^p(G)$ is a Banach algebra ([HR] 28.46). The converse implication is called L^p-*conjecture*. It holds for any locally compact group and $p \geq 2$. We make reference to Rajagopalan [445] in the case of $p > 2$, and to Rickert [464] in the case of $p = 2$. Lohoué [369] solved the L^p-conjecture for all almost connected, locally compact groups. It may be established easily for amenable groups.

Proposition 23.21. *Let G be a locally compact group.*
  (a) *If G is compact, G satisfies (KS).*
  (b) *Assume G to be amenable and let $p \in]1, 2[$. If G satisfies (KS_p) [resp. $L^p(G)$ is a Banach algebra], then G is compact.*

Proof. (a) For every $p \in [1, 2[$, we have $L^p(G) \subset L^1(G)$ ([HS] 13.17). For all $f \in L^p(G)$ and $g \in L^2(G)$, $f * g \in L^2(G)$ and (KS_p) holds.
  (b) Let $f \in L^p_+(G)$. Then $L_f \in Cv^2(G)$ [resp. $\|f * g\|_p \leq \|f\|_p \|g\|_p$ whenever $g \in L^p(G)$, hence $L_f \in Cv^p(G)$]. By Theorem 9.6, (Cv'_2) [resp. (Cv'_p)] holds. Then by duality $L^\infty(G) \subset L^{p'}(G)$ with $1/p + 1/p' = 1$; G is compact. □

Definition 23.22. *A locally compact group G is called a hermitian group if its group algebra $L^1(G)$ is hermitian, that is,*

$$(\forall f \in L^1(G)) f = f^* \Rightarrow \mathrm{sp}_{L^1(G)} f \subset \mathbf{R}.$$

General properties of hermitian groups are listed in [425]. Every open subgroup of a hermitian group is hermitian as it may be equipped with the induced Haar measure, and any closed $*$-subalgebra of a hermitian Banach algebra is hermitian ([461] 4.1.10). As every quotient of a hermitian Banach algebra is hermitian, Weil's formula shows that every quotient group of a hermitian group is hermitian. Jenkins proved that any hermitian, semisimple, connected Lie group with a finite center is compact ([284] 4.4).

Proposition 23.23. *Every hermitian almost connected locally compact group G is amenable.*

Proof. (a) We assume that G is connected. There exists a compact normal subgroup K of G such that $H = G/K$ is a connected Lie group. Consider $L = H/\operatorname{rad} H$ (Section 14B). Let Z be the center of L; L/Z is a hermitian, semisimple, connected Lie group with a trivial center and is thus compact. Then Proposition 13.4 implies amenability of L, H, and G.

(b) We now consider the general case; G/G_0 is compact, hence amenable. By (a), the hermitian, connected, open subgroup G_0 of G is amenable. Proposition 13.4 implies amenability of G. □

E. Amenable Semigroups

Let S be a *topological semigroup*, that is, a semigroup equipped with a Hausdorff topological structure such that the mapping

$$(x, y) \mapsto xy$$

$$S \times S \to S$$

is continuous. If f is a complex-valued function defined on S and $a \in S$, we put $_a f(x) = f(ax)$, $f_a(x) = f(xa)$ with $x \in S$. We also consider $l_a f : x \mapsto f(ax)$. The function $f : S \to \mathbf{C}$ is called *right uniformly continuous* if, for every net (a_i) in S converging to $a \in S$, we have

$$\lim_i \|l_{a_i} f - l_a f\| = 0.$$

We consider the Banach space $\mathscr{RUC}(S)$ of all bounded, right uniformly continuous, complex-valued functions defined on S. If $f \in \mathscr{RUC}(S)$ and $a \in S$, we have $l_a f \in \mathscr{RUC}(S)$. The topological semigroup S is called a (*left*) *amenable semigroup* if there exists a mean M on $\mathscr{RUC}(S)$ that is left invariant, that is,

$$(\forall f \in \mathscr{RUC}(S))(\forall a \in S) M(_a f) = M(f).$$

Similarly one defines right invariant means for bounded, left uniformly continuous functions on S.

To a large extent the theory of amenable groups may be carried over, in a straightforward way, to amenable semigroups, especially discrete ones. General references for amenable discrete semigroups are Day [115, 120] and Hewitt and Ross ([HR] Section 17).

Obviously if a topological semigroup is amenable with respect to the discrete topology, it is also amenable with respect to the given topology. For a topological semigroup S, $\overline{\operatorname{co}}\{\delta_x : x \in S\}$ is the set of all means on $\mathscr{RUC}(S)$ and is w*-compact in $\mathscr{RUC}(S)^*$. The set of left invariant means on $\mathscr{RUC}(S)$ is a right ideal in the set of all means on $\mathscr{RUC}(S)$ for the Arens multiplication. Amenability of topological semigroups may be characterized by a lot of equivalent properties reflecting those stated for amenable groups. Via the Markov–Kakutani fixed point theorem amenability of abelian semigroups is established. The class of amenable semigroups also admits stability properties that are the counterparts of those described for amenable groups.

We next study a stability property that is specific for amenability of semigroups.

Proposition 23.24. *If a topological semigroup S admits a nonvoid amenable open left ideal I, then S is amenable.*

Proof. Fix an element b in I. If $f \in \mathcal{RUC}(S)$ and $x \in I$, let $f' = (_b f)|_I$; $f' \in \mathcal{RUC}(I)$. If $a \in S$ and $x \in I$, we have $ab \in I$,

$$(_a f)'(x) = (_b(_a f))(x) = {}_{ab}f(x) = f(abx)$$

and, for a left invariant mean M on $\mathcal{RUC}(I)$,

$$M((_a f)') = M(_{ab}(f|_I)) = M(_b(f|_I)) = M(f').$$

We define

$$N(f) = M(f'),$$

$f \in \mathcal{RUC}(S)$; N is a mean on $\mathcal{RUC}(S)$. For all $f \in \mathcal{RUC}(S)$ and $a \in S$, $N(_a f) = N(f)$. □

Proposition 23.25. *Let S be an amenable semigroup and I a nonvoid right ideal. Then every left invariant mean on $\mathcal{RUC}(S)$ belongs to $\overline{\mathrm{co}}\{\delta_x : x \in I\}$.*

Proof. Let M be a left invariant mean on $\mathcal{RUC}(S)$. If $f \in \mathcal{RUC}(S)$ and $a \in I$, $_a f \in \mathcal{RUC}(S)$ and $M(f) = M(_a f)$. Hence $M \in \overline{\mathrm{co}}\{\delta_x : x \in I\}$ and $M(1_I) = 1$. □

We now examine some properties relating amenable semigroups to amenable groups.

Proposition 23.26. *Let G be an amenable group admitting a topological semigroup S (for the induced topology) that is a Borel subset of G. If there exists a left invariant mean M on $L^\infty(G)$ such that $M(1_S) > 0$, then S is an amenable semigroup.*

Proof. Given $f \in \mathcal{RUC}(S)$, we define

$$f'(x) = f(x) \quad \text{if } x \in S,$$
$$f'(x) = 0 \quad \text{if } x \in G \setminus S.$$

Then $f' \in L^\infty(G)$. We consider the mapping

$$F: f \mapsto f'$$

$$\mathscr{RUC}(S) \to L^\infty(G)$$

and let $\langle f, F^\sim(M) \rangle = \langle F(f), M \rangle$ for $f \in \mathscr{RUC}(S)$. Then $N = M(1_S)^{-1} F^\sim(M)$ is a mean on $\mathscr{RUC}(S)$. We show that N is left invariant.

If $f \in \mathscr{RUC}(S)$, $a \in S$, and $x \in G$, we put

$$g(x) = F(_a f)(x) -\,_a(F(f))(x);$$

$g \in L^\infty(G)$. As $aS \subset S$, we have $S \subset a^{-1}S$. Let E be the Borel subset $a^{-1}S \setminus S$. If $c \in E$, $g(c) = g(c) 1_E(c)$. If $c \in G \setminus E$, either $c \notin a^{-1}S$ or $c \in S$. In the first case, as $S \subset a^{-1}S$, we have $c \notin S$ and $F(_a f)(c) = 0$. As $ac \notin S$, $_a(F(f))(c) = F(f)(ac) = 0$. In the second case, $ac \in S$, $F(_a f)(c) =\,_a f(c) = f(ac)$ and $_a(F(f))(c) = F(f)(ac) = f(ac)$.

Thus in both cases $g(c) = 0$, and we have

$$g = g 1_E.$$

If there exist $b \in G$ and $j \in \mathbf{N}^*$ such that $a^j b \in a^{-1}S$, then $a^{j+1} b \in S$ and $a^{j+1} b \notin E$. Hence for any $b \in G$, $\{a^j b : j \in \mathbf{N}^*\}$ contains at most one element of E and, for every $n \in \mathbf{N}^*$,

$$\sum_{j=1}^n \,_{a^j} 1_E \leq 1_G.$$

By left invariance of M we obtain

$$n M(1_E) = M\left(\sum_{j=1}^n \,_{a^j} 1_E \right) \leq 1.$$

As $n \in \mathbf{N}^*$ may be chosen arbitrarily, we must have $M(1_E) = 0$. Then

$$|M(g)| = |M(g 1_E)| \leq \|g\|_\infty M(1_E) = 0,$$

that is, $M(F(_a f)) = M(_a(F(f)))$. As M is left invariant, $M(F(_a f)) = M(F(f))$. Finally

$$N(_a f) = M(1_S)^{-1} \langle _a f, F^\sim(M) \rangle = M(1_S)^{-1} \langle F(_a f), M \rangle$$

$$= M(1_S)^{-1} \langle F(f), M \rangle = N(f). \qquad \square$$

Corollary 23.27. *Let G be an amenable group admitting an open subsemigroup S. If there exists a left invariant mean M on $\mathscr{RUC}(G)$ that is concentrated on S, then S is an amenable semigroup.*

Proof. Let U be an open, relatively compact subset of G belonging to S^{-1}. For $x \in G$,

$$\xi_U * 1_S * \xi_U(x) = |U|^{-2} \int_G \int_G 1_U(z) 1_S(z^{-1}y) 1_U(y^{-1}x) \, dy \, dz$$

$$= |U|^{-2} \int_U |xU^{-1} \cap zS| \, dz = |U|^{-2} \int_U |z^{-1}xU^{-1} \cap S| \, dz.$$

If $z \in U$ and $x \in S$, we have $z^{-1}xU^{-1} \subset SxS \subset S$. Hence $\xi_U * 1_S * \xi_U(x) = |U|^{-1}|U^{-1}|$ whenever $x \in S$ and $\xi_U * 1_S * \xi_U \geq |U|^{-1}|U^{-1}|1_S$. We put

$$N(f) = |U||U^{-1}|^{-1}M(\xi_U * f * \xi_U),$$

$f \in L^\infty(G)$. By Proposition 4.17, $N \in \text{LIM}(L^\infty(G))$. Moreover

$$1 \geq N(1_S) = |U||U^{-1}|^{-1}M(\xi_U * 1_S * \xi_U) \geq M(1_S) = 1.$$

Therefore N induces a left invariant mean on $\mathscr{RUC}(S)$. □

Proposition 23.28. *If a locally compact group G is generated by an open amenable semigroup S, then G is amenable.*

Proof. Let M be a left invariant mean on $\mathscr{RUC}(S)$. For any $f \in \mathscr{RUC}(G)$, $f|_S \in \mathscr{RUC}(S)$. We may define a mean N on $\mathscr{RUC}(G)$ by putting

$$N(f) = M(f|_S)$$

for $f \in \mathscr{RUC}(G)$. For all $f \in \mathscr{RUC}(G)$ and $a \in S$, $_{a^{-1}}f \in \mathscr{RUC}(G)$. Also

$$N(_af) = M((_af)|_S) = M(f|_S) = N(f)$$

and

$$N(_{a^{-1}}f) = M((_{a^{-1}}f)|_S) = M(_a((_{a^{-1}}f)|_S))$$

$$= M((_a(_{a^{-1}}f))|_S) = M(f|_S) = N(f).$$

As S generates G, N is a left invariant mean on $\mathscr{RUC}(G)$. □

Proposition 23.29. *A discrete semigroup S admitting two disjoint, nonvoid, right ideals is not amenable. In particular, if S contains the free semigroup SF_2 on two generators, it is not amenable.*

Proof. Let I and J be two disjoint, nonvoid, right ideals in S and let $a \in I$ and $b \in J$. If there existed a left invariant mean M on S, we should have

$$M(1_S) \geq M(1_I) + M(1_J) \geq M(1_{aS}) + M(1_{bS}) = M(1_S) + M(1_S).$$

This inequality cannot hold as $M(1_S) = 1$. □

Proposition 23.30. *Every amenable, discrete, cancellative semigroup S is embeddable into an amenable discrete group.*

Proof. By Proposition 23.29, given a, $b \in S$, there exist $a', b' \in S$ such that $aa' = bb'$. This property suffices to conclude that the cancellative semigroup S is embeddable into a group ([364] Chap. X, 2.7) that may be considered to be generated by S. Then Proposition 23.28 establishes the statement. □

Lemma 23.31. *Let G be a group and let S be a subsemigroup generating G such that the class of all nonvoid right ideals of S has the finite intersection property. Then for every $a \in G$, $aS \cap S$ is nonvoid and hence constitutes a right ideal in S.*

Proof. If $x \in S$, $xS \subset S$ and $xS \cap S \neq \emptyset$. If $x \in S^{-1}$, we have $x^{-1} \in S$, $x^{-1}S \cap S \neq \emptyset$, and $S \cap xS \neq \emptyset$. Suppose now that, for $x_1, \ldots, x_n \in S \cup S^{-1}$ and $y = x_n \ldots x_1$, we have $yS \cap S \neq \emptyset$. We show that, for every $z \in S \cup S^{-1}$, $zyS \cap S \neq \emptyset$.

If $z \in S$, then $z(yS \cap S) \subset zyS \cap S$. If $z \in S^{-1}$, we consider the two right ideals $yS \cap S$ and $z^{-1}S$. By hypothesis $(yS \cap S) \cap (z^{-1}S) \neq \emptyset$. Therefore $(yS) \cap (z^{-1}S) \neq \emptyset$ and $(zyS) \cap S \neq \emptyset$. □

Proposition 23.32. *Let G be an amenable group admitting an open subsemigroup S. The following conditions are equivalent:*

  (i) *S is an amenable semigroup.*
  (ii) *The nonvoid open right ideals of S have the finite intersection property.*

Proof. (i) ⇒ (ii)
Let U be an open, relatively compact subset of G contained in S. If I is an open right ideal in G, for every $x \in G$,

$$1_I * \check{\xi}_U(x) = \frac{1}{|U|} \int_G 1_I(y) 1_U(x^{-1}y)\, dy = \frac{|I \cap xU|}{|U|} \leq 1.$$

We have $1_I * \check{\xi}_U \in \mathcal{RUC}(G)$ and, if $x \in I$, $xU \subset IS \subset I$. Hence $1_I * \check{\xi}_U(x) = |xU|/|U| = 1$ and $1_I \leq 1_I * \check{\xi}_U$. Moreover if $a \in S$ and $y \in U$, then $ay \in S$ and $1_S * \check{\xi}_U(a) = 1/|U| \int_G 1_S(ay) 1_U(y)\, dy = 1$. Hence $(1_S * \check{\xi}_U)|_S = 1_S|_S$.

If (ii) did not hold, there would exist two nonvoid, disjoint, open right ideals I' and I'' in S. Proposition 23.25 would imply the existence of a left invariant mean M on $\mathcal{RUC}(S)$ belonging to $\overline{\text{co}}\{\delta_x : x \in I'\} \cap \overline{\text{co}}\{\delta_y : y \in I''\}$. As

$I' \cap I'' = \emptyset$,

$$((1_{I'} + 1_{I''}) * \check{\xi}_U)|_S \leq (1_S * \check{\xi}_U)|_S = 1_S|_S.$$

So we should have

$$2 = M(1_{I'}|_S + 1_{I''}|_S) \leq M((1_{I'} * \check{\xi}_U)|_S + (1_{I''} * \check{\xi}_U)|_S) \leq M(1_S|_S) = 1,$$

and we would obtain a contradiction.

(ii) ⇒ (i)

Let H be the open amenable subgroup of G generated by S. By Lemma 23.31, for each $a \in H$, $I_a = aS \cap S$ is a nonvoid right ideal in S. By w*-compactness of the set of means on $\mathcal{RUC}(H)$, the hypothesis implies that $J = \cap_{a \in H} \{\delta_x : x \in I_a\}^- \neq \emptyset$. If $a \in H$ and $b \in I_{a^{-1}}$, then $ab \in aS \cap S$. For every $M \in J$, by (2.15) we have thus $M * 1_S(a) = M(_a 1_S) = 1$ and $M * 1_S = 1_H$. Now let M' be any left invariant mean on $\mathcal{RUC}(H)$. By Proposition 4.26 we have $N = M' \odot M \in \text{LIM}(\mathcal{RUC}(H))$. Moreover $N(1_S) = \langle M * 1_S, M' \rangle = \langle 1_H, M' \rangle = 1$. We may then conclude via Corollary 23.27. □

A fundamental result from abstract group theory due to Appel and Djorup asserts that a group generated by a two generator free semigroup SF_2 needs not be free ([13] Theorem). Therefore the following proposition is of interest.

Proposition 23.33. *A semigroup S of an amenable discrete group G inherits amenability if and only if S does not contain the free semigroup SF_2 on two generators.*

Proof. (a) If S is amenable, S does not contain SF_2 by Proposition 23.29.

(b) Assume S to be nonamenable. By Proposition 23.32 there exist two nonvoid, disjoint, right ideals I and J in S. Let $a \in I$ and $b \in J$. If the semigroup generated by a and b were not free, there would exist $s', s'' \in S \cup \{e\}$ such that $as' = bs''$, hence $I \cap J \neq \emptyset$, and we would obtain a contradiction. Therefore S must contain a copy of SF_2. □

Remark 23.34. There exist amenable groups admitting nonamenable subsemigroups.

Let $\{a_{m,n} : (m,n) \in \mathbf{Z}^2\}$ be a double-indexed sequence generating a free abelian group G. We denote the algebraic structure of G additively. The mappings

$$(m, n) \mapsto (m + 1, n)$$

$$(m, n) \mapsto (m, n + 1)$$

of $\mathbf{Z}^2$ onto $\mathbf{Z}^2$ induce automorphisms f and g on G. These mappings generate an abelian subgroup H of the automorphism group of G. Let Γ be the semidirect product of G by H ([HR] 2.6). For $(\alpha, \varphi), (\beta, \psi) \in \Gamma$,

$$(\alpha, \varphi)(\beta, \psi) = (\alpha + \varphi(\beta), \varphi \circ \psi).$$

As G and H are abelian, Γ is amenable.

Let $s = (a_{0,0}, f)$, $t = (a_{0,0}, g) \in \Gamma$. We show that s and t generate SF_2, thus Γ is nonamenable by Proposition 23.33. Consider a reduced word $w = u_1 \ldots u_d$, where $u_i = s$ or t for $i = 1, \ldots, d$. Suppose that s occurs p times and t occurs q times, hence $p + q = d$. The word may be written in the form $(\sum_{i=0}^{d-1} a_{m_i, n_i}, f^p \circ g^q)$ where $m_i, n_i \in \mathbf{N}$, $m_i + n_i = i$ whenever $i = 0, \ldots, d-1$. In particular if $d \geq 2$, $a_{m_1, n_1} = a_{1,0}$ or $a_{0,1}$ whether $u_1 = s$ or t. If there existed a reduced word $w' = u'_1 \ldots u'_{d'}$ equal to w, with respect to the form of the second coordinates in w and w', we should have $d = d'$. If $d \geq 2$, the first coordinates are sums of terms for which exactly one has its sum of indices equal to 1, this term being either $a_{1,0}$ or $a_{0,1}$ and being the same for w and w'. Hence $u_1 = u'_1$, and the words could be reduced. We conclude that s and t generate SF_2.

Proposition 23.35. *In the amenable discrete group G that is freely generated by two elements a and b of order 2, every subsemigroup is amenable.*

Proof. Assume that there exist two elements in G generating a free subsemigroup. These elements can only be given by

$$c = (ab) \cdots (ab) \qquad (m \text{ factors})$$

$$d = (ba) \cdots (ba) \qquad (n \text{ factors})$$

as all even powers of $ab \cdots aba$ and $ba \cdots bab$ are e and, if we had $c = (ab)^m$, $d = (ab)^n$, then the equality $c^n = d^m$ should hold.

If $m = n$ [resp. $m > n$, $m < n$], $cd = e$ [resp. $(cd)^m = c^{m-n}$, $(cd)^m = d^{n-m}$], and the elements c, d would not be independent. Hence G cannot contain a free subsemigroup on two generators, and we conclude via Proposition 23.33. □

A semigroup S is called an inverse semigroup if, for every $x \in S$, there exists a unique $x^* \in S$ satisfying $xx^*x = x^*$ and $x^*xx^* = x$. If S admits a left [resp. right] invariant mean for $l^\infty(S)$, it also admits a right [resp. left] invariant mean for $l^\infty(S)$. Let E_S be the set of all idempotent elements in S. It forms a commutative subsemigroup. Let $\approx$ be the equivalence relation defined on S in the following way: If $x, y \in S$, $x \approx y$ whenever there exists an idempotent u in E_S such that $ux = uy$. The corresponding quotient semigroup is a group G_S ([407] Theorem 1).

Proposition 23.36. *A discrete inverse semigroup S is amenable if and only if the associated discrete group G_S is an amenable group.*

Proof. (a) If a left invariant mean on $l^\infty(S)$ exists, then for the homomorphic image G_S of S there exists a left invariant mean on $l^\infty(G_S)$.

(b) Assume G_S to be amenable. Let $M \in \text{RIM}(l^\infty(G_S))$ and let M' be a right invariant mean on the abelian subsemigroup E_S of all idempotent elements in S.

Let $f \in l^\infty(S)$. If $s \in S$, consider $f^{(s)} = f_s|_{E_S}$. If s and t are equivalent elements in S, that is, there exists $u \in E_S$ such that $us = ut$, we have $f(jus) = f(jut)$ whenever $j \in E_S$ and

$$M'(f^{(s)}) = M'((f^{(s)})_u) = M'((f^{(t)})_u) = M'(f^{(t)}).$$

We may now define

$$f'(\dot{s}) = M'(f^{(s)}),$$

$\dot{s} \in G_S$; $f' \in l^\infty(G_S)$. We put

$$M''(f) = M(f'),$$

$f \in l^\infty(S)$; M'' is a mean on $l^\infty(S)$. Moreover for all $f \in l^\infty(S)$, $u \in S$, and $s \in S$, we have

$$(f_u)'(\dot{s}) = M'((f_u)^{(s)}) = M'(f^{(su)}) = f'(\dot{s}\dot{u}) = (f')_{\dot{u}}(\dot{s})$$

and

$$M''(f_u) = M((f')_{\dot{u}}) = M(f') = M''(f). \qquad \square$$

We study a particular type of amenable groups introduced by Jenkins. It associates amenability on groups and amenability on semigroups. If G is a locally compact group, we may consider $\Re(G)$ to be a discrete semigroup defined by $K_1 K_2 = \{x_1 x_2 : x_1 \in K_1, x_2 \in K_2\}$ for $K_1, K_2 \in \Re(G)$.

Definition 23.37. *A locally compact group G is said to be of type (J) if $\Re(G)$ is an amenable semigroup.*

If G is a locally compact abelian group, $\Re(G)$ is an abelian semigroup, hence an amenable semigroup and G is of type (J). If G is a compact group, let

$$m(f) = f(G),$$

$f \in l^\infty(\Re(G))$; m is a mean on $l^\infty(\Re(G))$. Then for $f \in l^\infty(\Re(G))$ and $K \in \Re(G)$, $m(_K f) = f(KG) = f(G) = m(f)$, that is, m is left invariant. Therefore every compact group is of type (J).

Proposition 23.38. *Let G be a group of type (J), H a locally compact group, and Φ an open continuous homomorphism of G onto H. Then H is of type (J).*

Proof. By continuity of Φ we may define

$$\Phi' : \Re(G) \to \Re(H)$$

$$K \mapsto \{\Phi(x) : x \in K\}.$$

As Φ is a homomorphism, for $K_1, K_2 \in \Re(G)$,

$$\Phi'(K_1 K_2) = \{\Phi(x_1 x_2) : x_1 \in K_1, x_2 \in K_2\}$$
$$= \{\Phi(x_1)\Phi(x_2) : x_1 \in K_1, x_2 \in K_2\}$$
$$= \Phi'(K_1)\Phi'(K_2),$$

that is, Φ' is a semigroup homomorphism. As Φ is open and surjective, Φ' is surjective ([B TG] Chap. I, Section 10, No. 4). The homomorphic image $\Re(H)$ of the amenable semigroup $\Re(G)$ under Φ' is an amenable semigroup, and H is of type (J). □

Proposition 23.39. *Every open subgroup H in a group G of type (J) is of type (J).*

Proof. If $K \in \Re(G)$, there exist $a_1, \ldots, a_n \in G$ such that $K \subset \cup_{i=1}^n Ha_i$ and $Ha_i \neq Ha_j$ whenever $i, j = 1, \ldots, n$ with $i \neq j$. For each $f \in l^\infty(\Re(H))$, we may define

$$f'(K) = \sum_{i=1}^n f(Ka_i^{-1} \cap H)/n.$$

For every $C \in \Re(H)$, we have $C \in \Re(G)$ and

$$(_C f)'(K) = \sum_{i=1}^n {}_C f(Ka_i^{-1} \cap H)/n$$
$$= \sum_{i=1}^n f(CKa_i^{-1} \cap H)/n = {}_C f'(K).$$

If M is a left invariant mean on $l^\infty(\Re(G))$, let

$$M'(f) = M(f'),$$

$f \in l^\infty(\Re(H))$; M is a mean on $l^\infty(\Re(H))$.

For all $f \in l^\infty(\Re(H))$ and $C \in \Re(H)$ we have

$$M'(_C f) = M(_C f') = M(f') = M'(f). \qquad \square$$

Proposition 23.40. *If H is a direct union $(H_i)_{i \in I}$ of compact open normal subgroups of a locally compact group G such that G/H is of type (J), then G itself is of type (J).*

Proof. As H is open, G/H is a discrete group. Let M be a left invariant mean on $l^\infty(\mathfrak{F}(G/H))$. We consider a system of representatives $E = \{x_l : l \in L\}$ for G/H. If $i \in I$, $f \in l^\infty(\Re(G))$ and $K \in \Re(G)$, let

$$f^{(i)}(K) = f(KH_i)$$

and define $f^{(i)'} \in l^\infty(\mathfrak{F}(G/H))$ by putting

$$f^{(i)'}(\{\dot{x}_l : l \in L_0\}) = f^{(i)}(\{x_l : l \in L_0\})$$

for every finite subset $\{\dot{x}_l : l \in L_0\}$ in G/H. For every $i \in I$, we may define a mean M_i on $l^\infty(\Re(G))$ by putting

$$M_i(f) = M(f^{(i)'}),$$

$f \in l^\infty(\Re(G))$. As H is open, for every $K \in \Re(G)$ there exist $l_1, \ldots, l_m \in L$ and compacta $K_1, \ldots, K_m$ in H such that $K = \cup_{k=1}^m x_{l_k} K_k$. Let $C = \{\dot{x}_{l_k} : k = 1, \ldots, m\}$. For $i \in I$, any finite subset $F = \{x_{l'_1}, \ldots, x_{l'_n}\}$ of E and any $f \in l^\infty(\Re(G))$, by normality we obtain

$$(_K f)^{(i)'}(\{\dot{x}_{l'_p} : p = 1, \ldots, n\}) = (_K f)^{(i)}(F) = {_K f}(FH_i)$$

$$= {_K f}(H_i F) = f\left(\bigcup_{p=1}^n KH_i x_{l'_p}\right)$$

$$= f\left(\bigcup_{p=1}^n \bigcup_{k=1}^m x_{l_k} H_i K_k x_{l'_p}\right).$$

There exists $i_0 \in I$ such that, for every $i \in I$ with $i \succ i_0$, $\cup_{k=1}^m K_k \subset H_i$. Therefore

$$(_K f)^{(i)'}(\{\dot{x}_{l'_p} : p = 1, \ldots, n\}) = f\left(\bigcup_{p=1}^n \bigcup_{k=1}^m x_{l_k} x_{l'_p} H_i\right)$$

$$= f^{(i)}(\{\dot{x}_{l_k} \dot{x}_{l'_p} : k = 1, \ldots, m; p = 1, \ldots, n\})$$

$$= f^{(i)}(\{\dot{x}_{l_k} \dot{x}_{l'_p} : k = 1, \ldots, m; p = 1, \ldots, n\})$$

$$= {_C f^{(i)}}(\{\dot{x}_{l'_p} : p = 1, \ldots, n\}),$$

that is,
$$({}_Kf)^{(i)'} = {}_cf^{(i)'}.$$

Hence if $K \in \Re(G)$, there exists $i_0 \in I$ such that, for every $i \in I$ with $i \succ i_0$ and every $f \in l^\infty(\Re(G))$,
$$M_i({}_Kf) = M\big(({}_Kf)^{(i)'}\big) = M\big({}_cf^{(i)'}\big) = M\big(f^{(i)'}\big) = M_i(f).$$

Any w*-cluster point of (M_i) is an invariant mean on $l^\infty(\Re(G))$. □

Proposition 23.41. *Every locally compact group G of type (J) is amenable.*

Proof. For $f \in \mathscr{C}(G)$, let
$$f'(K) = \frac{1}{|K|} \int_K f(x)\, dx \quad \text{if} \quad K \in \Re(G), \mathring{K} \neq \varnothing$$
$$= 0 \quad \text{if} \quad K \in \Re(G), \mathring{K} = \varnothing;$$

$f' \in l^\infty(\Re(G))$. Let $f \in \mathscr{C}(G)$ and $a \in G$. For every $K \in \Re(G)$ with $\mathring{K} \neq \varnothing$,
$$({}_af)'(K) = \frac{1}{|K|}\int_K {}_af(x)\, dx = \frac{1}{|aK|}\int_{aK} f(x)\, dx$$
$$= f'(aK) = {}_{\{a\}}f'(K)$$

hence $({}_af)' = {}_{\{a\}}f'$. Let M be a left invariant mean on $l^\infty(\Re(G))$. We define
$$N(f) = M(f'),$$

$f \in \mathscr{C}(G)$; N is a mean on $\mathscr{C}(G)$. For all $f \in \mathscr{C}(G)$, $a \in G$,
$$N({}_af) = M\big(({}_af)'\big) = M\big({}_{\{a\}}f'\big) = M(f') = N(f).$$

So $\text{LIM}(\mathscr{C}(G)) \neq \varnothing$ and G is amenable by Theorem 4.19. □

Recall that the discrete free group on two generators of order 2 is amenable by Proposition 14.2.

Proposition 23.42. *The discrete free group G on two generators a and b of order 2 is not of type (J).*

Proof. Every $x \in G$ may be written in the form $x = a^{\varepsilon_1} b^{\varepsilon_2} \cdots a^{\varepsilon_{n-1}} b^{\varepsilon_n}$ where $\varepsilon_2 = \cdots = \varepsilon_{n-1} = 1$ and $\varepsilon_1, \varepsilon_n = 0$ or 1. Let $l(x) = \sum_{i=1}^n \varepsilon_i$. If $F \in \mathfrak{F}(G)$, let $\delta(F) = \sup\{l(x) : x \in F\}$.

We consider $F_1 = \{a, ab\}$ and $F_2 = \{b, ba\}$. Suppose that $F = F_1 F_1' \cap F_2 F_2' \neq \emptyset$ with $F_1', F_2' \in \mathfrak{F}(G)$. There exists $x = a^{\varepsilon_1} \cdots b^{\varepsilon_n} \in F$ such that $\delta(F) = l(x)$. As $x \in F_1 F_1'$, necessarily $\varepsilon_1 = 1$. As $x \in F_2 F_2'$, we should also have $\varepsilon_1 = 0$. Therefore the right ideals generated by F_1 and F_2 are disjoint, and then Proposition 23.29 implies that $\mathfrak{F}(G)$ is a nonamenable semigroup. □

Remark 23.43. Not every extension of a type (J) group by a type (J) group is of type (J).

The discrete free group G on two generators a and b of order 2 considered in Proposition 23.42 contains the abelian subgroup generated by ab (or ba). This subgroup is normal, and the corresponding quotient group consisting of two elements is of type (J).

NOTES

Reiter [455] proved that the relativized forms of (P_1) and (P_1^*) characterize amenability for a closed subgroup of a locally compact group. He established general properties of the spaces $J_r(G, H), J_l(G, H)$ [453] and studied the subject in the more general setting of Segal algebras [455]. Properties of □ were established and widely used by Derighetti [129, 131]. The necessity of the condition given in 23.2 for amenability of the subgroup was shown by Reiter [453]. Derighetti [129] proved the sufficiency; also see Derighetti [130], Rindler [467], and Losert [371]. Rindler [467] showed 23.3. The discrete version of 23.4 is due to Flory [186]. Derighetti [131] demonstrated 23.5. Reiter [453] established 23.6–8. Rindler [467] proved a partial converse of 23.6: Let G be a locally compact group admitting a closed normal subgroup H for which G/H is infinite. If H is nonamenable, there exists a closed subspace X of $L^1(G/H)$ such that $f_a \in X$ whenever $f \in X$, $a \in H$, and $J_H(X)$ is not closed in $L^1(G/H)$. Derighetti [127, 128] obtained general relativization properties concerning in particular weak containment of the trivial representation of a closed subgroup in the restrictions of given unitary continuous representations of the group. Amenability of a closed subgroup H in a locally compact group G may be characterized in terms of approximate units properties. Reiter [454, 455] proved the following statements: If H is amenable, then $J_r(G, H)$ [resp. $J_l(G, H)$] has (multiple) approximate right [resp. left] units bounded by 2; if $J_r(G, H)$ [resp. $J_l(G, H)$] has bounded approximate left [resp. right] units, then H is amenable. In the case that H is normal, H is amenable if and only if

$J_l(G, H) = J_r(G, H)$ has multiple approximate left or right units; these units may be chosen to be hermitian and bounded by 2. As $J_l(G,G) = L^0(G)$, the latter statement generalizes Theorem 10.1. Derighetti [131] demonstrated that, in case H is nonnecessarily normal, amenability of H is insured if (a) $J_r(G, H)$ has bounded (multiple) approximate right units and (b) for every compact subset K of H and $\varepsilon > 0$, there exists $\varphi \in \mathscr{K}(G) \cap P^1(G)$ such that $\|J_{H,\rho}(_a\varphi - \varphi)\|_{L^1(G/H,\mu_\rho)} < \varepsilon$ whenever $a \in K$.

Properties 23.10–12 are due to Chou [86]. Rosenblatt [483] quoted example 23.13 and made further comments.

Von Neumann [414] had noticed that the rotation group of $\mathbf{R}^n$ contains F_2 in the nonamenable case $n = 3$, whereas it does not in the amenable cases $n = 1, 2$; the so-called von Neumann conjecture—not explicitly formulated by that author (it should more accurately be termed Day's conjecture [546])—says that a group is nonamenable if and only if it admits the closed subgroup F_2. Nonexistence of F_2 as a closed subgroup is a necessary condition for amenability of a locally compact group by Propositions 14.1 and 13.3. It is sufficient in case of G being almost connected as was established in Theorem 14.25. The sufficiency holds in other important particular instances thanks to the deep result that had been conjectured by Bass and Serre and established by Tits [538]: A linear group over a field of characteristic 0 either has a nonabelian free subgroup or admits a solvable subgroup of finite index. De la Harpe [251] gave interesting proofs for some special cases of Tits' results. One has the following fact. Let G be a subgroup of the group of all orthogonal linear transformations of $\mathbf{R}^n$ ($n = 1, 2$). Then G is amenable if and only if it does not contain F_2; see Wagon [546]. Making use of the dichotomy established by Tits, Deligne and Sullivan [124] showed that, for $n \in \mathbf{N}^*$, there exists a free nonabelian group of isometries acting freely on the sphere S_{2n+1}. See Borel [47] for a generalization to nontrivial, semisimple, connected compact Lie groups. Mycielski [408] had asserted that there would exist a nonelementary discrete group admitting no free nonabelian subgroup if the Burnside conjecture did hold, that is, there does not exist a finitely generated infinite group for which all nontrivial generators have prime order. We recall a precise formulation of Burnside's conjecture: Let $A_1, \ldots, A_m$ be a finite set of independent operations satisfying a system of relations $r^n = 1$ where $n \in \mathbf{N}^*$, r being any operation generated by $A_1, \ldots, A_m$. Is a group defined in this way necessarily finite? Adian and Novikov disproved the Burnside conjecture in case $m = 2$, and n admits an odd divisor that is at least 665: In the variety of all torsion groups of a given odd exponent of the form kn, where $k > 1$, $n \geq 665$, and k, n are relatively prime, there are infinitely many different finitely generated, nonabelian simple groups; see [2]. Mycielski [409] asked whether the nonamenable, infinite, free group on two generators in the variety defined by the equation $x^{665} = e$ fails to admit nonabelian free subgroups. Starting from the fact that exponentially bounded groups are necessarily amenable, Cohen [89] introduced a sharp growth condition: Let $\{x_1, \ldots, x_t : r_1, r_2, \ldots\}$ be a presentation of a discrete group G with $1 < t < \infty$; $G = F/H$ where F is the free group generated by $x_1, \ldots, x_t$ and H is its normal subgroup generated by $r_1, r_2, \ldots$. If $n \in \mathbf{N}$, let E_n be the set of all words of length n in F; let also $c_n = \text{card}(E_n \cap H)$. Then $c = \lim_{n \to \infty} (\sum_{i=0}^{n} c_i)^{1/n}$ exists and $c \in [1, 2t - 1]$. Cohen proved that G is amenable if and only if $c = 2t - 1$. By different methods Grigorčuk [231] independently obtained the same conclusion. Cohen formulated the property in the language of entropy. Ol'šanskiĭ proved the existence of an infinite group such that every proper subgroup is infinite cyclic [420]. He announced that Grigorčuk's criterion may be applied to obtain a nonamenable group that does not contain a free nonabelian subgroup and solved the von Neumann conjecture negatively [421].

Müller-Römer [406] studied contractible groups exhaustively and established 23.15; also see [405].

Rindler proved 23.17 [466] and 23.18 [469]. In [473] Rindler examined the property for compact groups and abelian groups. He gave more general versions in [471, 472].

Amenability of recurrent groups was established by Azencott [24]. Our proof of 23.19 is taken from Guivarc'h, Keane, and Roynette [243]; also see Derriennic and Guivarc'h [134], and Roynette [490]. An alternative demonstration is due to Zimmer [580]. The discrete case was studied by Avez [22]. He gave a direct proof of amenability for the discrete recurrent groups in [23]. For more information on recurrent groups we refer to Crépel [107]. A transient, that is, an aperiodic nonrecurrent probability measure μ on a locally compact completely separable group G is said to be of type I if 0 is the only cluster point of the set $\{\delta_x * \sum_{n=0}^{\infty} \mu^{*n} : x \in G\}$; it is called of type II in the opposite situation. Elie [156] established a characterization of the amenable, almost connected groups that are completely separable and admit a measure of type II.

As noted already in Section 12, growth conditions were considered thoroughly by Guivarc'h [241] and Jenkins [283, 285]. The locally compact group G is said to have polynomial growth of degree d if there exists a polynomial p of degree d such that, for every compact neighborhood V of e in G, there exists $\alpha_V > 0$ with $|V^n| \leq \alpha_V p(n)$ whenever $n \in \mathbf{N}^*$. The locally compact group G is said to be of exponential growth if, for every compact neighborhood V of e, there exists $t_V \in]1, \infty[$ such that $|V^n| \geq t_V^n$ whenever $n \in \mathbf{N}^*$; the group is not exponentially bounded. Hirsch and Thurston [264] gave a proof showing that, for discrete groups G, existence of a finite generating subset $S = S^{-1}$ satisfying $\liminf_{n \to \infty} (\text{card } S^{n+1}/\text{card } S^n) = 1$ implies existence of a left invariant mean on $l^{\infty}(G)$. Via considerations on proximal flows Glasner [203] established amenability of any finitely generated discrete group that is not of exponential growth. See Hulanicki [272] for earlier results on polynomial growth. Amenable groups may have exponential growth. A survey on these topics is Crépel and Lacroix [108]. The (solvable) affine group "$ax + b$" of the real line is amenable and has exponential growth. For a global study on the harmonic analysis of the affine group we refer to Khalill [313]. Consideration has been given to other properties that are related to growth conditions for connected locally compact groups G. The group G is said to be of type (NFS) if it does not admit elements a, b generating a free subsemigroup S that is uniformly discrete (i.e., there exists a neighborhood U of e such that $sU \cap tU = \emptyset$ whenever $s, t \in S$, $s \neq t$); G is said to be of type (R) if it admits a compact normal subgroup K such that G/K is a Lie group for which the eigenvalues of $Ad\ \dot{x}$ have absolute value 1 whenever $\dot{x} \in G/K$. Jenkins [283, 285] proved that both (NFS) and (R) characterize polynomial growth for connected completely separable locally compact groups. He also showed that these groups have either polynomial growth or exponential growth. We mention some important previous results. Wolf [559] had considered finitely generated solvable groups G that are polycyclic, that is, there exist normal subgroups G_i ($i = 0, 1, \ldots, n$) in G such that $G_0 = \{e\}$, $G_n = G$, $G_i \subset G_{i+1}$, and G_{i+1}/G_i is cyclic whenever $i = 0, 1, \ldots, n - 1$. Either G is almost nilpotent (i.e., admits a nilpotent subgroup of finite index) and has polynomial growth, or G is not almost nilpotent and has exponential growth. Wolf established the properties in his studies on the fundamental group of a riemannian manifold with nonpositive curvature. Milnor [390] proved that any finitely generated solvable group that is not polycyclic has exponential growth. From these statements Tits [538] deduced that every finitely generated linear group has either polynomial growth or exponential growth. Rosenblatt [479] could show that every finitely generated polycyclic group is almost nilpotent or admits a free subsemigroup on two generators. Chou [86] proved that a finitely generated group belonging to EG either is almost nilpotent or has exponential growth. Gromov [232] demonstrated that any finitely generated group with polynomial growth is almost nilpotent. Also see Gromov [233] for related results on riemannian manifolds. Jenkins' characterization of polynomial growth via (R) was established for connected Lie groups by Guivarc'h [241] who showed that these groups are of type (R) if and only if they as well as all their quotient groups are amenable unimodular. Baldi [27] proved that any connected Lie group is recurrent if and only if it has polynomial growth of order 2 at most. Jenkins [287] also examined growth conditions in relation to fixed point properties. In [289] he established interactions of growth conditions with existence properties of invariant linear functionals generalizing invariant means.

Jenkins [286] showed that a locally compact connected group G is exponentially bounded if and only if the following condition holds: For every compact subset K of G having nonvoid interior and every $\varepsilon > 0$, there exists $n \in \mathbf{N}^*$ such that $|K^{n+1} \triangle K^n| < \varepsilon |K^n|$. Emerson and Greenleaf [170] termed a locally compact group G to be strongly amenable in case, for every symmetric relatively compact open neighborhood K of e, $\lim_{n \to \infty} |K^{n+1}|/|K^n| = 1$. Compact groups are obviously strongly amenable. Emerson and Greenleaf proved all locally compact abelian groups to be strongly amenable via combinatorial arguments. Milnes [386] exhibited a first example of an amenable σ-compact group G admitting a compact symmetric neighborhood U of e with $(U^n) \notin \mathscr{A}_c(G)$. Jenkins [283] showed that, for amenable, connected, completely separable groups, strong amenability implies (R). He also considered the semidirect product of $\mathbf{R}$ with $\mathbf{R}^2$ given by the homomorphism $\varphi \colon \mathbf{R} \to \mathrm{Aut}(\mathbf{R}^2)$ where $\varphi(t)\,(x, y) = (e^t x, e^{-t} y)$ for $t \in \mathbf{R}$ and $(x, y) \in \mathbf{R}^2$. This group is amenable unimodular connected but not of type (R).

Lipsman proved (KS) for $SL(2, \mathbf{C})$ [359], $SL(n, \mathbf{C})$ [361], and also a Lorentz group [360]. Cowling [104] established (KS) for every semisimple, connected Lie group with finite center. Leinert [346] considered the property for F_2. Rajagopalan [446] proved the L^p-conjecture for amenable groups in the unimodular case, the general statement of which appears in Lohoué [369].

Palmer [425] gave 23.23. For related work see Hulanicki [269, 270, 273], Jenkins [288], and also Fountain, Ramsay, and Williamson [188]. Jenkins showed that any discrete group containing a free nonabelian subsemigroup on two generators is nonhermitian [278]; another proof was given in [280]. Related topics were studied in Jenkins [282]. Poguntke [438] proved that connected, nilpotent Lie groups are hermitian. Ludwig [373] established that connected groups having polynomial growth and compact extensions of nilpotent groups are hermitian. The first example of a nonhermitian amenable group was given by Jenkins [277]. Leptin [357] showed that the affine group "$ax + b$" is a hermitian group.

A locally compact group G is said to have the Wiener property (W) and G is said to be a Wiener group if, for every closed ideal I in $L^1(G)$, there exists a nondegenerate continuous unitary representation T of $L^1(G)$ such that $I \subseteq \mathcal{K}\!\mathit{er}\,T$. Wiener had established the property for $\mathbf{R}$ [555, 556]. The classical Wiener tauberian theorem due to Godement [206, 207] and Segal [510] asserts that any locally compact abelian group has property (W). Leptin [356] proved that all connected nilpotent Lie groups and all semidirect products of abelian groups have property (W). Gangolli [195] showed that every locally compact group that is a motion group, that is, a semidirect product of a compact subgroup and a closed normal abelian subgroup is a hermitian Wiener group. Nevertheless, Weit [553] proved that the one-sided version of Wiener's property (corresponding to right ideals) fails for the motion group of the plane. Ludwig [373] established that discrete solvable groups have property (W). Ehrenpreis and Mautner [155] demonstrated that $SL(2, \mathbf{R})$ is not a Wiener group. Duflo proved that no noncompact, semisimple, connected Lie group is a Wiener group; see [356]. Leptin and Poguntke [358] produced a connected solvable (amenable) Lie group that is neither a hermitian group nor a Wiener group; also see [425]. All known Wiener groups are amenable.

The fundamental properties of amenable semigroups were established by Day [115, 120] and Granirer [210]. An early reference is also Witz [558]. Invariant means on $l^\infty(\mathbf{N})$ were studied by Raimi [443, 444] and Chou [76]. For the proofs of 23.25–27, 32 we refer to Jenkins [279]. The demonstrations of

23.32 in the discrete case and 23.33 are due to Frey; also see Wilde and Witz [557]. The latter authors established 23.24, 30–31. The example 23.34 was considered by Hochster [265]. Jenkins [281] gave 23.35. Related topics were studied by Lau [335]. Let G be a Hausdorff topological group admitting a dense subsemigroup S. A bounded complex-valued function f on S is said to be uniformly continuous if $x \mapsto {}_x f$ and $x \mapsto f_x$ are continuous on S. Lau [334] showed that sufficient conditions for the existence of a left invariant mean on the space of uniformly continuous bounded functions on S are: (a) $\text{LIM}(\mathscr{C}(G)) \neq \varnothing$; (b) $\text{LIM}(\mathscr{L}\mathscr{U}\mathscr{C}(G)) \neq \varnothing$ and S has the finite intersection property for right ideals; and (c) $\text{LIM}(\mathscr{U}\mathscr{C}(G)) \neq \varnothing$ and S has the finite intersection property for right ideals and for left ideals. Duncan and Namioka [149] showed 23.36; also see Paterson [428] and Wordingham [574] for further properties concerning in particular inverse semigroups. Jenkins [281] introduced the class of groups which we call of type (J) and he termed σ-amenable groups. He proved 23.38–43. Ganesan [194] studied the class of all locally compact groups G, called P-amenable, for which the discrete semigroup $P^1(G)$ is amenable. The locally compact abelian groups and the compact groups belong to this class, whereas F_2 does not. Further important results on amenable semigroups were obtained by Junghenn [299, 300, 301], Lau [336], Wong [565], Berglund, Junghenn, and Milnes [34], Paterson [429], and Klawe [320]. Recent publication: Veršik [597].

24. GENERALIZED NOTIONS OF AMENABILITY

In this last section we give an outlook on various extensions of the notion of amenability; each of the three subsections is followed immediately by the corresponding notes. We first consider several types of invariant means, other than the classical ones characterizing amenable groups, and study their properties in connection with amenability. Many aspects of amenable groups may be generalized for actions of amenable transformation groups. Whereas we merely list different types of amenable actions studied by various authors, we give a few details on ergodicity in the context of amenable σ-compact groups. Finally we quote some of the outstanding results on amenable algebras; that important subject is probably far from having attained its final achievement.

A. Generalized Notions of Invariant Means

We outline the process leading to generalized notions of invariant means focusing on their relationship to amenability.

Let A be a Banach algebra and let $X = A^*$. On X we consider the canonical right Banach A-module structure given by

$$\langle uv, f \rangle = \langle v, fu \rangle$$

for $u, v \in A, f \in X$. The Banach subspace Y of X is said to be A-invariant if $fu \in Y$ whenever $u \in A$ and $f \in Y$. As in Section 2G, for $f \in Y$ and $F \in Y^*$, we then define $F \star f \in X$ by putting

$$\langle u, F \star f \rangle = \langle fu, F \rangle$$

for $u \in A$; $\|F \star f\|_X \le \|F\|_{Y^*}\|f\|_X$. We say that Y is A-introverted if $F \star f \in Y$ whenever $f \in Y$ and $F \in X^*$. As in (2.16) we have $F \star (fu) = (F \star f)u$ whenever $u \in A, f \in Y, F \in Y^*$. We consider an Arens multiplication $\odot$ in Y^* defined by

$$\langle f, F_1 \odot F_2 \rangle = \langle F_2 \star f, F_1 \rangle$$

for $F_1, F_2 \in Y^*, f \in Y$.

Let G be a locally compact group. We consider the Banach subspace $\mathscr{M}^1(G)^*$ of $L^\infty(G)$.

Definition 24.1. *Let G be a locally compact group. The Borel subset A of G is called topologically thick if, for every compact subset K of G and every $\alpha \in\,]0, 1[$, there exists $\mu \in M^1(G)$ such that $\nu * \mu(A) > \alpha$ whenever $\nu \in M^1_c(G)$ with $\operatorname{supp} \nu \subset K$.*

Proposition 24.2. *Let G be an amenable group and let A be a Borel subset of G. The following properties are equivalent:*

(i) *A is topologically thick.*
(ii) *There exists an $M^1(G)$-invariant mean N on $\mathscr{M}^1(G)^*$ [i.e., $\langle f\mu, N \rangle = \langle f, N \rangle$ whenever $f \in \mathscr{M}^1(G)^*, \mu \in M^1(G)$] such that $\langle \Theta_A, N \rangle = 1$ where $\Theta_A \in \mathscr{M}^1(G)^*$ is defined by $\langle \mu, \Theta_A \rangle = \langle 1_A, \mu \rangle = \mu(A), \mu \in \mathscr{M}^1(G)$.*
(iii) *There exists $M \in \operatorname{TLIM}(L^\infty(G))$ that is concentrated on A.*

Proof. (i) $\Rightarrow$ (ii)

By Proposition 6.7 there exists a net (μ_i) in $M^1(G)$ such that $\lim_i \|\mu * \mu_i - \mu_i\| = 0$ whenever $\mu \in M^1(G)$. The net (μ_i) admits a subnet (μ_{i_j}) converging to a mean N_1 on $\mathscr{M}^1(G)^*$ in the w^*-topology. For all $f \in \mathscr{M}^1(G)^*$ and $\mu \in M^1(G)$,

$$\langle f\mu, N_1 \rangle = \lim_j \langle f\mu, \mu_{i_j} \rangle = \lim_j \langle \mu_{i_j}, f\mu \rangle = \lim_j \langle \mu * \mu_{i_j}, f \rangle$$

$$= \lim_j \langle \mu_{i_j}, f \rangle = \lim_j \langle f, \mu_{i_j} \rangle = \langle f, N_1 \rangle,$$

that is, N_1 is $M^1(G)$-invariant.

If $\sigma_1 = (K_1, \varepsilon_1), \sigma_2 = (K_2, \varepsilon_2) \in \Re(G) \times]0,1[$, say $\sigma_1 \prec \sigma_2$ in case $K_1 \subset K_2$ and $\varepsilon_2 < \varepsilon_1$. As A is topologically thick, to every $\sigma = (K, \varepsilon) \in \Re(G) \times]0,1[$ we may associate $\nu_\sigma \in M^1(G)$ such that $\nu * \nu_\sigma(A) > 1 - \varepsilon$ whenever $\nu \in M_c^1(G)$ with $\operatorname{supp} \nu \subset K$. By density of $M_c^1(G)$ in $M^1(G)$, for every $\nu \in M^1(G)$, $\lim_\sigma \nu * \nu_\sigma(A) = 1$, hence $\lim_\sigma \langle \nu_\sigma, \Theta_A \nu \rangle = \lim_\sigma \langle \nu * \nu_\sigma, \Theta_A \rangle = 1$.

The net (ν_σ) admits a subnet (ν_{σ_τ}) converging to a mean N_2 in the w*-compact unit ball of $\mathcal{M}^1(G)^{**}$. As in Proposition 4.26(a), $N = N_1 \odot N_2$ is an $M^1(G)$-invariant mean on $\mathcal{M}^1(G)^*$. Moreover for every $\mu_0 \in M^1(G)$,

$$1 = \lim_\tau \langle \nu_{\sigma_\tau}, \Theta_A \mu_0 \rangle = \langle \Theta_A \mu_0, N_2 \rangle = \langle \mu_0, N_2 \star \Theta_A \rangle.$$

Therefore if $\mu = \alpha'\mu' - \alpha''\mu'' \in \mathcal{M}_\mathbf{R}^1(G)$ with $\alpha', \alpha'' \in \mathbf{R}_+$ and $\mu', \mu'' \in M^1(G)$, we have

$$\langle \mu, N_2 \star \Theta_A \rangle = \langle \alpha'\mu' - \alpha''\mu'', N_2 \star \Theta_A \rangle = \alpha' - \alpha'' = \mu(G) = \langle 1_G, \mu \rangle;$$

$N_2 \star \Theta_A = 1_G$ and

$$\langle \Theta_A, N \rangle = \langle \Theta_A, N_1 \odot N_2 \rangle = \langle N_2 \star \Theta_A, N_1 \rangle = \langle 1_G, N_A \rangle = 1.$$

(ii) $\Rightarrow$ (i)

Extend N to a mean N_1 on $L^\infty(G)^*$ via the Hahn–Banach theorem. By Proposition 3.3 there exists a net (φ_i) in $P^1(G)$ converging to N_1 in the w*-topology of $L^\infty(G)^*$. If A were not topologically thick, there would exist a compact subset K of G and $\alpha \in]0,1[$ such that to every $\mu \in M^1(G)$ there corresponds $\nu \in M_c^1(G)$ with $\operatorname{supp} \nu \subset K$ and

$$\langle \nu * \mu, \Theta_A \rangle = \nu * \mu(A) \leq \alpha.$$

But, for every $\nu \in M^1(G)$,

$$\lim_i |1 - \langle \nu * \varphi_i, \Theta_A \rangle| \leq \lim_i |1 - \langle \varphi_i, \Theta_A \rangle| + \lim_i |\langle \varphi_i - \nu * \varphi_i, \Theta_A \rangle|$$

$$= \lim_i |1 - \langle \Theta_A, \varphi_i \rangle| + \lim_i |\langle \Theta_A - \Theta_A \nu, \varphi_i \rangle|$$

$$= |1 - \langle \Theta_A, N_1 \rangle| + |\langle \Theta_A - \Theta_A \nu, N_1 \rangle|$$

$$= |1 - \langle \Theta_A, N \rangle| + |\langle \Theta_A - \Theta_A \nu, N \rangle| = 0;$$

so a contradiction would arise. Therefore A must be topologically thick.

$$(ii) \Rightarrow (iii)$$

There exists a net $(\nu_i)_{i \in I}$ in $M^1(G)$ that is w^*-converging to N. Choose $\varphi_0 \in P^1(G)$ and let $\varphi_i = \varphi_0 * \nu_i \in P^1(G), i \in I$. By Lemma 6.5 a subnet (φ_{i_j}) of (φ_i) converges to an element M of $\mathrm{TLIM}(L^\infty(G))$ in the w^*-topology of $L^\infty(G)^*$. Moreover

$$\langle 1_A, M \rangle = \lim_j \langle 1_A, \varphi_{i_j} \rangle = \lim_j \langle 1_A, \varphi_0 * \nu_{i_j} \rangle = \lim_j \langle \varphi_0 * \nu_{i_j}, \Theta_A \rangle$$

$$= \lim_j \langle \nu_{i_j}, \Theta_A \varphi_0 \rangle = \langle \Theta_A \varphi_0, N \rangle = \langle \Theta_A, N \rangle = 1.$$

$$(iii) \Rightarrow (ii)$$

Let $(\varphi_i)_{i \in I}$ be a net in $P^1(G)$ converging to M in the w^*-topology of $L^\infty(G)^*$. Choose $\varphi_0 \in P^1(G)$ and let $\psi_i = \varphi_0 * \varphi_i \in P^1(G), i \in I$. A subnet (ψ_{i_j}) of (ψ_i) is w^*-converging to a mean N on $L^\infty(G)$. For all $f \in \mathcal{M}^1(G)^*$ and $\mu \in M^1(G)$, by (2.12) we have

$$\langle f\mu, N \rangle = \lim_j \langle f\mu, \psi_{i_j} \rangle = \lim_j \langle \mu * \psi_{i_j}, f \rangle = \lim_j \langle \varphi_{i_j}, (\mu * \varphi_0)^* * f \rangle$$

$$= \langle (\mu * \varphi_0)^* * f, M \rangle = \langle \varphi_0^* * f, M \rangle$$

$$= \lim_j \langle \varphi_{i_j}, \varphi_0^* * f \rangle = \lim_j \langle \varphi_0 * \varphi_{i_j}, f \rangle$$

$$= \langle f, N \rangle.$$

Moreover

$$\langle \Theta_A, N \rangle = \lim_j \langle \Theta_A, \psi_{i_j} \rangle = \lim_j \langle \Theta_A, \varphi_0 * \varphi_{i_j} \rangle = \lim_j \langle \varphi_0 * \varphi_{i_j}, 1_A \rangle$$

$$= \lim_j \langle \varphi_{i_j}, \varphi_0^* * 1_A \rangle = \langle \varphi_0^* * 1_A, M \rangle = \langle 1_A, M \rangle = 1. \quad \square$$

Let G be a locally compact abelian group and let $\hat{G}$ be its dual group. Recall that $A(G) \simeq \{\hat{f}: f \in L^1(\hat{G})\}$ (Section 19A). If $u \in P(G) \cap A(G)$ such that $u(e) = 1$, by Bochner's theorem ([HR] 33.3) there exists a unique element $v \in L^1_+(\hat{G})$ such that $u = \hat{v}$ and $\|v\|_1 = \|u\| = u(e) = 1$, that is, $v \in P^1(\hat{G})$. We now formulate a general definition.

Definition 24.3. *Let G be a locally compact group. We define* $P^1(\hat{G}) = P(G) \cap \{u \in A(G) : u(e) = 1\}$ *and the set* $\mathfrak{M}$ (PM(G)) [*resp.* IM($\hat{G}$)] *of all states M on* PM(G) ($= A(G)^*$) [*such that* $\langle uT, M \rangle$ ($= \langle Tu, M \rangle$) $= \langle T, M \rangle$ *whenever* $T \in$ PM(G) *and* $u \in P^1(\hat{G})$].

If G is a locally compact group, for every $u \in P^1(\hat{G})$, $\|u\| = u(e) = 1$ and $P^1(\hat{G})$ is the set of all ultraweak states on PM(G) ([171] 3.15; 2.6.2)). This set is w*-dense in $\mathfrak{M}$(PM(G)) ([218] Corollary 4; [410] Theorem 4.3). In particular, if G is a locally compact abelian group, $\mathscr{UC}(\hat{G})$ coincides with $A(G)$PM(G) and IM($\hat{G}$) coincides with TLIM($\mathscr{UC}(\hat{G})$), that is, with IM($\mathscr{UC}(\hat{G})$).

Proposition 24.4. *If G is any locally compact group,* IM($\hat{G}$) $\neq \emptyset$.

Proof. For $u \in A(G)$, we consider the continuous operator

$$t_u : \quad T \mapsto uT$$

$$\text{PM}(G) \to \text{PM}(G).$$

As $A(G)$ is a commutative algebra, $\{t_u^{\sim} : u \in P^1(\hat{G})\}$ is a commutative semigroup. For every $u \in P^1(\hat{G})$, $uP^1(\hat{G}) \subset P^1(\hat{G})$ and, as $P^1(\hat{G})$ is w*-dense in $\mathfrak{M}$(PM(G)), we have $t_u^{\sim}(\mathfrak{M}(\text{PM}(G))) \subset \mathfrak{M}(\text{PM}(G))$. Now the Markov–Kakutani fixed point theorem ([B EVT] Chap. IV, App. 1, théorème 1) implies the existence of $M \in \mathfrak{M}(\text{PM}(G))$ such that $t_u^{\sim} M = M$ whenever $u \in P^1(\hat{G})$, that is, $M \in$ IM($\hat{G}$). □

Definition 24.5. *Let G be a locally compact group. A pseudomeasure T on G is called weakly almost periodic if the operator*

$$t : \quad u \mapsto uT$$

$$A(G) \to \text{PM}(G)$$

is w-compact, that is, $\{uT : u \in A(G), \|u\|_{A(G)} \leq 1\}^-$ *is compact in the topology* $\sigma(\text{PM}(G), \text{PM}(G)^*)$. *We denote by* $\mathscr{W}(\hat{G})$ *the set of all weakly almost periodic pseudomeasures on G*.

Proposition 24.6. *If G is an amenable group,* $\mathscr{W}(\hat{G}) \subset \mathscr{UC}(\hat{G})$.

Proof. Let $T \in \mathscr{W}(\hat{G})$. As G is an amenable group, by Theorem 10.4 there exist approximate units (u_i) bounded by 1 in $A(G)$. Hence (u_iT) converges to T in the w*-topology of $\mathrm{PM}(G)$. As T is weakly almost periodic, a subnet $(u_{i_j}T)$ of (u_iT) converges to T in the w-topology of the von Neumann algebra $\mathrm{PM}(G)$, hence $T \in A(G)\mathrm{PM}(G) = \mathscr{U}\mathscr{C}(\hat{G})$. □

Proposition 24.7. *If G is any locally compact group, the restrictions of all elements in $\mathrm{IM}(\hat{G})$ to $\mathscr{W}(\hat{G})$ coincide.*

Proof. Let $M_0, M_1 \in \mathrm{IM}(\hat{G})$ and $T_0 \in \mathscr{W}(\hat{G})$. We put $c = \langle T_0, M_0 \rangle$. As $P^1(\hat{G})$ is w*-dense in $\mathfrak{M}(\mathrm{PM}(G))$, we may determine a net (u_i) in $P^1(\hat{G})$ such that $\lim_i \langle u_i, T \rangle = \langle T, M_0 \rangle$ whenever $T \in \mathrm{PM}(G)$. For every $u \in P^1(\hat{G})$,

$$\lim_i \langle u, u_iT_0 \rangle = \lim_i \langle u_i, uT_0 \rangle = \langle uT_0, M_0 \rangle = \langle T_0, M_0 \rangle = c.$$

So (u_iT_0) converges to $c 1_{A(G)}$ in the w*-topology of $\mathrm{PM}(G)$. As T_0 is weakly almost periodic, there exists a subnet $(u_{i_j}T_0)$ of (u_iT_0) that converges to $c 1_{A(G)}$ in the $\sigma(\mathrm{PM}(G), \mathrm{PM}(G)^*)$-topology. Finally notice that, for every $u \in P^1(\hat{G}), \langle T_0, M_1 \rangle = \langle uT_0, M_1 \rangle$; thus $\langle T_0, M_1 \rangle = c \langle 1_{A(G)}, M_1 \rangle = \langle T_0, M_0 \rangle$. □

We add some related results applying to amenable groups. These topics illustrate possibilities of generalizing standard methods. We freely transpose the notations of Section 2G. Let G be any locally compact group. If $u \in A(G)$ and $v \in B(G)$, we have $uv \in A(G)$. If $T \in \mathrm{PM}(G) = A(G)^*$, we may consider uT to belong to $B(G)^*$ with $\langle v, uT \rangle = \langle uv, T \rangle$; therefore $A(G)\mathrm{PM}(G) \subset B(G)^*$, so $\mathscr{U}\mathscr{C}(\hat{G}) \subset B(G)^*$ and $B(G)^{**} \subset \mathscr{U}\mathscr{C}(\hat{G})^*$. If $F \in \mathscr{U}\mathscr{C}(\hat{G})^*$, $T \in \mathrm{PM}(G)$, $u \in A(G)$, and $v \in A(G)$, we have $\langle v, F \star (uT) \rangle = \langle v(uT), F \rangle = \langle uv, F \star T \rangle = \langle v, u(F \star T) \rangle$. Thus $F \star (uT) = u(F \star T) \in A(G)\mathrm{PM}(G)$ and, by density, $F \star g \in \mathscr{U}\mathscr{C}(\hat{G})$ whenever $F \in \mathscr{U}\mathscr{C}(\hat{G})^*, g \in \mathscr{U}\mathscr{C}(\hat{G})$, that is, $\mathscr{U}\mathscr{C}(\hat{G})$ is introverted.

Proposition 24.8. *Let G be an amenable group and let i be the canonical embedding of $B(G)$ into $B(G)^{**}$.*

(a) *If $v \in B(G)$, the mapping*

$$F \mapsto i(v) \odot F$$

$$\mathscr{U}\mathscr{C}(\hat{G})^* \to \mathscr{U}\mathscr{C}(\hat{G})^*$$

is $w^* - w^*$ continuous.

(b) *If $v_1, v_2 \in B(G), i(v_1v_2) = i(v_1) \odot i(v_2)(= i(v_2) \odot i(v_1))$.*

Proof. (a) Let $u_0 \in A(G)$, $T_0 \in \text{PM}(G)$ and $T = u_0 T_0$. For every $F \in \mathcal{UC}(\hat{G})^*$,

$$\langle T, i(v) \odot F \rangle = \langle F \star T, i(v) \rangle = \langle u_0(F \star T_0), i(v) \rangle = \langle v, u_0(F \star T_0) \rangle$$
$$= \langle u_0 v, F \star T_0 \rangle = \langle u_0 v T_0, F \rangle = \langle vT, F \rangle.$$

Let now (F_j) be a net in $\mathcal{UC}(\hat{G})^*$ that is w*-converging to F. As $vT \in B(G)A(G)\text{PM}(G) \subset A(G)\text{PM}(G)$, we have

$$\lim_j \langle T, i(v) \odot F_j \rangle = \lim_j \langle vT, F_j \rangle = \langle vT, F \rangle = \langle T, i(v) \odot F \rangle.$$

(b) Let $u_0 \in A(G)$, $T_0 \in \text{PM}(G)$, and $T = u_0 T_0$. Then we have

$$\langle T, i(v_1) \odot i(v_2) \rangle = \langle i(v_2) \star T, i(v_1) \rangle = \langle u_0(i(v_2) \star T_0), i(v_1) \rangle$$
$$= \langle v_1, u_0(i(v_2) \star T_0) \rangle = \langle u_0 v_1, i(v_2) \star T_0 \rangle$$
$$= \langle u_0 v_1 T_0, i(v_2) \rangle$$
$$= \langle v_2, (u_0 v_1) T_0 \rangle = \langle u_0 v_1 v_2, T_0 \rangle = \langle v_1 v_2, u_0 T_0 \rangle = \langle v_1 v_2, T \rangle$$
$$= \langle T, i(v_1 v_2) \rangle. \qquad \square$$

Proposition 24.9. *Let G be an amenable group and let X be a closed introverted subspace of $\text{PM}(G)$. Then $Y = A(G)X$ is a closed introverted subspace of $\text{PM}(G)$.*

Proof. By Proposition 19.5, Y is a closed subspace of $\text{PM}(G)$. If $F \in Y^*$, extend F to an element F_1 of $\text{PM}(G)^*$ via the Hahn–Banach theorem. As X is introverted, for all $u \in A(G), T \in X$, we have

$$F_1 \star (uT) = u(F_1 \star T) \in A(G)X = Y. \qquad \square$$

Proposition 24.10. *If G is an amenable group and X is a closed introverted subspace of $\text{PM}(G)$, there exists an isometric isomorphism of $(A(G)X)^*$ onto the subspace of $\mathcal{L}(X)$ formed by all $A(G)$-multipliers.*

Proof. (a) By Proposition 24.9, $Y = A(G)X$ is a closed introverted subspace of $\text{PM}(G)$. We may define

$$\Phi : F \mapsto \Phi(F)$$

$$Y^* \to \mathcal{L}(X)$$

by putting $\langle u, \Phi(F)(T) \rangle = \langle uT, F \rangle$ for $T \in X, u \in A(G)$; $\|\Phi\| \leq 1$.

Let (u_i) be a net of approximate units bounded by 1 in $A(G)$. For all $u \in A(G)$, $T \in X$, and $F \in Y^*$, we have $\lim_i \|u_i(uT) - uT\|_{PM(G)} = 0$ and

$$\|\Phi(F)(uT)\| \geq \lim_i |\langle u_i, \Phi(F)(uT)\rangle|$$

$$= \lim_i |\langle u_i uT, F\rangle| = |\langle uT, F\rangle|.$$

Hence $\|\Phi(F)\| \geq \|F\|$ and then $\|\Phi\| = 1$.

(b) Let $F_1, F_2 \in Y^*$. For all $u \in A(G), T \in X$,

$$\langle u, \Phi(F_1 \odot F_2)(T)\rangle = \langle uT, F_1 \odot F_2\rangle = \langle F_2 \star (uT), F_1\rangle$$

$$= \langle u\Phi(F_2)(T), F_1\rangle = \langle u, \Phi(F_1) \circ \Phi(F_2)(T)\rangle,$$

that is, $\Phi(F_1 \odot F_2) = \Phi(F_1) \circ \Phi(F_2)$; Φ is a homomorphism of Y^* into the space of all $A(G)$-multipliers in $\mathscr{L}(X)$.

(c) If $S \in \mathscr{L}(X)$, for all $u \in A(G), T \in X$, we consider $\langle S^\sim(u), T\rangle = \langle u, S(T)\rangle$. If S is an $A(G)$-multiplier and F_0 is a w*-cluster point of $(S^\sim(u_i))$, then for all $u \in A(G)$, $T \in X$,

$$\langle u, S(T)\rangle = \lim_i \langle uu_i, S(T)\rangle = \lim_i \langle u_i, uS(T)\rangle$$

$$= \lim_i \langle u_i, S(uT)\rangle = \lim_i \langle S^\sim(u_i), uT\rangle$$

$$= \langle uT, F_0\rangle = \langle u, \Phi(F_0)(T)\rangle,$$

that is, $S = \Phi(F_0)$ and Φ is surjective. □

We state a general version of Proposition 19.17.

Corollary 24.11. *Let G be an amenable group and let X be a closed introverted subspace of $\mathscr{UC}(\hat{G})$. Then X^* is isometrically isomorphic to the space of $A(G)$-multipliers in $\mathscr{L}(X)$.*

Proof. As G is amenable, $A(G)$ has approximate units; hence $A(G)X = X$, X being closed. It suffices then to apply Proposition 24.10. □

NOTES

Mitchell [391] had considered thickness in discrete semigroups S characterizing subsets A that are large enough to support a left invariant mean: For every finite subset F in S there exists $a \in S$ such that $Fa \subset A$. Wong [563] established 24.2. If S is a locally compact semitopological semigroup and A is a Borel subset, Day [123] investigated 90 formally distinct thickness properties for A linked to left amenability conditions of the semigroup. The properties are variations of five fundamental notions that we list in increasing order of restriction on A: (a) A is thick in the sense of Mitchell; (b) A is topologically lumpy, that is, for every $\alpha \in \,]0, 1[$ and every $\nu \in M_c^1(S)$, there exists $a \in S$ such that $\nu \otimes \delta_a(A) > \alpha$; (c) A is topologically thick; (d) A is topologically substantial, that is, for every compact subset K in S, there exists $\mu \in M^1(S)$ such that $\mu \otimes \nu\, (A) = 1$ for every $\nu \in M_c^1(S)$ with $\operatorname{supp} \nu \subset K$; and (e) A is lumpy, that is, for every compact subset K in S, there exists $a \in S$ such that $Ka \subset A$. Lumpiness had been defined by Day in [122] whereas Wong had introduced (d) in [564]. Wong studied the subject extensively in [566, 567, 569, 570, 571]. Further statements concerning discrete semigroups are due to Klawe [319, 320] and Junghenn [299].

The study of invariant means on pseudomeasures was inaugurated by Dunkl and Ramirez [150, 151] who established explicit proofs for compact groups. The properties given in 24.4 and 24.6–7 were proved for general amenable groups by Granirer [218]. He elaborated interesting related properties that remain valid for arbitrary locally compact groups; also see [219]. The results stated in 24.8–11 are taken from Lau [340]. He made further comments concerning the set of invariant means on $\mathcal{UC}(\hat{G})$ in case G is amenable. Renaud [459] obtained general properties concerning invariant means on pseudomeasures. He showed that the mean is unique if and only if the group is discrete. See Chou [87] for a general study of $\operatorname{IM}(\hat{G})$ and also for the particular case of noncompact amenable σ-compact groups.

If X is a topological space, let $C(X)$ denote the vector space of all complex-valued (nonnecessarily bounded) functions defined on X. The space X is termed pseudocompact if $C(X)$ coincides with $\mathcal{C}(X)$, that is, every continuous function defined on X is bounded. Properties of pseudocompact groups may be found in Comfort and Ross [92]. Every countably compact space X is pseudocompact. Otherwise there would exist a continuous function f defined on X and a sequence (x_n) in X such that $|f(x_n)| > n$ whenever $n \in \mathbf{N}^*$; a subsequence of (x_n) would converge to an element of X and then f could not be continuous. For a topological group G, Argabright [15] considered invariant means on $C_\mathbf{R}(G)$. He reported that Glicksberg had shown the existence of a noncompact abelian topological group that is pseudocompact: Consider the subgroup G of an uncountable product of copies of $\mathbf{Z}_2$ that consists of all points having only countably many nonzero coordinates; then G is noncompact, but it is obviously countably compact and therefore also pseudocompact. So it admits a left invariant mean on $C(G) \simeq \mathcal{C}(G)$. Argabright proved that in the case that G is a locally compact group that is realcompact or discrete or abelian, $C(G)$ admits a left invariant mean if and only if G is compact. This statement was generalized by Moran [401]. No example of a nonpseudocompact group G admitting an invariant mean on $C(G)$ seems to be known.

Dixmier [138] had inaugurated the study of invariant means on spaces of vector-valued functions defined on a semigroup. Let S be a semigroup and E

a real Banach space. We denote by $l_\mathbf{R}^\infty(S, E^*)$ the Banach space of all bounded functions mapping S into E^*. In accordance with Corollary 3.4 a mean on $l_\mathbf{R}^\infty(S, E^*)$ is a linear mapping $N: l_\mathbf{R}^\infty(S, E^*) \to E^*$ such that $N(f)$ belongs to the w*-closure of $\text{co}\{f(x): x \in S\}$ whenever $f \in l_\mathbf{R}^\infty(S, E^*)$. If $f \in l_\mathbf{R}^\infty(S, E^*)$ and $a, x \in S$, we still write $_af(x) = f(ax)$. The mean N is called invariant if $N(_af) = N(f)$ whenever $f \in l_\mathbf{R}^\infty(S, E^*)$ and $a \in S$. If $f \in l_\mathbf{R}^\infty(S)$ and $\xi \in E$, we consider

$$f_\xi': x \mapsto \langle \xi, f(x) \rangle$$

$$S \to \mathbf{R}.$$

If M is a left invariant mean on $l_\mathbf{R}^\infty(S)$, one may define an invariant mean M' on $l_\mathbf{R}^\infty(S, E^*)$ by putting

$$\langle \xi, M'(f) \rangle = \langle f_\xi', M \rangle,$$

$f \in l_\mathbf{R}^\infty(S, E^*), \xi \in E$. The theory of invariant means was extended systematically to vector-valued functions on a semigroup by Husain and Wong [274]. They considered E more generally to be a locally convex Hausdorff space that is quasi-barrelled, that is, strongly bounded subsets of E^* are equicontinuous. This generalization was carried out especially in the case of a locally compact group G. Denote by $\mathscr{L}^\infty(G, E^*)$ the set of all functions $f \in l^\infty(G, E^*)$ such that, for every $\xi \in E$, the mapping $x \mapsto \langle \xi, f(x) \rangle$ is λ-measurable. Let $N(G, E^*)$ be the subspace of $\mathscr{L}^\infty(G, E^*)$ formed by all elements f such that, for every $\xi \in X$, the mapping $x \mapsto \langle \xi, f(x) \rangle$ is locally λ-null. A mean on $L^\infty(G, E^*) = \mathscr{L}^\infty(G, E^*)/N(G, E^*)$ is a linear mapping M from $L^\infty(G, E^*)$ into E^* such that, for every $f \in \mathscr{L}^\infty(G, E^*)$, $M(f)$ lies in the w*-closure of co $\{g(x): x \in G\}$ whenever $g \in l^\infty(G, E^*)$ with $g - f \in N(G, E^*)$. If $f \in L^\infty(G, E^*)$ and $\varphi \in L^1(G)$, one may define $\varphi \cdot f \in L^\infty(G, E^*)$ by

$$\langle \xi, \varphi \cdot f(x) \rangle = \int_G \varphi(y) \langle \xi, f(y^{-1}x) \rangle \, dy,$$

$x \in G, \xi \in E$. The mean M on $L^\infty(G, E^*)$ is called topologically invariant if $M(\varphi \cdot f) = M(f)$ whenever $f \in L^\infty(G, E^*)$ and $\varphi \in P^1(G)$. Husain and Wong showed that topological invariance implies invariance. They generalized the localization property 20.2 and the stationarity property 20.3. A similar theory was developed by Bombal and Vera [42, 43] for locally convex Hausdorff spaces. Bombal [41] studied the general problem of the existence of invariant means taking their values in a normed algebra. Invariant means for vector-valued functions defined on semigroups were also considered by Bottesch [48]. For a study on continuous linear functionals defined on Banach spaces that are invariant under a semigroup of operators generalizing the notions of invariant means see W. Woodward and Chivukula [573]. Other abstract generalizations of the notion of translation-invariant mean were studied by Fremlin and Talagrand [189].

If G is a locally compact group and $(K, |\cdot|)$ is a complete valued field with ultrametric valuation, that is, $|\alpha + \beta| \leq \sup\{|\alpha|, |\beta|\}$ whenever $\alpha, \beta \in K$, let $\mathscr{C}(G, K)$ be the Banach space over K formed by all continuous bounded functions defined on G with values in K equipped with the norm defined by $\|f\| = \sup\{|f(x)|: x \in G\}$ for $f \in \mathscr{C}(G, K)$. The group G is said to be K-amenable if there exists a left invariant mean on $\mathscr{C}(G, K)$, that is, a K-linear functional M on $\mathscr{C}(G, K)$ such that (a) $M(1_{G,K}) = 1$, where $1_{G,K}$ is the function mapping G onto the unit of K; (b) $|M(f)| \leq \|f\|$ whenever $f \in \mathscr{C}(G, K)$; and (c) $M(_a f) = M(f)$ whenever $f \in \mathscr{C}(G, K)$, $a \in G$ with $_a f(x) = f(ax)$ for $x \in G$. As the ultrametric valued field K is totally disconnected ([396] Chapitre I, 3.3, corollaire), a natural isomorphism of $\mathscr{C}(G, K)$ onto $\mathscr{C}(G/G_0, K)$ exists, so it suffices to consider K-amenability in case G is locally compact totally disconnected, hence 0-dimensional ([HR] 3.5). Schickhof [503] developed the theory for locally compact 0-dimensional groups. In particular, he studied Reiter's condition and investigated weak K-amenability for G, that is, the existence of invariant means for the subspace of $\mathscr{C}(G, K)$ consisting of the functions f for which $f(G)$ has compact closure.

Lengagne [348] studied invariant means for the vector spaces of bounded real-valued functions defined on a category and for compact topological categories.

The relevance of the theory of invariant means on vector-valued or operator-valued functions for the studies of quantum statistical mechanics and physical systems is emphasized by Doplicher, Kadison, Kastler, and Robinson [141] and Emch, Knops, and Verboven [160].

B. Amenable Actions

Actions of amenable groups in connection with *fixed point conditions* were considered in Section 5. *Ergodic* properties concern an important application of (FP). A detailed historical account on the notion of ergodicity is Lo Bello [365]. The term "ergodic" had been introduced by Boltzmann in a physical context. Consider a gas consisting of N molecules assumed to be spherical and enclosed in a container. Each molecule is determined by its three coordinates and the projections of the momentum vector onto the three coordinate axes. The state of the whole system is therefore described by a point P (the phase point) on a $6N - 1$ dimensional surface Γ of constant energy in a $6N$ dimensional phase space. The evolution of the system is given by the motion of P on Γ. If a first observation locates P at $\zeta \in \Gamma$, the next observation locates it at a point $T(\zeta)$. Gibbs assumed that, for any μ-measurable subset A of Γ and any measure-preserving transformation T, there exists a time average $\lim_{n \to \infty} \sum_{i=0}^{n-1} 1_A(T^i(\zeta))/n$ for P in A. He believed it to be equal to the space average $\mu(A)$. Studying this question Poincaré was led to the recurrence theorem that may be formulated in the following way: If T is a measure-preserving transformation on a space Z admitting a probability measure and A is a measurable subset of Z, then for almost every $\zeta \in Z$ there exists $n \in \mathbf{N}^*$ such

GENERALIZED NOTIONS OF AMENABILITY

that $T^n(\zeta) \in A$. Hence the subset of A consisting of those elements ζ, for which there exists no n, has measure zero. Currently a general ergodic transformation T is defined to be a transformation for which every T-invariant subset has measure 0 or is almost all of the space. The classical versions of ergodic theorems are (a) von Neumann's mean ergodic theorem and (b) Birkhoff's individual ergodic theorem. We state their results here:

Let T be a measurable, measure-preserving transformation on $[0, 1]$.

(a) If $f \in L^2([0, 1])$, there exists $f^\natural \in L^2([0, 1])$ such that

$$\lim_{n \to \infty} \int_0^1 \left| \sum_{i=0}^{n-1} f(T^i(\zeta))/n - f^\natural(\zeta) \right|^2 d\zeta = 0.$$

(b) If f is measurable on $[0, 1]$, $\lim_{n \to \infty} \sum_{i=0}^{n-1} f(T^i(\zeta))/n$ exists almost everywhere and defines $f^\natural \in L^1([0, 1])$; $f^\natural$ is T-invariant, that is, $f^\natural = f^\natural \circ T$.

We study *ergodicity* in the framework of amenability considering the action of an amenable group on a measure space. Although various types of general versions are available, we adopt a special point of view giving insight into the techniques of standard methods.

We consider a locally compact group G and a left Banach G_d-module Z for the action

$$\Lambda_x : \zeta \mapsto x\zeta$$

$$Z \to Z.$$

Assume that the action is weakly measurable, that is, for all $\zeta \in Z, \eta \in Z^*$, the mapping $x \mapsto \langle \Lambda_x \zeta, \eta \rangle$ is measurable. If $\zeta \in Z$, let $C_\zeta = \overline{\mathrm{co}} \{\Lambda_x \zeta : x \in G\}$.
$$G \to \mathbf{C}$$
For $\zeta \in Z, \eta \in Z^*, x \in G$, we define

$$\varphi_{\zeta,\eta}(x) = \langle \Lambda_x \zeta, \eta \rangle;$$

$\varphi_{\zeta,\eta} \in L^\infty(G)$. If $M \in \mathfrak{M}(L^\infty(G)), \zeta \in Z, \eta \in Z^*$, let

$$\langle \eta, T_{M,\zeta} \rangle = M(\varphi_{\zeta,\eta}).$$

Lemma 24.12. *Let G be a locally compact group having weak measurable action on the left Banach G_d-module Z and let $M \in \mathfrak{M}(L^\infty(G))$. For $\zeta \in Z$ such that C_ζ is w-compact, we have $T_{M,\zeta} \in C_\zeta$. In particular, if G is amenable and $M \in \mathrm{LIM}(L^\infty(G))$, then $T_{M,\zeta}$ is a fixed point in C_ζ.*

Proof. As Z is a left Banach G_d-module, there exists $k \in \mathbf{R}_+^*$ such that $|\langle x\zeta, \eta \rangle| \leq k \|\zeta\|_Z \|\eta\|_{Z^*}$ whenever $x \in G, \zeta \in Z, \eta \in Z^*$; also $|M(\varphi_{\zeta,\eta})| \leq$

$k\|\zeta\|_Z\|\eta\|_{Z^*} < \infty$, hence $T_{M,\zeta} \in Z^{**}$. We may consider C_ζ to be w*-closed in Z^{**}. If we had $T_{M,\zeta} \notin C_\zeta$, by the Hahn–Banach theorem there would exist $\eta_0 \in Z^{***} = Z^*$ such that

$$\mathcal{R}e\,\langle \eta_0, T_{M,\zeta}\rangle > \sup \mathcal{R}e\,\{\langle \eta_0, z\rangle : z \in C_\zeta\} = \alpha.$$

For every $x \in G$,

$$\mathcal{R}e\,\varphi_{\zeta,\eta_0}(x) = \mathcal{R}e\,\langle x\zeta, \eta_0\rangle = \mathcal{R}e\,\langle \eta_0, x\zeta\rangle \le \alpha$$

and

$$\mathcal{R}e\,\langle \eta_0, T_{M,\zeta}\rangle = M(\mathcal{R}e\,\varphi_{\zeta,\eta_0}) \le \alpha,$$

so we would come to a contradiction. We must have $T_{M,\zeta} \in C_\zeta$.

Suppose that $M \in \mathrm{LIM}(L^\infty(G))$. For all $\zeta \in Z$, $\eta \in Z^*$, $a \in G$, and $x \in G$, we have

$$\varphi_{\zeta,\eta a}(x) = \langle x\zeta, \eta a\rangle = \langle ax\zeta, \eta\rangle.$$

Hence $M(\varphi_{\zeta,\eta a}) = M(\varphi_{\zeta,\eta})$ and

$$\langle \eta, aT_{M,\zeta}\rangle = \langle \eta a, T_{M,\zeta}\rangle = \langle \eta, T_{M,\zeta}\rangle,$$

that is, $T_{M,\zeta}$ is a fixed point. $\square$

If G is a locally compact group and Z is a left Banach G_d-module, for $f \in L^1(G)$ and $\zeta \in Z$, let

$$f\zeta = \int_G f(x)\Lambda_{x^{-1}}(\zeta)\,dx = \int_G f(x)(x^{-1}\zeta)\,dx.$$

If $f, g \in L^1(G)$ and $\zeta \in Z$,

$$f(g\zeta) = \int_G f(y)\left(\int_G g(x)(y^{-1}x^{-1}\zeta)\,dx\right)dy$$

$$= \int_G f(y)\left(\int_G g(xy^{-1})\Delta(y^{-1})x^{-1}\zeta\,dx\right)dy = \int_G g*f(x)x^{-1}\zeta\,dx$$

$$= (g*f)\zeta.$$

If $f \in L^1(G), a \in G, \zeta \in Z$, we have

$$f\Lambda_a(\zeta) = \int_G f(x)\Lambda_{x^{-1}}\Lambda_a(\zeta)\,dx = \int_G f(x)\Lambda_{x^{-1}a}(\zeta)\,dx$$

$$= \int_G f(ax)\Lambda_{x^{-1}}(\zeta)\,dx = (_a f)\zeta. \tag{1}$$

We now establish an *abstract ergodic theorem*.

Theorem 24.13. *Let G be an amenable σ-compact group. Let Z be a left Banach G_d-module and assume that the action $\Lambda_x : \zeta \to x\zeta (x \in G)$ on Z is weakly measurable. Let $(U_n) \in \mathscr{A}(G)$.*
  (a) *If $\zeta \in Z$ such that $C_\zeta = \overline{\mathrm{co}}\,\{x\zeta : x \in G\}$ is w-compact, define*

$$E_n(\zeta) = \xi_{U_n}\zeta = \frac{1}{|U_n|}\int 1_{U_n}(x)\Lambda_{x^{-1}}(\zeta)\,dx = \frac{1}{|U_n|}\int_{U_n}\Lambda_{x^{-1}}(\zeta)\,dx \in Z.$$

  Then $(E_n(\zeta))$ converges to a fixed point of C_ζ that is the unique fixed point in C_ζ and is therefore independent of the choice of (U_n).
  (b) *Let $Z_1 = \{\zeta \in Z : (\forall x \in G)\Lambda_x(\zeta) = \zeta\}$ and let Z_2 be the closed vector subspace of Z generated by $\{\zeta - \Lambda_x(\zeta) : x \in G, \zeta \in Z\}$. Then $Z = Z_1 \oplus Z_2$ and (E_n) converges strongly to the projection of Z onto Z_1.*

Proof. As Z is a left Banach G-module, $k = \sup\{\|\Lambda_x\| : x \in G\} \in \mathbf{R}_+$.
  (a) Choose $\zeta \in Z$. By Lemma 24.12 there exists a fixed point ζ_0 in C_ζ. If $\varepsilon > 0$, we determine $\alpha_1, \ldots, \alpha_m > 0$ and $x_1, \ldots, x_m \in G$ such that $\sum_{i=1}^m \alpha_i = 1$ and $\|\zeta' - \zeta_0\| < \varepsilon/(2k+1)$ with $\zeta' = \sum_{i=1}^m \alpha_i \Lambda_{x_i^{-1}}(\zeta)$. For every $n \in \mathbf{N}^*$, let $\varphi_n = \xi_{U_n}$. By (1) we have

$$\varphi_n \zeta' = \varphi_n \left(\sum_{i=1}^m \alpha_i \Lambda_{x_i^{-1}}(\zeta)\right) = \left(\sum_{i=1}^m \alpha_{i\,x_i^{-1}}\varphi_n\right)\zeta.$$

As

$$\left\|\sum_{i=1}^m \alpha_{i\,x_i^{-1}}\varphi_n - \varphi_n\right\|_1 = \left\|\sum_{i=1}^m \alpha_i(_{x_i^{-1}}\varphi_n - \varphi_n)\right\|_1 \leq \sum_{i=1}^m \alpha_i \|_{x_i^{-1}}\varphi_n - \varphi_n\|_1$$

$$= \sum_{i=1}^m \alpha_i \frac{|x_i U_n \triangle U_n|}{|U_n|},$$

there exists $n_0 \in \mathbf{N}^*$ such that, for every $n \geq n_0$,

$$\sup\{\|\Lambda_x(\zeta)\| : x \in G\} \cdot \left\|\sum_{i=1}^m \alpha_{i\,x_i^{-1}}\varphi_n - \varphi_n\right\|_1 < \frac{\varepsilon}{2}.$$

As ζ_0 is a fixed point, for every $n \in \mathbf{N}^*$,

$$\varphi_n \zeta_0 = \int_G \varphi_n(x) \Lambda_{x^{-1}}(\zeta_0) \, dx = \int_G \varphi_n(x) \zeta_0 \, dx = \zeta_0.$$

So finally, for $n \geq n_0$,

$$\|\zeta_0 - E_n(\zeta)\| = \|\varphi_n \zeta_0 - \varphi_n \zeta\| \leq \|\varphi_n \zeta_0 - \varphi_n \zeta'\| + \|\varphi_n \zeta' - \varphi_n \zeta\|$$

$$\leq \|\varphi_n\|_1 k \|\zeta_0 - \zeta'\| + \left\|\left(\sum_{i=1}^m \alpha_i \,_{x_i^{-1}}\varphi_n - \varphi_n\right)\zeta\right\|$$

$$\leq k\|\zeta_0 - \zeta'\| + \left\|\sum_{i=1}^m \alpha_i \,_{x_i^{-1}}\varphi_n - \varphi_n\right\|_1 \sup\{\|\Lambda_x(\zeta)\| : x \in G\}$$

$$< \frac{\varepsilon}{2} + \frac{\varepsilon}{2} = \varepsilon.$$

Hence $\lim_{n \to \infty} \|\zeta_0 - E_n(\zeta)\| = 0$ and then also $\lim_{n \to \infty} \|\zeta_0 - E_n(\xi)\| = 0$ for every $\xi \in C_\zeta$.

Now if there existed another fixed point ζ_1 in C_ζ, then for every $n \in \mathbf{N}^*$ we should have

$$E_n(\zeta_1) = \int_G \varphi_n(x) \Lambda_{x^{-1}}(\zeta_1) \, dx = \zeta_1$$

and

$$\zeta_1 = \lim_{n \to \infty} E_n(\zeta_1) = \zeta_0.$$

(b) Let $P: \zeta \mapsto \lim_{n \to \infty} E_n(\zeta); \|P\| \leq k$ and $P \in \mathscr{L}(Z)$.
$\quad\quad\quad Z \to Z$
For every $\zeta \in Z$, $P\zeta \in Z_1$ and

$$P(P\zeta) = \lim_{n \to \infty} E_n(P\zeta) = P\zeta$$

so $P^2 = P$. If $\zeta \in Z_1$, then $P\zeta = \lim_{n \to \infty} E_n(\zeta) = \zeta \in Z_1$. We conclude that Z_1 is the closed subspace $P(Z)$ of Z.

Let $\zeta \in Z$ and $x \in G$. By (1) we have

$$\|P(\zeta - \Lambda_x(\zeta))\| \leq \limsup_{n \to \infty} \|\varphi_n(\zeta - \Lambda_x(\zeta))\| = \limsup_{n \to \infty} \|(\varphi_n - {}_x\varphi_n)\zeta\|$$

$$\leq \lim_{n \to \infty} \|\varphi_n - {}_x\varphi_n\|_1 \sup\{\|\Lambda_x(\zeta)\| : x \in G\} = 0$$

so $\zeta - \Lambda_x(\zeta) \in Z_1^\perp$. Conversely if $\zeta \in Z_1^\perp$, $P\zeta = 0$ is the unique fixed point in C_ζ and, for every $\varepsilon > 0$, there exist $\alpha_1, \ldots, \alpha_m > 0$ and $x_1, \ldots, x_m \in G$ such that $\sum_{i=1}^m \alpha_i = 1$ and $\|\sum_{i=1}^m \alpha_i \Lambda_{x_i}(\zeta)\| < \varepsilon$. Hence

$$\left\| \zeta - \sum_{i=1}^m \alpha_i(\zeta - \Lambda_{x_i}(\zeta)) \right\| = \left\| \sum_{i=1}^m \alpha_i \Lambda_{x_i}(\zeta) \right\| < \varepsilon,$$

that is, $\zeta \in Z_2$. □

The next results constitute (a) a *mean ergodic theorem* and (b) a *pointwise ergodic theorem*.

Theorem 24.14. *Let G be an amenable σ-compact group and $(U_n) \in \mathscr{A}(G)$. Suppose that G acts on a measure space (X, μ) equipped with the finitely additive G-invariant nonnegative measure μ. Assume also that the mapping $(x, \xi) \mapsto x\xi$, $G \times X \to X$ is measurable. Let $p \in {]}1, \infty{[}$ or $p = 1$ in case μ is bounded. If $f \in L^p(X), n \in \mathbf{N}^*$ and $\xi \in X$, let $A_n f(\xi) = (1/|U_n|) \int_{U_n} f(x\xi) \, dx$.*

(a) *If $f \in L^p(X)$, then $(A_n f)$ converges to an element $f^\natural$ in $L^p(X)$ that is G-invariant, that is, $f^\natural(x\xi) = f^\natural(\xi)$ for all $x \in G$ and almost all $\xi \in X$; $f^\natural$ is independent of the choice of (U_n).*

(b) *Let $f \in L^p_\mathbf{R}(X)$. Then $(A_n f)$ converges μ-almost everywhere to $f^\natural$ if and only if $\sup \{A_n f(\xi) : n \in \mathbf{N}^*\} < \infty$ for μ-almost every $\xi \in X$.*

Proof. (a) We apply Theorem 24.13 to the Banach space $Z = L^p(X, \mu)$. If $f \in L^p(X), x \in G$, let

$$(xf)(\xi) = f(x^{-1}\xi),$$

$\xi \in X$. As μ is G-invariant, for $p \in [1, \infty[$, $\|xf\|_p = \|f\|_p$ whenever $f \in L^p(X)$ and $x \in G$, so Z is a left Banach G_d-module.

In case $p \in {]}1, \infty{[}$, for every $f \in L^p(X), C_f = \overline{\mathrm{co}}\{xf : x \in G\}$ is a w*-closed subset of the closed ball of radius $\|f\|_p$ in $L^{p'}(X)^* = L^p(X)$. This ball is w*-compact by Alaoglu's theorem, hence C_f is w-compact in the reflexive space Z. In the case that $p = 1$ and the measure μ is bounded, $C_f = \overline{\mathrm{co}}\{xf : x \in G\}$ is also w-compact ([DS] IV. 8. 9).

Let $n \in \mathbf{N}^*, f \in L^p(X)$. For $\xi \in X$,

$$A_n f(\xi) = \frac{1}{|U_n|} \int_{U_n} (x^{-1}f)(\xi) \, dx.$$

So Theorem 24.13 asserts that, for every $f \in L^p(X), (A_n f)$ converges to an element $f^\natural$ in $L^p(X)$ that is fixed; $x^{-1}f^\natural = f^\natural$ and $f^\natural(x\xi) = f^\natural(\xi)$ for every $x \in G$ and almost every $\xi \in X$.

(b) Let $f \in L^p_\mathbf{R}(X)$. If $(A_n f)$ converges μ-almost everywhere to $f^\natural$, then obviously $\sup \{A_n f(\xi) : n \in \mathbf{N}^*\} < \infty$ for μ-almost every $\xi \in X$. Assume now that the latter condition holds. Consider the projection $P: f \to f^{\natural}$ of $L^p(X)$ onto $Z_1 = \text{Im } P$. If $f \in Z_1$, for every $n \in \mathbf{N}^*$, $A_n f = f$ and $(A_n f)$ is trivially μ-almost everywhere convergent to f. Consider then $f \in Z_2 = \mathcal{K}\!\mathit{er}\, P$. In order to establish the implication in that case, by Banach's theorem ([DS] IV. 11.2) it suffices to establish μ-convergence for a dense subset of Z_2. By Proposition 24.13 applied to $L^p(X)$, Z_2 admits a dense subset consisting of functions of the form $f = g - ag$ where $g \in L^p(X), a \in G$. We may also assume that $g \in L^p(X) \cap L^{\infty}(X)$. For $n \in \mathbf{N}^*$ and μ-almost every $\xi \in X$,

$$|A_n f(\xi)| = \frac{1}{|U_n|} \int 1_{U_n}(x)|g(x\xi) - g(a^{-1}x\xi)|\, dx \le \frac{|U_n \triangle a^{-1}U_n|}{|U_n|} \|g\|_{\infty}.$$

Therefore $\lim_{n \to \infty} |A_n f(\xi)| = 0$, that is, $(A_n f)$ converges to the null function for μ-almost every point of X. □

NOTES

Amenable groups appeared as transformation groups in Section 5. Now we mention some further fixed point properties related to amenability. Kallman [302] proved that if G is a locally compact group admitting closed subgroups H and K such that H is amenable, G/K is compact, and K normalizes H, then $\overline{KH}/K$ admits a $\overline{KH}$-invariant probability measure. Wong [561] showed that amenability of a locally compact group G may be characterized by the existence of fixed points for actions of $P^1(G)$ and $M^1(G)$: If E is any locally convex Hausdorff real vector space admitting an action of the semigroup $L^1(G)$ [resp. $\mathcal{M}^1(G)$] that is continuous and Z is any compact convex $P^1(G)$ [resp. $M^1(G)$]-invariant subset of E, then Z admits a point that is invariant under the action of $P^1(G)$ [resp. $M^1(G)$]. A boundary of a locally compact group G is a compact homogeneous space B of G having the following property: For every $\mu \in M^1(B)$, there exist a sequence (a_n) in G and $\dot{a} \in B$ such that, for the induced action of G on $M^1(G), (a_n \mu)$ converges to the measure $\delta_{\dot{a}}$. This definition was introduced by Furstenberg [191] who considered mainly Lie groups. A group is said to be unbounded if it admits only the boundary reduced to a singleton. If G is amenable, by (FP) it is unbounded. Furstenberg showed that unbounded connected Lie groups are compact extensions of solvable groups; they are therefore amenable by Theorem 13.4. As one more formulation of a fixed point property showing the existence of an invariant mean via a Hahn–Banach type argument we quote the following result due to Boissier [40]: Let G be a lusinian ([B TG] IX, Section 6, No. 4, définition 7) group acting continuously and nontrivially on a lusinian space Z admitting a G-invariant nonnegative measure μ; if there exists $\zeta_0 \in Z$ such that $\{\mu(x\zeta_0): x \in G\} \ne \{0\}$ and $\sup\{\mu(x\zeta_0): x \in G\} < \infty$, then there exists a left invariant mean on $\mathscr{C}_{\mathbf{R}}(G)$.

For general fixed point properties of semigroups one should consult Mitchell [392, 393, 394], Argabright [16], Grossman [235, 236, 237], Junghenn [298], and Iversen [275]. Day [116] proved that if S is a topological semigroup, then the existence of a left invariant mean on $\mathscr{C}(S)$ implies the existence of a common fixed point in each compact convex subset of each locally convex, Hausdorff, real vector space for all continuous affine actions of S. For a wide-ranging account on fixed point properties concerning topological semigroups we refer to Berglund, Junghenn, and Milnes [34].

Huff [266] examined uniqueness of fixed points for actions of amenable semigroups. If the locally compact group G acts continuously and affinely on a compact convex subset Z of a locally convex, Hausdorff real vector space and $0 \in Z$, then 0 is a trivial fixed point and amenability of G does not guarantee the existence of a nonzero fixed point. Important results concerning the existence of nonzero fixed points are due to Jenkins [290]. He considered a group G that is connected, locally compact or discrete, finitely generated, solvable. He showed that polynomial growth of G is characterized by the following property: If G acts continuously and affinely on a compact convex subset Z of a locally convex, Hausdorff real vector space E and there exists $F \in E^*$ such that $\langle \zeta, F \rangle > 0$ and $(\sup\{\langle x\zeta, F \rangle : x \in G\})^{-1}\zeta \in Z$ whenever $\zeta \in Z \setminus \{0\}$, then Z admits a nonzero fixed point. The latter property is not shared by all groups having polynomial growth.

If G is a locally compact group admitting a closed subgroup H, we say that H admits the fixed point property with respect to G if, for every continuous and affine action of G on a compact convex subset Z of a locally convex, Hausdorff real vector space, there exists a fixed point for the action of H in Z. This property states that there exists a mean M on $\mathscr{RUC}(G)$ such that $M(_a f) = M(f)$ whenever $f \in \mathscr{RUC}(G)$ and $a \in H$. Let G be a locally compact, completely separable group. If $\mu \in M^1(G)$, denote by H_μ the space of μ-harmonic functions belonging to $\mathscr{RUC}(G)$. Furstenberg [191, 192] considered the compact structure space Π_μ of H_μ; he called it the Poisson space of G with respect to μ. Azencott [24] proved the following results. Let G be a locally compact, completely separable group, and let $\mu \in M^1(G)$. The stabilizer of Π_μ in G has the fixed point property with respect to G. Hence if Π_μ is a singleton, that is, H_μ is reduced to the space of constant functions on G, then G is amenable. In particular, recurrent groups are amenable. Let G be an amenable, completely separable group and let μ be a probability measure on G that is étalée, that is, there exists $n \in \mathbf{N}^*$ such that μ^{*n} is not singular with respect to a right Haar measure; if G acts transitively on Π_μ, then Π_μ is finite. Azencott established properties characterizing, among the completely separable Lie groups, the groups G that are amenable and satisfy the following property: For every étalée measure μ in $M^1(G)$, G acts transitively on Π_μ. One instance is the finiteness of any Poisson space corresponding to an étalée probability measure. The characterization was extended to general locally compact completely separable groups by Frémond, and Sueur-Pontier [190]. For amenable, almost connected, completely separable groups G admitting an aperiodic étalée probability measure μ, Birgé and Raugi [35] gave a necessary and sufficient condition in order that all μ-harmonic functions are constant functions (Choquet–Deny property); the condition is fulfilled for instance in the case that μ is hermitian. As all groups for which the Poisson space is a singleton are amenable, these authors were able to answer a conjecture formulated by Furstenberg: An almost connected, locally compact, completely separable group G is amenable if and only if there exists an aperiodic étalée probability measure μ on G for which all μ-harmonic functions defined on G are constant functions. Relying on Azencott's results, Elie and Raugi [157] proved that, for an amenable, almost connected, completely separable group G and an étalée probability measure μ of moment 1, the following conditions are equivalent: Π_μ is finite, Π_μ is a homogeneous space, and Π_μ is metrizable. For further refinements of these properties see Raugi [449].

Glasner [203] showed that to any topological group G there may be associated a minimal strongly proximal flow (G, Z) that is universal, that is, admits every minimal proximal flow as a factor; C. Moore [398] called $M^1(Z)$ the Glasner space of the group G. Glasner studied the class of all topological groups admitting only trivial minimal and proximal flows (he termed them strongly amenable groups). These groups satisfy the conditions of 5.4. A compact extension of a nilpotent group is of that type, but a solvable group may be not. Considering connected semisimple groups, C. Moore proved amenability for the stabilizer groups of the elements of the corresponding Glasner spaces; also see Guivarc'h [242].

Let G be a discrete group, (G, Z) an affine flow, and, for $x \in G$, let S_x be the subset of Z consisting of all points that are fixed for the action $\Lambda_x \colon \zeta \mapsto x\zeta$. Call any point a in G weakly attractive if, for every weak neighborhood $U \overset{Z \to Z}{\text{of}}$ S_a and every compact convex subset C in $Z \setminus U$, there exists $k \in \mathbf{N}^*$ such that $\Lambda_{a^{kn}}(C) \subset U$ whenever $n \in \mathbf{Z}^*$. The connection of the nonamenability of a group with the nonexistence of a free subgroup on two generators is illustrated by the following fact due to S. Chen [68]: If G is a nonamenable discrete group and, for an affine flow (G, Z), all points of G are weakly attractive, then G admits the subgroup F_2.

Actions of amenable σ-compact groups seem to constitute the appropriate framework for ergodic properties. Important forerunners of ergodic theorems were established by Calderón [62, 63], mainly for the case $\mathbf{R}$; Day [111, 114, 115, 120], Dixmier [138], Witz [558], and Tempel'man [536, 537] for semigroups; and Chou [79] for amenable, discrete, countable infinite groups. Chatard [67] proved a mean ergodic theorem for an amenable σ-compact, completely separable group G acting continuously on a compact metrizable space Z that admits a G-invariant measure. Previously [66] she had studied the problem for σ-compact, locally compact abelian groups G investigating properties of almost periodic functions in $L^1(G)$ that are defined to be elements admitting relatively compact orbits in $L^1(G)$. Aribaud [19] obtained an ergodic theorem following from his general study on the action of a locally compact group G by continuous automorphisms on a locally convex topological vector space for which G is weakly almost periodic, that is, the orbit of each element is weakly relatively compact; also see [18] for earlier work of that author. As a matter of fact we should mention a long list of ergodic theorems concerning dynamical systems defined by a measure space (X, μ) equipped with a probability measure μ and an amenable transformation group having continuous action for which μ is invariant. A reference showing the importance of the subject is Guichardet [239].

Early attempts to obtain general ergodic theorems had been made by Renaud [456]. Greenleaf [228] established 24.12–13. He proved the mean ergodic theorem 24.14(a) whereas Emerson [164] gave the pointwise ergodic theorem 24.14(b).

With the notations of 24.14, $(U_n) \in \mathcal{A}(G)$ is said to be p-admissible if, for any $f \in L^p(X)$, the sequence $(A_n f)$ converges μ-almost everywhere to an element in $L^p(X)$. In the case that μ is bounded, $(U_n) \in \mathcal{A}(G)$ is said to be admissible if it is p-admissible for every $p \in [1, \infty[$. Emerson [164] established a transfer principle that is typical in ergodic theory and goes back to Calderón [63]. It reduces admissibility properties to considerations on the functions

$$x \mapsto \sup\left\{ |U_n|^{-1} \left| \int_{U_n} f(yx)\, dy \right| : n \in \mathbb{N}^* \right\}$$

($f \in L^p(G)$) defined on the group. Not every averaging sequence in an amenable σ-compact group is admissible. Let α be a fixed irrational number and consider the action of $\mathbb{Z}$ on $I = \mathbb{R} \pmod 1$ defined by $n \cdot \xi = \xi + n\alpha \pmod 1$, $n \in \mathbb{Z}, \xi \in I$. Emerson produced $(A_n) \in \mathcal{A}(\mathbb{Z})$ for which

$$\limsup_{n \to \infty} \frac{1}{\operatorname{card} A_n} \sum_{x \in A_n} (x \cdot \xi)^{-1/2} = \infty$$

whenever $\xi \in I$. In an arbitrary amenable σ-compact group he proved admissibility for any nondecreasing left regular averaging sequence. Ornstein, and B. Weiss [423] gave a proof of 1-admissibility for left regular averaging sequences applying to amenable discrete countable groups. Chatard [67] had shown that if a locally compact group admits a right regular averaging sequence and is thus unimodular, 1-admissibility holds in the situation considered in her study. Relying on the structural properties of amenable connected groups Greenleaf and Emerson [230] succeeded ingeniously in establishing the existence of p-admissible ($1 < p < \infty$), nonnecessarily left regular, averaging sequences in these groups. From these results Conze and Dang Ngoc [99, 100] deduced general ergodic theorems for amenable connected groups acting on von Neumann algebras. Jenkins [283] studied ergodicity for locally compact groups in connection with growth conditions. Rosenblatt [478] obtained ergodic properties in case G is an amenable, discrete, finitely generated group having a measurable action on a probability space; also see del Junco and Rosenblatt [297].

As a generalization of ergodic properties, convergence of averaging rates was studied for a measurable partition with finite entropy of a probability space on which a discrete amenable group acts. Important recent results are due to Pickel [434], Kieffer [315, 316], Tagi-Zade [526], Ornstein and B. Weiss [422, 423], and Šujan [523].

If G is a discrete group acting on a compact space Z, by Theorem 5.4 amenability of G implies the existence of a G-invariant probability measure on Z. The converse may obviously fail as for a trivial action of the group G on Z there exists a G-invariant probability measure on Z whether G is amenable or not. We proceed now to compare amenability with amenable actions. The most important study is due to Greenleaf [227], who relied on general properties elaborated by Gulick, Liu, and van Rooij [244]. We summarize the principal points of that theory.

Let G be a locally compact group acting continuously on a locally compact space Z, that is, there exists a continuous mapping

$$(x, \zeta) \mapsto x\zeta$$

$$G \times Z \to Z$$

such that $x(y\zeta) = xy\zeta$ whenever $x, y \in G$, $\zeta \in Z$, and $e\zeta = \zeta$ whenever $\zeta \in Z$. We may define a Banach $\mathcal{M}^1(G)$-module $\mathscr{C}(Z)$ by putting, for $f \in \mathscr{C}(Z)$ and $\mu \in \mathcal{M}^1(G)$, $\mu f(\zeta) = \int_G f(x\zeta) d\mu(x)$, $\zeta \in Z$. Let $f \in \mathscr{C}(Z), \mu \in \mathcal{M}^1(G)$. If $(\zeta_i)_{i \in I}$ converges to ζ in Z, for a given $\varepsilon > 0$, let K be a compact subset of G such that $\|f\| |\mu|(G \setminus K) < \varepsilon/3$ and $i_0 \in I$ such that $\|\mu\| |f(x\zeta_i) - f(x\zeta)| < \varepsilon/3$ whenever $x \in K, i \succ i_0$. Then also

$$|\mu f(\zeta_i) - \mu f(\zeta)| \leq \int_K |f(x\zeta_i) - f(x\zeta)| d|\mu|(x) + 2\|f\| |\mu|(G \setminus K) < \varepsilon.$$

Therefore we have $\mu f \in \mathscr{C}(Z)$. We denote by $\mathscr{UC}(Z)$ the space of all continuous bounded complex-valued functions f defined on Z that are uniformly continuous for the action of G, that is, the mapping $x \mapsto \delta_x f$ is
$$G \to \mathscr{C}(Z)$$
continuous. Let $X = \mathscr{C}(Z), \mathscr{UC}(Z)$ or $X = L^\infty(Z, \nu)$ for a nonnegative, quasi-invariant measure ν on Z. If $f \in X$, $a \in G$, and $\varphi \in L^1(G)$, we have $\delta_x f \in X$ and $\varphi f \in X$. The mean M on X is called invariant mean [resp. topologically invariant mean] if $M(\delta_x f) = M(f)$ whenever $f \in X, x \in G$ [resp. $M(\varphi f) = M(f)$ whenever $f \in X, \varphi \in P^1(G)$]. Consider the following properties:

(i) [resp. (i′)] There exists an invariant [resp. topologically invariant] mean on $\mathscr{C}(Z)$.

(ii) [resp. (ii′)] There exists an invariant [resp. topologically invariant] mean on $\mathscr{UC}(Z)$.

(iii) [resp. (iii')] For some real, nonnegative, nontrivial quasi-invariant measure ν on Z, there exists an invariant [resp. topologically invariant] mean on $L^\infty(Z, \nu)$.

(iv) [resp. (iv')] For every real, nonnegative, nontrivial, quasi-invariant measure ν on Z, there exists an invariant [resp. topologically invariant] mean on $L^\infty(Z, \nu)$.

Making use of the techniques of Section 4, one shows that (i), (i'), (ii), and (ii') are equivalent, that they are implied by (iii), and that (iii') $\Rightarrow$ (iii), (iv') $\Rightarrow$ (iv). If the equivalent conditions hold, we say that G admits an *amenable action* on Z.

If G is amenable, fix $\zeta_0 \in Z$. The mapping $x \mapsto x\zeta_0$ induces a mapping $G \to Z$ from $\mathscr{C}(Z)$ to $\mathscr{C}(G)$ and one may lift a topologically left invariant mean on $\mathscr{C}(G)$ to a topologically left invariant mean on $\mathscr{C}(Z)$. Hence if G is amenable, G acts amenably on Z. The locally compact group G is amenable if and only if it acts amenably on every homogeneous space G/H via the natural mapping. As a matter of fact, in case $H = \{e\}$, amenable action signifies amenability of the group. In the particular case where Z is a homogeneous space G/H, H being a closed subgroup of the locally compact group G, the eight properties considered by Greenleaf are equivalent: the implication (i') $\Rightarrow$ (iv') is established.

Consider the compact group $G = SO(3, \mathbf{R})$, and let H be the stabilizer of an element in the unit sphere of $\mathbf{R}^3$. The compact homogeneous space $Z = G/H$ admits a unique, normalized, quasi-invariant measure ν, and ν gives rise to an invariant mean on $L^\infty(Z, \nu)$. Therefore G_d acts amenably on Z, but by Corollary 14.3 the group G_d is nonamenable. Thus in general (iv') does not imply amenability of the group.

Anker [11] noticed that the equivalence $(P_1) \Leftrightarrow (P_1^*)$ given in 6.10 may be carried over to the continuous action of a locally compact group on a locally compact space admitting a quasi-invariant measure. We mention a property showing the possibility of characterizing amenability of a locally compact group in terms of amenable actions. Greenleaf [227] established the following result generalizing 17.10: A locally compact group G is amenable if and only if, for every closed subgroup H, every $T \in \hat{G}$ is weakly contained in $\operatorname{ind}_{H \uparrow G} T|_H$. As a matter of fact, if the property holds, let $H = \{e\}$. Amenability of G follows from the considerations made in Section 17A. Conversely, let G be an amenable group. As G acts amenably on the homogeneous space G/H equipped with the corresponding quasi-invariant measure, one has a Reiter-type condition. The proof of 8.4 then yields weak containment of i_G in $\operatorname{ind}_{H \uparrow G} i_H$. Hence $i_G \otimes T$ is weakly contained in $(\operatorname{ind}_{H \uparrow G} i_H) \otimes T$, that is, T is weakly contained in $\operatorname{ind}_{H \uparrow G} T|_H$. From this statement Beiglböck [31] deduced that, if H_1, H_2 are closed subgroups of an amenable group G such that $H_1 \subset H_2$, $T_1 \in \hat{H}_1$, $T_2 \in \hat{H}_2$ and $T_2|_{H_1}$ is weakly equivalent to T_1, then $\operatorname{ind}_{H_2 \uparrow G} T_2$ is weakly contained in $\operatorname{ind}_{H_1 \uparrow G} T_1$. Anker [12] established characterizations in terms of weak containment properties for the amenability of the continuous action of a locally compact group on a locally compact space equipped with a quasi-invariant measure. He also examined convolutors in this context; for earlier related results see Coifman and G. Weiss [90, 91].

Greenleaf's observations [227] concerning the nonequivalence of amenable actions with (iii), (iv) emphasize the nonequivalence of amenability for a locally compact group G and the existence of G-invariant measures. Recently Rosenblatt studied the subject for discrete groups [485, 487]. Previously he had examined the problem in connection with generalized Følner conditions [477] and growth properties [478, 479]. Also see [486].

K. Sakai widely transposed classical properties of amenable groups to amenable transformation groups in [495], and to amenable transformation semigroups in [496, 497, 499, 500]; for previous work of that author see [492, 493, 494]. He also studied linear representations of a topological group on various topological vector spaces and obtained generalized versions of (GR), cohomological properties, and ergodic theorems [498]. General forms of Day's criterion and convolution properties were given by Guivarc'h [242]. Losert [371] proved 15.3 in this general setup.

Eymard [173, 174, 176] introduced and studied extensively *amenable homogeneous spaces*. Let G be a Hausdorff topological group and let H be a closed subgroup. The homogeneous space G/H is called amenable if the following equivalent conditions hold:

(a) There exists an invariant mean on $\mathcal{UC}(G/H)$;

(b) if E is any locally convex, Hausdorff, real vector space, Z is any compact, convex subset of E, G acts continuously affinely on E, and Z admits a point that is invariant under the action of H, then Z admits a point that is invariant under the action of G. Eymard mainly considered the case where G is locally compact. He established that amenability of G/H is characterized by a lot of other properties, for instance existence of invariant means on $L^\infty(G/H), \mathcal{C}(G/H)$, analogues of the Reiter and Glicksberg–Reiter properties, a weak containment property, and convolution properties. An example of an amenable homogeneous space that is not a group is given by $SL(2,\mathbf{R})/D^{(2)}(SL(2,\mathbf{Z}))$. Toure [541, 542, 543] transposed several other characterizations of amenable groups to amenable homogeneous spaces; he also obtained interesting general statements. Derighetti [126] established refinements for (P_1). Berg and Christensen [32] extended the properties of Section 18A; also see the comments of Derriennic and Guivarc'h [134]. Rindler [466] considered uniformly distributed sequences in amenable homogeneous spaces. Schott [506, 507, 508] studied growth conditions on homogeneous spaces in connection with amenability. He proved that a nonamenable homogeneous space has exponential growth. Riemersma [465] obtained various generalizations and abstract versions.

Zimmer [577, 580, 581, 586] developed a very important theory involving amenability of an ergodic action. It is defined by a condition generalizing the fixed point property considered by Eymard. Let G be a locally compact, completely separable group. An ergodic G-space is defined by a standard Borel space Z (i.e., a σ-algebra of subsets, Borel subsets, of Z determine a polish topology) admitting a Borel measurable action $G \times Z \to Z$ and a probability measure μ on Z that is G-quasi-invariant and ergodic. If B is a Borel group (i.e., the group admits a topology for which multiplication and inversion are Borel measurable), one considers a cocycle $\alpha: G \times Z \to B$, that is, α is Borel measurable and $\alpha(xy, \zeta) = \alpha(x, \zeta)\alpha(y, x\zeta)$ for all $x, y \in G$ and μ-almost all $\zeta \in Z$. In particular, if π is a homomorphism of G into B, the corresponding cocycle is called restriction of π to $G \times Z$. Let E be a separable Banach space and let E_1^* be the unit ball of its dual. The isometric automorphisms of E form a Borel group $\mathrm{Iso}(E)$ for the strong topology. A Borel field of compact, convex subsets of E_1^* is defined by a mapping Φ of Z into the set of compact, convex subsets of E_1^* such that $\{(\zeta, u): u \in \Phi(\zeta)\}(\zeta \in Z)$ are Borel subsets of $Z \times E_1^*$. If α is a cocycle of $G \times Z \to \mathrm{Iso}(E)$, there exists an adjoint cocycle $\alpha^\sim$ of $G \times Z$ into the group of homeomorphisms of E_1^* defined by $\alpha^\sim(x, \zeta) = (\alpha(x, \zeta)^{-1})^\sim$, $x \in G, \zeta \in Z$. A Borel field is called α-invariant if $\alpha^\sim(x, \zeta) \Phi(x\zeta) = \Phi(\zeta)$ for every $x \in G$ and almost every $\zeta \in Z$. Any Borel measurable mapping $\varphi: Z \to E_1^*$ is termed an α-invariant section if $\alpha^\sim(x, \zeta)\varphi(x\zeta) = \varphi(x)$ for every $x \in G$ and μ-almost every $\zeta \in Z$. An ergodic G-space Z is defined as amenable if, for every separable Banach space E, every cocycle $\alpha: G \times Z \to \mathrm{Iso}(E)$, every α-invariant field Φ, there exists an α-invariant section φ such that $\varphi(\zeta) \in \Phi(\zeta)$ for almost every $\zeta \in Z$. If Z is an ergodic G-space, (G, Z) is called an amenable pair in case for every homomorphism $\pi: G \to \mathrm{Iso}(E)$ and every G-invariant, compact, convex subset K in E_1^*, the existence of any α-invariant section φ, where α is the restriction of π to $G \times Z$, implies the existence of a fixed point for G in K.

Every ergodic action of an amenable group is ergodic, that is, if G is an amenable group and Z is an ergodic G-space, then (G, Z) is an amenable pair. Let H be a closed subgroup of the locally compact, completely separable group G; then G/H is an amenable G-space if and only if H is amenable. Also $(G, G/H)$ is an amenable pair if and only if G/H is an amenable homogeneous space [580]. Whereas every ergodic action of a group that is induced from an amenable subgroup is an amenable action, the converse is generally false. Zimmer showed that the converse holds nevertheless for connected, semisimple Lie groups with finite center [582] and more generally for all connected groups [584]. A subsequent question is the determination of the amenable subgroups. C. Moore [398] elaborated a classification of the maximal amenable subgroups in a semisimple group.

Further studies on the subject are Zimmer [578, 579, 583, 585, 587, 588], C. Moore and Zimmer [399], and P. Hahn [248]. One should also consult Furstenberg's [193] expository report concerning results due to Margulis and Zimmer. Anantharaman-Delaroche [7, 8, 9] introduced and studied a general notion of amenability for an action of a locally compact group on a von Neumann algebra, generalizing Zimmer's concepts. Ocneanu [419] considered actions of amenable discrete countable groups on von Neumann algebras.

The C^*-algebra A is called nuclear (Takesaki property) if, for every C^*-algebra B, $A \otimes B$ admits a unique C^*-norm. If E_1 and E_2 are Banach spaces, a crossnorm p on $E_1 \otimes E_2$ satisfies $p(x_1 \otimes x_2) = \|x_1\| \|x_2\|$ for all $x_1 \in E_1, x_2 \in E_2$. Let G be a locally compact group acting continuously on a von Neumann algebra A, that is, there exists a continuous homomorphism of G into the group of automorphisms of A. Consider the generalized group algebra $L^1(G, A)$. The crossed product of A by G is the enveloping C^*-algebra $C^*(G, A)$ of $L^1(G, A)$. Similarly one defines $C_L^*(G, A)$. If A is any C^*-algebra, the completion of $C^*(G) \otimes A$ for the largest crossnorm is $C^*(G, A)$ [153, 238, 239]. Guichardet [238] proved that the completion of $C_L^*(G) \otimes A$ for the smallest crossnorm is $C_L^*(G, A)$. Hence if G is amenable, $C_L^*(G)$ is nuclear. Making use of the proof of 20.17, Lance [330] established amenability of any discrete group G for which $C_L^*(G)$ is nuclear. Haagerup [246] sharpened the nonnuclearity of $C_L^*(F_2)$. For an amenable, completely separable group G acting continuously on a separable C^*-algebra A, Sauvageot [502] investigated the primitive ideal space of the crossed product $C^*(G, A)$. Gootman and Rosenberg [209] solved the "generalized Effros–F. Hahn conjecture" [153]: Every primitive ideal of $C^*(G, A)$ is the kernel of a representation that is induced from a representation of $C^*(H, A)$, H being a closed subgroup of G.

The utilization of fixed point characterizations of amenability in the mathematical theory of statistical experiments is shown by Heyer [263].

Recent publications: Moulin Ollagnier [595], Schmidt [596].

C. Amenable Algebras

The following definition is suggested by Theorem 11.8.

Definition 24.15. *The Banach algebra A is called amenable if, for every Banach A-module X, $H_1(A, X^*) = \{0\}$.*

This important class of Banach algebras admits natural stability properties. We indicate the proofs for two of them.

Proposition 24.16. *Let A and B be Banach algebras and let Φ be a continuous homomorphism of A onto a dense subset of B. If A is amenable, then B is amenable.*

Proof. If X is a Banach B-module, X is a Banach A-module for the structure defined by $a \cdot \xi = \Phi(a)\xi, \xi \cdot a = \xi\Phi(a)$, $a \in A, \xi \in X$. If D is a continuous derivation mapping B into X^*, then $D\Phi$ is a continuous derivation of A into X^*, and by hypothesis there exists $\eta \in X^*$ such that $D\Phi = d_1\eta$. Hence $D = d_1\eta$ on $\Phi(A)$, and then also on B by density. □

Proposition 24.17. *Let A be a Banach algebra and let I be a closed ideal of A. If I and A/I are amenable algebras, then A is amenable.*

Proof. Let X be a Banach A-module and let D be a continuous derivation of A into X^*. Then $D_1 = D|_I$ is a continuous derivation of I into X^*, and by hypothesis there exists $\eta \in X^*$ such that $D_1 = d_1\eta$. For all $a \in A$ and $b \in I$, we have

$$0 = (D - d_1\eta)(ab) = (D - d_1\eta)(a)b + a(D - d_1\eta)(b) = (D - d_1\eta)(a)b.$$

Hence for every $\xi \in X$, $\langle \xi, (D - d_1\eta)(a)b \rangle = 0$, that is,

$$\langle b\xi, (D - d_1\eta)(a) \rangle = 0. \tag{2}$$

Similarly we have

$$\langle \xi b, (D - d_1\eta)(a) \rangle = 0. \tag{3}$$

Let X_I be the closed vector space generated by $IX \cup XI$; X_I is a Banach A-submodule of X, and (2), (3) assert that $D - d_1\eta$ maps A into $X_I^\perp \simeq (X/X_I)^*$. Moreover one may consider $D_2 = D - d_1\eta$ to be a continuous derivation of A/I into $(X/X_I)^*$. Then by hypothesis there exists $\zeta \in X_I^\perp$ such that $D_2 = d_1\zeta$. Therefore we have $D = d_1(\eta + \zeta)$. □

We now consider more specifically amenable C^*-algebras. Any derivation on a C^*-algebra is automatically continuous ([501] 4.1.3). If A is a C^*-algebra, $A \hat{\otimes} A$ becomes a Banach A-module if one defines

$$a(b \otimes c) = ab \otimes c, \quad (b \otimes c)a = b \otimes ca$$

for $a, b, c \in A$. We also consider the Banach A-module structure defined by

$$a \cdot (b \otimes c) = b \otimes ac, \quad (b \otimes c) \cdot a = ba \otimes c$$

for $a, b, c \in A$. These two structures commute. As for groups, amenability of C^*-algebras may be characterized by invariance properties.

Proposition 24.18. *Let A be a C^*-algebra with unit u. The following properties are equivalent*:

(i) *A is amenable.*

(ii) *There exists a continuous linear operator Φ of $(A \hat{\otimes} A)^*$ into $C = \{f \in (A \hat{\otimes} A)^* : (\forall a \in A) \; af = fa\}$ such that $\Phi|_C = \mathrm{id}_C$ and $\Phi(a \cdot f) = a \cdot \Phi(f), \Phi(f \cdot a) = \Phi(f) \cdot a$ whenever $f \in (A \hat{\otimes} A)^*$ and $a \in A$.*

(iii) *If Y is a Banach A-module, X is a Banach A-submodule, and $F \in X^*$ with $F(v\xi v^*) = F(\xi)$ whenever $\xi \in X$ and $v \in A_u$, then there exists $F_1 \in Y^*$ extending F such that $F_1(v\xi v^*) = F_1(\xi)$ whenever $\xi \in Y$ and $v \in A_u$.*

Proof. (i) $\Rightarrow$ (ii)
We consider the Banach A-module $Y = (A \hat{\otimes} A)^* \hat{\otimes} (A \hat{\otimes} A)$ with

$$a(f \otimes t) = f \otimes (at), \quad (f \otimes t)a = f \otimes (ta)$$

for $f \in (A \hat{\otimes} A)^*$, $t \in A \hat{\otimes} A$, and $a \in A$. Let Z be the closed vector subspace of Y generated by the elements $(a \cdot f) \otimes t - f \otimes (t \cdot a)$, $(f \cdot a) \otimes t - f \otimes (a \cdot t)$, where $f \in (A \hat{\otimes} A)^*$, $t \in A \hat{\otimes} A$, and $a \in A$. Let W be the closed vector subspace of Y generated by the elements $f \otimes t$, where $f \in C$, $t \in A \hat{\otimes} A$. As Z and W are A-submodules of Y, the closed vector subspace X of Y generated by $Z \cup W$ is an A-submodule of Y; moreover X/Z is an A-submodule of Y/Z.

Consider $F \in Y^*$ defined by

$$\langle f \otimes t, F \rangle = f(t),$$

for $f \in (A \hat{\otimes} A)^*$ and $t \in A \hat{\otimes} A$. Let $D = d_1 F$. For all $a \in A$, $f \in (A \hat{\otimes} A)^*$, and $t \in A \hat{\otimes} A$, we have

$$D(a)(f \otimes t) = (aF - Fa)(f \otimes t) = F((f \otimes t)a - a(f \otimes t))$$

$$= F(f \otimes (ta) - f \otimes (at)) = \langle ta - at, f \rangle = \langle t, af - fa \rangle. \quad (4)$$

Therefore $D(a)|_W = 0$. As the module structures commute, for all $a, b \in A$, $f \in (A \hat{\otimes} A)^*$, and $t \in A \hat{\otimes} A$, we also have

$$D(a)((b \cdot f) \otimes t - f \otimes (t \cdot b))$$

$$= (aF)((b \cdot f) \otimes t) - (aF)(f \otimes (t \cdot b))$$

$$- (Fa)((b \cdot f) \otimes t) + (Fa)(f \otimes (t \cdot b))$$

$$= F(((b \cdot f) \otimes t)a) - F((f \otimes (t \cdot b))a)$$

$$- F(a((b \cdot f) \otimes t)) + F(a(f \otimes (t \cdot b)))$$

$$= F((b \cdot f) \otimes (ta)) - F(f \otimes ((t \cdot b)a))$$

$$- F((b \cdot f) \otimes at) + F(f \otimes a(t \cdot b))$$

$$= \langle ta, b \cdot f \rangle - \langle (t \cdot b)a, f \rangle - \langle at, b \cdot f \rangle + \langle a(t \cdot b), f \rangle$$

$$= \langle t, b \cdot (af - fa) \rangle - \langle t \cdot b, af - fa \rangle = 0.$$

This result allows us to conclude that $D(a)|_Z = 0$. Hence $D(a)|_X = 0$, and there exists a continuous derivation D' of A into $(Y/X)^*$ defined by $D'(a)(\dot{\eta}) = D(a)\eta$ where, as usual, $\dot{\eta}$ is the conjugacy class of $\eta \in Y$. As A is amenable, there exists $F' \in (Y/X)^*$ such that $D' = d_1 F'$. We now define $F_1 \in Y^*$ by putting $F_1(\eta) = F'(\dot{\eta})$, $\eta \in Y$. We consider $T \in \mathscr{L}((A \hat{\otimes} A)^*)$, where

$$T(f)(t) = \langle f \otimes t, F_1 \rangle$$

for $f \in (A \hat{\otimes} A)^*$ and $t \in A \hat{\otimes} A$. For every $a \in A$, we have

$$D'(a)((f \otimes t)^{\cdot}) = (aF' - F'a)((f \otimes t)^{\cdot}) = \langle (f \otimes t)a - a(f \otimes t), F_1 \rangle$$
$$= \langle f \otimes (ta - at), F_1 \rangle = T(f)(ta - at)$$
$$= (aT(f) - T(f)a)(t). \tag{5}$$

Relations (4) and (5) imply that, for every $t \in A \hat{\otimes} A$,

$$(af - fa)(t) = (aT(f) - T(f)a)(t),$$
$$(a(f - T(f)) - (f - T(f))a)(t) = 0,$$

that is, $f - Tf \in C$. Finally, consider the continuous linear operator Φ of $(A \hat{\otimes} A)^*$ into C defined by $\Phi = id - T$. If $f \in C$ and $t \in A \hat{\otimes} A$, as $F_1|_W = 0$, we have $T(f)(t) = \langle f \otimes t, F_1 \rangle = 0$, and therefore $\Phi|_C = id_C$. As $F_1|_Z = 0$,

$$(\Phi(a \cdot f))(t) - (a \cdot \Phi(f))(t) = -\langle (a \cdot f) \otimes t - f \otimes (t \cdot a), F_1 \rangle = 0$$

whenever $f \in (A \hat{\otimes} A)^*$, $t \in A \hat{\otimes} A$ and $a \in A$. Therefore if $f \in (A \hat{\otimes} A)^*$ and $a \in A$, we have $\Phi(a \cdot f) = a \cdot \Phi(f)$. Similarly $\Phi(f \cdot a) = \Phi(f) \cdot a$.

(ii) $\Rightarrow$ (i)

Let X be a Banach A-module and consider a continuous derivation D of A into X^*. For every $\xi \in X$, define a bilinear functional f_ξ on $A \hat{\otimes} A$ by

$$f_\xi(b \otimes c) = D(b)(\xi c)$$

for $b, c \in A$; f_ξ may be identified with an element of $(A \hat{\otimes} A)^*$. Also, define $F \in X^*$ by

$$F(\xi) = \Phi(f_\xi)(u \otimes u),$$

$\xi \in X$. If $a \in A$ and $\xi \in X$, we consider the bilinear functional $g_{\xi,a}$ on $A \hat{\otimes} A$, where

$$g_{\xi,a}(b \otimes c) = D(a)(\xi cb)$$

for $b, c \in A$. For every $d \in A$,

$$dg_{\xi,a}(b \otimes c) = g_{\xi,a}((b \otimes c)d) = g_{\xi,a}(b \otimes cd) = D(a)(\xi(cd)b)$$
$$= D(a)(\xi c(db)) = g_{\xi,a}(db \otimes c)$$
$$= g_{\xi,a}(d(b \otimes c)) = (g_{\xi,a}d)(b \otimes c)$$

hence $g_{\xi,a} \in C$.

For $a, b, c \in A$ and $\xi \in X$, we have

$$(f_\xi \cdot a)(b \otimes c) = f_\xi(b \otimes ac) = D(b)(\xi ac)$$

and

$$f_{\xi a}(b \otimes c) = D(b)(\xi ac)$$

hence

$$f_\xi \cdot a = f_{\xi a}. \tag{6}$$

Moreover

$$a \cdot f_\xi(b \otimes c) = f_\xi(ba \otimes c) = D(ba)(\xi c)$$
$$= (D(b)a)(\xi c) + (bD(a))(\xi c) = D(b)(a\xi c) + D(a)(\xi cb)$$
$$= f_{a\xi}(b \otimes c) + g_{\xi, a}(b \otimes c)$$

hence

$$a \cdot f_\xi = f_{a\xi} + g_{\xi, a}. \tag{7}$$

For all $a \in A$ and $\xi \in X$, by (6), (7), and the fact that $\Phi(f_\xi) \in C$, we obtain

$$d_1 F(a)(\xi) = (aF - Fa)(\xi) = F(\xi a - a\xi) = \Phi(f_{\xi a - a\xi})(u \otimes u)$$
$$= \Phi(f_\xi \cdot a - a \cdot f_\xi + g_{\xi, a})(u \otimes u)$$
$$= \Phi(f_\xi)(u \otimes a - a \otimes u) + \Phi(g_{\xi, a})(u \otimes u)$$
$$= (a\Phi(f_\xi) - \Phi(f_\xi)a)(u \otimes u) + g_{\xi, a}(u \otimes u)$$
$$= D(a)(\xi),$$

that is, $D(a) = d_1 F(a)$.

$$\text{(ii)} \Rightarrow \text{(iii)}$$

Let F' be any extension of F to an element of Y^*. For every $\eta \in Y$, we define $F_\eta \in (A \hat{\otimes} A)^*$ by putting

$$F_\eta(a \otimes b) = F'(a\eta b),$$

$a, b \in A$. Define also $F'' \in Y^*$ by putting

$$\langle \eta, F'' \rangle = \Phi(F_\eta)(u \otimes u),$$

$\eta \in Y$.

If $v \in A_u$ and $a, b \in A$, we have

$$F_{v^*\eta v}(a \otimes b) = F'(av^*\eta vb) = F_\eta(av^* \otimes vb) = v^* \cdot F_\eta \cdot v(a \otimes b),$$

that is, $F_{v^*\eta v} = v^* \cdot F_\eta \cdot v$. With respect to the properties of Φ, we obtain

$$\langle v^*\eta v, F''\rangle = \Phi(v^* \cdot F_\eta \cdot v)(u \otimes u) = v^* \cdot \Phi(F_\eta) \cdot v(u \otimes u)$$

$$= \Phi(F_\eta)(v^* \otimes v) = \Phi(F_\eta)(v^*(u \otimes u)v)$$

$$= \Phi(F_\eta)((u \otimes u)vv^*) = \Phi(F_\eta)(u \otimes u) = \langle \eta, F''\rangle.$$

Furthermore as $\Phi|_C = id_C$, for every $\xi \in X$,

$$\langle \xi, F''\rangle = \Phi(F_\xi)(u \otimes u) = F_\xi(u \otimes u) = F'(\xi) = F(\xi),$$

that is, $F''|_X = F$.

$$\text{(iii)} \Rightarrow \text{(ii)}$$

Consider X, Y, and Z as in the proof of (i) $\Rightarrow$ (ii).

For $f \in (A \hat{\otimes} A)^*$ and $t \in A \hat{\otimes} A$, let $F(f \otimes t) = f(t)$; $F \in X^*$. Necessarily $F|_Z = 0$ and F induces $F' \in (X/Z)^*$.

As $F(v\xi v^*) = F(\xi)$ whenever $\xi \in X$ and $v \in A_u$, we also have $F'(v\dot{\xi}v^*) = F'(\dot{\xi})$ whenever $\dot{\xi} \in X/Z$ and $v \in A_u$. Then by assumption there exists $F_1' \in (Y/Z)^*$ extending F' and it is sufficient to put

$$\Phi(f)(t) = F_1'((f \otimes t)\dot{\,})$$

for $t \in A \hat{\otimes} A$ and $f \in (A \hat{\otimes} A)^*$. $\square$

The following proposition proves the existence of nonamenable C^*-algebras. Recall that if G is a locally compact group, then $PM(G) = A(G)^* \supset B(G)^* = C^*(G)^{**} \supset C^*(G)$ (Section 19A).

Proposition 24.19. *Let G be a discrete group.*
  (a) *If A is an amenable C^*-algebra such that $C_L^*(G) \subset A \subset PM(G)$, then G is an amenable group.*
  (b) *If G is an amenable group, then $C_L^*(G)$ is an amenable C^*-algebra.*

Proof. (a) Note that $\mathscr{L}(l^2(G))$ is a Banach A-module admitting A as a submodule. We consider $F \in A^*$ defined by

$$\langle a, F\rangle = (a\delta_e | \delta_e),$$

$a \in A$. We have $F(vav^*) = F(a)$ whenever $a \in A$ and v is a unitary element in

A. By Proposition 24.18 there exists $F' \in \mathscr{L}(l^2(G))^*$ extending F such that $F'(v\varphi v^*) = F'(\varphi)$ whenever $\varphi \in \mathscr{L}(l^2(G))$ and $v \in A_u$. We may assume that F' is hermitian. For a fixed unitary element v in A, let

$$\Phi_1(\varphi) = F'_+(v\varphi v^*), \qquad \Phi_2(\varphi) = F'_-(v\varphi v^*),$$

$\varphi \in \mathscr{L}(l^2(G))$. Then Φ_1, Φ_2 are positive-definite and, for $\varphi \in \mathscr{L}(l^2(G))$, we have

$$F'(\varphi) = F'(v\varphi v^*) = \Phi_1(\varphi) - \Phi_2(\varphi),$$

$$\|F'\| \leq \|\Phi_1\| + \|\Phi_2\| = \|F'\|.$$

Therefore necessarily $F'_\pm(v\varphi v^*) = F'_\pm(\varphi)$ ([D] 12.3.4). As F' extends F, F'_+ and F'_- cannot be zero simultaneously. Hence there exists a state Φ on $\mathscr{L}(l^2(G))$ such that $\Phi(\varphi) = \Phi(L_x\varphi L_x^*)$ whenever $\varphi \in \mathscr{L}(l^2(G))$ and $x \in G$.

If $\Theta \in l^\infty(G)$, define $M_\Theta \in \mathscr{L}(l^2(G))$ by $M_\Theta(f) = \Theta f, f \in l^2(G)$. For all $f \in l^2(G)$ and $x, z \in G$,

$$L_{x^{-1}}M_\Theta L_{x^{-1}}^* f(z) = L_{x^{-1}}M_\Theta L_x f(z) = L_{x^{-1}}\big(\Theta(z)f(x^{-1}z)\big)$$

$$= \Theta(xz)f(z) = {}_x\Theta(z)f(z)$$

hence $L_{x^{-1}}M_\Theta L_{x^{-1}}^* = M_{x\Theta}$. We now define $M \in l^\infty(G)^*$ by

$$M(\Theta) = \Phi(M_\Theta),$$

$\Theta \in l^\infty(G)$. As Φ is a state, M is a mean on $l^\infty(G)$. For every $x \in G$ and every $\Theta \in l^\infty(G)$,

$$M({}_x\Theta) = \Phi(M_{x\Theta}) = \Phi\big(L_{x^{-1}}M_\Theta L_{x^{-1}}^*\big) = \Phi(M_\Theta) = M(\Theta).$$

Therefore $M \in \mathrm{LIM}(l^\infty(G))$.

(b) We verify (iii) of Proposition 24.18.

Let Y be a Banach C_L^*-module and let X be a submodule. Consider $F \in X^*$ such that $F(v\xi v^*) = F(\xi)$ whenever $\xi \in X$ and $v \in C_L^*(G)_u$. Let F' be an extension of F to an element of Y^* and put $F_x(\eta) = F'(L_x\eta L_x^*)$, for $x \in G, \eta \in Y$. We may then consider $F'_\eta \in l^\infty(G)$ defined by $F'_\eta(x) = F_x(\eta)$, $x \in G$. If $M \in \mathrm{RIM}(l^\infty(G))$, let also

$$\Phi: \eta \mapsto M\big(F'_\eta\big);$$

$\Phi \in Y^*$. For every $\xi \in X$ and every $x \in G$,

$$F_x(\xi) = F'(L_x\xi L_x^*) = F(L_x\xi L_x^*) = F(\xi).$$

Hence $F'_\xi(x) = F(\xi)$ and $\Phi(\xi) = F(\xi)$, that is, Φ extends F.

If $\eta \in Y$ and $x \in G$, $a \in G$, we have

$$F_{L_a\eta L_a^*}^{\prime(x)} = F_x(L_a\eta L_a^*) = F'(L_x L_a \eta L_a^* L_x^*) = F'(L_{xa}\eta L_{xa}^*).$$

As M is right invariant,

$$\Phi(L_a\eta L_a^*) = M(F'_{L_a\eta L_a^*}) = M(F'_\eta) = \Phi(\eta).$$

Therefore $\Phi(v\eta v^*) = \Phi(\eta)$ whenever $\eta \in Y$ and $v = C_L^*(G)_u$. □

Let A be a C^*-algebra and let $\tilde{A}$ be the associated C^*-algebra with unit u. To the Banach A-module X we associate the Banach $\tilde{A}$-module X for which $u\xi = \xi = \xi u$ whenever $\xi \in X$. We extend any derivation $D: A \to X^*$ to $\tilde{A}$ by putting $D(u) = 0$. Also, let

$$\Phi_D: x \mapsto D(x)x^*$$

$$\tilde{A} \to X^*.$$

Definition 24.20. *The C^*-algebra A is called strongly amenable if, for every Banach A-module X and every continuous derivation D of A into X^*, there exists $\eta \in \overline{\mathrm{co}}\,\Phi_D(\tilde{A}_u)$ such that $D = -d_1\eta$.*

Obviously every strongly amenable C^*-algebra is amenable. All postliminal C^*-algebras are strongly amenable ([291] theorem 7.9).

Many properties of amenable groups admit counterparts for algebras. The following property, a transposition of Proposition 17.5, may serve as an illustration of that fact.

Proposition 24.21. *Every continuous representation T of a strongly amenable C^*-algebra A on a Hilbert space $\mathcal{H}$ is equivalent to a unitary representation.*

Proof. It suffices to consider the case of a C^*-algebra with unit. We equip $\mathcal{L}(\mathcal{H})$ with the Banach A-module structure defined by

$$aV = T_a V, \qquad Va = V(T_{a^*})^*,$$

$a \in A$, $V \in \mathcal{L}(\mathcal{H})$ and consider $D \in \mathcal{L}(A, \mathcal{L}(\mathcal{H}))$ defined by

$$D(a) = T_a - (T_{a^*})^*,$$

$a \in A$. For all $a, b \in A$,

$$aD(b) + D(a)b = T_a(T_b - (T_{b^*})^*) + (T_a - (T_{a^*})^*)(T_{b^*})^*$$

$$= T_{ab} - (T_{a^*})^*(T_{b^*})^* = T_{ab} - (T_{b^*a^*})^* = T_{ab} - (T_{(ab)^*})^* = D(ab);$$

D is a continuous derivation. As A is strongly amenable, there exists $V \in \overline{\mathrm{co}}\{D(v)v^* : v \in A_u\}$ such that $D = -d_1 V$. Now for every $a \in A$,

$$T_a - (T_{a^*})^* = D(a) = -aV + Va = -T_a V + V(T_{a^*})^*;$$

therefore

$$T_a(id_{\mathscr{H}} + V) = (id_{\mathscr{H}} + V)(T_{a^*})^*. \qquad (8)$$

Let $U = id_{\mathscr{H}} + V$. For every $v \in A_u$, we have

$$D(v)v^* = (T_v - (T_{v^*})^*)T_v^* = T_v T_v^* - id_{\mathscr{H}}$$

and therefore $U \in \overline{\mathrm{co}}\{T_v T_v^* : v \in A_u\}$.

If $\xi \in \mathscr{H}$ and $v \in A_u$, $(T_v T_v^* \xi | \xi) \geq 0$; T and U are positive-definite operators. Let S be the square root of U. For all $\xi \in \mathscr{H}$ and $v \in A_u$, we have

$$\|\xi\| = \|(T_{v^*})^* T_v^* \xi\| \leq \|T\| \|T_v^* \xi\|. \qquad (9)$$

For every $\varepsilon > 0$, there exist $\alpha_1, \ldots, \alpha_n \in \mathbf{R}_+^*$ and $v_1, \ldots, v_n \in A_u$ such that $\sum_{i=1}^n \alpha_i = 1$ and $\|U - \sum_{i=1}^n \alpha_i T_{v_i} T_{v_i}^*\| < \varepsilon$. If $\sum_{i=1}^n \alpha_i T_{v_i} T_{v_i}^* \xi = 0$, we have $\sum_{i=1}^n \alpha_i (T_{v_i} T_{v_i}^* \xi | \xi) = 0$, $\sum_{i=1}^n \alpha_i \|T_{v_i}^* \xi\| = 0$, and, for every $i = 1, \ldots, n$, $\|T_{v_i}^* \xi\| = 0$. We deduce from (9) that $\xi = 0$. Therefore U^{-1} and S^{-1} exist. The relation (8) says that, for every $a \in A$,

$$T_a S^2 = S^2 (T_{a^*})^*,$$

$$S^{-1} T_a S = S(T_{a^*})^* S^{-1}.$$

The representation $T' = S^{-1} T S$ is equivalent to T. Finally note that, as the positive-definite S is hermitian, for every $a \in A$,

$$(T_a')^* = (S^{-1} T_a S)^* = (S(T_{a^*})^* S^{-1})^* = S^{-1} T_{a^*} S = T_{a^*}'. \qquad \square$$

NOTES

Amenable algebras were introduced by Johnson [291]. A general survey on amenable Banach algebras is to be found in Bonsall and Duncan [46]. Stegmeir [518] extensively studied general properties and examples of amenable Banach algebras. Johnson [291] established 24.16–17 and further stability properties for the class of amenable Banach algebras. Johnson [294] also proved the following "perturbation property": Let A be an amenable Banach algebra with multiplication π; there exists $\varepsilon > 0$ such that, for any multiplication π' on A with $\|\pi - \pi'\| < \varepsilon$, (A, π') is again an amenable Banach algebra. More developments on the subject are given in [295]. Johnson [292] defined an approximate diagonal for a Banach algebra A to be a bounded net (m_i) in $A \hat{\otimes} A$ such that $\lim_i (am_i - m_i a) = 0$ and $\lim_i \theta(m_i) a = a$ whenever $a \in A$, θ being the continuous linear mapping $A \hat{\otimes} A \to A$ defined by $\theta(a \otimes b) = ab$ for $a, b \in A$; a virtual diagonal is a point M in $(A \hat{\otimes} A)^{**}$ such that $aM = Ma$ and $\theta^{\sim\sim}(M)a = a$ whenever $a \in A$. He showed that A is amenable if and only if it admits an approximate diagonal or, equivalently, a virtual diagonal. Racher [440] expressed for amenability of a locally compact group G a necessary and sufficient condition in terms of existence of a virtual diagonal for $L^1(G)$.

The notion of strong amenable C^*-algebras was also introduced by Johnson [291]. Bunce [55, 57] established 24.18–19; Lau [338] obtained general statements. Paschke [427] studied crossed products of strongly amenable unital C^*-algebras by endomorphisms. Rosenberg [476] showed that the class of amenable (resp. strongly amenable) C^*-algebras is closed under the process of taking crossed products by amenable discrete groups; he also proved the existence of amenable C^*-algebras that are not strongly amenable. Johnson [291] noticed that if G is an amenable discrete group and T is a unitary representation of G on a Hilbert space $\mathcal{H}$, then $\{T_f : f \in l^1(G)\}^-$ is a strongly amenable C^*-subalgebra of $\mathcal{L}(\mathcal{H})$. Therefore 24.19 implies that, for any discrete group G, $C^*(G)$ is amenable if and only if it is strongly amenable. Lau [343] introduced a special type of Banach algebras that are preduals of nonnecessarily unique von Neumann algebras; for this class including the Fourier–Stieltjes algebras of locally compact groups and the measure algebras of locally compact semigroups he defined a notion of amenability that may be characterized in terms of existence of bounded approximate units. Bunce [56] proved 24.21; Barnes [30] and Bunce [58] established general statements concerning this equivalence problem.

The important class of amenable von Neumann algebras may be characterized by a lot of remarkable properties. We can only refer to the outstanding achievements of Connes [93, 94] that are summarized in [95] and developed in [96]. An early reference is Johnson, Kadison, and Ringrose [296]. We list the main characterizations of amenability for a von Neumann algebra A over a Hilbert space $\mathcal{H}$ admitting an infinite countable orthonormal basis: (i) A has property (P), that is, for every $T \in \mathcal{L}(\mathcal{H})$, the commutant of A intersects the closed convex hull of $\{VTV^* : V \in A_u\}$; (ii) A is approximately finite-dimensional, that is, A is generated by an increasing sequence of finite-dimensional subalgebras; (iii) A is semidiscrete, that is, the mapping $\Sigma_i a_i \otimes b_i \mapsto \Sigma_i a_i b_i$ of $A \otimes_{\min} A^{\sim\sim}$ into the C^*-algebra of $\mathcal{L}(\mathcal{H})$ generated by A and its commutant $A^{\sim\sim}$ is an isometry, $A \otimes_{\min} A^{\sim\sim}$ being $A \otimes A^{\sim\sim}$ equipped with the minimal norm for which the completion is a C^*-algebra; (iv) A is injective (Hakeda–Tomiyama property), that is, there exists a projection with norm 1 of $\mathcal{L}(\mathcal{H})$ onto A. The implication (iii) $\Rightarrow$ (iv) was established by Effros and Lance [154]. Bunce and Paschke [59] gave a proof showing that amenability implies injectivity. By Proposition 20.17 a discrete group G is amenable if and only if the von Neumann algebra PM(G) is amenable. Effros and Lance [154] proved that, for an amenable discrete group G, PM(G) is semidiscrete. See also de la Harpe [250] and Strătilă [521].

Connes showed that every amenable separable C^*-algebra is nuclear [94]. For considerations on the converse implication see Choi and Effros [73, 74], and Bunce and Paschke [60]. Haagerup [247] established that all nuclear C^*-algebras are amenable.

Amenable von Neumann algebras are counterparts of discrete amenable groups. Comparing amenability for a von Neumann algebra A over a Hilbert space $\mathcal{H}$ and amenability for a discrete group G, one has to replace $l^\infty(G)$ by $\mathscr{L}(\mathcal{H})$ and invariant means by hypertraces, that is, states on $\mathscr{L}(\mathcal{H})$ that are invariant under the action of unitary elements. We end up by stating a formal general notion of amenability in terms of existence of invariant means. Let X be a standard Borel space and $\mathscr{R}$ a Borel subset of $X \times X$ that is an equivalence relation. On X one considers a measure μ that is quasi-invariant on $\mathscr{R}$, that is, for each μ-null Borel subset A in X, the set $\{x \in X: (\exists y \in A) (x, y) \in \mathscr{R}\}$ is μ-null; one endows $\mathscr{R}$ with the measure $m = \int v^x d\mu(x)$ where v^x ($x \in X$) is the counting measure on $\mathscr{R}^x = \{(x, y): y \sim x\}$. A measurable bijection Φ of a measurable subset of X onto a measurable subset of X is called partial transformation; for any $f: \mathscr{R} \to \mathbf{R}$ and any partial transformation Φ one defines

$$f^\Phi(y, x) = 0 \text{ if } x \in X, y \notin \mathscr{I}m \, \Phi,$$

$$f^\Phi(y, x) = f(\Phi^{-1}(y), x) \text{ if } x \in X, y \in \mathscr{I}m \, \Phi.$$

Connex, Feldman and B. Weiss [97] call $\mathscr{R}$ amenable if it posseses a left invariant mean, that is, there exists $P: L_\mathbf{R}^\infty(\mathscr{R}, m) \to L_\mathbf{R}^\infty(X, \mu)$ such that $P \geq 0$, $P(1_\mathscr{R}) = 1_X$ and $P(f^\Phi) = (Pf)^\Phi$ whenever $f \in L_\mathbf{R}^\infty(\mathscr{R}, m)$ and Φ is a partial transformation. The existence of such a left invariant mean is equivalent to the amenability of an associated von Neumann algebra.

Recent publications: Cuntz [591], Gootman [592], C. Moore [594], and Schmidt [596].

BIBLIOGRAPHY

BASIC REFERENCES

Bourbaki N. *Élements de mathématique*.
[B A]   *Algèbre*. Chap. 1–3. Hermann, Paris, 1970.
[B TG]   *Topologie générale*. Chap. 1–4, 1971; Chap. 5–10, 1974. Hermann, Paris.
[B EVT]   *Espaces vectoriels topologiques*. Chap. 1–5. Masson, Paris, 1981.
[B INT]   *Intégration*. Chap. 1–4, 1965; Chap. 5, 1967; Chap. 6, 1959; Chap. 7–8, 1963; Chap. 9, 1969. Hermann, Paris.
[B LIE]   *Groupes et algèbres de Lie*. Chap. 1, 1971; Chap. 2–3, 1972; Chap. 4–6, 1968; Chap. 7–8, 1975. Hermann, Paris.
[B TS]   *Théories spectrales*. Chap. 1–2. Hermann, Paris, 1967.
[D]   Dixmier, Jacques. *Les C*-algèbres et leurs représentations*, 2nd ed. Gauthier–Villars, Paris, 1969.
[DS]   Dunford, Nelson, and Jacob T. Schwartz. *Linear Operators*. Part I. General theory, 3rd ed. Interscience, New York, 1966.
   Hewitt, Edwin, and Kenneth A. Ross. *Abstract Harmonic Analysis*.
[HR I]   Vol. I, 2nd ed. Springer-Verlag, Berlin, 1979.
[HR II]   Vol. II. Springer-Verlag, Berlin, 1970.
[HS]   Hewitt, Edwin, and Karl Stromberg. *Real and Abstract Analysis*. 2nd ed. Springer-Verlag, Berlin, 1969.

CITED REFERENCES

1. Adel'son-Vel'skiĭ, G. M., and Ju. A. Šreider. Banach means on groups (In Russian). *Usp. Mat. Nauk* (N. S.) **12**, 131–136 (1957).
2. Adian, S. I. Classifications of periodic words and their application in group theory. *Burnside Groups*. Lecture Notes in Mathematics, Vol. 806, 1–40. Springer-Verlag, Berlin, 1980.
   Adian, S. I. See [416].
3. Adler, Andrew. On continuity of invariant measures. *Proc. Amer. Math. Soc.* **41**, 487–491 (1973).
4. Akemann, Charles A. Operator algebras associated with fuchsian groups. *Houston J. Math.* **7**, 295–301 (1981).
5. Akemann, Charles A., and Martin E. Walter. The Riemann–Lebesgue property for arbitrary locally compact groups. *Duke Math. J.* **43**, 225–236 (1976).

6. Alfsen, Erik M., and Per Holm. A note on compact representations and almost periodicity in topological groups. *Math. Scand.* **10**, 127–136 (1962).
7. Anantharaman-Delaroche, Claire. Sur la moyennabilité des actions libres d'un groupe localement compact dans une algèbre de von Neumann. *C. R. Acad. Sci. Paris*, Sér. A **289**, 605–607 (1979).
8. Anantharaman-Delaroche, Claire. Action moyennable d'un groupe localement compact sur une algèbre de von Neumann. *Math. Scand.* **45**, 289–304 (1979).
9. Anantharaman-Delaroche, Claire. Action moyennable d'un groupe localement compact sur une algèbre de von Neumann II. *Math. Scand.* **50**, 251–268 (1982).
10. Anker, Jean-Philippe. *Contenance faible au sens de Fell et moyennabilité*. Faculté des Sciences, University of Lausanne, 1978.
11. Anker, Jean-Philippe. Sur la propriété P_*. *Monatsh. Math.* **90**, 87–90 (1980).
12. Anker, Jean-Philippe. *Aspects de la p-induction en analyse harmonique*. Payot, Lausanne, 1982.
13. Appel, K. I., and F. M. Djorup. On the group generated by a free semigroup. *Proc. Amer. Math. Soc.* **15**, 838–840 (1964).
14. Arens, Richard. The adjoint of a bilinear operation. *Proc. Amer. Math. Soc.* **2**, 839–848 (1951).
15. Argabright, Loren N. Invariant means on topological semigroups. *Pacific J. Math.* **16**, 193–203 (1966).
16. Argabright, Loren N. Invariant means and fixed points: a sequel to Mitchell's paper. *Trans. Amer. Math. Soc.* **130**, 127–130 (1968).
17. Argabright, Loren N., and Carroll O. Wilde. Semigroups satisfying a strong Følner condition. *Proc. Amer. Math. Soc.* **18**, 587–591 (1967).
18. Aribaud, François. Un théorème ergodique pour les espaces L^1. *J. Functional Analysis* **5**, 395–411 (1970).
19. Aribaud, François. Sur la moyenne temporelle d'un système dynamique. *Illinois J. Math.* **17**, 90–110 (1973).
20. Arsac, Gilbert. Sur l'espace de Banach engendré par les coefficients d'une représentation unitaire. *Publ. Dép. Math.* (Lyon) **13**, 1–101 (1976).
21. Avez, André. Limite de quotients pour des marches aléatoires sur des groupes. *C. R. Acad. Sci. Paris*, Sér. A **276**, 317–320 (1973).
22. Avez, André. Théorème de Choquet-Deny pour les groupes à croissance non exponentielle. *C. R. Acad. Sci. Paris*, Sér. A **279**, 25–28 (1974).
23. Avez, André. Croissance des groupes de type fini et fonctions harmoniques. *Théorie ergodique*. Lecture Notes in Mathematics, Vol. 532, 35–49. Springer-Verlag, Berlin, 1976.
24. Azencott, Robert. *Espaces de Poisson des groupes localement compacts*. Lecture Notes in Mathematics, Vol. **148**. Springer-Verlag, Berlin, 1970.
25. Baggett, Lawrence. A weak containment theorem for groups with a quotient R-group. *Trans. Amer. Math. Soc.* **128**, 277–290 (1967).
26. Baggett, Lawrence. A description of the topology on the dual spaces of certain locally compact groups. *Trans. Amer. Math. Soc.* **132**, 175–215 (1968).
27. Baldi, P. Caractérisation des groupes de Lie connexes récurrents. *Ann. Inst. H. Poincaré*, Sect. B **XVII**, 281–308 (1981).
28. Banach, Stefan. Sur le problème de la mesure. *Fund. Math.* **4**, 7–33 (1923).
29. Banach, Stefan, and Alfred Tarski. Sur la décomposition des ensembles de points en parties respectivement congruentes. *Fund. Math.* **6**, 244–277 (1924).
30. Barnes, Bruce A. When is a representation of a Banach $*$-algebra Naimark-related to a $*$-representation? *Pacific J. Math.* **72**, 5–25 (1977).

CITED REFERENCES

31. Beiglböck, W. D. Über einen Satz von L. Pukanszky. *Manuscripta Math.* **7**, 113–123 (1972).
32. Berg, Christian, and Jens Peter Reus Christensen. On the relation between amenability of locally compact groups and the norms of convolution operators. *Math. Ann.* **208**, 149–153 (1974).
33. Berg, Christian, and Jens Peter Reus Christensen. Sur la norme des opérateurs de convolution. *Invent. Math.* **23**, 173–178 (1974).
34. Berglund, John F., Hugo D. Junghenn, and Paul Milnes. *Compact Right Topological Semigroups and Generalizations of Almost Periodicity*. Lecture Notes in Mathematics, Vol. 663. Springer-Verlag, Berlin, 1978.
35. Birgé, Lucien, and Albert Raugi. Fonctions harmoniques sur les groupes moyennables. *C. R. Acad. Sci. Paris*, Sér. A **278**, 1287–1289 (1974).
36. Blattner, Robert J. On induced representations. *Amer. J. Math.* **83**, 79–98 (1961).
37. Blum, Julius R., and Bennet Eisenberg. Generalized summing sequences and the mean ergodic theorem. *Proc. Amer. Math. Soc.* **42**, 423–429 (1974).
38. Blum, Julius R., Bennet Eisenberg, and L. S. Hahn. Ergodic theory and the measure of sets in the Bohr group. *Acta Sci. Math.* (Szeged) **34**, 17–24 (1973).
39. Bohr, Harald. *Fastperiodische Funktionen*. Springer-Verlag, Berlin, 1932.
40. Boissier, Monique. Sur des conditions de moyennabilité d'un groupe. *Nant. Math.* **1**, 36–39 (1974).
41. Bombal, Fernando. Medidas invariantes con valores en A-modulos normados. *Rev. Mat. Hisp.-Amer.* **32**, 223–238 (1972).
42. Bombal, Fernando, and G. Vera. Medias en espacios localmente convexos y semi-reflexividad. *Collect. Math.* **XXIV**, 267–295 (1973); **XXVI**, 3–4 (1975).
43. Bombal, Fernando, and G. Vera. Funciones vectoriales casi convergentes. *Collect. Math.* **XXVI**, 141–156 (1975).
44. Bondar, James V., and Paul Milnes. Amenability: A survey for statistical applications of Hunt-Stein and related conditions on groups. *Wahrscheinlichkeitstheorie und Verw. Gebiete* **57**, 103–128 (1981).
    Bondar, James V. See [389].
45. Bonic, Robert A. Symmetry in group algebras of discrete groups. *Pacific J. Math.* **11**, 73–94 (1961).
46. Bonsall, F. F., and J. Duncan. *Complete Normed Algebras*. Springer-Verlag, Berlin, 1973.
47. Borel, Armand. On free subgroups of semisimple groups. *Enseignement Math.* **29**, 151–164 (1983).
48. Bottesch, Martin. Invariante Mittel für Vektorfunktionen. *Mathematica* **19**, 129–136 (1977).
49. Bożejko, Marek. Some aspects of harmonic analysis on free groups. *Colloq. Math.* **XLI**, 265–271 (1979).
50. Bożejko, Marek. Uniformly amenable discrete groups. *Math. Ann.* **251**, 1–6 (1980).
51. Bożejko, Marek. Remark on Herz-Schur multipliers on free groups. *Math. Ann.* **258**, 11–15 (1981).
52. Bożejko, Marek. A new group algebra and lacunary sets in discrete noncommutative groups. *Studia Math.* **LXX**, 165–175 (1981).
53. Brooks, Robert. Amenability and the spectrum of the laplacian. *Bull. Amer. Math. Soc.* **6**, 87–89 (1982).
54. Brown, Arlen, and Carl Pearcy. Structure of commutators of operators. *Ann. of Math.* **82**, 112–127 (1965).
55. Bunce, John W. Characterizations of amenable and strongly amenable C^*-algebras. *Pacific J. Math.* **43**, 563–572 (1972).

56. Bunce, John W. Representations of strongly amenable C^*-algebras. *Proc. Amer. Math. Soc.* **32**, 241–246 (1972).

57. Bunce, John W. Finite operators and amenable C^*-algebras. *Proc. Amer. Math. Soc.* **56**, 145–151 (1976).

58. Bunce, John W. The similarity problem for representations of C^*-algebras. *Proc. Amer. Math. Soc.* **81**, 409–414 (1981).

59. Bunce, John W., and William L. Paschke. Quasi-expectations and amenable von Neumann algebras. *Proc. Amer. Math. Soc.*, **71**, 232–236 (1978).

60. Bunce, John W., and William L. Paschke. Derivations on a C^*-algebra and its double dual. *J. Functional Analysis* **37**, 235–247 (1980).

61. Burckel, Robert B. *Weakly Almost Periodic Functions on Semigroups.* Gordon and Breach, New York, 1970.

62. Calderón, A. P. A general ergodic theorem. *Ann. of Math.* **58**, 182–191 (1953).

63. Calderón, A. P. Ergodic theory and translation-invariant operators. *Proc. Nat. Acad. Sci. U.S.A.* **59**, 349–353 (1968).

64. Carling, L. N. On the restriction map of the Fourier–Stieltjes algebra $B(G)$ and $B_p(G)$. *J. Functional Analysis* **25**, 236–243 (1977).

65. Cecchini, Carlo. Operators on $VN(G)$ commuting with $A(G)$. *Colloq. Math.* **XLIII**, 137–142 (1980).

66. Chatard, Jacqueline. Moyenne de fonctions presque périodiques intégrables. *C. R. Acad. Sci. Paris*, Sér. A **270**, 259–262 (1970).

67. Chatard, Jacqueline. Applications des propriétés de moyenne d'un groupe localement compact à la théorie ergodique. *Ann. Inst. H. Poincaré*, Sect. B **VI**, 307–326 (1970).

Chen, Ing-Sheun. See [329].

68. Chen, Su-Shing. On nonamenable groups. *Internat. J. Math. and Math. Sci.* **1**, 529–532 (1978).

Chivukula, R. Rao. See [573].

69. Choda, Hisashi, and Marie Choda. Fullness, simplicity and inner amenability. *Math. Japon.* **24**, 235–246 (1979).

70. Choda, Marie. The factors of inner amenable groups. *Math. Japon.* **24**, 145–152 (1979).

71. Choda, Marie. Effect of inner amenability on strong ergodicity. *Math. Japon.* **28**, 109–115 (1983).

72. Choda, Marie, and Yasuo Watatani. Conditions for inner amenability of groups. *Math. Japon.* **24**, 401–402 (1979).

Choda, Marie. See [69].

73. Choi, Man-Duen, and Edward G. Effros. Separable nuclear C^*-algebras and injectivity. *Duke Math. J.* **43**, 309–322 (1976).

74. Choi, Man-Duen, and Edward G. Effros. Nuclear C^*-algebras and injectivity: the general case. *Indiana Univ. Math. J.* **26**, 443–446 (1977).

75. Chou, Ching. On the size of the set of left invariant means on a semigroup. *Proc. Amer. Math. Soc.* **23**, 199–205 (1969).

76. Chou, Ching. Minimal sets and ergodic measures for $\beta\mathbf{N} \setminus \mathbf{N}$. *Illinois J. Math.* **13**, 777–788 (1969).

77. Chou, Ching. On a conjecture of E. Granirer concerning the range of an invariant mean. *Proc. Amer. Math. Soc.* **26**, 105–107 (1970).

78. Chou, Ching. On topologically invariant means on a locally compact group. *Trans. Amer. Math. Soc.* **151**, 443–456 (1970).

79. Chou, Ching. On a geometric property of the set of invariant means on a group. *Proc. Amer. Math. Soc.* **30**, 296–301 (1971).

CITED REFERENCES

80. Chou, Ching. The multipliers of the space of almost convergent sequences. *Illinois J. Math.* **16**, 687–694 (1972).
81. Chou, Ching. Weakly almost periodic functions with zero mean. *Bull. Amer. Math. Soc.* **80**, 297–299 (1974).
82. Chou, Ching. Weakly almost periodic functions and almost convergent functions on a group. *Trans. Amer. Math. Soc.* **206**, 175–200 (1975).
83. Chou, Ching. The exact cardinality of the set of invariant means on a group. *Proc. Amer. Math. Soc.* **55**, 103–106 (1976).
84. Chou, Ching. Locally compact groups which are amenable as discrete groups. *Proc. Amer. Math. Soc.* **76**, 46–50 (1979).
85. Chou, Ching. Minimally weakly almost periodic groups. *J. Functional Analysis* **36**, 1–17 (1980).
86. Chou, Ching. Elementary amenable groups. *Illinois J. Math.* **24**, 396–407 (1980).
87. Chou, Ching. Topological invariant means on the von Neumann algebra $VN(G)$. *Trans. Amer. Math. Soc.* **273**, 207–229 (1982).
88. Chou, Ching. Weakly almost periodic functions and Fourier-Stieltjes algebras of locally compact groups. *Trans. Amer. Math. Soc.* **274**, 141–157 (1982).

Christensen, Jens Peter Reus. See [32, 33].

89. Cohen, Joel M. Cogrowth and amenability of discrete groups. *J. Functional Analysis* **48**, 301–309 (1982).
90. Coifman, Ronald R., and Guido Weiss. Operators transferred by representations of an amenable group. *Proceedings of Symposia in Pure Mathematics*, Vol. **26**, 369–372. American Mathematical Society, Providence, 1973.
91. Coifman, Ronald R., and Guido Weiss. Operators associated with representations of amenable groups, singular integrals induced by ergodic flows, the rotation method and multipliers. *Studia Math.* **XLVII**, 285–303 (1973).
92. Comfort, W. W., and Kenneth A. Ross. Pseudocompactness and uniform continuity in topological groups. *Pacific J. Math.* **16**, 483–496 (1966).
93. Connes, Alain. Classification of injective factors. *Ann. of Math.* **104**, 73–115 (1976).
94. Connes, Alain. On the cohomology of operator algebras. *J. Functional Analysis* **28**, 248–253 (1978).
95. Connes, Alain. Von Neumann algebras. *Proceedings of the International Congress of Mathematicians*, Helsinki, 1978, Vol. **1**, 97–109. University of Helsinki, 1980.
96. Connes, Alain. Classification des facteurs. *Proceedings of Symposia in Pure Mathematics*, Vol. **38/2**, 43–109. American Mathematical Society, Providence, 1982.
97. Connes, Alain, Jacob Feldman, and Benjamin Weiss. An amenable equivalence relation is generated by a single transformation. *Ergod. Th. and Dynam. Sys.* **1**, 431–450 (1981).
98. Converse, George, Isaac Namioka, and R. R. Phelps. Extreme invariant positive operators. *Trans. Amer. Math. Soc.* **137**, 375–385 (1969).
99. Conze, Jean-Pierre, and N. Dang Ngoc. Noncommutative ergodic theorems. *Bull. Amer. Math. Soc.* **83**, 1297–1299 (1977).
100. Conze, Jean-Pierre, and N. Dang Ngoc. Ergodic theorems for noncommutative dynamical systems. *Invent. Math.* **46**, 1–15 (1978).
101. Corduneanu, Constantin. *Almost Periodic Functions*. Interscience Tracts in Pure and Applied Mathematics. Wiley, New York, 1968.
102. Cowling, Michael. La synthèse des convoluteurs de L^p de certains groupes pas moyennables. *Boll. Un. Mat. Ital.* (5) **14A**, 551–555 (1977); **6**, 317 (1982).
103. Cowling, Michael. Some applications of Grothendieck's theory of topological tensor products in harmonic analysis. *Math. Ann.* **232**, 273–285 (1978).

104. Cowling, Michael. The Kunze–Stein phenomenon. *Ann. of Math.* **107**, 209–234 (1978).
105. Cowling, Michael. An application of the Littlewood-Paley theory in harmonic analysis. *Math. Ann.* **241**, 83–86 (1979).
106. Cowling, Michael, and John J. F. Fournier. Inclusions and noninclusion of spaces of convolution operators. *Trans. Amer. Math. Soc.* **221**, 59–95 (1976).
107. Crépel, Pierre. Récurrence des marches aléatoires sur les groupes de Lie. *Théorie ergodique.* Lecture Notes in Mathematics, Vol. **532**, 50–69. Springer-Verlag, Berlin, 1976.
108. Crépel, Pierre, and J. Lacroix. Théoremes de renouvellement pour les marches aléatoires sur les groupes localement compacts. *Proceedings*: *Sém. Théorie du potentiel, No. 2.* Lecture Notes in Mathematics, Vol. **563**, 27–42. Springer-Verlag, Berlin, 1976.

Dang Ngoc, N. See [99, 100].

109. Darsow, W. F. Positive definite functions and states. *Ann. of Math.* **60**, 447–453 (1954).
110. Davis, Henry W. On the mean value of Haar measurable almost periodic functions. *Duke Math. J.* **34**, 201–214 (1967).
111. Day, Mahlon M. Means and ergodicity of semigroups. *Bull. Amer. Math. Soc.* **55**, 1054 (1949).
112. Day, Mahlon M. Means on semigroups and groups. *Bull. Amer. Math. Soc.* **55**, 1054–1055 (1949).
113. Day, Mahlon M. Amenable groups. *Bull. Amer. Math. Soc.* **56**, 46 (1950).
114. Day, Mahlon M. Means for the bounded functions and ergodicity of the bounded representations of semigroups. *Trans. Amer. Math. Soc.* **69**, 276–291 (1950).
115. Day, Mahlon M. Amenable semigroups. *Illinois J. Math.* **1**, 509–544 (1957).
116. Day, Mahlon M. Fixed-point theorems for compact convex sets. *Illinois J. Math.* **5**, 585–590 (1961).
117. Day, Mahlon M. Correction to my paper "Fixed-point theorems for compact convex sets". *Illinois J. Math.* **8**, 713 (1964).
118. Day, Mahlon M. Convolutions, means, and spectra. *Illinois J. Math.* **8**, 100–111 (1964).
119. Day, Mahlon M. Amenability and equicontinuity. *Studia Math.* **XXXI**, 481–494 (1968).
120. Day, Mahlon M. Semigroups and amenability. *Proceedings*: *Symp. Semigroups*, 5–53. Academic, New York, 1969.
121. Day, Mahlon M. Invariant renorming. *Proceedings*: *Symp. Fixed Point Theory and Applications*, 51–62. Academic, New York, 1976.
122. Day, Mahlon M. Lumpy subsets in left-amenable locally compact semigroups. *Pacific J. Math.* **62**, 87–92 (1976).
123. Day, Mahlon M. Left thick to left lumpy — a guided tour. *Pacific J. Math.* **101**, 71–92 (1982).
124. Deligne, Pierre, and Dennis Sullivan. Division algebras and the Hausdorff–Banach–Tarski paradox. *Enseignement Math.* **29**, 145–150 (1983).
125. Derighetti, Antoine. Some results on the Fourier–Stieltjes algebra of a locally compact group. *Comment. Math. Helv.* **45**, 219–228 (1970).
126. Derighetti, Antoine. On the property (P_1) of locally compact groups. *Comment. Math. Helv.* **46**, 226–239 (1971).
127. Derighetti, Antoine. Sur certaines propriétés des représentations unitaires des groupes localement compacts. *Comment. Math. Helv.* **48**, 328–339 (1973).
128. Derighetti, Antoine. Sulla nozione di contenimento debole e la proprietà di Reiter. *Rend. Sem. Mat. Fis. Milano* **44**, 47–54 (1974).
129. Derighetti, Antoine. Sur la propriété P_1. *C. R. Acad. Sci. Paris*, Sér. A **283**, 317–319 (1976).

130. Derighetti, Antoine. Sur les représentations unitaires des sous-groupes fermés d'un groupe localement compact. *Symposia Mathematica* **XXII**: *Analisi armonica e spazi di funzioni su gruppi localmente compatti*, 133–143. Academic, London, 1977.

131. Derighetti, Antoine. Some remarks on $L^1(G)$. *Math. Z.* **146**, 189–194 (1978).

132. Derighetti, Antoine. Relations entre les convoluteurs d'un groupe localement compact et ceux d'un sous-groupe fermé. *Bull. Sci. Math.* **106**, 69–84 (1982).

133. Derighetti, Antoine. A propos des convoluteurs d'un groupe quotient. *Bull. Sci. Math.* **107**, 3–23 (1983).

134. Derriennic, Yves, and Yves Guivarc'h. Théorème de renouvellement pour les groupes non moyennables. *C. R. Acad. Sci. Paris*, Sér. A **277**, 613–615 (1973).

135. Dhombres, Jean. Sur une classe de moyennes. *Ann. Inst. Fourier* (Grenoble) **17**, 135–156 (1967).

136. Dieudonné, Jean. Sur le produit de composition (II). *J. Math. Pures Appl.* **39**, 275–292 (1960).

137. Dieudonné, Jean. *Éléments d'analyse*, Tome V. Gauthier–Villars, Paris, 1975.

138. Dixmier, Jacques. Les moyennes invariantes dans les semi-groupes et leurs applications. *Acta Sci. Math.* (Szeged) **12**, 213–227 (1950).

139. Dixmier, Jacques. Opérateurs de rang fini dans les représentations unitaires. *Inst. Hautes Etudes Sci. Publ. Math.* **6**, 305–317 (1960).

140. Dixmier, Jacques. *Les algèbres d'opérateurs dans l'espace hilbertien*. 2nd ed. Gauthier–Villars, Paris, 1969.

Djorup, F. M. See [13].

141. Doplicher, Sergio, Richard V. Kadison, Daniel Kastler, and Derek W. Robinson. Asymptotically abelian systems. *Comm. Math. Phys.* **6**, 101–120 (1967).

142. Doran, Robert S., and Josef Wichmann. *Approximate Identities and Factorization in Banach Modules*. Lecture Notes in Mathematics, Vol. **768**. Springer-Verlag, Berlin, 1979.

143. Doss, Raouf. On the Fourier–Stieltjes transforms of singular or absolutely continuous measures. *Math. Z.* **97**, 77–84 (1967).

144. Doss, Raouf. On the transform of a singular or an absolutely continuous measure. *Proc. Amer. Math. Soc.* **19**, 361–363 (1968).

145. Douglas, R. G. On the inversion invariance of invariant means. *Proc. Amer. Math. Soc.* **16**, 642–644 (1965).

146. Douglass, Steven A. Summing sequences for amenable semigroups. *Michigan Math. J.* **20**, 169–179 (1973).

147. Dreseler, Bernd, and Walter Schempp. Spaces of test functions and theorems of the Bochner–Schoenberg–Eberlein type. *Math. Z.* **148**, 207–213 (1976).

148. Dubins, L. E. Le paradoxe de Hausdorff–Banach–Tarski. *Gazette des Mathématiciens* **12**, 71–76 (1979).

149. Duncan, J., and I. Namioka. Amenability of inverse semigroups and their semigroup algebras. *Proc. Roy. Soc. Edinburgh* **80A**, 309–321 (1978).

Duncan, J. See [46].

150. Dunkl, Charles F., and Donald E. Ramirez. Existence and nonuniqueness of invariant means on $\mathscr{L}^\infty(\hat{G})$. *Proc. Amer. Math. Soc.* **32**, 525–530 (1972).

151. Dunkl, Charles F., and Donald E. Ramirez. Weakly almost periodic functionals on the Fourier algebra. *Trans. Amer. Math. Soc.* **185**, 501–514 (1973).

152. Effros, Edward G. Property Γ and inner amenability. *Proc. Amer. Math. Soc.* **47**, 483–486 (1975).

153. Effros, Edward G., and Frank Hahn. Locally compact transformation groups and C^*-algebras. *Mem. Amer. Math. Soc.* **75** (1967).

154. Effros, Edward G., and E. Christopher Lance. Tensor products of operator algebras. *Advances in Math.* **25**, 1–34 (1977).
Effros, Edward G. See [73, 74].
155. Ehrenpreis, L., and F. I. Mautner. Some properties of the Fourier transform on semisimple Lie groups III. *Trans. Amer. Math. Soc.* **90**, 431–484 (1959).
Eisenberg, Bennet. See [37, 38].
156. Elie, Laure. Renouvellement sur les groupes moyennables. *C. R. Acad. Sci. Paris*, Sér. A **284**, 555–558 (1977).
157. Elie, Laure, and Albert Raugi. Fonctions harmoniques sur certains groupes résolubles. *C. R. Acad. Sci. Paris*, Sér. A **280**, 377–379 (1975).
158. Ellis, Robert. Locally compact transformation groups. *Duke Math. J.* **24**, 119–125 (1957).
159. Ellis, Robert. *Lectures on Topological Dynamics*. Benjamin, New York, 1969.
160. Emch, G. G., H. J. F. Knops, and E. J. Verboven. *On Partial Weakly Clustering States with an Application to the Ising Model*. Instituut voor Theoretische Fysica, Faculteit der Wiskunde en Natuurwetenschappen, Katholieke Univ. Nijmegen, 1967.
161. Emerson, William R. Asymptotic results for a class of integral operators over groups. *Journal of Mathematics and Mechanics* **17**, 737–758 (1968).
162. Emerson, William R. Ratio properties in locally compact amenable groups. *Trans. Amer. Math. Soc.* **133**, 179–204 (1968).
163. Emerson, William R. Large symmetric sets in amenable groups and the individual ergodic theorem. *Amer. J. Math.* **96**, 242–247 (1974).
164. Emerson, William R. The pointwise ergodic theorem for amenable groups. *Amer. J. Math.* **96**, 472–487 (1974).
165. Emerson, William R. Averaging strongly subadditive set functions in unimodular amenable groups I. *Pacific J. Math.* **61**, 391–400 (1975).
166. Emerson, William R. Averaging strongly subadditive set functions in unimodular amenable groups II. *Pacific J. Math.* **64**, 353–368 (1976).
167. Emerson, William R. Characterizations of amenable groups *Trans. Amer. Math. Soc.* **241**, 183–194 (1978).
168. Emerson, William R. The Hausdorff paradox for general group actions. *J. Functional Analysis* **32**, 213–227 (1979).
169. Emerson, William R., and Frederick P. Greenleaf. Covering properties and Følner conditions for locally compact groups. *Math. Z.* **102**, 370–384 (1967).
170. Emerson, William R., and Frederick P. Greenleaf. Asymptotic behavior of products $C^p = C + \cdots + C$ in locally compact abelian groups. *Trans. Amer. Math. Soc.* **145**, 171–204 (1969).
Emerson, William R. See [230].
171. Eymard, Pierre. L'algèbre de Fourier d'un groupe localement compact. *Bull. Soc. Math. France*, **92**, 181–236 (1964).
172. Eymard, Pierre. Algèbres A_p et convoluteurs de $L^p(G)$. *Séminaire Bourbaki, Vol.* 1969/70. Lecture Notes in Mathematics, Vol. **180**, 364–381. Springer-Verlag, Berlin, 1971.
173. Eymard, Pierre. Sur les moyennes invariantes et les représentations unitaires. *C. R. Acad. Sci. Paris*, Sér. A **272**, 1649–1652 (1971).
174. Eymard, Pierre. *Moyennes invariantes et représentations unitaires*. Lecture Notes in Mathematics, Vol. **300**. Springer-Verlag, Berlin, 1972.
175. Eymard, Pierre. Initiation à la théorie des groupes moyennables. *Analyse harmonique sur les groupes de Lie*. Lecture Notes in Mathematics, Vol. **497**, 89–107. Springer-Verlag, Berlin, 1975.
176. Eymard, Pierre. Invariant means and unitary representations. *Proceedings of Symposia in Pure Mathematics*, Vol. **26**, 373–376. Academic, New York, 1976.

CITED REFERENCES

177. Fairchield, Lonnie. Extreme invariant means without minimal support. *Trans. Amer. Math. Soc.* **172**, 83–93 (1972).
178. Faraut, Jacques. Moyennabilité et normes d'opérateurs de convolution. *Analyse harmonique sur les groupes de Lie.* Lecture Notes in Mathematics, Vol. **497**, 153–163. Springer-Verlag, Berlin, 1975.
179. Faraut, Jacques. Semi-groupes de Feller invariants sur les espaces homogènes non moyennables. *Analyse harmonique sur les groupes de Lie.* Lecture Notes in Mathematics, Vol. **497**, 164–171. Springer-Verlag, Berlin, 1975.
180. Feichtinger, Hans G. Multipliers of Banach spaces of functions on groups. *Math. Z.* **152**, 47–58 (1976).

    Feldman, Jacob. See [97].

181. Fell, J. M. G. The dual spaces of C^*-algebras. *Trans. Amer. Math. Soc.* **94**, 365–403 (1960).
182. Fell, J. M. G. Weak containment and induced representations of groups. *Canad. J. Math.* **14**, 237–268 (1962).
183. Figà-Talamanca, Alessandro, and Massimo Picardello. Multiplicateurs de $A(G)$ qui ne sont pas dans $B(G)$. *C. R. Acad. Sci. Paris*, Sér. A **277**, 117–119 (1973).
184. Flory, Volker. On the Fourier algebra of a locally compact amenable group. *Proc. Amer. Math. Soc.* **29**, 603–606 (1971).
185. Flory, Volker. *Eine Lebesgue–Zerlegung und funktorielle Eigenschaften der Fourier–Stieltjes Algebra.* Inaugural Dissertation, University of Heidelberg, 1972.
186. Flory, Volker. Estimating norms in C^*-algebras of discrete groups. *Math. Ann.* **224**, 41–52 (1976).
187. Følner, Erling. On groups with full Banach mean value. *Math. Scand.* **3**, 243–254 (1955).
188. Fountain, J. B., R. W. Ramsay, and John H. Williamson. Functions of measures on compact groups. *Proc. Roy. Irish Acad.*, Sect. A **76**, 235–251 (1976).

    Fournier, John J. F. See [106].

189. Fremlin, David D., and Michel Talagrand. A decomposition theorem for additive set-functions, with applications to Pettis integrals and ergodic means. *Math. Z.* **168**, 117–142 (1979).
190. Frémond, C., and M. Sueur-Pontier. Caractérisation des groupes localement compacts de type (T) ayant la propriété du point fixe. *Ann. Inst. H. Poincaré*, Sect. B **VII**, 293–298 (1971).
191. Furstenberg, Harry. A Poisson formula for semi-simple Lie groups. *Ann. of Math.* **77**, 335–386 (1963).
192. Furstenberg, Harry. Random Walks on Lie Groups. *Harmonic Analysis and Representations of Semisimple Lie Groups*, 467–489. D. Reidel Publishing Company, London, 1980.
193. Furstenberg, Harry. Rigidity and cocycles for ergodic actions of semi-simple Lie groups. *Séminaire Bourbaki, Vol. 1979/80.* Lecture Notes in Mathematics, Vol. **842**, 273–292. Springer-Verlag, Berlin, 1981.
194. Ganesan, S. *P-amenable Locally Compact Groups.* (To be published) (1983).
195. Gangolli, Ramesh. On the symmetry of L_1 algebras of locally compact motion groups, and the Wiener tauberian theorem. *J. Functional Analysis* **25**, 244–252 (1977).
196. Gerl, Peter. Diskrete, mittelbare Gruppen. *Monatsh. Math.* **77**, 307–318 (1973).
197. Gerl, Peter, Gleichverteilung auf lokalkompakten Gruppen. *Math. Nachr.* **71**, 249–260 (1976).
198. Gerl, Peter. Wahrscheinlichkeitsmaße auf diskreten Gruppen. *Arch. Math.* (Basel) **31**, 605–610 (1978).
199. Gilbert, John. Convolution operators on $L^p(G)$ and properties of locally compact groups. *Pacific J. Math.* **24**, 257–268 (1968).

200. Gilbert, John. Tensor products of Banach spaces and weak type (p,p) multipliers. *J. Approx. Theory* **13**, 136–145 (1975).

201. Gilbert, John. L^p-convolution operators and tensor products of Banach spaces. *Bull. Amer. Math. Soc.* **80**, 1127–1132 (1974).

202. Gillman, Leonard, and Meyer Jerison. *Rings of continuous functions*. 2nd ed. Springer-Verlag, Berlin, 1976.

203. Glasner, Shmuel. *Proximal flows*. Lecture Notes in Mathematics, Vol. **517**. Springer – Verlag, Berlin, 1976.

204. Glasner, Shmuel. On Choquet–Deny measures. *Ann. Inst. H. Poincaré*, Sect. B **XII**, 1–10 (1976).

205. Glicksberg, Irving. On convex hulls of translates. *Pacific J. Math.* **13**, 97–113 (1963).

206. Godement, Roger. Extension à un groupe abélien quelconque des théorèmes taubériens de N. Wiener et d'un théorème de A. Beurling. *C. R. Acad. Sci. Paris*, Sér. A **223**, 16–18 (1946).

207. Godement, Roger. Théorèmes taubériens et théorie spectrale. *Ann. Sci. Ecole Normale Sup.* **64**, 119–138 (1947).

208. Godement, Roger. Les fonctions de type positif et la théorie des groupes. *Trans. Amer. Math. Soc.* **63**, 1–84 (1948).

209. Gootman, Elliot C., and Jonathan Rosenberg. The structure of crossed product C^*-algebras: a proof of the generalized Effros–Hahn conjecture. *Invent. Math.* **52**, 283–298 (1979).

210. Granirer, Edmond E. On the invariant mean on topological semigroups and on topological groups. *Pacific J. Math.* **XV**, 107–140 (1965).

211. Granirer, Edmond E. Extremely amenable semigroups. *Math. Scand.* **17**, 177–197 (1965).

212. Granirer, Edmond E. On the range of an invariant mean. *Trans. Amer. Math. Soc.* **125**, 384–394 (1966).

213. Granirer, Edmond E. Extremely amenable semigroups II. *Math. Scand.* **20**, 93–113 (1967).

214. Granirer, Edmond E. On finite equivalent invariant measures for semigroups of transformations. *Duke Math. J.* **38**, 395–408 (1971).

215. Granirer, Edmond E. Exposed points of convex sets and weak sequential convergence. *Mem. Amer. Math. Soc.* **123** (1972).

216. Granirer, Edmond E. Criteria for compactness and for discreteness of locally compact amenable groups. *Proc. Amer. Math. Soc.* **40**, 615–624 (1973).

217. Granirer, Edmond E. The radical of $L^\infty(G)^*$. *Proc. Amer. Math. Soc.* **41**, 321–324 (1973).

218. Granirer, Edmond E. Weakly almost periodic and uniformly continuous functionals on the Fourier algebra of any locally compact group. *Trans. Amer. Math. Soc.* **189**, 371–382 (1974).

219. Granirer, Edmond E. Properties of the set of topologically invariant means on P. Eymard's W^*-algebra $VN(G)$. *Nederl. Akad. Wetensch. Proc.*, Ser. A **77**, 116–121 (1974).

220. Granirer, Edmond E. Density theorems for some linear subspaces and some C^*-subalgebras of $VN(G)$. *Symposia Mathematica* **XXII**. *Analisi armonica e spazi di funzioni su gruppi localmente compatti*, 61–70. Academic, New York, 1977.

221. Granirer, Edmond E. On group representations whose C^*-algebra is an ideal in its von Neumann algebra. *Ann. Inst. Fourier* (Grenoble) **29**, 4, 37–52 (1979).

222. Granirer, Edmond E. An application of the Radon Nikodym property in harmonic analysis. *Boll. Un. Mat. Ital.* **18B**, 663–671 (1981).

223. Granirer, Edmond E. On group representations whose C^*-algebra is an ideal in its von Neumann algebra II. *J. London Math. Soc.* **26**, 308–316 (1982).

CITED REFERENCES

224. Granirer, Edmond E., and Anthony T. Lau. A characterization of locally compact amenable groups. *Illinois J. Math.* **15**, 249–257 (1971).
225. Granirer, Edmond E., and Michael Leinert. On some topologies which coincide on the unit sphere of the Fourier–Stieltjes algebra $B(G)$ and of the measure algebra $M(G)$. *Rocky Mount. J. Math.* **11**, 459–472 (1981).
226. Greenleaf, Frederick P. *Invariant Means on Topological Groups*. Van Nostrand, New York, 1969.
227. Greenleaf, Frederick P. Amenable actions of locally compact groups. *J. Functional Analysis* **4**, 295–315 (1969).
228. Greenleaf, Frederick P. Ergodic theorems and the construction of summing sequences in amenable locally compact groups. *Comm. Pure Appl. Math.* **26**, 29–46 (1973).
229. Greenleaf, Frederick P. Concrete methods for summing almost periodic functions and their relation to uniform distribution of semigroup actions. *Colloq. Math.* **XLI**, 105–116 (1979).
230. Greenleaf, Frederick P., and William R. Emerson. Group structure and the pointwise ergodic theorem for connected amenable groups. *Advances in Math.* **14**, 153–172 (1974).
    Greenleaf, Frederick P. See [169, 170].
231. Grigorčuk, R. I. Symmetric random walks on discrete groups (In Russian). *Usp. Mat. Nauk* **35**, 132–152 (1978).
232. Gromov, Mikhael. Groups of polynomial growth and expanding maps. *Inst. Hautes Etudes Sci. Publ. Math.* **53**, 53–78 (1981).
233. Gromov, Mikhael. *Structures métriques pour les variétés riemaniennes*. Cedic/Fernand Nathan, Paris, 1981.
234. Grosser, Siegfried, and Martin Moskowitz. Compactness conditions in topological groups. *J. Reine Angew. Math.* **246**, 1–40 (1971).
235. Grossman, Marvin W. A categorical approach to invariant means and fixed point properties. *Semigroup Forum* **5**, 14–44 (1972).
236. Grossman, Marvin W. Uniqueness of invariant means on certain introverted spaces. *Bull. Austr. Math. Soc.* **9**, 109–120 (1973).
237. Grossman, Marvin W. Invariant means and fixed point properties on completely regular spaces. *Bull. Austr. Math. Soc.* **16**, 203–212 (1977).
238. Guichardet, Alain. *Tensor Products of C^*-Algebras*. Aarhus Universitet Lecture Notes Series, **12**, 1969.
239. Guichardet, Alain. Systèmes dynamiques non commutatifs. *Astérisque* **13–14** (1974).
240. Guichardet, Alain. *Cohomologie des groupes topologiques et des algèbres de Lie*. Cedic/Fernand Nathan, Paris, 1980.
241. Guivarc'h, Yves. Croissance polynomiale et périodes des fonctions harmoniques. *Bull. Soc. Math. France* **101**, 333–379 (1973).
242. Guivarc'h, Yves. Quelques propriétés asymptotiques des produits de matrices aléatoires. *Ecole d'été de probabilités de Saint-Flour*. Lecture Notes in Mathematics, Vol. **774**, 177–250. Springer-Verlag, Berlin, 1980.
243. Guivarc'h, Yves, Michael Keane, and Bernard Roynette. *Marches aléatoires sur les groupes de Lie*. Lecture Notes in Mathematics, Vol. **624**. Springer-Verlag, Berlin, 1977.
    Guivarc'h, Yves. See [134].
244. Gulick, S. I., Teng-sun Liu, and Arnoud C. M. van Rooij. Group algebra modules II. *Canad. J. Math.* **19**, 151–173 (1967).
245. Haagerup, Uffe. An example of a non nuclear C^*-algebra, which has the metric approximation property. *Invent. Math.* **50**, 279–293 (1979).

246. Haagerup, Uffe. The reduced C^*-algebra of the free group on two generators. 18*th Scandinavian Congress of Mathematicians*. Birkhäuser, Boston, 321–335 (1981).

247. Haagerup, Uffe. All nuclear C^*-algebras are amenable. *Invent. Math.* **74**, 305–319 (1983).

Hahn, Frank. See [153].

Hahn, L. S. See [38].

248. Hahn, Peter. The σ-representations of amenable groupoids. *Rocky Mountain J. Math.* **9**, 631–639 (1979).

249. Harpe, Pierre de la. Moyennabilité de quelques groupes topologiques de dimension infinie. *C. R. Acad. Sci. Paris*, Sér. A **277**, 1037–1040 (1973).

250. Harpe, Pierre de la. Moyennabilité du groupe unitaire et propriété P de Schwartz des algèbres de von Neumann. *Proceedings: Algèbres d'opérateurs*. Lecture Notes in Mathematics, Vol. **725**, 220–227. Springer-Verlag, Berlin, 1979.

251. Harpe, Pierre de la. Free groups in linear groups. *Enseignement Math.* **29**, 129–144 (1983).

252. Hauenschild, Wilfried. Zur Darstellungstheorie von SIN-Gruppen. *Math. Ann.* **210**, 257–276 (1974).

253. Hausdorff, Felix. *Grundzüge der Mengenlehre*. Veit, Leipzig, 1914. Reedited by Chelsea Publishing Company, New York, 1978.

254. Helgason, Sigurdur. *Differential geometry and symmetric spaces*. Academic, New York, 1962.

255. Herz, Carl. Remarques sur la note précédente de M. Varopoulos. *C. R. Acad. Sci. Paris*, Sér. A **260**, 6001–6004 (1965).

256. Herz, Carl. Le rapport entre l'algèbre A_p d'un groupe et d'un sous-groupe. *C. R. Acad. Sci. Paris*, Sér. A **271**, 244–246 (1970).

257. Herz, Carl. Synthèse spectrale pour les sous-groupes par rapport aux algèbres A_p. *C. R. Acad. Sci. Paris*, Sér. A **271**, 316–318 (1970).

258. Herz, Carl. The theory of p-spaces with an application to convolution operators. *Trans. Amer. Math. Soc.* **154**, 69–82 (1971).

259. Herz, Carl. Harmonic synthesis for subgroups. *Ann. Inst. Fourier* (Grenoble) **XXIII**, 3, 91–123 (1973).

260. Herz, Carl. Une généralisation de la notion de transformée de Fourier–Stieltjes. *Ann. Inst. Fourier* (Grenoble) **XXIV**, 3, 145–157 (1974).

261. Herz, Carl. On the asymmetry of norms of convolution operators, I. *J. Functional Analysis* **23**, 11–22 (1976).

262. Herz, Carl, and Nestor Rivière. Estimates for translation-invariant operators on spaces with mixed norms. *Stud. Math.* **XLIV**, 511–515 (1972).

263. Heyer, Herbert. Theory of Statistical Experiments. Springer-Verlag, New York, 1982.

264. Hirsch, Morris W., and William P. Thurston. Foliated bundles, invariant measures and flat manifolds. *Ann. of Math.* **101**, 369–390 (1975).

265. Hochster, M. Subsemigroups of amenable groups. *Proc. Amer. Math. Soc.* **21**, 363–364 (1969).

Holm, Per. See [6].

266. Huff, Robert E. Existence and uniqueness of fixed-points for semigroups of affine maps. *Trans. Amer. Math. Soc.* **152**, 99–106 (1970).

267. Hulanicki, Andrzej. Groups whose regular representation weakly contains all unitary representations. *Studia Math.* **XXIV**, 37–59 (1964).

268. Hulanicki, Andrzej. Means and Følner condition on locally compact groups. *Studia Math.* **XXVII**, 87–104 (1966).

269. Hulanicki, Andrzej. On the spectral radius of hermitian elements in group algebras. *Pacific J. Math.* **18**, 277–287 (1966).

CITED REFERENCES

270. Hulanicki, Andrzej. On the spectral radius in group algebras. *Studia Math.* **XXXIV**, 209–214 (1970).
271. Hulanicki, Andrzej. On positive functionals on a group algebra multiplicative on a subalgebra. *Studia Math.* **XXXVII**, 163–171 (1971).
272. Hulanicki, Andrzej. On the spectrum of convolution operators on groups with polynomial growth. *Invent. Math.* **17**, 135–142 (1972).
273. Hulanicki, Andrzej. Subalgebra of $L_1(G)$ associated with Laplacian on a Lie group. *Colloq. Math.* **XXXI**, 259–287 (1974).
274. Husain, Taqdir, and James C. S. Wong. Invariant means on vector-valued functions. *Ann. Scuola Norm. Sup. Pisa Sci. Fis. Mat.* **27**, 717–727, 729–742 (1973).
275. Iversen, P. Mean value of compact convex sets, Day's fixed point theorem and invariant subspaces. *J. Math. Anal. Appl.* **57**, 1–11 (1977).
276. Iwasawa, Kenkishi. On some types of topological groups. *Ann. of Math.* **50**, 507–558 (1949).
277. Jenkins, Joe W. An amenable group with a nonsymmetric group algebra. *Bull. Amer. Math. Soc.* **75**, 357–360 (1969).
278. Jenkins, Joe W. Symmetry and nonsymmetry in the group algebras of discrete groups. *Pacific J. Math.* **32**, 131–145 (1970).
279. Jenkins, Joe W. Amenable subsemigroups of a locally compact group. *Proc. Amer. Math. Soc.* **25**, 766–770 (1970).
280. Jenkins, Joe W. On the spectral radius of elements in a group algebra. *Illinois J. Math.* **15**, 551–554 (1971).
281. Jenkins, Joe W. Sigma-amenable locally compact groups. *Proc. Amer. Math. Soc.* **28**, 621–626 (1971).
282. Jenkins, Joe W. Free semigroups and unitary group representations. *Studia Math.* **XLIII**, 27–39 (1972).
283. Jenkins, Joe W. A characterization of growth in locally compact groups. *Bull. Amer. Math. Soc.* **79**, 103–106 (1973).
284. Jenkins, Joe W. Nonsymmetric group algebras. *Studia Math.* **XLV**, 295–307 (1973).
285. Jenkins, Joe W. Growth of connected locally compact groups. *J. Functional Analysis* **12**, 113–127 (1973).
286. Jenkins, Joe W. Følner's condition for exponentially bounded groups. *Math. Scand.* **35**, 165–174 (1974).
287. Jenkins, Joe W. A fixed point theorem for exponentially bounded groups. *J. Functional Analysis* **22**, 346–353 (1976).
288. Jenkins, Joe W. Representations of exponentially bounded groups. *Amer. J. Math.* **98**, 29–38 (1976).
289. Jenkins, Joe W. Invariant functionals and polynomial growth. *Astérisque* **74**, 171–181 (1980).
290. Jenkins, Joe W. On group actions with nonzero fixed points. *Pacific J. Math.* **91**, 363–371 (1980).
     Jerison, Meyer. See [202].
291. Johnson, Barry E. Cohomology in Banach algebras. *Mem. Amer. Math. Soc.* **127** (1972).
292. Johnson, Barry E. Approximate diagonals and cohomology of certain annihilator Banach algebras. *Amer. J. Math.* **94**, 685–698 (1972).
293. Johnson, Barry E. Introduction to cohomology in Banach algebras. *Algebras in Analysis*, 84–100. Academic, New York, 1975.

294. Johnson, Barry E. Perturbations of Banach algebras. *Proc. London Math. Soc.* **34**, 439–458 (1977).

295. Johnson, Barry E. Stability on Banach algebras. *General Topology and Its Relations to Modern Analysis and Algebra IV.* Lecture Notes in Mathematics, Vol. **609**, 109–114. Springer-Verlag, Berlin, 1977.

296. Johnson, Barry E., Richard V. Kadison, and John R. Ringrose. Cohomology of operator algebras, III. Reduction to normal cohomology. *Bull. Soc. Math. France* **100**, 73–96 (1972).

297. Junco, Andres del, and Joseph M. Rosenblatt. Counterexamples in ergodic theory and number theory. *Math. Ann.* **245**, 185–197 (1979).

298. Junghenn, Hugo D. Some general results on fixed points and invariant means. *Semigroup Forum* **11**, 153–164 (1975).

299. Junghenn, Hugo D. Amenability of function spaces on thick subsemigroups. *Proc. Amer. Math. Soc.* **75**, 37–41 (1979).

300. Junghenn, Hugo D. Topological left amenability of semidirect products. *Canad. Math. Bull.* **24**, 79–85 (1981).

301. Junghenn, Hugo D. Amenability induced by amenable homomorphic images. *Semigroup Forum* **24**, 11–23 (1982).

Junghenn, Hugo D. See [34].

Kadison, Richard V. See [141, 296].

302. Kallman, Robert R. The existence of invariant measures on certain quotient spaces. *Advances in Math.* **11**, 387–391 (1973).

303. Kaniuth, Eberhard. A note on reduced duals of certain locally compact groups. *Math. Z.* **150**, 189–194 (1976).

304. Kaniuth, Eberhard. On separation in reduced duals of groups with compact invariant neighbourhoods of the identity. *Math. Ann.* **232**, 177–182 (1978).

305. Kaniuth, Eberhard. Primitive ideal spaces of groups with relatively compact conjugacy classes. *Archiv Math.* (Basel) **32**, 16–24 (1979).

306. Kaniuth, Eberhard. Weak containment and tensor products of group representations. *Math. Z.* **180**, 107–117 (1982).

307. Kaniuth, Eberhard, and Günter Schlichting. Zur harmonischen Analyse klassenkompakter Gruppen II. *Invent. Math.* **10**, 332–345 (1970).

Kastler, Daniel. See [141].

308. Kaufman, Robert. Remarks on invariant means. *Proc. Amer. Math. Soc.* **18**, 120–122 (1965).

309. Kaur, Gurnam, and J. N. Manocha. Invariant means on topological groups. *Proc. Nat. Inst. Sci. India*, Part A **32**, 391–394 (1968).

Keane, Michael. See [243].

310. Keller, Gordon. Amenable groups and varieties of groups. *Illinois J. Math.* **16**, 257–269 (1972).

311. Kesten, Harry. Full Banach mean values on countable groups. *Math. Scand.* **7**, 146–156 (1959).

312. Kesten, Harry. Symmetric random walks on groups. *Trans. Amer. Math. Soc.* **92**, 336–354 (1959).

313. Khalill, Idriss. Sur l'analyse harmonique du groupe affine de la droite. *Studia Math.* **LI**, 139–167 (1974).

314. Khelemskiĭ, A. Ya., and M. V. Sheinberg. Amenable Banach algebras. *Funkcional Anal. i Priložen* **13**, 42–48 (1979). English translation: *Functional Anal. Appl.* **13**, 32–37 (1979).

315. Kieffer, J. C. An entropy equidistribution property for a measurable partition under the action of an amenable group. *Bull. Amer. Math. Soc.* **81**, 464–466 (1975).

CITED REFERENCES

316. Kieffer, J. C. A generalized Shannon–McMillan theorem for the action of an amenable group on a probability space. *The Annals of Probability* **3**, 1031–1037 (1975).
317. Kieffer, J. C. A ratio limit theorem for a strongly subadditive set function in a locally compact amenable group. *Pacific J. Math.* **61**, 183–190 (1975).
318. Klawe, Maria M. Semidirect product of semigroups in relation to amenability, cancellation properties, and strong Følner conditions. *Pacific J. Math.* **73**, 91–106 (1977).
319. Klawe, Maria M. On the dimension of left invariant means and left thick subsets. *Trans. Amer. Math. Soc.* **231**, 507–518 (1977).
320. Klawe, Maria M. Dimensions of the sets of invariant means of semigroups. *Illinois J. Math.* **24**, 233–243 (1980).

Knops, H. J. F. See [160].

321. Kotzmann, Ernst, Viktor Losert and Harald Rindler. Dense ideals of group algebras. *Math. Ann.* **246**, 1–14 (1979).
322. Kotzmann, Ernst, and Harald Rindler. Central approximate units in a certain ideal of $L^1(G)$. *Proc. Amer. Math. Soc.* **57**, 155–158 (1976).
323. Krom, Melven, and Myren Krom. Groups with free nonabelian subgroups. *Pacific J. Math.* **35**, 425–427 (1970).
324. Krom, Melven, and Myren Krom. Free subgroups and Følner's conditions. *Math. Scand.* **32**, 242–244 (1973).

Krom, Myren. See [323, 324].

325. Kugler, Werner. Über die Einfachheit gewisser verallgemeinerter L^1-Algebren. *Arch. Math.* (Basel) **26**, 82–88 (1975).
326. Kuipers, L., and H. Niederreiter. *Uniform Distribution of Sequences*. Interscience Tracts in Pure and Applied Mathematics. Wiley, New York, 1974.
327. Kunze, Ray A., and Elias M. Stein. Uniformly bounded representations and harmonic analysis of the 2 × 2 real unimodular group. *Amer. J. Math.* **82**, 1–62 (1960).

Lacroix, J. See [108].

328. Lai, Hang Chin. On some properties of $A^p(G)$-algebras. *Proc. Japan Acad.* **45**, 572–576 (1969).
329. Lai, Hang Chin, and Ing-Sheun Chen. Harmonic analysis on the Fourier algebras $A_{1,p}(G)$. *J. Austr. Math. Soc.*, Ser. A **30**, 438–452 (1981).
330. Lance, E. Christopher. On nuclear C^*-algebras. *J. Functional Analysis* **12**, 157–176 (1973).

Lance, E. Christopher. See [154].

331. Landstad, Magnus B. On the Bohr topology in amenable topological groups. *Math. Scand.* **28**, 207–214 (1971).
332. Lau, Anthony To-Ming. Functional analytic properties of topological semigroups and n-extreme amenability. *Trans. Amer. Math. Soc.* **152**, 431–439 (1970).
333. Lau, Anthony To-Ming. Topological semigroups with invariant means in the convex hull of multiplicative means. *Trans. Amer. Math. Soc.* **148**, 69–84 (1970).
334. Lau, Anthony To-Ming. Invariant means on dense subsemigroups of topological groups. *Canad. J. Math.* **XXIII**, 797–801 (1971).
335. Lau, Anthony To-Ming. Amenability and invariant subspaces. *J. Austr. Math. Soc.*, Ser. A **18**, 200–204 (1974).
336. Lau, Anthony To-Ming. Some fixed point theorems and their applications to W^*-algebras. *Fixed Point Theory and Its Applications*, 121–129. Academic, New York, 1976.
337. Lau, Anthony To-Ming. Closed convex invariant subsets of $L_p(G)$. *Trans. Amer. Math. Soc.* **232**, 131–142 (1977).
338. Lau, Anthony To-Ming. Characterizations of amenable Banach algebras. *Proc. Amer. Math. Soc.* **70**, 156–160 (1978).

339. Lau, Anthony To-Ming. Operators which commute with convolutions on subspaces of $L_\infty(G)$. Colloq. Math. **XXXIX**, 351-359 (1978).

340. Lau, Anthony To-Ming. Uniformly continuous functionals on the Fourier algebra of any locally compact group. Trans. Amer. Math. Soc. **251**, 39-59 (1979).

341. Lau, Anthony To-Ming. The second conjugate algebra of the Fourier algebra of a locally compact group. Trans. Amer. Math. Soc. **267**, 53-63 (1981).

342. Lau, Anthony To-Ming. Invariantly complemented subspaces of $L_\infty(G)$ and amenable locally compact groups. Illinois J. Math. **26**, 226-235 (1982).

343. Lau, Anthony To-Ming. Analysis on a class of Banach algebras with applications to harmonic analysis on locally compact groups and semigroups. Fund. Math. **118**, 161-175 (1983).

    Lau, Anthony To-Ming. See [224].

344. Lebesgue, Henri. Leçons sur l'intégration et la recherche des fonctions primitives. Gauthier-Villars, Paris, 1904: Reedited by Chelsea Publishing Company, Bronx, 1973.

345. Lehmann, E. L. Testing Statistical Hypotheses. Wiley, New York, 1959.

346. Leinert, Michael. Convoluteurs de groupes discrets. C. R. Acad. Sci. Paris, Sér. A **271**, 630-631 (1971).

347. Leinert, Michael. Abschätzung von Normen gewisser Matrizen und eine Anwendung. Math. Ann. **240**, 13-19 (1979).

    Leinert, Michael. See [225].

348. Lengagne, Guy. Moyennes invariantes dans une catégorie. C. R. Acad. Sci. Paris, Sér. A **273**, 704-707 (1971).

349. Leptin, Horst. Faltungen von Borelschen Maßen mit L^p-Funktionen auf lokal kompakten Gruppen. Math. Ann. **163**, 111-117 (1966).

350. Leptin, Horst. On a certain invariant of a locally compact group. Bull. Amer. Math. Soc. **72**, 870-874 (1966).

351. Leptin, Horst. Verallgemeinerte L^1-Algebren und projektive Darstellungen lokal kompakter Gruppen I. Invent. Math. **3**, 257-281 (1967).

352. Leptin, Horst. Verallgemeinerte L^1-Algebren und projektive Darstellungen lokal kompakter Gruppen II. Invent. Math. **4**, 68-86 (1967).

353. Leptin, Horst. On locally compact groups with invariant means. Proc. Amer. Math Soc. **19**, 489-494 (1968).

354. Leptin, Horst. Sur l'algèbre de Fourier d'un groupe localement compact. C. R. Acad. Sci. Paris, Sér. A **266**, 1180-1182 (1968).

355. Leptin, Horst. Zur harmonischen Analyse klassenkompaketer Gruppen. Invent. Math. **5**, 249-254 (1968).

356. Leptin, Horst. Ideal theory in group algebras of locally compact groups. Invent. Math. **31**, 259-278 (1976).

357. Leptin, Horst. Lokal kompakte Gruppen mit symmetrischen Algebren. Symposia Mathematica **XXII**: Analisi Armonica e Spazi di Funzioni su Gruppi Localmente Compatti, 267-281. Academic, New York, 1977.

358. Leptin, Horst, and Detlev Poguntke. Symmetry and nonsymmetry for locally compact groups. J. Functional Analysis **33**, 119-134 (1979).

359. Lipsman, Ronald L. Uniformly bounded representations of $SL(2,\mathbf{C})$. Amer. J. Math. **91**, 47-66 (1969).

360. Lipsman, Ronald L. Uniformly bounded representations of the Lorentz groups. Amer. J. Math. **91**, 938-962 (1969).

361. Lipsman, Ronald L. Harmonic analysis on $SL(n,\mathbf{C})$. J. Functional Analysis **3**, 126-155 (1969).

CITED REFERENCES

362. Liu, Teng-sun, and Arnoud C. M. van Rooij. Invariant means on a locally compact group. *Monatsh. Math.* **78**, 356-359 (1974).
  Liu, Teng-sun. See [244].
363. Liukkonen, John. Characters and centroids of [SIN] amenable groups. Recent advances in the representation theory of rings and C^*-algebras by continuous sections. *Mem. Amer. Math. Soc.* **148**, 145-151 (1974).
364. Ljapin, E. S. *Semigroups*. 3rd ed. Translations of Mathematical Monographs. American Mathematical Society, Providence, 1974.
365. Lo Bello, Anthony. *On the Origin and History of Ergodic Theory*. Allegheny College, Pa., 1981.
366. Locher-Ernst, L. Wie man aus einer Kugel zwei zu ihr kongruente Kugeln herstellen kann. *Elem. Math.* **XI**, 25-48 (1956).
367. Lohoué, Noël. Sur certaines propriétés remarquables des algèbres $A_p(G)$. *C. R. Acad. Sci. Paris*, Sér. A **273**, 893-896 (1971).
368. Lohoué, Noël. Sur les convoluteurs de $L^p(G)$. *C. R. Acad. Sci. Paris*, Sér. A **278**, 1543-1545 (1974).
369. Lohoué, Noël. Estimations L^p des coefficients de représentation et opérateurs de convolution. *Advances in Math.* **38**, 178-221 (1980).
370. Loomis, Lynn H. *An Introduction to Abstract Harmonic Analysis*. Van Nostrand, New York, 1953.
371. Losert, Viktor. Some properties of groups without property (P_1). *Comment. Math. Helv.* **54**, 133-139 (1979).
372. Losert, Viktor, and Harald Rindler. Almost invariant sets. *Bull. London Math. Soc.* **13**, 145-148 (1981).
  Losert, Viktor. See [321].
373. Ludwig, Jean. A class of symmetric and a class of Wiener group algebras. *J. Functional Analysis* **31**, 187-194 (1979).
374. Lust-Piquard, Françoise. Eléments ergodiques et totalement ergodiques dans $L^\infty(\Gamma)$. *Studia Math.* **LXIX**, 191-225 (1981).
375. Luthar, Indar S. Uniqueness of the invariant mean on abelian topological semigroups. *Trans. Amer. Math. Soc.* **104**, 403-411 (1962).
376. Maak, Wilhelm, *Fastperiodische Funktionen*. 2nd ed. Springer-Verlag, Berlin, 1967.
377. Mackey, George W. Induced representations of locally compact groups I. *Ann. of Math.* **55**, 101-139 (1952).
378. Mackey, George W. *The Theory of Unitary Group Representations*. University of Chicago Press, Chicago, 1976.
  Manocha J. N. See [309].
379. Margulis, G. A. Factor groups of discrete subgroups. *Soviet Math. Dokl.* **19**, 1145-1149 (1978).
380. Margulis, G. A. Some remarks on invariant means. *Monatsh. Math.* **90**, 233-235 (1980).
  Mautner, F. I. See [155].
381. Maxones, Walter, and Harald Rindler. Asymptotisch gleichverteilte Maßfolgen in Gruppen vom Heisenberg-Typ. *Österreich. Akad. Wiss. Math.-Naturwiss. Kl. S.-B. II* **185**, 485-504 (1977).
382. Maxones, Walter, and Harald Rindler. Asymptotisch gleichverteilte Netze von Wahrscheinlichkeitsmaßen auf lokalkompakten Gruppen. *Colloq. Math.* **XL**, 131-145 (1978).
383. Maxones, Walter, and Harald Rindler. Bemerkungen zu einer Arbeit von P. Gerl „Gleichverteilung auf lokalkompakten Gruppen". *Math. Nachr.* **79**, 193-199 (1977).

384. McKennon, Kelly. Multipliers, positive functionals, positive-definite functions, and Fourier–Stieltjes transforms. *Mem. Amer. Math. Soc.* **111** (1971).

385. Michele, Leonede de, and Paolo Soardi. Symbolic calculus in $A_p(G)$. *Atti Accad. Naz. Lincei Rend. Cl. Sci. Fis. Natur.* **LVII**, 24–30, 31–35 (1974).

386. Milnes, Paul. Counterexample to a conjecture of Greenleaf. *Canad. Math. Bull.* **13**, 497–499 (1970).

387. Milnes, Paul. Left mean-ergodicity, fixed points, and invariant means. *J. Math. Anal. Appl.* **65**, 32–43 (1978).

388. Milnes, Paul. Amenable groups for which every topologically left invariant mean is right invariant. *Rocky Mount. J. Math.* **11**, 261–266 (1981).

389. Milnes, Paul, and James V. Bondar. A simple proof of a covering property of locally compact groups. *Proc. Amer. Math. Soc.* **73**, 117–118 (1979).
Milnes, Paul. See [34, 44].

390. Milnor, John. Growth of finitely generated solvable groups. *J. Differential Geometry* **2**, 447–449 (1968).

391. Mitchell, Theodore. Constant functions and left invariant means on semigroups. *Trans. Amer. Math. Soc.* **119**, 244–261 (1965).

392. Mitchell, Theodore. Fixed points and multiplicative left invariant means. *Trans. Amer. Math Soc.* **122**, 195–202 (1966).

393. Mitchell, Theodore. Function algebras, means and fixed points. *Trans. Amer. Math. Soc.* **130**, 117–126 (1968).

394. Mitchell, Theodore. Topological semigroups and fixed points. *Illinois J. Math.* **14**, 630–641 (1970).

395. Mocanu, Constanța. A generalization of an invariant of H. Leptin (In Romanian). *Studia Babeș-Bolyai Math.* **XXIII**, 45–49 (1978).

396. Monna, A. F. *Analyse non-archimédienne*. Springer-Verlag, Berlin, 1970.

397. Montgomery, Deane, and Leo Zippin. *Topological Transformation Groups*. Interscience Tracts in Pure and Applied Mathematics. Interscience, New York, 1955.

398. Moore, Calvin C. Amenable subgroups of semisimple groups and proximal flows. *Israel J. Math.* **34**, 121–138 (1979).

399. Moore, Calvin C., and Robert J. Zimmer. Groups admitting ergodic actions with generalized discrete spectrum. *Invent. Math.* **51**, 171–188 (1979).

400. Moore, Gregory H. *Zermelo's Axiom of Choice. Its Origins, Development, and Influence*. Springer-Verlag, Berlin, 1982.

401. Moran, William. Invariant means on $\mathscr{C}(G)$ *J. London Math. Soc.* **2**, 133–138 (1970),

402. Moran, William, and John H. Williamson. Isotone measures on groups. *Math. Proc. Cambridge Philos. Soc.* **84**, 89–107 (1978).
Moskowitz, Martin. See [234].

403. Moulin Ollagnier, Jean, and Didier Pinchon. Une nouvelle démonstration du théorème de E. Følner. *C. R. Acad. Sci. Paris*, Sér. A **287**, 557–560 (1978).

404. Moulin Ollagnier, Jean, and Didier Pinchon. Filtre moyennant et valeurs moyennes des capacités invariantes. *Bull. Soc. Math. France* **110**, 259–277 (1982).

405. Müller-Römer, Peter. Contracting extensions and contractible groups. *Bull. Amer. Math. Soc.* **79**, 1264–1269 (1973).

406. Müller-Römer, Peter. Kontrahierende Erweiterungen und kontrahierbare Gruppen. *J. Reine Angew. Math.* **283/284**, 238–264 (1976).

407. Munn, W. D. A class of irreducible matrix representations of an arbitrary inverse semigroup. *Proceedings of the Glasgow Mathematical Society* **5**, 41–48 (1961).

408. Mycielski, Jan. Note on: S. Banach. Sur le problème de la mesure. *Stefan Banach; Oeuvres, Vol. One,* 318–322. Institut mathématique de l'Académie polonaise des sciences, Warsaw, 1967.
409. Mycielski, Jan. Finitely additive invariant measures I. *Colloq. Math.* **XLII**, 309–318 (1967).
410. Namioka, I. Partially ordered linear topological spaces. *Mem. Amer. Math. Soc.* **24**, 1–50 (1957).
411. Namioka, I. Følner's conditions for amenable semigroups. *Math. Scand.* **15**, 18–28 (1964).
412. Namioka, I. On a recent theorem by H. Reiter. *Proc. Amer. Math. Soc.* **17**, 1101–1102 (1966).

Namioka, I. See [98, 149].

413. Nebbia, Claudio. Multipliers and asymptotic behaviour of the Fourier algebra of nonamenable groups. *Proc. Amer. Math. Soc.* **84**, 549–554 (1982).
414. Neumann, John von. Zur allgemeinen Theorie des Maßes. *Fund. Math.* **13**, 73–116 (1929).
415. Neumann, John von. Almost periodic functions in a group I. *Trans. Amer. Math. Soc.* **36**, 445–492 (1934).

Niederreiter, H. See [326].

416. Novikov, P. S., and S. I. Adian. Infinite periodic groups. (In Russian). *Izv. Akad. Nauk SSSR,* Ser. Mat. **32**, 212–244, 251–254, 709–731 (1968).
417. Oberlin, Daniel M. Translation-invariant operators on $L^p(G)$, $0 < p < 1$ (II). *Canad. J. Math.* **XXIX**, 626–630 (1977).
418. Oberlin, Daniel M. Translation-invariant operators of weak type. *Pacific J. Math.* **85**, 155–164 (1979).
419. Ocneanu, Adrian. Action des groupes moyennables sur les algèbres de von Neumann. *C. R. Acad. Sci. Paris,* Sér. A **291**, 399–401 (1980).
420. Ol'sanskiĭ, A. Ju. Infinite groups with cyclic subgroups. *Soviet Math. Dokl.* **20**, 343–346 (1979).
421. Ol'sanskiĭ, A. Ju. On the question of the existence of an invariant mean on a group. *Usp. Mat. Nauk* **35**, 199–200 (1980). Russian Math. Surveys **35**, 180–181 (1980).
422. Ornstein, Donald S., and Benjamin Weiss. Ergodic theory of amenable group actions I: The Rohlin lemma. *Bull. Amer. Math. Soc.* **2**, 161–164 (1980).
423. Ornstein, Donald S., and Benjamin Weiss. The Shannon–McMillan–Breiman theorem for a class of amenable groups. *Israel J. Math.* **44**, 53–60 (1983).
424. Pachl, Jan K. Uniform measures on topological groups. *Compositio Math.* **45**, 385–392 (1982).
425. Palmer, T. W. Classes of nonabelian, noncompact, locally compact groups. *Rocky Mount. J. Math.* **8**, 683–741 (1978).
426. Paschke, William L. Inner amenability and conjugation operators. *Proc. Amer. Math. Soc.* **71**, 117–118 (1978).
427. Paschke, William L. The crossed product of a C^*-algebra by an endomorphism. *Proc. Amer. Math. Soc.* **80**, 113–118 (1980).

Paschke, William. See [59, 60].

428. Paterson, Alan L. T. Weak containment and Clifford semigroups. *Proc. Roy. Soc. Edinburgh,* Sect. A **81**, 23–30 (1978).
429. Paterson, Alan L. T. Amenability and locally compact semigroups. *Math. Scand.* **42**, 271–288 (1978).
430. Paterson, Alan L. T. Amenable groups for which every topological left invariant mean is invariant. *Pacific J. Math.* **84**, 391–397 (1979).
431. Paterson, Alan L. T. On invariant means which are not inversion invariant. *J. London Math. Soc.* **19**, 312–318 (1979).

432. Paterson, Alan L. T. Amenability and translation experiments. *Canad. J. Math.* **XXXV**, 49–58 (1983).
    Pearcy, Carl. See [54].
433. Phelps, R. R. Extreme positive operators and homomorphisms. *Trans. Amer. Math. Soc.* **108**, 265–274 (1963).
    Phelps, R. R. See [98].
    Picardello, Massimo. See [183].
434. Pickel, B. S. Informational futures of amenable groups (In Russian). *Dokl. Akad. Nauk SSSR* **223**, 1067–1070 (1975).
435. Pier, Jean-Paul. Operators and functionals on Banach spaces that are topologically invariant under the action of a σ-compact unimodular amenable group. *An. Şti. Univ. "Al. I. Cuza" Iaşi*, Seçt. Ia Mat. **XXI**, 25–32 (1975).
436. Pier, Jean-Paul. La théorie des groupes moyennables et ses prolongements. *Actas del V Congreso de la Agrupación de Matematicas de Expresión Latina*, 79–103. Instituto Jorge Juan de Matematicas, Madrid, 1978.
437. Pier, Jean-Paul. Quasi-invariance intérieure sur les groupes localement compacts. *Actualités mathématiques*, 431–436. Gauthier-Villars, Paris, 1982.
    Pinchon, Didier. See [403, 404].
438. Poguntke, Detlev. Nilpotente Liesche Gruppen haben symmetrische Gruppenalgebren. *Math. Ann.* **227**, 51–59 (1977).
    Poguntke, Detlev. See [358].
439. Promislow, David. Nonexistence of invariant measures. *Proc. Amer. Math. Soc.* **88**, 89–92 (1983).
440. Racher, Gerhard. On amenable and compact groups. *Monats. Math.* **92**, 305–311 (1981).
441. Racher, Gerhard. A proposition of A. Grothendieck revisited. *Banach Theory and its Applicaations*. Lecture Notes in Mathematics, Vol. 991, 215–227. Springer-Verlag, Berlin, 1983.
442. Raimi, Ralph A. On a theorem of E. Følner. *Math. Scand.* **6**, 47–49 (1958).
443. Raimi, Ralph A. Invariant means and invariant matrix methods of summability. *Duke Math. J.* **30**, 81–94 (1963).
444. Raimi, Ralph A. Homeomorphisms and invariant measures for $\beta N \setminus N$. *Duke Math. J.* **33**, 1–12 (1966).
445. Rajagopalan, M. L^p-conjecture for locally compact groups I. *Trans. Amer. Math. Soc.* **125**, 216–222 (1966).
446. Rajagopalan, M. L^p-conjecture for locally compact groups II. *Math. Ann.* **169**, 331–339 (1967).
447. Rajagopalan, M., and P. V. Ramakrishnan. On a conjecture of Granirer and strong Følner condition. *J. Indian Math. Soc.* **37**, 85–92 (1973).
448. Rajagopalan, M., and Klaus G. Witz. On invariant means which are not inverse invariant. *Canad. J. Math.* **20**, 222–224 (1968).
    Ramakrishnan, P. V. See [447].
    Ramirez, Donald E. See [150, 151].
    Ramsay, R. W. See [188].
    Randow, R. von. See [520].
449. Raugi, Albert. Fonctions harmoniques et théorèmes limites pour les marches aléatoires sur les groupes. *Bull. Soc. Math. France*, mémoire **54**, 5–118 (1977).
    Raugi, Albert. See [35, 157].
450. Reiter, Hans. Investigations in harmonic analysis. *Trans. Amer. Math. Soc.* **73**, 401–427 (1952).

451. Reiter, Hans. Sur la propriété (P_1) et les fonctions de type positif. *C. R. Acad. Sci. Paris*, Sér. A **258**, 5134–5135 (1964).
452. Reiter, Hans. On some properties of locally compact groups. *Nederl. Wetensch. Proc.*, Ser. A **68**, 697–701 (1965).
453. Reiter, Hans. *Classical Harmonic Analysis and Locally Compact Groups*. Oxford University Press, Oxford, 1968.
454. Reiter, Hans. Sur certains idéaux dans $L^1(G)$. *C. R. Acad. Sci. Paris*, Sér. A **267**, 882–885 (1968).
455. Reiter, Hans. L^1-*Algebras and Segal Algebras*. Lecture Notes in Mathematics, Vol. 231. Springer-Verlag, Berlin, 1971.
456. Renaud, Pierre François. General ergodic theorems for locally compact groups. *Amer. J. Math.* **93**, 52–64 (1971).
457. Renaud, Pierre François. Centralizers of the Fourier algebra of an amenable group. *Proc. Amer. Math. Soc.* **32**, 539–542 (1972).
458. Renaud, Pierre François. Equivalent types of invariant means on locally compact groups. *Proc. Amer. Math. Soc.* **31**, 495–498 (1972).
459. Renaud, Pierre François. Invariant means on a class of von Neumann algebras. *Trans. Amer. Math. Soc.* **170**, 285–291 (1972).
460. Revuz, Daniel. *Markov Chains*. North-Holland Publishing Company, Amsterdam, 1975.
461. Rickart, Charles E. *General Theory of Banach Algebras*. Van Nostrand, New York, 1960.
462. Rickert, Neil W. Amenable groups and groups with the fixed point property. *Trans. Amer. Math. Soc.* **127**, 221–232 (1967).
463. Rickert, Neil W. Some properties of locally compact groups. *J. Austr. Math. Soc.*, Ser. A **VII**, 433–454 (1967).
464. Rickert, Neil W. Convolution of L^2 functions. *Colloq. Math.* **XIX**, 301–303 (1968).
465. Riemersma, Martinus. Amenability of homogeneous spaces. *Nederl. Akad. Wetensch. Proc.*, Ser. A **76**, 376–380 (1973).
466. Rindler, Harald. Ein Gleichverteilungsbegriff für mittelbare Gruppen. *Österreich. Akad. Wiss. Math.-Natur. Kl. S.-B. II* **182**, 107–119 (1973).
467. Rindler, Harald. Über ein Problem von Reiter und ein Problem von Derighetti zur Eigenschaft P_1 lokalkompakter Gruppen. *Comment. Math. Helv.* **48**, 492–497 (1973).
468. Rindler, Harald. Zur Eigenschaft P_1 lokalkompakter Gruppen. *Nederl. Akad. Wetensch. Proc.*, Ser A **76**, 142–147 (1973).
469. Rindler, Harald. Gleichverteilte Folgen von Operatoren. *Compositio Math.* **29**, 201–212 (1974).
470. Rindler, Harald. Approximierende Einheiten in Idealen von Gruppenalgebren. *Anz. Österreich. Akad. Wiss. Math.-Naturwiss. Kl.* **5**, 37–39 (1976).
471. Rindler, Harald. Gleichverteilte Folgen in lokalkompakten Gruppen. *Monatsh. Math.* **82**, 207–235 (1976).
472. Rindler, Harald. Uniformly distributed sequences in quotient groups. *Acta Sci. Math.* (Szeged) **38**, 153–156 (1976).
473. Rindler, Harald. Zur $L^1(G)$-Gleichverteilung auf abelschen und kompakten Gruppen. *Arch. Math.* (Basel) **26**, 209–213 (1975).
474. Rindler, Harald. Approximate units in ideals of group algebras. *Proc. Amer. Math. Soc.* **71**, 62–64 (1978).

Rindler, Harald. See [321, 322, 372, 381, 382, 383].
Ringrose, John R. See [296].
Rivière, Nestor. See [262].

475. Robert, Alain. Exemples de groupes de Fell. *C. R. Acad. Sci. Paris*, Sér. A **287**, 603–606 (1978).
    Robinson, Derek W. See [141].
    Rooij, Arnoud C. M. van. See [244, 362].
476. Rosenberg, Jonathan. Amenability of crossed products of C^*-algebras. *Comm. Math. Phys.* **57**, 187–191 (1977).
    Rosenberg, Jonathan. See [209].
477. Rosenblatt, Joseph M. A generalization of Følner's condition. *Math. Scand.* **33**, 153–170 (1973).
478. Rosenblatt, Joseph M. Equivalent invariant measures. *Israel J. Math.* **17**, 261–270 (1974).
479. Rosenblatt, Joseph M. Invariant measures and growth conditions. *Trans. Amer. Math. Soc.* **193**, 33–53 (1974).
480. Rosenblatt, Joseph M. Invariant means and invariant ideals in $L_\infty(G)$ for a locally compact group G. *J. Functional Analysis* **21**, 31–51 (1976).
481. Rosenblatt, Joseph M. Invariant means for the bounded measurable functions on a non-discrete locally compact group. *Math. Ann.* **220**, 219–228 (1976).
482. Rosenblatt, Joseph M. The number of extensions of an invariant mean. *Compositio Math.* **33**, 147–159 (1976).
483. Rosenblatt, Joseph M. Uniform distribution in compact groups. *Mathematika* **23**, 198–207 (1976).
484. Rosenblatt, Joseph M. Invariant means on the continuous bounded functions. *Trans. Amer. Math. Soc.* **236**, 315–324 (1978).
485. Rosenblatt, Joseph M. Finitely additive invariant measures II. *Colloq. Math.* **XLII**, 361–363 (1979).
486. Rosenblatt, Joseph M. Strongly equivalent invariant measures. *Math. Proc. Cambridge Philos. Soc.* **88**, 33–43 (1980).
487. Rosenblatt, Joseph M. Uniqueness of invariant means for measure-preserving transformations. *Trans. Amer. Math. Soc.* **265**, 623–636 (1981).
488. Rosenblatt, Joseph M. Ergodic and mixing random walks on locally compact groups. *Math. Ann.* **257**, 31–42 (1981).
489. Rosenblatt, Joseph M., and Michel Talagrand. Different types of invariant means. *J. London Math. Soc.* **24**, 525–532 (1981).
    Rosenblatt, Joseph M. See [297].
    Ross, Kenneth A. See [92].
490. Roynette, Bernard. Marches aléatoires sur les groupes de Lie. *Ecole d'été de probabilités de Saint-Flour VII*. Lecture Notes in Mathematics, Vol. 678, 237–379. Springer-Verlag, Berlin, 1978.
    Roynette, Bernard. See [243].
491. Rudin, Walter. Invariant means on L^∞. *Studia Math.* **XLIV**, 219–227 (1972).
492. Sakai, Kôkichi. Amenable transformation groups. *Sci. Rep. Kagoshima Univ.* **22**, 1–7 (1973).
493. Sakai, Kôkichi. Amenable transformation groups II. *Proc. Japan Acad.* **49**, 428–431 (1973).
494. Sakai, Kôkichi. Følner's conditions for amenable transformation semigroups. *Sci. Rep. Kagoshima Univ.* **23**, 7–13 (1974).
495. Sakai, Kôkichi. Some remarks on locally compact transformation groups with invariant means. *Sci. Rep. Kagoshima Univ.* **24**, 11–15 (1975).
496. Sakai, Kôkichi. On amenable transformation semigroups I, II. *J. Math. Kyoto Univ.* **16**, 555–595, 597–626 (1976).

CITED REFERENCES

497. Sakai, Kôkichi. On amenable transformation semigroups III. *Sci. Rep. Kagoshima Univ.* **25**, 31–51 (1976).
498. Sakai, Kôkichi. On linear representations of amenable groups. *Sci. Rep. Kagoshima Univ.* **25**, 53–68 (1976).
499. Sakai, Kôkichi. On *J*-unitary representations of amenable groups. *Sci. Rep. Kagoshima Univ.* **26**, 33–41 (1977).
500. Sakai, Kôkichi. On amenable transformation semigroups IV. *Sci. Rep. Kagoshima Univ.* **31**, 1–19 (1982).
501. Sakai, Shôichirô. *C*-Algebras and W*-Algebras.* Springer-Verlag, Berlin, 1971.
502. Sauvageot, Jean-Luc. Idéaux primitifs induits dans les produits croisés. *J. Functional Analysis* **32**, 381–392 (1979).

Schempp, Walter. See [147].

503. Schickhof, W. H. Non-archimedean invariant means. *Compositio Math.* **30**, 169–180 (1975).

Schlichting, Günter. See [307].

504. Schochetman, I. Nets of subgroups and amenability. *Proc. Amer. Math. Soc.* **29**, 397–403 (1971).
505. Schochetman, I. The dual topology of certain group extensions. *Advances in Math.* **35**, 113–128 (1980).
506. Schott, René. Irrfahrten auf nicht mittelbaren homogenen Räumen. *Arbeitsberichte*, Institut für Mathematik, Univ. Salzburg **1–2**, 64–76 (1981).
507. Schott, René. Croissance et moyennabilité des espaces homogènes. *C. R. Acad. Sci. Paris*, Sér. A **293**, 373–376 (1981).
508. Schott, René. Marches aléatoires sur les espaces homogènes. Loi du logarithme itéré pour certaines intégrales stochastiques. Thèse, University of Nancy, 1982.
509. Schwartz, Jacob T. Two finite, non-hyperfinite, non-isomorphic factors. *Comm. Pure Appl. Math.* **16**, 19–26 (1963).
510. Segal, Irving E. The group algebra of a locally compact group. *Trans. Amer. Math. Soc.* **61**, 69–105 (1947).

Sheinberg, M. V. See [314].

511. Sherman, Jon. A new characterization of amenable groups. *Trans. Amer. Math. Soc.* **254**, 365–390 (1979).
512. Silverman, Robert J. Means on semigroups and the Hahn-Banach extension property. *Trans. Amer. Math. Soc.* **83**, 222–237 (1956).
513. Skudlarek, H. L. On a two-sided version of Reiter's condition and weak containment. *Arch. Math.* (Basel) **31**, 605–610 (1978).
514. Snell, Roy C. The range of invariant means on locally compact groups and semigroups. *Proc. Amer. Math. Soc.* **37**, 441–447 (1973).
515. Snell, Roy C. The range of invariant means on locally compact abelian groups. *Canad. Math. Bull.* **17**, 567–573 (1974).

Soardi, Paolo. See [385].

516. Spector, René. Sur la structure locale des groupes abéliens localement compacts. *Bull. Soc. Math. France*, mémoire **24** (1970).

Šreider, Ju. A. See [1].

517. Stegeman, Jan D. On a property concerning locally compact groups. *Nederl. Akad. Wetensch. Proc.*, Ser. A **68**, 702–703 (1965).
518. Stegmeir, Ulrich. *Approximierende Einsen in Idealen von Gruppenalgebren.* Dissertation, Technische Univ. Munich, 1978.

Stein, Elias. See [327].

519. Stewart, James. Positive definite functions and generalizations, an historical survey. *Rocky Mount. J. Math.* **6**, 409–434 (1976).

520. Stone, M., and R. von Randow. Statistically inspired conditions on the group structure of invariant experiments and their relationships with other conditions on locally compact topological groups. *Z. Wahrscheinlichkeitstheorie und Verw. Gebiete* **10**, 70–80 (1968).

521. Strătilă, Şerban. *Modular Theory in Operator Algebras*. Editura Academiei, Bucarest; Abacus Press, Turnbridge Wells, England, 1981.

522. Stromberg, Karl. The Banach–Tarski paradox. *Amer. Math. Monthly* **86**, 151–161 (1979). Sueur-Pontier, M. See [190].

523. Šujan, Štefan. Generators for amenable group actions. *Monatsh. Math.* **95**, 67–79 (1983).

524. Sullivan, Dennis. For $n > 3$ there is only one finitely additive rotationally invariant measure on the n-sphere defined on all Lebesgue measurable subsets. *Bull. Amer. Math. Soc.* **4**, 121–123 (1981).
Sullivan, Dennis. See [124].

525. Sz.-Nagy, Béla. On uniformly bounded linear transformations in Hilbert space. *Acta Sci. Math.* (Szeged) **11**, 152–157 (1947).

526. Tagi-Zade, A. T. Entropy characteristics of amenable groups (In Russian). *Akad. Nauk. Azerbaĭdžan SSR Dokl.* **34**, 11–14 (1978).

527. Takenouchi, Osamu. Sur une classe de fonctions continues de type positif sur un groupe localement compact. *Math. J. Okayama Univ.* **4**, 143–173 (1954).

528. Talagrand, Michel. Moyennes de Banach extrémales. *C. R. Acad. Sci. Paris*, Sér. A **282**, 1359–1362 (1976).

529. Talagrand, Michel. Géométrie des simplexes de moyennes invariantes. *J. Functional Analysis* **34**, 304–337 (1979).

530. Talagrand, Michel. Capacités invariantes: le cas alterné *C. R. Acad. Sci. Paris*, Sér. A **284**, 543–546 (1977).

531. Talagrand, Michel. Capacités invariantes extrémales. *Ann. Inst. Fourier* (Grenoble) **XXVIII**, 4, 79–146 (1978).

532. Talagrand, Michel. Moyennes invariantes s'annulant sur des idéaux. *Compositio Math.* **42**, 213–216 (1981).

533. Talagrand, Michel. Some functions with a unique invariant mean. *Proc. Amer. Math. Soc.* **82**, 253–256 (1981).
Talagrand, Michel. See [189, 489].

534. Tarski, Alfred. Sur l'équivalence des polygones. *Przegląd mat.-fiz.* **2**, 47–60 (1924).

535. Tarski, Alfred. Algebraische Fassung des Maßproblems. *Fund. Math.* **31**, 47–66 (1938).
Tarski, Alfred. See [29].

536. Tempel'man, A. A. Ergodic theorems for general dynamic systems. *Soviet Math. Dokl.* **8**, 1213–1216 (1967).

537. Tempel'man, A. A. Ergodic functions and averaging sequences. *Soviet Math. Dokl.* **24**, 78–82 (1981).
Thurston, William P. See [264].

538. Tits, Jacques. Free subgroups in linear groups. *J. Algebra* **20**, 250–270 (1972).

539. Tits, Jacques. Travaux de Margulis sur les sous-groupes discrets de groupes de Lie. *Séminaire Bourbaki. Vol. 1975 / 76*. Lecture Notes in Mathematics, Vol. **567**, 174–180. Springer-Verlag, Berlin, 1977.

540. Tomter, Per. A new proof for Følner's condition for maximally almost periodic groups. *Math. Scand.* **18**, 25–34 (1966).

541. Toure, Saliou. Sur quelques propriétés des espaces homogènes moyennables. *C. R. Acad. Sci. Paris*, Sér. A **273**, 717–719 (1971).

CITED REFERENCES

542. Toure, Saliou. Espaces homogènes moyennables et représentations des produits semi-directs. *Analyse harmonique sur les groupes de Lie II*. Lecture Notes in Mathematics, Vol. 739, 568-617. Springer-Verlag, Berlin, 1979.
543. Toure, Saliou. Quelques applications associées à l'application de Reiter. *Afrika Mat.* **III**, 87-114 (1981).
544. Van Dijk, G. The property P_1 for semisimple $\mathfrak{p}$-adic groups. *Nederl. Akad. Wetensch. Proc.*, Ser. A **75**, 82-85 (1972).
545. Vasilescu, F.-H., and L. Zsidó. Uniformly bounded groups in finite W^*-algebras. *Acta Sci. Math.* (Szeged) **36**, 189-192 (1974).

Vera, G. See [42, 43].

Verboven, E. J. See [160].

546. Wagon, Stanley. Invariance properties of finitely additive measures in R^n. *Illinois J. Math.* **25**, 74-86 (1981).
547. Wagon, Stanley. Circle-squaring in the twentieth century. *The Mathematical Intelligencer* **3**, 176-181 (1981).
548. Wallach, Nolan R. *Harmonic Analysis on Homogeneous Spaces*. Marcel Dekker, New York, 1973.

Walter, Martin E. See [5].

549. Wang, S. P. On the Mautner phenomenon and groups with property (T). *Amer. J. Math.* **104**, 1191-1210 (1982).
550. Warner, Garth. *Harmonic Analysis on Semisimple Lie Groups*. Volume I. Springer-Verlag, Berlin, 1972.
551. Watatani, Yasuo. The character groups of amenable group C^*-algebras. *Math. Japon.* **24**, 141-144 (1979).

Watatani, Yasuo. See [72].

552. Weil, André. *L'Intégration dans les groupes topologiques et ses applications*. 2nd ed. Hermann, Paris, 1953.

Weiss, Benjamin. See [97, 422, 423].

Weiss, Guido. See [90, 91].

553. Weit, Yitzhak. On the one-sided Wiener's theorem for the motion group. *Ann. of Math.* **111**, 415-422 (1980).
554. Wichmann, Josef. Bounded approximate units and bounded approximate identities. *Proc. Amer. Math. Soc.* **41**, 547-550 (1973).

Wichmann, Josef. See [142].

555. Wiener, Norbert. Tauberian theorems. *Ann. of Math.* **33**, 1-100 (1932).
556. Wiener, Norbert. *The Fourier Integral and Certain of its Applications*. Cambridge University Press, Cambridge, 1933: Reprinted by Dover Publications, New York, 1958.
557. Wilde, Carroll O., and Klaus G. Witz. Invariant means and the Stone-Čech compactification. *Pacific J. Math.* **21**, 577-586 (1967).

Wilde, Carroll O. See [17].

Williamson, John H. See [188, 402].

558. Witz, Klaus G. Applications of a compactification for bounded operator semigroups. *Illinois J. Math.* **8**, 685-696 (1964).

Witz, Klaus G. See [448, 557].

559. Wolf, Joseph A. Growth of finitely generated solvable groups and curvature of Riemannian manifolds. *J. Differential Geometry* **2**, 421-445 (1968).
560. Wong, James C. S. Topologically stationary locally compact groups and amenability. *Trans. Amer. Math. Soc.* **144**, 351-363 (1969).

561. Wong, James C. S. Topological invariant means on locally compact groups and fixed points. *Proc. Amer. Math. Soc.* **27**, 572–578 (1971).
562. Wong, James C. S. An ergodic property of locally compact amenable semigroups. *Pacific J. Math.* **48**, 615–619 (1973).
563. Wong, James C. S. A characterization of topological left thick subsets in locally compact left amenable semigroups. *Pacific J. Math.* **62**, 295–303 (1976).
564. Wong, James C. S. Amenability and substantial semigroups. *Canad. Math. Bull.* **19**, 231–234 (1976).
565. Wong, James C. S. Abstract harmonic analysis of generalised functions on locally compact semigroups with applications to invariant means. *J. Austral. Math. Soc.*, Ser. A **23**, 84–94 (1977).
566. Wong, James C. S. A characterisation of locally compact amenable subsemigroups. *Canad. Math. Bull.* **23**, 305–312 (1979).
567. Wong, James C. S. On topological analogues of left thick subsets in semigroups. *Pacific J. Math.* **83**, 571–585 (1979).
568. Wong, James C. S. Characterisations of extremely amenable semigroups. *Math. Scand.* **48**, 101–108 (1981).
569. Wong, James C. S. On left thickness of subsets in semigroups. *Proc. Amer. Math. Soc.* **84**, 403–407 (1982).
570. Wong, James C. S. On the relation between left thickness and topological left thickness in semigroups. *Proc. Amer. Math. Soc.* **86**, 471–476 (1982).
571. Wong, James C. S. Left thickness, left amenability and extreme left amenability. *Semigroup Forum* **26**, 307–316 (1983).
  Wong, James C. S. See [274].
572. Woodward, Gordon S. Translation-invariant linear forms on $C_0(G)$, $C(G)$, $L^p(G)$ for noncompact groups. *J. Functional Analysis* **16**, 205–220 (1974).
573. Woodward, W. Randolph, and R. Rao Chivukula. Invariant linear functionals. *J. Austral. Math. Soc.*, Ser. A **14**, 293–303 (1972).
574. Wordingham, J. R. The left regular $*$-representation of an inverse semigroup. *Proc. Amer. Math. Soc.* **86**, 55–58 (1982).
575. Yeadon, F. J. Fixed points and amenability: a counterexample. *J. Math. Anal. Appl.* **45**, 718–720 (1974).
576. Yoshizawa, H. Some remarks on unitary representations of the free group. *Osaka Math. J.* **3**, 55–63 (1951).
577. Zimmer, Robert J. Amenable ergodic actions, hyperfinite factors, and Poincaré flows. *Bull. Amer. Math. Soc.* **83**, 1078–1080 (1977).
578. Zimmer, Robert J. On the von Neumann algebra of an ergodic group action. *Proc. Amer. Math. Soc.* **66**, 289–293 (1977).
579. Zimmer, Robert J. Hyperfinite factors and amenable ergodic actions. *Invent. Math.* **41**, 23–31 (1977).
580. Zimmer, Robert J. Amenable ergodic group actions and an application to Poisson boundaries of random walks. *J. Functional Analysis* **27**, 350–372 (1978).
581. Zimmer, Robert J. Amenable pairs of groups and ergodic actions and the associated von Neumann algebras. *Trans. Amer. Math. Soc.* **243**, 271–286 (1978).
582. Zimmer, Robert J. Induced and amenable ergodic actions of Lie groups. *Ann. Sci. Ecole Norm. Sup.* **11**, 407–428 (1978).
583. Zimmer, Robert J. On the cohomology of ergodic group actions. *Israel J. Math.* **35**, 289–300 (1980).

RECENT PUBLICATIONS

584. Zimmer, Robert J. Strong rigidity for ergodic actions of semisimple Lie groups. *Ann. of Math.* **112**, 511–529 (1980).
585. Zimmer, Robert J. On the cohomology of ergodic actions of semisimple Lie groups and discrete subgroups. *Amer. J. Math.* **103**, 937–951 (1981).
586. Zimmer, Robert J. Ergodic theory, group representations, and rigidity. *Bull. Amer. Math. Soc.* **6**, 383–416 (1982).
587. Zimmer, Robert J. Ergodic theory, semisimple Lie groups, and foliations by manifolds of negative curvature. *Inst. Hautes Etudes Sci. Publ. Math.* **55**, 37–62 (1982).
588. Zimmer, Robert J. Curvature of leaves in amenable foliations. *Amer. J. Math.* **105**, 1011–1022 (1983).
    Zimmer, Robert J. See [399].
    Zippin, Leo. See [397].
    Zsidó, L. See [545].

RECENT PUBLICATIONS

589. Anker, Jean-Philippe. Applications de la p-induction en analyse harmonique. *Comment. Math. Helv.* **58**, 622–645 (1983).
590. Boidol, Joachim. Group algebras with a unique C^*-norm. *J. Functional Analysis* (To be published).
591. Cuntz, Joachim. K-theoretic amenability for discrete groups. *J. Reine Angew. Math.* **344**, 180–195 (1983).
592. Gootman, Elliot C. On certain properties of crossed products. *Proceedings of Symposia in Pure Mathematics*, Vol. **38**/1, 311–321. American Mathematical Society, Providence, 1982.
593. Lance, E. Christopher. K-theory for certain group C^*-algebras. *Acta Math.* **151**, 209–230 (1983).
594. Moore, Calvin C. Ergodic theory and von Neumann algebras. *Proceedings of Symposia in Pure Mathematics*, Vol. **38**/2, 179–226. American Mathematical Society, Providence, 1982.
595. Moulin Ollagnier, Jean. Théorème ergodique presque sous-additif et convergence en moyenne de l'information. *Ann. Inst. H. Poincaré*, Sect. B **XIX**, 257–266 (1983).
596. Schmidt, Klaus. Amenability, Kazhdan's property T, strong ergodicity and invariant means for ergodic group-actions. *Ergod. Th. and Dynam. Sys.* **1**, 223–236 (1981).
597. Veršik, A. Amenability and approximation of infinite groups. *Sel. Math. Sov.* **2**, 311–330 (1982).

NOTATION INDEX

A^1	30	$C(X)$	350
$\tilde{A}$	8	$\mathscr{C}(S)$	6
$\hat{A}$	8	$\mathscr{C}_0(S)$	6
$A^{\sim}, A^{\sim\sim}$	239	$C_p(G)$	93
(A)	114	$C^*(G)$	78, 208
a^0	10	$C^*_{\mathscr{S}}(G)$	78
A^T	239	$C^*(G, A),$	
A_h	8	$\quad C^*_L(G, A)$	366
A_u	8	$(Cv_p), (Cv_p^*)$	85
$A^{(2)}$	290	$(Cv'_p), (Cv'^*_p)$	85
$A(G)$	94	$Cv^p(G)$	82
$A_p(G)$	91	card S	6
$A_p^1(G)$	223	$\operatorname{co}(E)$	8
$AP(G)$	284	$\overline{\operatorname{co}}(E)$	8
$\mathscr{A}(G)$	166		
$\mathscr{A}_c(G)$	167	$(D), (D'), (D_p)$	48
$\mathscr{A}\mathscr{N}(G)$	179	d_1, d_2	98
		$dx, d\lambda(x)$	11
(B)	114	$\mathscr{D}(G)$	243
B_G	284	$D^1(X)$	9
$B(G)$	208	$D^{(n)}(G)$	120
$B_p(G)$	93	$d_l(f), d_r(f)$	58
$B^{(1)}(G), B^{(2)}(G)$	210	$d_1^c(f)$	152
$\beta(S)$	6	$d_{1,G,H}(f),$	
		$\quad d_{r,G,H}(f)$	310
$\mathfrak{c}$	6		
C_ξ	353	e	10
$(C_p), (C'_p)$	85	E_S	332

EG	120	$\text{IM}(\hat{G})$	346
EG_0, EG_Λ	318	$\text{IM}(X)$	32
(EB)	114	$\text{IIM}(X)$	289
$\check{f}, \tilde{f}$	10	$\underset{H \uparrow G}{\text{ind}} T$	184
f^*	13		
$\hat{f}$	209	(J)	333
$f^\natural$	353, 357	$\mathscr{J}(G)$	249
$(F), (F^*)$	62	$J_H, J_{h,\rho}$	127
F_2	122	$J^\rho, J^\rho_{H,\rho}$	183
$\mathfrak{F}(S)$	6	$J_1(G, H),$	
(FP)	45	$J_r(G, H)$	310
$[FC]^-$	116	K_a	116
$_af, f_a$	10	k^E, k_c^E	221
f_β	285	$\mathscr{K}(S)$	6
f_μ	20	$\mathfrak{K}(S)$	6
$f\zeta$	354	$\mathfrak{K}_0(G)$	11
		$(KS), (KS_p)$	325
$\hat{G}$	78, 79	$(K_\mu), (K_{\mu,x})$	202
$(G), (G^*)$	80		
G_0	133	$(L), (L^*)$	62
G_d	11	L_a	10
G_S	332	l_a	326
$\hat{G}_r$	78	L_f	77
$(G_p), (G_p^*)$	194	L_μ	27, 82
$(GR_1), (GR_r)$	59	$L_G, L_G(f)$	27
(GR_1^c)	153	$\mathscr{L}_G(f)$	27
$GL(n, \mathbf{R}),$		$L_{G,H}$	310
$GL(n, \mathbf{C})$	125	$\mathscr{L}(X)$	7
$GL(n, \mathbf{Q}_p)$	132	$\mathscr{L}(X, Y)$	7
		$\mathscr{L}_\mathbf{R}(X, Y)$	7
H_μ	359	$\mathscr{L}_+(X, Y)$	8
$h(I)$	253	$\mathscr{L}^2(A, X)$	98
$\mathfrak{H}(G)$	11	$l^p(X)$	10
$H_1(A, X)$	98	$\mathscr{L}^p(X, \mu)$	10
$H_1(G, X^*)$	99	$L^p(X, \mu)$	10
		$L^0(X, \mu)$	10
I, I_{X, A^1}	255	$l^\infty(X)$	10
i_G	80	$\mathscr{L}^\infty(X, \mu)$	10
$[IN]$	275		

NOTATION INDEX

$L^\infty(X,\mu)$	10		$P^1(\hat{G})$	346
$\mathrm{LIM}(X)$	32		p_f^ν	185
$\mathscr{LUC}(G)$	11		$PM(G)$	94
$L^p(G) \hat{\otimes} L^{p'}(G)$	91		$PM^p(G)$	94
$L^p(G, L^q(H))$	203		$PF^p(G)$	216
			$P^p(X,\mu)$	10
$\hat{M}$	40			
$m(G)$	297		$\mathbf{Q}_p$	132
$M_{U,A}$	270		$Q(G)$	216
$\mathscr{M}(X)$	9			
$\mathscr{M}^1(X)$	9		(R)	340
$\mathscr{M}_c^1(X)$	9		$r(x)$	8
$\mathscr{M}_f^1(X)$	9		$R_G, R_G(f)$	58
$M^0(X)$	9		$R_a(f)$	12
$M^1(X)$	9		$R_{G,H}$	310
$\mathscr{M}_a^1(G)$	12		R_a^H	314
$\mathscr{M}_s^1(G)$	12		$\mathscr{R}_1\mathscr{A}(G,k)$,	
$\mathscr{M}_d^1(G)$	12		$\mathscr{R}_r\mathscr{A}(G,k)$	177
$\mathfrak{M}(X)$	23		$\mathrm{RIM}(X)$	32
$\mathfrak{M}(X, A^1)$	255		$\mathscr{RUC}(G)$	11
$\mathfrak{M}(PM(G))$	346		$\mathscr{RUC}(S)$	326
			rad G	133
NF	318			
N_X	242		S, S_{X,A^1}	255
$N(A)$	74		s_Q	235
$\mathscr{N}(X)$	10		S_n	6
$N(G, X^*)$	99		sp x	8
$\mathscr{N}(X,P)$	255		$(SF), (SF^*)$	62
(NFS)	340		SF_2	329
			$\mathfrak{S}(G)$	235
O_ζ	43		$(SG_p), (SG_p')$	194
			[SIN]	190
(P)	240, 375		$\mathscr{S}(X,\mu)$	10
p'	10		$SL(2,\mathbf{R})$	125
P_x, P_f	323		$SL(n,\mathbf{R}), SL(n,\mathbf{C})$	125
$(P_p), (P_p^*)$	51		$SL(n,\mathbf{Q}_p)$	132
(PG)	114		$\mathscr{SA}(G)$	166
$P(G)$	78		$\mathscr{SA}_c(G)$	167
$\mathfrak{P}(S)$	6		supp f	6
$P_L(G)$	186			

NOTATION INDEX

supp M	253	βS	7
supp μ	9	$\beta(M, X)$	33
$\mathbf{T}$	6	γ_p	51
$T^\sim$	7	γ_p^*	52
T^*	9		
(T)	186	Γ, Γ_1	91
t_Q	235	$\Gamma_{X,U}$	270
$\mathcal{T}_\mathcal{U}$	293		
$\mathcal{T}(X), \mathcal{TN}(X)$	270	Δ (symmetric difference of subsets)	
$\mathcal{T}_s(X), \mathcal{T}_s\mathcal{N}(X)$	271	Δ (modular function)	11
T_G^H	203	$(\Delta), (\Delta_1), (\Delta')$	49
$T_{M,\zeta}$	353	δ_a	9
TIM(X), TLIM(X), TRIM(X)	32	ϑ, ϑ^*	157
(U_X)	188	ι	289
$\mathcal{UC}(G)$	11		
$\mathcal{UC}(\hat{G})$	218	κ, κ^*	157
$V, V(X, A^1)$	258	λ	11
$\mathcal{V}(X, A^1)$	258	Λ_x	353
w_β	285	$\hat{\mu}$	209
(W)	341	μ^*	13
(WF)	62	μ^{*n}	12
$\mathcal{W}(\hat{G})$	346	μ^a, μ^s, μ^d	12
$W^p(G)$	216	μ_ρ	126
WAP(G)	307		
		ξ_A	11
$\hat{x}$	8		
x^*	8	Π_μ	359
X'	7	π^ν	185
X^*	7		
$X_\mathbf{R}^*$	7	ρ	126
X_+^*	8		
$\mathcal{X}_T$	184	σ, σ^*	80
		σ_p, σ_p^*	194
$Z(G, X^*)$	99	τ_p, τ_p'	194
$\alpha(M, X)$	33	Υ	32

NOTATION INDEX

Θ_A	343	$\hat{\mu}$	209
		$\langle\cdot,\cdot\rangle$	7
$\varphi_{\xi,\eta}$	353	$(\cdot\|\cdot)$	10
Φ_D	373		
		$\|f\|$	6
χ (character)	8, 208	$\|T\|$	7
χ, associated		$\|\cdot\|_\infty$	10
to ρ	183	$\|\cdot\|_p$	10
χ_S	243	$\|\cdot\|_{\mathscr{S}}$	78
		$\|\cdot\|_{\hat{G}}$	208
$\psi_{\xi,\eta}$	188	$\|\cdot\|_{A_p}$	91
		$\|\cdot\|_{B_p}$	93
ω_{I_0}	297	$\|\cdot\|_{C_p}$	93
$1'$	10	$\|\cdot\|_{Cv^p}$	82
1_A	6	$\|\cdot\|_{p,q}$	203
$|A|$	11	$\|\cdot\|_{1,p}$	223
$\Phi^{\frac{1}{2}}$	73	$\|\cdot\|_{\hat{\otimes}}$	91
$\prec$	6		
$\ll$	243	$*$	12
$\sim$	243	$\star$	20
$\check{f}, \tilde{f}$	10	$\odot$	21
f^*	13	$\otimes$	91, 189
$\hat{f}$	209	$\circledast$	293
$f^\natural$	353, 357	$\square$ associated	
$_af, f_a$	10	to a pair (G, H)	310
$\check{F}$	32		
μ^*	13	$\vee, \bigvee_n, \bigvee_1$	243

AUTHOR INDEX

Adel'son-Vel'skiĭ, 117
Adian, 338
Adler, 264
Akemann, 226, 227, 304
Alfsen, 284
Anantharaman-Delaroche, 365
Anker, 61, 156, 191, 207, 363
Appel, 331
Arens, 23
Argabright, 182, 350, 359
Aribaud, 360
Arsac, 191, 226
Avez, 117, 207, 339
Azencott, 339, 359, 360

Baggett, 190, 191
Baldi, 340
Banach, 3, 245, 305, 306
Barnes, 228, 375
Bass, 338
Beiglböck, 363
Berg, 206, 364
Berglund, 245, 342, 359
Birgé, 360
Birkhoff, 353
Blattner, 183
Blum, 181
Bohr, 4, 284
Boidol, 191
Boissier, 358
Boltzmann, 352
Bombal, 351
Bondar, 6, 72, 157
Bonic, 181
Bonsall, 111, 117, 374
Borel, 338
Bottesch, 351
Bourbaki, 122

Bozejko, 181, 208, 227, 304
Brooks, 181
Bunce, 375, 376
Burckel, 284, 307

Calderón, 360, 361
Carling, 227, 360
Cecchini, 227
Chatard, 181, 360, 361
Chen, I., 227
Chen, S., 360
Chivukula, 351
Choda, H., 304
Choda, M., 303, 304
Choi, 376
Chou, 263, 264, 297, 302, 303, 304, 305, 307,
    308, 338, 340, 341, 350, 360
Christensen, 206, 364
Cohen, 338, 339
Coifman, 364
Comfort, 350
Connes, 375, 376
Converse, 302
Conze, 361
Corduneanu, 284
Cowling, 207, 216, 227, 341
Crépel, 339
Cuntz, 376

Dang Ngoc, 361
Darsow, 139
Davis, 181
Day, 5, 23, 29, 42, 47, 61, 89, 117, 122, 207,
    245, 263, 264, 303, 326, 341, 350, 359, 360
Deligne, 338
Derighetti, 61, 98, 122, 156, 207, 226, 304,
    337, 338, 364
Derriennic, 206, 339, 364

Dhombres, 284
Dieudonné, 61, 89, 122, 134, 139
Dixmier, 42, 72, 81, 117, 122, 139, 190, 240, 245, 350, 360
Djorup, 331
Doplicher, 352
Doran, 22
Doss, 226
Douglas, 263, 303
Douglass, 181
Dreseler, 227
Dubins, 139
Duflo, 341
Duncan, 111, 117, 342, 374
Dunkl, 228, 350

Effros, 303, 375, 376
Ehrenpreis, 341
Eisenberg, 181
Elie, 339, 360
Ellis, 43, 48
Emch, 352
Emerson, 61, 72, 180, 181, 245, 246, 340, 361
Eymard, 6, 89, 91, 98, 122, 157, 190, 208, 227, 364

Fairchield, 302
Faraut, 206, 208
Feichtinger, 207
Feldman, 376
Fell, 72, 81
Figà-Talamanca, 227
Flory, 207, 226, 337
Følner, 72, 122, 181
Fountain, 341
Fournier, 207
Fremlin, 351
Frémond, 359
Frey, 182, 342
Furstenberg, 47, 139, 190, 358, 359, 360, 365

Ganesan, 342
Gangolli, 341
Gelfand, 81
Gerl, 156, 207
Gibbs, 352
Gilbert, 81, 89, 207, 227
Gillman, 285
Glasner, 48, 303, 304, 339, 360
Glicksberg, 61, 350
Godement, 72, 78, 81, 293, 341
Gootman, 366, 376
Granirer, 42, 156, 226, 227, 228, 245, 264, 302, 303, 304, 305, 308, 341, 350

Greenleaf, 6, 42, 72, 122, 139, 180, 181, 184, 190, 302, 303, 340, 361, 362, 363, 364
Grigorčuk, 339
Gromov, 340
Grosser, 275
Grossman, 304, 359
Guichardet, 81, 361, 366
Guivarc'h, 117, 139, 206, 339, 340, 360, 364
Gulick, 362

Haagerup, 156, 227, 366, 376
Hahn, L., 181
Hahn, P., 365
Harpe, de la, 305, 338, 376
Hauenschild, 190
Hausdorff, 2, 3, 4, 139, 243, 245, 303
Helgason, 134
Herz, 91, 98, 156, 203, 207, 227
Hewitt, 122, 245, 303, 326
Heyer, 366
Hirsch, 339
Hochster, 342
Holm, 284
Huff, 359
Hulanicki, 29, 42, 61, 72, 81, 117, 122, 139, 181, 339, 341
Husain, 351

Iversen, 359
Iwasawa, 133, 136

Jenkins, 117, 181, 325, 333, 339, 340, 341, 342, 359, 361
Jerison, 285
Johnson, 23, 98, 111, 156, 374, 375
Junco, del, 182, 306, 361
Junghenn, 245, 342, 350, 359

Kadison, 352, 375
Kakutani, 139, 211
Kallman, 358
Kaniuth, 190, 275
Kastler, 352
Kaufman, 303
Kaur, 303
Keane, 339
Keller, 182
Kesten, 207
Khalill, 340
Khelemskiĭ, 111
Kieffer, 180, 181, 362
Klawe, 182, 342, 350
Kneser, 180
Knops, 352

AUTHOR INDEX

Kotzmann, 156
Krom, Melven, 181, 245
Krom, Myren, 181, 245
Kugler, 228
Kuipers, 320
Kunze, 325

Lacroix, 339
Lai, 227
Lance, 191, 366, 375, 376
Landstad, 304
Lau, 29, 156, 227, 228, 245, 304, 308, 342, 350, 375
Lebesgue, 1, 2, 305, 306
Lehmann, 157
Leinert, 156, 227, 341
Lengagne, 352
Leptin, 72, 89, 98, 117, 139, 180, 228, 341
Lipsman, 341
Liu, 302, 362
Liukkonen, 190
Lo Bello, 352
Locher-Ernst, 139
Lohoué, 207, 227, 325, 341
Loomis, 22, 284
Losert, 156, 306, 337, 364
Ludwig, 341
Lust-Piquard, 264
Luthar, 303

Maak, 284
Mackey, 183, 184
Manocha, 303
Margulis, 190, 191, 306, 365
Mautner, 341
Maxones, 156
McKennon, 226, 227
Michele, de, 227
Milnes, 6, 72, 157, 181, 245, 302, 303, 341, 342, 359
Milnor, 340
Mitchell, 245, 263, 304, 349, 350, 359
Mocanu, 180
Montgomery, 133
Moore, C., 360, 365, 376
Moore, G., 246, 365
Moran, 275, 350
Moskowitz, 275
Moulin Ollagnier, 72, 181, 302, 366
Müller Römer, 339
Mycielski, 72, 305, 338

Namioka, 29, 42, 61, 72, 182, 302, 342
Nebbia, 227

Neumann, von, 4, 5, 29, 42, 117, 122, 139, 246, 303, 307, 338, 353
Niederreiter, 320
Novikov, 338

Oberlin, 206, 207
Ocneanu, 365
Ol'sanskiĭ, 339
Ornstein, 361, 362

Pachl, 305
Palmer, 6, 139, 275, 341
Paschke, 303, 375, 376
Paterson, 157, 190, 303, 304, 342
Phelps, 302, 304
Picardello, 227
Pickel, 362
Pier, 6, 302, 304
Pinchon, 72, 181, 302
Poguntke, 341
Poincaré, 352
Promislow, 307

Racher, 227, 375
Raĭkov, 81
Raimi, 245, 341
Rajagopalan, 182, 303, 325, 341
Ramakrishnan, 182
Ramirez, 350
Ramsay, 341
Randow, von, 181
Raugi, 360
Reiter, 22, 29, 42, 61, 81, 98, 117, 122, 139, 309, 337
Renaud, 226, 302, 350, 361
Rickert, 47, 122, 139, 325
Riemersma, 364
Rindler, 29, 98, 156, 306, 337, 339, 364
Ringrose, 111, 375
Rivière, 156, 203, 207
Robert, 190
Robinson, 352
Rooij, van, 302, 362
Rosenberg, 366, 375
Rosenblatt, 156, 182, 208, 263, 264, 303, 304, 305, 306, 307, 338, 340, 361, 364
Ross, 122, 245, 303, 326, 350
Roynette, 339
Rudin, 263
Ruziewicz, 306

Saeki, 22
Sakai, K., 190, 246, 364
Sakai, S., 240, 245

Sauvageot, 366
Schempp, 227
Schickhof, 352
Schlichting, 275
Schmidt, 366, 376
Schochetman, 122, 190
Schott, 364
Schwartz, 245
Segal, 341
Serre, 338
Sheinberg, 111
Sherman, 246
Silverman, 245
Skudlarek, 156, 190
Snell, 264
Soardi, 227
Spector, 227
Šreider, 117
Stegeman, 61
Stegmeir, 156, 374
Stein, 325
Stewart, 81
Stone, 181
Strătilă, 240, 376
Stromberg, 139
Sueur-Pontier, 359
Šujan, 362
Sullivan, 306, 338
Sz. Nagy, 190

Tagi-Zade, 362
Takenouchi, 122, 139
Talagrand, 302, 303, 305, 307, 351
Tarski, 3, 245, 246
Tempel'man, 360
Thurston, 339
Tits, 191, 338, 340
Tomter, 304
Toure, 364

Van Dijk, 139
Vasilescu, 190
Vera, 351
Verboven, 352
Veršik, 342
Von Neumann, *see* Neumann, von

Wagon, 246, 307, 338
Wallach, 134
Walter, 226, 227
Wang, 190
Warner, 134
Watatani, 304
Weil, 4, 81, 284
Weiss, B., 361, 362, 376
Weiss, G., 364
Weit, 341
Wichmann, 22
Wiener, 341
Wilde, 182, 342
Williamson, 275, 341
Witz, 303, 341, 342, 360
Wolf, 340
Wong, 23, 29, 42, 61, 156, 245, 264, 304, 342, 349, 350, 351, 358
Woodward, G., 245
Woodward, W., 351
Wordingham, 342

Yeadon, 245
Yoshizawa, 139

Zimmer, 339, 364, 365
Zippin, 133
Zsidó, 190

SUBJECT INDEX

Abelian group, 113, 115
Action of a semigroup, 17
Adjoint operator, 7, 9
Admissible sequence, 361
Affine flow, 44
Almost connected group, 133
  periodic function, 4, 284
  solvable group, 134
Amenable action, 363
Amenable Banach algebra, 366
Amenable group, 31, 38
σ-Amenable group, 342
Amenable homogeneous space, 364
Amenable pair, 365
Amenable semigroup, 38, 326
Amenable von Neumann algebra, 375
Aperiodic probability measure, 192
Approximately finite-dimensional von Neumann algebra, 375
Approximate units, 14
  bounded, 14
  left, 14
  multiple, 14
  right, 14
Arens multiplication, 22
Asymptotically invariant net, 48
Averaging net, 179
  sequence, 166
    regular, 177
    strongly, 166

Birkhoff's individual ergodic theorem, 353
Boundary of a locally compact group, 358
Bruhat function, 118
Burnside's conjecture, 338

(C)-group, 136
C*-algebra, 8

(CCR)-group, 191
Centralizer, 19
Character, 8, 79
  group, 79
Choquet-Deny property, 323
Cocycle, 365
Cohomology group, first, 98
Commutant, 239
Compact group, 112, 115
Completely separable topological space, 7
Concentrated mean, 248
Conjugacy class, 116
Connected component, 133
Contractible group, 320
Convolution operator, 27
  product, 12
Convolutor, 82
Countable set, 6
Covering lemma, 66
Crossed product, 366
Cyclic vector, 73

Day's asymptotical invariance properties, 48
  conjecture, 338
Degree of polynomial growth, 339
Derivation, 98, 99
  inner, 99
Dirac measure, 9
Ditkin's conditions, 221
Dixmier's criterion, 41
Dual:
  of an involutive algebra, 77
  reduced, 78
Dual object of a locally compact group, 78
Dual space:
  algebraic, 7
  topological, 7

415

Effros-F. Hahn conjecture, generalized, 366
Elementary group, 120
Entropy, 206
Enveloping C*-algebra, 78
Ergodic G-space, 365
Ergodicity, 352
Ergodic theorem:
  abstract, 355
  mean, 357
  pointwise, 357
Essential Banach module, 100
Étalée probability measure, 359
Exponential growth, 339
Exponentially bounded group, 114
Exposed point, 308
Extremely amenable semigroup, 304

Fell group, 190
  topology, 78
Fixed point property, 45
Flow, 43
  affine, 44
  irreducible affine, 44
  minimal, 43
  proximal, 43
  strongly proximal, 44
Følner's conditions, 62
  strong, 62
  weak, 62
Fourier algebra, 94
Fourier-Gelfand transform, 8
Fourier-Stieltjes algebra, 208
Fourier-Stieltjes transform, 209
Fourier transform, 209
Free group:
  on a finite number of generators, 123
  of order 2, 115
  on 2 generators, 122

Gelfand topology, 9
Gelfand transform, 8
G-invariant probability measure, 40, 126
Glasner space, 360
Glicksberg-Reiter property, 58
Godement's conditions, 79
Group algebra, 13
Growth conditions, 114

Haar measure, 11
  normalized, 11
Hakeda-Tomiyama property, 375
Harmonic function, 323
Hausdorff-Banach-Tarski paradox, 3
Hermitian element, 8

Hermitian group, 325
Hull of an ideal, 253

Induced representation, 183
Injective von Neumann algebra, 375
Inner amenable group, 292
Inner mean, 292
Introverted space, 21
Invariant function space, 17, 19
  left, 19
  right, 20
  topologically left, 19
  topologically right, 20
Invariant mean, 30, 32
  left, 31
  right, 31
  topologically, 32
  topologically left, 31
  topologically right, 31
Invariant measure, 126
Inverse semigroup, 332
Inversion invariant mean, 289
Involution, 8
Iwasawa decomposition, 134

Kazhdan's property, 186
Kernel in a commutative Banach function
  algebra, 221

L^p-conjecture, 325
Left invariant functional, 237
Left invariant mean, 31
Left regular representation:
  of G, 76
  of $L^1(G)$, 77
Leptin's conditions, 62
Locally finite group, 121
Lumpiness, 350

Mackey-Bruhat formula, 127
Marcienkiewicz interpolation theorem-type
  result, 203
Markov-Kakutani fixed point theorem, 113
Maximally almost periodic group, 190
Mean, 23
  invariant, 32
  left invariant, 31
  right invariant, 31
  topologically invariant, 32
  topologically left invariant, 31
  topologically right invariant, 31
  trivial, 292
Modular function, 11

SUBJECT INDEX

Module;
  Banach A-, 18
  Banach G-, 18
  left Banach, 17
  right Banach, 18
  two-sided Banach, 18
Multiplier, 19
Multiplier function, 93
Multiset, 243
Multiset function, 243
Mutually singular means, 253

Nested family, 258
Neumann algebra, von, 239
  approximately finite-dimensional, 375
  injective, 375
Neumann's conjecture, von, 338
Neumann's mean ergodic theorem, von, 353
Nontrivial mean, 292
Nuclear C*-algebra, 366

Orbit, 43

P-amenable group, 342
Paradoxical decomposition, 3, 243
Poisson space, 359
Polynomial growth, 114
  degree of, 339
Positive-definite function, 78
  associated to a representation, 78
Positive-definite measure, 78
Positive-definite operator, 73
Positive functional, 72
  associated to a representation, 73
Prehilbert structure, 9
Probability measure, 9
Proximal elements, 43
Proximal flow, 43
Pseudocompact group, 350
Pseudofunction, 216
Pseudomeasure, 94

Quasi-invariant measure, 126
Quasi-regular representation, 184
Quotient group, 118

Radical, 133
Recurrent element, 323
Recurrent group, 323
Recurrent random walk, 323
Reduced dual, 78
Regular algebra, 221
Regular averaging sequence:
  left, 177
  right, 177

Reiter's conditions, 51
Relativization problem, 309
Representation:
  of an algebra, 76
  of a group, 73
  nondegenerate, 73
Representations, equivalent, 73
Riemann measurable function, 307
Right invariant mean, 31

Scattered sequence, 252
Semidiscrete von Neumann algebra, 375
Semigroup:
  action of a, 17
  topological, 326
Semisimple group, 134
Set of spectral synthesis, 221
σ-amenable group, 342
$\sigma(X,X^*)$-topology, 7
$\sigma(X^*,X)$-topology, 7
Singular mean, 253
Solvable group, 120
Spectral radius, 8
Spectral set, 221
Spectral synthesis, 220
Spectrum, 8
Square root of an operator, 73
State, 23
Stationarity, 232
Stone-Čech compactification, 7, 285
Strongly amenable C*-algebra, 373
Strongly amenable group:
  Emerson-Greenleaf, 340
  Glasner, 360
Strongly bounded representation, 186
Strongly proximal flow, 44
Strong topology, 9
Structure space, 8
*-Subalgebra, 8
Subflow, 43
Superharmonic function, 323
Support of a mean, 253
Support of a representation, 76, 78
Sz. Nagy-Dixmier property, 186

Takesaki property, 366
Tauberian algebra, 221
Thick, topologically, 343
Thickness properties, 350
Topologically invariant mean, 32
Topologically irreducible representation, 77
Topologically left invariant mean, 31
Topologically right invariant mean, 31
Topological semigroup, 326

W-Topology, 7
W*-Topology, 7
σ (X,X*)-Topology, 7
σ (X*,X)-Topology, 7
Transformation semigroup, 17
Transient probability measure, 339
Trivial mean, 292

Ultrastrong topology, 9
Ultraweak topology, 9, 94
Unbounded group, 358
Uniform amenability, 181, 182
Uniformly bounded representation, 186
Uniformly continuous function, 11
  left, 11
  right, 11, 326
Uniformly distributed group, 320
Unimodular group, 12
Unital Banach module, 100
Unitary element, 8
Unitary group, 8
Unitary representation, 76

Von Neumann algebra, *see* Neumann algebra, von

Weak containment, 76, 78
Weak direct product, 121
Weakly almost periodic function, 307
Weakly almost periodic pseudomeasure, 346
Weakly attractive element, 360
Weakly equivalent representations, 189
Weakly measurable action, 353
Weak (p,p)-type operator, 205
Weak topology, 7, 9
Weak*-topology, 7
Weil's formula, 127
Wiener group, 341
Wiener property, 341
w-topology, 7
w*-topology, 7

Zero:
  left, 99
  right, 99